P9-AFP-579

Optimal Combining and Detection

Statistical Signal Processing for Communications

With signal combining and detection methods now representing a key application of signal processing in communication systems, this book provides a range of key techniques for receiver design when multiple received signals are available. Various optimal and suboptimal signal combining and detection techniques are explained in the context of multiple-input multiple-output (MIMO) systems, including successive interference cancellation (SIC) based detection and lattice reduction (LR) aided detection. The techniques are then analyzed using performance analysis tools. The fundamentals of statistical signal processing are also covered, with two chapters dedicated to important background material. With a carefully balanced blend of theoretical elements and applications, this book is ideal for both graduate students and practicing engineers in wireless communications.

JINHO CHOI is currently a Professor in the School of Engineering, and Chair of the Wireless Group, at Swansea University, UK. He is the author of *Adaptive and Iterative Signal Processing in Communications* (Cambridge University Press, 2006) and the recipient of the 1999 Best Paper Award for Signal Processing from EURASIP. A Senior Member of the IEEE, his current research interests include wireless communications and array/statistical signal processing.

Optimal Combining and Detection

Statistical Signal Processing for Communications

JINHO CHOI

Swansea University, UK

CAMBRIDGE UNIVERSITY PRESS
Cambridge, New York, Melbourne, Madrid, Cape Town, Singapore,
São Paulo, Delhi, Dubai, Tokyo

Cambridge University Press
The Edinburgh Building, Cambridge CB2 8RU, UK

Published in the United States of America by Cambridge University Press, New York

www.cambridge.org
Information on this title: www.cambridge.org/9780521517607

First published 2010

Printed in the United Kingdom at the University Press, Cambridge

A catalog record for this publication is available from the British Library

Library of Congress Cataloging in Publication data
Choi, Jinho.
Optimal combining and detection : statistical signal processing for communications / Jinho Choi.
p. cm.
Includes bibliographical references and index.
ISBN 978-0-521-51760-7 (hardback)
1. Signal processing. 2. Antenna arrays. 3. Mixing circuits. I. Title.
TK5102.9.C4842 2010
621.382′2 – dc22 2009043214

ISBN 978-0-521-51760-7 Hardback

Contents

Preface

Statistical signal processing is a set of statistical techniques that have been developed to deal with random signals in a number of applications. Since it is rooted in detection and estimation theory, which are well established in statistics, the fundamentals are not changed although new applications have emerged. Thus, I did not have any strong motivation to write another book on statistical signal processing until I was convinced that there was a sufficient amount of new results to be put together with fundamentals of detection and estimation theory in a single book.

These new results have emerged in applying statistical signal processing techniques to wireless communications since 1990. We can consider a few examples here. The first example is smart antenna. Smart antenna is an application of array signal processing to cellular systems to exploit spatial selectivity for improving spectral efficiency. Using antenna arrays, the spatial selectivity can be used to mitigate incoming interfering signals at a receiver or control the transmission direction of signals from a transmitter to avoid any interference with the receivers which do not want to receive the signal. The second example is based on the development of code division multiple access (CDMA) systems for cellular systems. In CDMA systems, multiple users are allowed to transmit their signals simultaneously with different signature waveforms. The matched filter can be employed to detect a desired signal with its signature waveform. This detector is referred to as the single-user detector as it only detects one user's signal. Although this single-user detector is able to provide a reasonable performance, it is also possible to improve the performance to detect multiple signals simultaneously. This detector is called the multiuser detector. The third example is multiple-input multiple-output (MIMO) systems. In MIMO systems, multiple signals are transmitted and multiple signals are received. Thus, it is required to detect multiple signals simultaneously. These new applications promote advances of statistical signal processing. In particular, new and advanced techniques for signal combining and detection have emerged.

This book is intended to provide fundamentals of signal detection and estimation together with new results that have been developed for the new applications mentioned above.

I would like to thank many people for supporting this work, in particular: I. M. Kim (Queens University), C. Ling (Imperial College), and F. Adachi (Tohoku University). They helped me by providing constructive comments and proofreading. Needless to say the responsibility for the remaining errors, typos, unclear passages, and weaknesses is mine. I would also like to thank those people who inspire and encourage me all the time:

F. Adachi (Tohoku University) for encouragement as my mentor, J. Ritcey (University of Washington) for long-term friendship, and many others including my students for useful discussions.

Special thanks go to J. Ha, Y. Han, and H. J. Lee (Korea Advanced Institute of Science and Technology) who hosted me and offered an opportunity to teach a summer course with most of the materials in this book in 2008 at Information Communications University which became part of Korea Advanced Institute of Science and Technology in 2009. It was my great pleasure to teach young and talented students at Information Communications University. Their comments were very helpful in shaping this book.

Finally, I would like to offer very special thanks to my wife, Kila, and children, Seji and Wooji, for their generous support, understanding, and love.

Symbols

General

$j = \sqrt{-1}$

$\mathbb{F}_2$: binary field

$\mathbb{Z}$: set of integer numbers

$\mathbb{R}^n$: real-valued n-dimensional vector space

$\mathbb{C}^n$: complex-valued n-dimensional vector space

$\times$: Cartesian product (if it does not mean the product)

$|\mathcal{A}|$: cardinality of set $\mathcal{A}$ or the number of the elements in $\mathcal{A}$

$\emptyset$: empty set

$\cup$: set union

$\cap$: set intersection

$\setminus$: set difference or set-minus

$\mathcal{A}^c$: the complementary set of set $\mathcal{A}$

$u(x)$: step function

$\delta(x)$: Dirac delta function

Statistics related symbols

$f_X(x)$: pdf of random variable X

$F_X(x)$: cdf of random variable X

$\Pr(\mathcal{A})$: probability of random event $\mathcal{A}$

$\mathcal{E}[X]$: statistical expectation of X

$\mathrm{Var}(X)$: variance of X

$\mathcal{Q}(x)$: Q-function, $\mathcal{Q}(x) = \int_x^{\infty} \frac{1}{\sqrt{2\pi}} \mathrm{e}^{-\frac{z^2}{2}} \mathrm{d}z$

$\mathcal{N}(\mathbf{x}, \mathbf{R})$: Gaussian probability density function with mean $\mathbf{x}$ and covariance $\mathbf{R}$

$\mathcal{CN}(\mathbf{x}, \mathbf{R})$: circularly symmetric complex Gaussian probability density function with mean $\mathbf{x}$ and covariance $\mathbf{R}$

Vector/Matrix related symbols

$||\cdot||_p$: p-norm

$||\cdot||_{\mathrm{F}}$: Frobenius norm

$(\cdot)^{\mathrm{T}}$: transpose

$(\cdot)^{\mathrm{H}}$: Hermitian transpose

$\det(\cdot)$: determinant of a square matrix

$\mathrm{tr}(\cdot)$: trace of a square matrix

$\mathrm{Diag}(a_1, a_2, \ldots, a_N)$: $N \times N$ diagonal matrix whose elements are $a_1, a_2, \ldots, a_N$

$[\mathbf{a}]_n$: nth element of a vector $\mathbf{a}$
$[\mathbf{A}]_{m,n}$: (m, n)th element of a matrix $\mathbf{A}$
$[\mathbf{A}]_{m_1:m_2,n_1:n_2}$: a submatrix of $\mathbf{A}$ obtained by taking the elements in the m_1th to m_2th columns and the n_1th to n_2th rows
$[\mathbf{A}]_{:,n}$: nth column vector of $\mathbf{A}$
$[\mathbf{A}]_{n,:}$: nth row vector of $\mathbf{A}$

Abbreviations

AR	autoregressive
ARV	array response vector
ASK	amplitude shift keying
AWGN	additive white Gaussian noise
BER	bit error rate
BSC	binary symmetric channel
cdf	cumulative distribution function
CDMA	code division multiple access
CLT	central limit theorem
CRB	Cramer–Rao Bound
CSCG	circularly symmetric complex Gaussian
CVP	closed vector problem
DFE	decision feedback equalizer
DMC	discrete memoryless channel
DMI	direct matrix inversion
DPSK	differential phase shift keying
EGC	equal gain combining
FA	false alarm
GLR	generalized likelihood ratio
GLRT	generalized likelihood ratio test
GSDC	generalized selection diversity combining
iid	independent and identically distributed
ISI	intersymbol interference
LCMV	linearly constrained minimum variance
LLR	log-likelihood ratio
LMS	least mean square
LR	lattice reduction or likelihood ratio
LS	least square
MAC	multiple access channel
MAP	maximum a posteriori probability
MIMO	multiple-input multiple-output
MISO	multiple-input single-output
ML	maximum likelihood
MLE	maximum likelihood estimate

MMSE	minimum mean square error
MRC	maximal ratio combining
MSE	mean square error
MSNR	maximum signal to noise ratio
MUSIC	multiple signal classification
MVDR	minimum variance distortionless response
MVUE	minimum variance unbiased
PAM	pulse amplitude modulation
pdf	probability density function
PEP	pairwise error probability
QAM	quadrature amplitude modulation
QPSK	quadrature phase shift keying
RLS	recursive least square
ROC	receiver operating characteristics
SD	selection diversity
SDR	software defined radio
SLLN	strong law of large numbers
SMI	sample matrix inversion
SIC	successive interference cancellation
SIMO	single-input multiple-output
SISO	single-input single-output
SINR	signal to interference-plus-noise ratio
SNR	signal to noise ratio
SVP	shortest vector problem
ULA	uniform linear array
WLLN	weak law of large numbers
WLS	weighted least square
WSS	wide-sense stationary
ZF	zero-forcing

1 Introduction

Statistical signal processing is a set of tools for dealing with random signals. As a set of tools, statistical signal processing has a broad range of applications from radars and sonars to speech and image processing. There are a number of books on this topic (e.g. (Scharf 1991) and (Orfanidis 1988)). In this book, instead of providing a comprehensive description of statistical signal processing with a broad range of applications, we focus on key approaches for communications. In particular, we attempt to present mainly signal detection and combining techniques in the context of wireless communications.

1.1 Applications in digital communications

The main aim of digital communications is to transmit a sequence of bits over a given channel to a receiver with minimum errors. In implementing digital communication systems, however, there are various constraints to be taken into account. For example, the transmission power is usually limited and the complexity of receiver is also limited. With practical implementation constraints including computational complexity, statistical signal processing plays a crucial role in designing a receiver for digital communications. Although there are a number of different roles that statistical signal processing can play, we confine ourselves to two main topics in this book: one is signal detection and the other is signal combining.

Signal detection has been well established as the main topic in communications. However, advances in multiuser detection have opened up a whole new approach for joint detection (Verdu 1998). In this book, we focus on optimal and suboptimal approaches for joint detection in the context of multiple-input multiple-output (MIMO) communications.

Signal combining is to combine multiple observations and plays a crucial role in both array signal processing and wireless communications. In particular, in wireless communications, signal combining is essential to mitigate fading at the receiver equipped with multiple receive antennas. Furthermore, signal combining can be generalized to mitigate interfering signals as it can provide spatial selectivity in smart antennas. In this book, we discuss signal combining based on this generalized view.

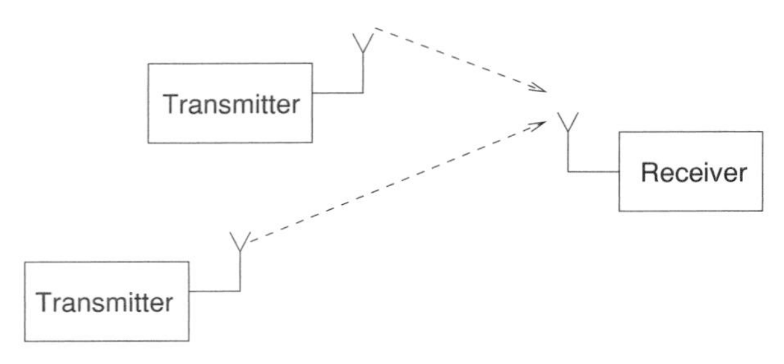

Figure 1.1 Multiuser communication.

1.2 Detection problems

Signal detection (Whalen 1971) (Kay 1998) is an application of statistical hypothesis testing in statistics. In statistical hypothesis testing, there are a finite number of hypotheses. With a set of observations, statistical hypothesis testing attempts to choose a hypothesis that explains observations best under a certain performance criterion. In signal detection, the set of hypotheses is decided by the signal alphabet for transmitted signals. Let $\mathcal{S}$ denote the signal alphabet. Then, the received signal over a memoryless channel at a receiver is given by

$$r = hs + n, \tag{1.1}$$

where $s \in \mathcal{S}$ is the transmitted signal, h is the channel gain, and n is the background noise. The receiver is to detect s from r provided that h is known and the statistical properties of n are also known. In general, the size of $\mathcal{S}$ is finite in digital communications. According to statistical hypothesis testing, signal detection is to choose a signal from $\mathcal{S}$, which can most likely generate r according to a certain criterion. Thus, it is important to define a detection criterion. Various decision criteria, which can be possibly employed for signal detection, have been proposed in statistical hypothesis testing. Some examples are the maximum likelihood (ML), Bayesian, maximum a posteriori probability (MAP) decision criteria. Since the theory of statistical hypothesis testing is well established in statistics, its application to signal detection is straightforward. In Chapter 2, we present an overview of detection theory based on the theory of statistical hypothesis testing.

While detection theory heavily relies on the theory of statistical hypothesis testing, the issues related to implementation with practical constraints are not quite covered by the theory. Most implementation issues are subject to a computational complexity constraint. To illustrate this issue, we can consider the signal detection problem in a multiuser communication system. If there are two users who transmit signals simultaneously as shown in Fig. 1.1, the received signal is given by

$$r = h_1 s_1 + h_2 s_2 + n, \tag{1.2}$$

where h_k and s_k are the channel gain and transmitted symbol from the kth user, respectively. For the signal detection in this case, which is called multiuser detection (Verdu 1998), we need to extend the signal alphabet as $\mathcal{S} \times \mathcal{S}$, where $\times$ denotes the

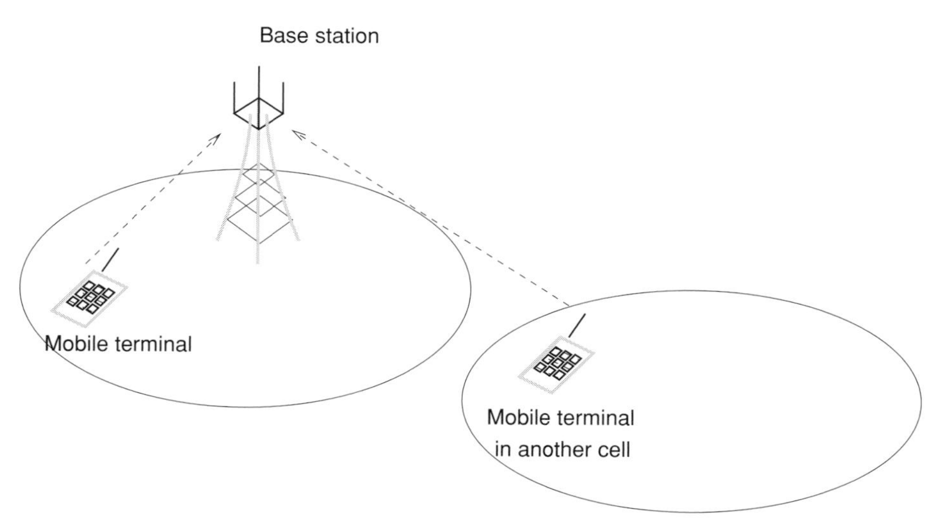

Figure 1.2 Multiuser detection in a cellular system.

Cartesian product, to detect s_1 and s_2 jointly. Then, it can be shown that the size of the signal alphabet becomes $|\mathcal{S} \times \mathcal{S}| = |\mathcal{S}|^2$. Since the size of the extended signal alphabet grows *exponentially* with the number of users who transmit signals simultaneously, the complexity of signal detection can grow *exponentially* with the number of users if an exhaustive search is used for detection. Thus, the computational complexity becomes one of the major issues as the number of users increases. Usually, a suboptimal approach is required to reduce the complexity at the expense of performance loss.

Since multiuser detection has become an important issue in wireless communications, various low complexity suboptimal detectors have been proposed for multiuser detection over the last two decades. An example of multiuser detection can be found in cellular systems. As shown in Fig. 1.2, a base station can receive the signals from a user in its cell as well as a user in another cell if both users transmit signals through a common channel. Another example arises when a transmitter transmits different signals through multiple antennas. In this case, s_1 and s_2 in (1.2) can be considered the signals transmitted from the first and second antennas, respectively. In Chapters 7–10, we present optimal and suboptimal detectors for multiuser detection in the context of MIMO, which allows a general description of multiuser detection problems and optimal/suboptimal detectors.

1.3 Combining problems

While a signal or symbol is chosen from a pre-determined signal set or alphabet in signal detection (the transmitted signal should be one of the signals in a signal alphabet), a signal can be estimated in signal estimation without the constraint that the signal belongs to the signal alphabet. Consider the received signal in (1.2). The estimated signal can be a function of the received signal, r, as follows:

$$\hat{s}_k = g_k(r), \quad k = 1, 2, \tag{1.3}$$

where $g_k(r)$ is the estimator for s_k. If a linear estimator is employed, we have

$$\hat{s}_k = g_k r, \quad k = 1, 2, \tag{1.4}$$

where g_k is the coefficient of a linear estimator for s_k, which is a real or complex number. Exploiting statistical properties of s_k and r, an estimator can be derived to minimize the difference between s_k and $\hat{s}_k$. To measure the difference, the minimum mean square error (MMSE) criterion is widely adopted (Wiener 1949) and the resulting estimator is called the MMSE estimator.

Signal estimation can be considered as an application of statistical estimation theory in statistics (Kay 1993). In Chapter 3, we present an overview of estimation theory. Various estimators and their properties are discussed in Chapter 3. In particular, we emphasize the linear MMSE estimator as it can be easily obtained for a number of applications and becomes optimal when signals and noise are Gaussian.

When multiple observations are available, it is necessary to combine them to estimate a certain signal or parameter. Signal combining is a generalization of signal estimation with multiple observations. Suppose that we have L observations or received signals as follows:

$$r_l = h_l s + n_l, \quad l = 0, 1, \ldots, L-1, \tag{1.5}$$

where h_l and n_l are the channel coefficient and background noise of the lth received signal, respectively, and s is the signal of interest. A combiner to estimate s is a function of $r_0, r_1, \ldots, r_{L-1}$, denoted by $g(\cdot)$, and its estimate is given by

$$\hat{s} = g(r_0, r_1, \ldots, r_{L-1}). \tag{1.6}$$

If a linear combiner is used, an estimate of s is given by

$$\hat{s} = \sum_{l=0}^{L-1} g_l r_l. \tag{1.7}$$

The coefficients, $g_l, l = 0, 1, \ldots, L-1$, can be found under various performance criteria. In particular, if the MMSE criterion is employed, these coefficients are readily found by applying the orthogonality principle.

In Chapter 4, we present optimal combining methods and discuss performance issues when a receiver can have multiple received signals in wireless communications.

An important application of signal combining in wireless communications is smart antenna. Smart antenna often refers to *signal processing algorithms* for antenna arrays (Van Trees 2002) in wireless communications. Thus, another name for smart antenna is antenna array processing for wireless communications. The main idea of smart antenna is to exploit that spatial selectivity that has been created by antenna arrays. Spatial selectivity becomes crucial in wireless communication environments where the performance is degraded by interfering signals. Exploiting spatial selectivity to mitigate interfering signals, the performance of wireless communication systems can be improved. In Chapter 5, we present key ideas of smart antenna with array processing algorithms for wireless communications.

A further generalization of signal combining is presented in Chapter 6 when multiple signals coexist. It is shown that the MMSE combiner plays a key role in joint signal combining.

1.4 Background

Appendices provide overviews of the necessary background knowledge for this book: namely, signals and systems, information theory, and vectors and matrices. Furthermore, since we deal with random signals in statistical signal processing, probability theory is essential as background knowledge. In this section, we briefly review key definitions and results of probability and random processes. For a detailed account, the reader is referred to (Papoulis 1984) (Porat 1994) (Leon-Garcia 1994).

1.4.1 Review of probability

A sample space Ω is the set of all possible outcomes (or events) of an experiment. An outcome A is a subset of Ω. A probability measure $\Pr(\cdot)$ is a mapping from Ω to the real line with the following properties:

(i) $\Pr(A) \geq 0,\ A \in \Omega$

(ii) $\Pr(\Omega) = 1$

(iii) For a countable set of events, $\{A_m\}$, if $A_l \cup A_m = \emptyset$, for $l \neq m$, then

$$\Pr\left(\bigcup_{l=1}^{\infty} A_l\right) = \sum_{l=1}^{\infty} \Pr(A_l).$$

This property is called the addition law of probabilities.

The joint probability of two events A and B is $\Pr(A \cap B)$. The conditional probability of A given B is

$$\Pr(A|B) = \frac{\Pr(A \cap B)}{\Pr(B)}, \quad \Pr(B) > 0.$$

A and B are independent if and only if

$$\Pr(A \cap B) = \Pr(A)\Pr(B)$$

and this implies

$$\Pr(A|B) = \Pr(A).$$

1.4.2 Random variables

A random variable is a mapping from an event ω in Ω to a real number, denoted by $X(\omega)$. The *cumulative distribution function* (cdf) of X is defined as

$$\begin{aligned} F_X(x) &= \Pr(\{\omega \mid X(\omega) \leq x\}) \\ &= \Pr(X \leq x), \end{aligned}$$

for any real number x, and the *probability density function* (pdf) is defined as

$$f_X(x) = \frac{\mathrm{d}}{\mathrm{d}x} F_X(x),$$

where the subscript X on F and f identifies the random variable. It can be shown that

$$F_X(x) = \int_{-\infty}^{x} f_X(z)\mathrm{d}z.$$

We have $F_X(\infty) = 1$, because $\Pr(X \leq \infty) = 1$. Furthermore, $\int_{-\infty}^{\infty} f_X(z)\mathrm{d}z = F_X(\infty) = 1$. The subscript X is often omitted if the context is clear.

Note that we use capital letters to denote random variables in this book if possible. For example, X is a random variable, while x is a variable. Random vectors will be written in boldface letters. For example, $\mathbf{x}$ is a random vector. This could lead to confusion, because a vector (not a random vector) will also be written in boldface letters. To avoid this confusion, we will make clear indication if necessary.

There are some well-known pdfs as follows:

(i) Gaussian pdf with mean μ and variance σ^2:

$$f(x) = \frac{1}{\sqrt{2\pi\sigma^2}} \exp\left(-\frac{1}{2}(x-\mu)^2\right).$$

As the Gaussian pdf is frequently used in this book, it is denoted by $\mathcal{N}(\mu, \sigma^2)$. That is, if X is a Gaussian random variable with mean μ and variance σ^2, we write $X \sim \mathcal{N}(\mu, \sigma^2)$.

(ii) Exponential pdf ($a > 0$):

$$f(x) = \begin{cases} a\mathrm{e}^{-ax}, & \text{if } x \geq 0; \\ 0, & \text{otherwise.} \end{cases}$$

(iii) Rayleigh pdf ($b > 0$):

$$f(x) = \begin{cases} \frac{x}{b}\mathrm{e}^{-\frac{x^2}{b}}, & \text{if } x \geq 0; \\ 0, & \text{otherwise} \end{cases}$$

(iv) Chi-square pdf of n degrees of freedom:

$$f(x) = \begin{cases} \frac{x^{\frac{n-2}{2}}\mathrm{e}^{-\frac{x}{2}}}{2^{\frac{n}{2}}\Gamma(\frac{n}{2})}, & \text{if } x \geq 0; \\ 0, & \text{otherwise,} \end{cases}$$

where $\Gamma(x)$ is the gamma function.

A joint cdf of random variables, $X_1, X_2, \ldots, X_n$, is defined as

$$F_{X_1,X_2,\ldots,X_n}(x_1, x_2, \ldots, x_n) = \Pr(X_1 \leq x_1, X_2 \leq x_2, \ldots, X_n \leq x_n).$$

The joint pdf is defined as

$$f_{X_1,X_2,\ldots,X_n}(x_1, x_2, \ldots, x_n) = \frac{\partial^n}{\partial x_1 \partial x_2 \ldots \partial x_n} F_{X_1,X_2,\ldots,X_n}(x_1, x_2, \ldots, x_n).$$

The conditional pdf of X_1 given $X_2 = x_2$ is defined as

$$f_{X_1|X_2}(x_1|x_2) = \frac{f_{X_1,X_2}(x_1, x_2)}{f_{X_2}(x_2)}, \quad f_{X_2}(x_2) > 0.$$

The expectation of X is defined as

$$\mathcal{E}[X] = \int x f_X(x) \mathrm{d}x.$$

In addition, the expectation of $g(X)$, a function of X, is given by

$$\mathcal{E}[g(X)] = \int g(x) f_X(x) \mathrm{d}x.$$

The variance of X is defined as

$$\begin{aligned} \mathrm{Var}(X) &= \mathcal{E}[(X - \mathcal{E}[X])^2] \\ &= \int (x - \mathcal{E}[X])^2 f_X(x) \mathrm{d}x. \end{aligned}$$

The conditional mean of X given $Y = y$ is defined as

$$\mathcal{E}[X|Y = y] = \int x f_{X|Y}(x|Y = y) \mathrm{d}x.$$

The pdf of a real-valued joint Gaussian random vector, $\mathbf{x}$, is given by

$$f(\mathbf{x}) = \frac{1}{\sqrt{\det(2\pi \mathbf{C}_\mathbf{x})}} \exp\left(-\frac{1}{2}(\mathbf{x} - \bar{\mathbf{x}})^{\mathrm{T}} \mathbf{C}_\mathbf{x}^{-1} (\mathbf{x} - \bar{\mathbf{x}})\right),$$

where $\bar{\mathbf{x}} = \mathcal{E}[\mathbf{x}]$ is the mean vector and $\mathbf{C}_\mathbf{x} = \mathcal{E}[(\mathbf{x} - \bar{\mathbf{x}})(\mathbf{x} - \bar{\mathbf{x}})^{\mathrm{T}}]$ is the covariance matrix. For convenience, if $\mathbf{x}$ is a Gaussian random vector with mean $\bar{\mathbf{x}}$ and covariance matrix $\mathbf{C}_\mathbf{x}$, we write $\mathbf{x} \sim \mathcal{N}(\bar{\mathbf{x}}, \mathbf{C}_\mathbf{x})$.

If $\mathbf{x}$ is a circularly symmetric complex Gaussian (CSCG) random vector,

$$f(\mathbf{x}) = \frac{1}{\det(\pi \mathbf{C}_\mathbf{x})} \exp\left(-(\mathbf{x} - \bar{\mathbf{x}})^{\mathrm{H}} \mathbf{C}_\mathbf{x}^{-1} (\mathbf{x} - \bar{\mathbf{x}})\right),$$

where $\bar{\mathbf{x}} = \mathcal{E}[\mathbf{x}]$ and $\mathbf{C}_\mathbf{x} = \mathcal{E}[(\mathbf{x} - \bar{\mathbf{x}})(\mathbf{x} - \bar{\mathbf{x}})^{\mathrm{H}}]$. For a CSCG random vector, we can show that

$$\mathcal{E}[(\mathbf{x} - \bar{\mathbf{x}})(\mathbf{x} - \bar{\mathbf{x}})^{\mathrm{T}}] = \mathbf{0}.$$

If $\mathbf{x}$ is a CSCG random vector with mean $\bar{\mathbf{x}}$ and covariance matrix $\mathbf{C}_\mathbf{x}$, we write $\mathbf{x} \sim \mathcal{CN}(\bar{\mathbf{x}}, \mathbf{C}_\mathbf{x})$.

A sequence of random variables, X_l, $l = 1, 2, \ldots$, is called an independent and identically distributed (iid) sequence if the X_l's are independent and their distributions are identical.

Let $X_1, X_2, \ldots, X_k$ be an iid sequence. Denote by μ and σ^2 the mean and variance of X_l, respectively. For given any $\delta > 0$ and $\epsilon > 0$, there is an integer k that satisfies

$$\Pr\left(\mu - \epsilon \le \frac{X_1 + X_2 + \cdots + X_k}{k} \le \mu + \epsilon\right) > 1 - \delta.$$

This behavior of an iid sequence is called the weak law of large numbers (WLLN).

If $X_l, l = 1, 2, \ldots$, are iid with a finite mean μ, then

$$\Pr\left(\lim_{k\to\infty} \frac{X_1 + X_2 + \cdots + X_k}{k} = \mu\right) = 1.$$

This is called the strong law of large numbers (SLLN).

Let X_l be a sequence of independent random variables with $\bar{x}_l = \mathcal{E}[X_l]$ and $\sigma_l^2 = \mathrm{Var}(X_l)$. Consider the normalized sum as follows:

$$\begin{aligned} Y_k &= \frac{S_k - \mathcal{E}[S_k]}{\mathrm{Var}(S_k)} \\ &= \frac{\sum_{l=1}^k X_l - \bar{x}_l}{\sum_{l=1}^k \sigma_l^2}, \end{aligned}$$

where $S_k = \sum_{l=1}^k X_l$. If

$$\lim_{k\to\infty} \Pr(a \le Y_k \le b) = \frac{1}{\sqrt{2\pi}} \int_a^b \mathrm{e}^{-\frac{y^2}{2}} \mathrm{d}y, \quad (a < b)$$

the sequence X_l, $l = 1, 2, \ldots$, is said to satisfy the central limit theorem (CLT). If a sequence satisfies the following condition, called the Lyapunov condition:

$$\lim_{k\to\infty} \frac{\sum_{l=1}^k \mathcal{E}[|X_l - \bar{x}_l|^3]}{\left(\sum_{l=1}^k \sigma_l^2\right)^{\frac{3}{2}}} = 0,$$

this sequence satisfies the CLT.

For any iid sequence, $X_l, l = 1, 2, \ldots$, if $\mathcal{E}[|X_l - \mu|^3] = \alpha < \infty$, where $\mu = \mathcal{E}[X_l]$, the Lyapunov condition is satisfied. This implies that this sequence satisfies the CLT. To see whether or not the Lyapunov condition is satisfied, we can show that

$$\frac{\sum_{l=1}^k \mathcal{E}[|X_l - \bar{x}_l|^3]}{\left(\sum_{l=1}^k \sigma_l^2\right)^{\frac{3}{2}}} = \frac{k\alpha}{k^{\frac{3}{2}}\sigma^3} = \frac{1}{\sqrt{k}} \frac{\alpha}{\sigma^3},$$

where σ^2 is the variance of X_l. From this, we can see that the Lyapunov condition is satisfied as $\lim_{k\to\infty} \frac{1}{\sqrt{k}} \frac{\alpha}{\sigma^3} = 0$.

1.4.3 Random processes

A (discrete-time) random process is a sequence of random variables, $\{x_m\}$. The mean and autocorrelation function of $\{x_m\}$ are denoted by $\mathcal{E}[x_m]$ and $R_x(l, m) = \mathcal{E}[x_l x_m^*]$, respectively. A random process is called wide-sense stationary (WSS) if

$$\mathcal{E}[x_l] = \mu, \quad \text{for all } l;$$

$$\mathcal{E}[x_l x_m^*] = \mathcal{E}[x_{l+p} x_{m+p}^*], \quad \text{for all } l, m, p.$$

For a WSS random process, we have

$$R_x(l, m) = R_x(l - m).$$

The power spectrum of a WSS random process, $\{x_m\}$, is defined as

$$S_x(z) = \sum_{l=-\infty}^{\infty} z^{-l} R_x(l).$$

In addition, we can show that

$$R_x(l) = \frac{1}{2\pi} \int_{-\pi}^{\pi} S_x(e^{j\omega}) e^{jl\omega} d\omega,$$

where $z = e^{j\omega}$, $-\pi \le \omega < \pi$.

A zero-mean WSS random process is called white if

$$R_x(l) = \begin{cases} \sigma_x^2, & \text{if } l = 0; \\ 0, & \text{otherwise,} \end{cases}$$

where $\sigma_x^2 = \mathcal{E}[|x_l|^2]$. It can be shown that the power spectrum of a zero-mean white WSS random process is flat, i.e. $S_x(e^{j\omega}) = \sigma_x^2$ for all ω.

If x_l is an output of the linear system whose impulse response is $\{h_p\}$ with a zero-mean white random process input, n_l, the autocorrelation function of $\{x_m\}$ is given by

$$R_x(l) = \sigma_n^2 \sum_l h_l h_{l-m},$$

where $\sigma_n^2 = \mathcal{E}[|n_l|^2]$. In addition, its power spectrum is given by

$$S_x(z) = \sigma_n^2 H(z) H(z^{-1})$$

or

$$S_x(e^{j\omega}) = \sigma_n^2 |H(e^{j\omega})|^2, \quad z = e^{j\omega}.$$

2 Fundamentals of detection theory

Statistical hypothesis testing is a process to accept or reject a hypothesis based on observations, where multiple hypotheses are proposed to characterize observations. Taking observations as realizations of a certain random variable, each hypothesis can be described by a different probability distribution of the random variable. Under a certain criterion, a hypothesis can be accepted for given observations. Signal detection is an application of statistical hypothesis testing.

In this chapter, we present an overview of signal detection and introduce key techniques for performance analysis. We mainly focus on fundamentals of signal detection in this chapter, while various signal detection problems and detection algorithms will be discussed in later chapters (e.g. Chapters 7, 8, and 9).

2.1 Elements of hypothesis testing

There are three key elements in statistical hypothesis testing, which are (i) observation(s); (ii) set of hypotheses; and (iii) prior information. With these key elements, the decision process or hypothesis testing can be illustrated as in Fig. 2.1.

In statistical hypothesis testing, observations and prior information are all important and should be taken into account. However, in some cases, no prior information is available or prior information could be useless. In this case, statistical hypothesis testing relies only on observations.

Suppose that there are M (≥ 2) hypotheses. Then, we have an M-ary hypothesis testing in which we will choose one of the M hypotheses that are proposed to characterize observations and prior information under a certain performance criterion. There are various hypothesis tests or decision rules depending on criteria. Some examples are given as follows.

- Maximum a posteriori probability (MAP) decision rule
- Bayesian decision rule
- Maximum likelihood (ML) decision rule

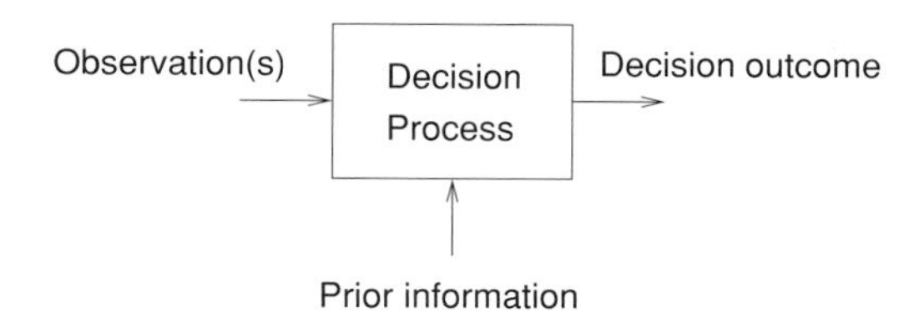

Figure 2.1 Block diagram for decision process.

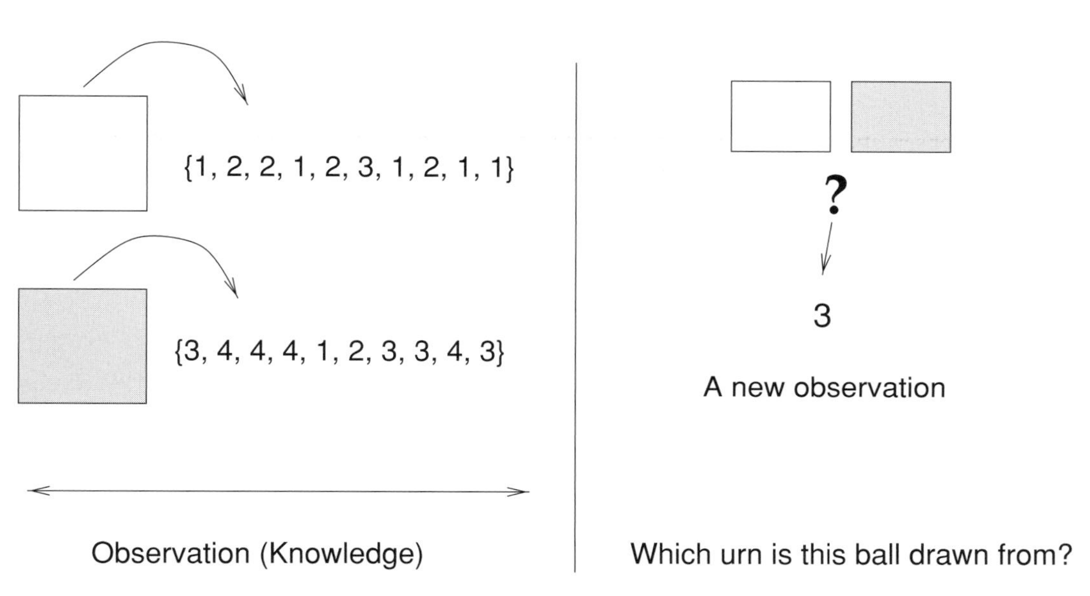

Figure 2.2 Two urns of balls.

These well-known decision rules can be generalized by the likelihood ratio (LR) based decision rule.

2.1.1 MAP decision rule

Consider an example to illustrate how the MAP decision rule works. Suppose that there are two urns (white and gray) in which there are a number of balls. A certain number is marked on each ball. It is assumed that the distribution of the numbers on balls is different for each urn. A ball is drawn from one of the urns and we want to decide which urn the ball is drawn from based on the number on the drawn ball. We can build the following two hypotheses:

$$\begin{aligned} H_0 &: \text{ the ball is drawn from the white urn;} \\ H_1 &: \text{ the ball is drawn from the gray urn.} \end{aligned}$$

To know the distribution of the numbers on balls for each urn, suppose that 10 balls are drawn from each urn as shown in Fig. 2.2.

Based on the empirical distribution results in Fig. 2.2, we can build conditional distributions of the numbers on balls as follows:

$$\Pr(1|H_0) = \frac{5}{10};$$
$$\Pr(2|H_0) = \frac{4}{10};$$
$$\Pr(3|H_0) = \frac{1}{10};$$

and

$$\Pr(1|H_1) = \frac{1}{10};$$
$$\Pr(2|H_1) = \frac{1}{10};$$
$$\Pr(3|H_1) = \frac{4}{10};$$
$$\Pr(4|H_1) = \frac{4}{10}.$$

In addition, we assume that the probability that white (H_0) or gray urn (H_1) is chosen is the same, i.e.

$$\Pr(H_0) = \Pr(H_1) = \frac{1}{2}.$$

Then, we can have

$$\Pr(1) = \frac{6}{20};\ \Pr(2) = \frac{5}{20};\ \Pr(3) = \frac{5}{20};\ \Pr(4) = \frac{4}{20},$$

where $\Pr(k)$ denotes the probability that the ball on which number k is marked is drawn. Taking $\Pr(H_m)$ as the a priori probability of H_m, we can find the a posteriori probability of H_m as follows:

$$\Pr(H_0|1) = \frac{5}{6};$$
$$\Pr(H_1|1) = \frac{1}{6};$$
$$\Pr(H_0|2) = \frac{4}{5};$$
$$\Pr(H_1|2) = \frac{1}{5};$$
$$\Pr(H_0|3) = \frac{1}{5};$$
$$\Pr(H_1|3) = \frac{4}{5};$$
$$\Pr(H_0|4) = 0;$$
$$\Pr(H_1|4) = 1,$$

where $\Pr(H_m|k)$ can also be interpreted as the conditional probability that hypothesis H_m is true provided that the number on the drawn ball is k. As in Fig. 2.2, if the number on the ball is $k = 3$, we decide that the ball is drawn from the "gray" urn (in other words, we accept the hypothesis H_1) since $\Pr(H_1|3) = \frac{4}{5}$ is greater than $\Pr(H_0|3) = \frac{1}{5}$. As in this example, if we choose the hypothesis that maximizes the a posteriori probability, the resulting decision process is called the MAP hypothesis testing.

We now build formally the MAP decision rule for binary hypothesis testing. In general, H_0 and H_1 are referred to as the null hypothesis and the alternative hypothesis, respectively, in the binary hypothesis testing. We assume that the a priori probabilities, $\Pr(H_0)$ and $\Pr(H_1)$, are known. In addition, the conditional probability, $\Pr(X|H_m)$, is given, where X denotes the random variable for an observation. Then, the MAP decision rule is given by

$$\Pr(H_0|X = x) \underset{H_1}{\overset{H_0}{\gtrless}} \Pr(H_1|X = x), \tag{2.1}$$

where H_0 is chosen if $\Pr(H_0|X = x) > \Pr(H_1|X = x)$ and vice versa. Here, x denotes the realization of X. Note that in (2.1), we do not consider the case where $\Pr(H_0|X = x) = \Pr(H_1|X = x)$. In this case, a decision can be made arbitrarily (we will discuss this issue later). As in (2.1), the decision outcome can be considered as a function of x. Using Bayes' rule, we can also show that

$$\frac{\Pr(X = x|H_0)}{\Pr(X = x|H_1)} \underset{H_1}{\overset{H_0}{\gtrless}} \frac{\Pr(H_1)}{\Pr(H_0)}.$$

Note that if X is a continuous random variable, $\Pr(X = x|H_m)$ is replaced by $f(X = x|H_m)$, where $f(X|H_m)$ denotes the conditional pdf of X given H_m.

Example 2.1.1 Consider a random signal, denoted by X, with the following hypothesis pair:

$$\begin{aligned} H_0 &: X = N; \\ H_1 &: X = v + N, \end{aligned} \tag{2.2}$$

where N is a Gaussian random variable with zero mean and variance σ^2 and v is a positive constant. The decision rule associated with the hypothesis pair in (2.2) is to decide whether or not a constant signal (i.e. v) is present when the observation, X, is corrupted by the noise, N. Let $\mathcal{N}(\mu, \sigma^2)$ denote the pdf of a Gaussian random variable where μ and σ^2 denote the mean and variance, respectively. That is,

$$\mathcal{N}(\mu, \sigma^2) = \frac{1}{\sqrt{2\pi\sigma^2}} \exp\left(-\frac{(x-\mu)^2}{2\sigma^2}\right).$$

Then, we have

$$\begin{aligned} f(x|H_0) &= \mathcal{N}(0, \sigma^2); \\ f(x|H_1) &= \mathcal{N}(v, \sigma^2). \end{aligned}$$

Table 2.1 Four possible cases after decision.

Accept	H_0 is true	H_1 is true
H_0	Correct	Type II error (miss)
H_1	Type I error (false alarm)	Correct (detection)

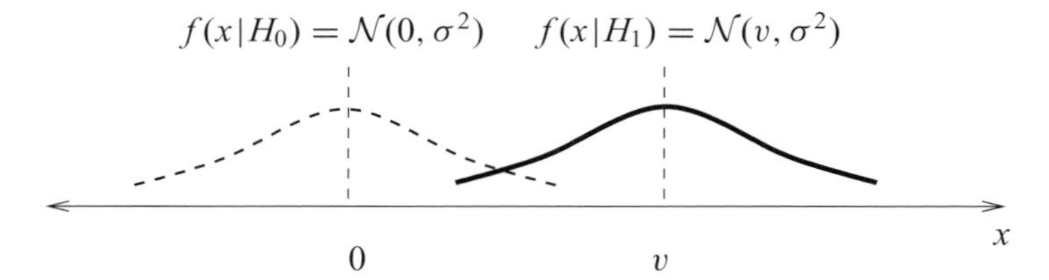

Figure 2.3 Illustration of the hypothesis pair in Eq. (2.2) when $v > 0$.

Suppose that $v > 0$. The two pdfs for the hypothesis pair in (2.2) are illustrated in Fig. 2.3. Since

$$\frac{f(X = x \mid H_0)}{f(X = x \mid H_1)} = \exp\left(-\frac{v(2x - v)}{2\sigma^2}\right),$$

the MAP decision rule is simplified as follows:

$$x \underset{H_0}{\overset{H_1}{\gtrless}} \frac{v}{2} - \frac{\sigma^2 \log \tau}{v}, \tag{2.3}$$

where $\tau = \frac{\Pr(H_0)}{\Pr(H_1)}$.

As mentioned earlier, the decision rule is a function of x. The decision rule $t(x)$ can be expressed as

$$t(x) = \begin{cases} 0, & \text{if } x \in \mathcal{X}_0; \\ 1, & \text{if } x \in \mathcal{X}_1, \end{cases}$$

where $\mathcal{X}_0$ and $\mathcal{X}_1$ are the decision regions of H_0 and H_1, respectively. For example, the decision regions for the MAP decision rule in (2.3) are given by

$$\mathcal{X}_0 = \left\{ x \mid x < \frac{v}{2} - \frac{\sigma^2 \ln \tau}{v} \right\};$$

$$\mathcal{X}_1 = \left\{ x \mid x > \frac{v}{2} - \frac{\sigma^2 \ln \tau}{v} \right\}.$$

Note that $\mathcal{X}_1$ is also called the rejection region or critical region, while $\mathcal{X}_0$ is called the acceptance region in the binary hypothesis testing.

As shown in Table 2.1, there are the four possible cases after decision. Among them, the following two types of decision errors are usually considered to characterize the performance:

Type I error (false alarm):	Accept H_1 when H_0 is true;
Type II error (miss):	Accept H_0 when H_1 is true.

Using the decision rule $t(x)$, the probability of type I error or false alarm (FA) probability can be given by

$$\begin{aligned} P_{\mathrm{FA}} &= \Pr(X \in \mathcal{X}_1 | H_0) \\ &= \int t(x) f(x | H_0) \mathrm{d}x \\ &= \mathcal{E}[t(X) | H_0]. \end{aligned} \tag{2.4}$$

The probability of type II error or miss probability is given by

$$\begin{aligned} P_{\mathrm{MS}} &= \Pr(X \in \mathcal{X}_0 | H_1) \\ &= \int (1 - t(x)) f(x | H_1) \mathrm{d}x \\ &= \mathcal{E}[1 - t(X) | H_1], \end{aligned} \tag{2.5}$$

while the probability of detection is given by

$$\begin{aligned} P_{\mathrm{D}} &= 1 - P_{\mathrm{MS}} \\ &= \mathcal{E}[t(X) | H_1]. \end{aligned} \tag{2.6}$$

The reason why type I and II error probabilities are called the FA and miss probabilities, respectively, is that the null hypothesis, H_0, often represents the case that no signal is present, while the alternative hypothesis, H_1, represents the case that a signal is present.

2.1.2 Bayesian decision rule

The Bayesian decision rule is to minimize the cost associated with the decision. Let D_m denote the decision that accepts H_m. In addition, denote by C_{im} the associated cost of D_i when the hypothesis H_m is true. Assume that

$$\begin{aligned} C_{10} &> C_{00}; \\ C_{01} &> C_{11}. \end{aligned}$$

That is, the cost of erroneous decision is higher than the cost of correct decision. It is possible to minimize the average cost $\mathcal{E}[C_{im}]$ by properly deciding the acceptance regions, $\mathcal{X}_0$ and $\mathcal{X}_1$. The average cost is given by

$$\begin{aligned}\bar{C} &= \mathcal{E}[C_{im}] \\ &= \sum_i \sum_m C_{im} \Pr(D_i, H_m) \\ &= \sum_i \sum_m C_{im} \Pr(D_i | H_m) \Pr(H_m). \end{aligned} \tag{2.7}$$

To find the optimal decision rule that minimizes the cost, we can consider the following problem:

$$\min_{\mathcal{X}_0, \mathcal{X}_1} \bar{C}. \tag{2.8}$$

For convenience, assume that $\mathcal{X}_1 = \mathcal{X}_0^c$, where $\mathcal{X}_0^c$ denotes the complementary set of $\mathcal{X}_0$. Since

$$\Pr(D_1 | H_m) = 1 - \Pr(D_0 | H_m),$$

the average cost in (2.7) can be expressed as

$$\bar{C} = C_{10} \Pr(H_0) + C_{11} \Pr(H_1) + \int_{\mathcal{X}_0} g_1(x) - g_0(x) \mathrm{d}x,$$

where

$$\begin{aligned} g_1(x) &= \Pr(H_1)(C_{01} - C_{11}) f(x | H_1); \\ g_0(x) &= \Pr(H_0)(C_{10} - C_{00}) f(x | H_0). \end{aligned}$$

It follows that

$$\bar{C} = \text{Constant} + \int_{\mathcal{X}_0} g_1(x) - g_0(x) \mathrm{d}x.$$

Thus, we have

$$\min_{\mathcal{X}_0} \bar{C} \Leftrightarrow \min_{\mathcal{X}_0} \left\{ \int_{\mathcal{X}_0} g_1(x) - g_0(x) \mathrm{d}x \right\}.$$

The optimal region that minimizes the cost is now given by

$$\mathcal{X}_0 = \{x \mid g_1(x) \leq g_0(x)\}.$$

In addition, we have

$$\mathcal{X}_1 = \{x \mid g_1(x) > g_0(x)\}.$$

Consequently, the Bayesian decision rule that minimizes the cost is given by

$$\frac{g_0(x)}{g_1(x)} \underset{H_1}{\overset{H_0}{\gtrless}} 1$$

or

$$\frac{f(x|H_0)}{f(x|H_1)} \underset{H_1}{\overset{H_0}{\gtrless}} \frac{\Pr(H_0)}{\Pr(H_1)} \frac{C_{01} - C_{11}}{C_{10} - C_{00}}. \tag{2.9}$$

From (2.9), we can see that the ratio of the cost differences, $\frac{C_{01}-C_{11}}{C_{10}-C_{00}}$, can characterize the Bayesian decision rule rather than the values of individual costs, C_{im}'s, for binary hypothesis testing. It is noteworthy that $(C_{10} - C_{00})$ and $(C_{01} - C_{11})$ are positive. In addition, if $C_{10} - C_{00} = C_{01} - C_{11}$, then the Bayesian decision rule becomes the MAP hypothesis test.

2.1.3 ML decision rule

If the a priori probability is not available, the MAP decision rule may not be applicable. In this case, we can consider another decision rule based on likelihood functions. For a given value of observation, x, the likelihood function is defined as

$$f_m(x) = f(x|H_m), \ m = 0, 1. \tag{2.10}$$

The likelihood function is a function of the hypothesis or m, not a function of x as x is given. The maximum likelihood (ML) decision rule is to choose the hypothesis that maximizes the likelihood. That is,

$$f_0(x) \underset{H_1}{\overset{H_0}{\gtrless}} f_1(x)$$

or

$$\frac{f_0(x)}{f_1(x)} \underset{H_1}{\overset{H_0}{\gtrless}} 1. \tag{2.11}$$

The ratio on the left-hand side in (2.11), $f_0(x)/f_1(x)$ is called the likelihood ratio (LR). It is often convenient to take the logarithm (base e) and the log-likelihood ratio (LLR) is defined as follows:

$$\text{LLR}(x) = \log \frac{f_0(x)}{f_1(x)}. \tag{2.12}$$

Now, the ML decision rule can be given by

$$\text{LLR}(x) \underset{H_1}{\overset{H_0}{\gtrless}} 0. \tag{2.13}$$

Note that the MAP decision rule is reduced to the ML decision rule if $\Pr(H_0) = \Pr(H_1)$. In other words, the ML decision rule is a special case of the MAP decision rule when the a priori probabilities are the same.

2.1.4 Likelihood ratio (LR) based decision rule

To generalize the ML decision rule, we can consider the following LR-based decision rule:

$$\frac{f_0(x)}{f_1(x)} \underset{H_1}{\overset{H_0}{\gtrless}} \tau, \tag{2.14}$$

where τ is a threshold that is pre-determined.

Example 2.1.2 Consider the following hypothesis pair:

$$H_0 : X = \mu_0 + N;$$
$$H_1 : X = \mu_1 + N,$$

where $\mu_0 < \mu_1$ and $N \sim \mathcal{N}(0, \sigma^2)$. Then, we have

$$f_0(x) = \mathcal{N}(\mu_0, \sigma^2);$$
$$f_1(x) = \mathcal{N}(\mu_1, \sigma^2).$$

The LLR is given by

$$\text{LLR}(x) = -\frac{(\mu_1 - \mu_0)}{\sigma^2}\left(x - \frac{\mu_0 + \mu_1}{2}\right).$$

The LR-based decision rule becomes

$$\left(x - \frac{\mu_0 + \mu_1}{2}\right) \underset{H_0}{\overset{H_1}{\gtrless}} \frac{\sigma^2}{\mu_1 - \mu_0} \log \tau.$$

Let

$$\bar{\tau} = \frac{\sigma^2}{\mu_1 - \mu_0} \log \tau + \frac{\mu_0 + \mu_1}{2}.$$

Then, the LR-based decision rule is reduced to

$$x \underset{H_0}{\overset{H_1}{\gtrless}} \bar{\tau}. \tag{2.15}$$

The LR-based decision rule can be considered as a generalization of the MAP, Bayesian, or ML decision rule. In particular, from (2.14), if we let

$$\tau = \frac{\Pr(H_0)(C_{10} - C_{00})}{\Pr(H_1)(C_{01} - C_{11})},$$

then the LR-based decision rule becomes the Bayesian decision rule. Figure 2.4 shows the relationship of various decision rules for binary hypothesis testing.

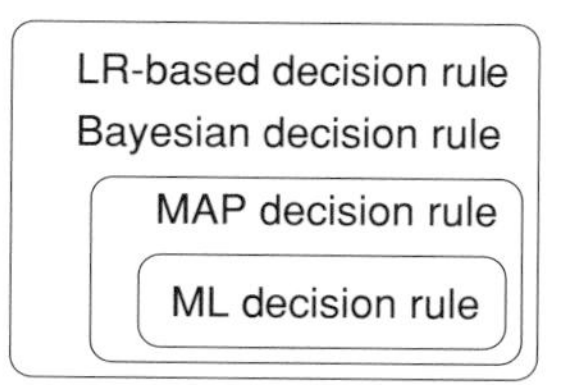

Figure 2.4 Relationship of various decision rules for binary hypothesis testing.

2.2 Neyman–Pearson lemma

For each decision rule, we can find the detection probability and FA probability. Now, we consider the reverse approach: for a given target decision probability or FA probability, derive a decision rule. To derive an optimal decision rule, consider the following optimization problem:

$$\max_t P_{\mathrm{D}}(t) \quad \text{subject to } P_{\mathrm{FA}}(t) \leq \alpha, \tag{2.16}$$

where t denotes a decision rule and α is the maximum FA probability. In (2.16), the solution is the optimal decision rule that maximizes the detection probability subject to the maximum FA probability constraint.

Lemma 2.2.1 (*Neyman–Pearson Lemma*) *Define the decision rule t' as*

$$t'(x) = \begin{cases} 1, & \text{if } f_1(x) > \eta f_0(x); \\ \gamma(x), & \text{if } f_1(x) = \eta f_0(x); \\ 0, & \text{if } f_1(x) < \eta f_0(x), \end{cases} \tag{2.17}$$

where $\eta > 0$ and $\gamma(x) \in [0, 1]$ is a randomized rule for the case of $f_1(x) = \eta f_0(x)$. For example,

$$\gamma(x) = \begin{cases} 1, & \text{with probability } q; \\ 0, & \text{with probability } 1 - q, \end{cases}$$

where q is decided such that $P_{\mathrm{FA}} = \alpha$. Then, the decision rule in (2.17) is the solution of the problem in (2.16) and called the Neyman–Pearson (NP) rule.

Proof: Suppose that $\tilde{t}$ is any decision rule that satisfies $P_{\mathrm{FA}} \leq \alpha$. In order to show the optimality of t', we need to show that

$$P_{\mathrm{D}}(t') \geq P_{\mathrm{D}}(\tilde{t}) \quad \text{for any } \tilde{t}.$$

Let $\mathcal{X}$ denote the observation set, i.e. $x \in \mathcal{X}$. For any $x \in \mathcal{X}$, (by the definition of t') it follows that

$$(t'(x) - \tilde{t}(x))(f_1(x) - \eta f_0(x)) \geq 0.$$

Then, we have

$$\int_{x \in \mathcal{X}} (t'(x) - \tilde{t}(x))(f_1(x) - \eta f_0(x))\mathrm{d}x \geq 0$$

or

$$\int_{x \in \mathcal{X}} t'(x) f_1(x)\mathrm{d}x - \int_{x \in \mathcal{X}} \tilde{t}(x) f_1(x)\mathrm{d}x \\ \geq \eta \left(\int_{x \in \mathcal{X}} t'(x) f_0(x)\mathrm{d}x - \int_{x \in \mathcal{X}} \tilde{t}(x) f_0(x)\mathrm{d}x \right).$$

From this, we can show that

$$P_{\mathrm{D}}(t') - P_{\mathrm{D}}(\tilde{t}) \geq \eta(P_{\mathrm{FA}}(t') - P_{\mathrm{FA}}(\tilde{t})) \geq 0.$$

This implies that $P_{\mathrm{D}}(t') \geq P_{\mathrm{D}}(\tilde{t})$, which completes the proof. □

As shown in (2.17), the NP decision rule is identical to the LR-based decision rule with the threshold $\tau = 1/\eta$ except the randomized rule (when $f_1(x) = \eta f_0(x)$).

Example 2.2.1 Consider the following hypothesis pair:

$$H_0 : X = N;$$
$$H_1 : X = v + N,$$

where $v > 0$ and $N \sim \mathcal{N}(0, \sigma^2)$. According to (2.15), the FA probability is given by

$$\begin{aligned} \alpha &= \int_{-\infty}^{\infty} t(x) f_0(x)\mathrm{d}x \\ &= \int_{\bar{\tau}}^{\infty} f_0(x)\mathrm{d}x \\ &= \int_{\bar{\tau}}^{\infty} \frac{1}{\sqrt{2\pi\sigma^2}} \exp\left(-\frac{x^2}{2\sigma^2}\right) \mathrm{d}x. \end{aligned}$$

Define the *Q-function* as

$$\mathcal{Q}(x) = \int_x^{\infty} \frac{1}{\sqrt{2\pi}} \mathrm{e}^{-t^2/2}\mathrm{d}t,$$

where $\mathcal{Q}(x)$, $x \geq 0$, is the tail of the normalized Gaussian pdf (i.e., $\mathcal{N}(0, 1)$) from x to ∞. Then, we can show that

$$\alpha = \mathcal{Q}\left(\frac{\bar{\tau}}{\sigma}\right)$$

or

$$\bar{\tau} = \sigma \mathcal{Q}^{-1}(\alpha).$$

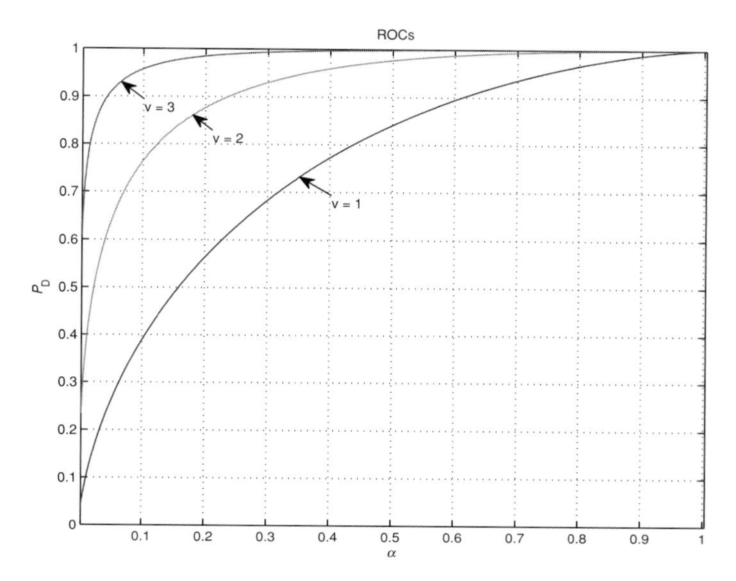

Figure 2.5 Receiver operating characteristics (ROCs) of the NP decision rule in Example 2.2.1.

The detection probability is

$$\begin{aligned} P_{\mathrm{D}} &= \int_{\bar{\tau}}^{\infty} f_1(x)\mathrm{d}x \\ &= \int_{\bar{\tau}}^{\infty} \frac{1}{\sqrt{2\pi\sigma^2}} \exp\left(-\frac{(x-v)^2}{2\sigma^2}\right) \mathrm{d}x \\ &= \mathcal{Q}\left(\frac{\bar{\tau}-v}{\sigma}\right) \\ &= \mathcal{Q}\left(\mathcal{Q}^{-1}(\alpha) - \frac{v}{\sigma}\right). \end{aligned}$$

This result shows that the detection probability is a function of the FA probability.

A parametric plot of the relationship between the detection and FA probabilities is called the receiver operating characteristics (ROCs). In Fig. 2.5, the ROCs of the NP decision rule in Example 2.2.1 are shown for different values of v.

2.3 Symmetric signal detection

In digital communications, symmetric signal detection problems are frequently considered. In this section, we focus on a symmetric signal detection problem.

Suppose that the hypotheses of interest are given by

$$\begin{aligned} H_0 &: X = V + N; \\ H_1 &: X = -V + N, \end{aligned} \tag{2.18}$$

where X is the received signal, $V > 0$, and $N \sim \mathcal{N}(0, \sigma^2)$. Due to the symmetry in this case, the probabilities of type I and type II errors are the same:

$$P_{\text{err}} = \Pr(\text{Accept } H_0 \mid H_1) = \Pr(\text{Accept } H_1 \mid H_0).$$

For a given $X = x$, the LLR function becomes

$$\begin{aligned} \text{LLR}(x) &= \log\left(\frac{f_0(x)}{f_1(x)}\right) \\ &= -\frac{1}{2\sigma^2}\left((x - V)^2 - (x + V)^2\right) \\ &= \frac{2V}{\sigma^2}x. \end{aligned}$$

Based on the ML decision rule, we have

$$\text{LLR}(x) \underset{H_1}{\overset{H_0}{\gtrless}} 0 \tag{2.19}$$

or the ML decision rule is simplified as

$$x \underset{H_1}{\overset{H_0}{\gtrless}} 0.$$

That is, the ML detection is simply a hard-decision of the observation $X = x$.

2.3.1 Error probability

Due to the symmetry in (2.18), the error probability is given by

$$P_{\text{err}} = \Pr(\text{LLR}(X) > 0 | H_1) = \Pr(\text{LLR}(X) < 0 | H_0).$$

It can be shown that

$$\begin{aligned} P_{\text{err}} &= \Pr(\text{LLR}(X) > 0 | H_1) \\ &= \Pr(X > 0 | H_1) \\ &= \int_0^{\infty} \frac{1}{\sqrt{2\pi}\sigma} \mathrm{e}^{-\frac{1}{2\sigma^2}(x+V)^2} \mathrm{d}x. \end{aligned} \tag{2.20}$$

We can also show that

$$\begin{aligned} P_{\text{err}} &= \int_{\frac{V}{\sigma}}^{\infty} \frac{1}{\sqrt{2\pi}} e^{-\frac{x^2}{2}} dx \\ &= \mathcal{Q}\left(\frac{V}{\sigma}\right). \end{aligned}$$

If we define the signal-to-noise ratio (SNR) as

$$\text{SNR} = \frac{V^2}{\sigma^2}.$$

Then, $P_{\text{err}} = \mathcal{Q}(\sqrt{\text{SNR}})$. Since $\mathcal{Q}(x)$ is a decreasing function, we can see that the error probability decreases with the SNR.

2.3.2 Bounds on error probability for Gaussian noise

For some cases, it is useful to derive bounds on the error probability. The complementary error function is defined as

$$\begin{aligned} \text{erfc}(x) &= \frac{2}{\sqrt{\pi}} \int_{x}^{\infty} e^{-t^2} dt \\ &= 1 - \text{erfc}(x), \quad \text{for } x > 0, \end{aligned}$$

where the error function is defined as $\text{erf}(x) = \frac{2}{\sqrt{\pi}} \int_0^x e^{-x^2} dx$. The relationship between the Q-function, $\mathcal{Q}(\cdot)$, and the complementary error function is given by

$$\begin{aligned} \text{erfc}(x) &= 2\mathcal{Q}\left(\sqrt{2}x\right); \\ \mathcal{Q}(x) &= \frac{1}{2}\text{erfc}\left(\frac{x}{\sqrt{2}}\right). \end{aligned}$$

The complementary error function has the following bounds:

$$\left(1 - \frac{1}{2x^2}\right) \frac{e^{-x^2}}{\sqrt{\pi}x} < \text{erfc}(x) < \frac{e^{-x^2}}{\sqrt{\pi}x},$$

where the lower bound is valid if $x > 1/\sqrt{2}$. Thus, $\mathcal{Q}(\cdot)$ is bounded as follows:

$$\left(1 - \frac{1}{x^2}\right) \frac{e^{-x^2/2}}{\sqrt{2\pi}x} < \mathcal{Q}(x) < \frac{e^{-x^2/2}}{\sqrt{2\pi}x},$$

where the lower bound is valid if $x > 1$. For the error probability, we can have the following upper bound:

$$\begin{aligned} P_{\text{err}} &= \mathcal{Q}(\sqrt{\text{SNR}}) \\ &< \frac{e^{-\text{SNR}/2}}{\sqrt{2\pi}\sqrt{\text{SNR}}}. \end{aligned}$$

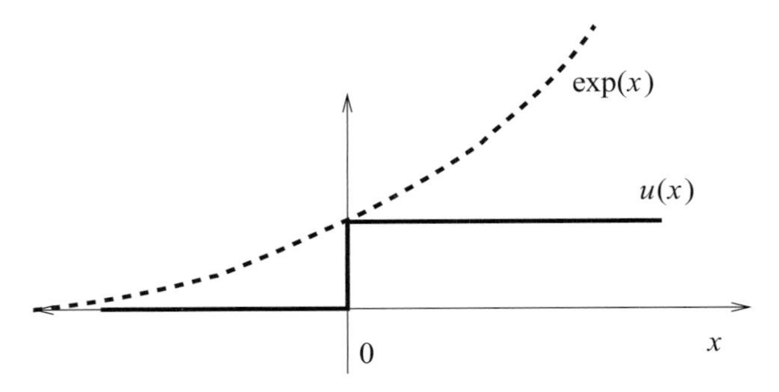

Figure 2.6 An upper bound on the step function.

2.3.3 Chernoff bound

The Chernoff bound is very useful in providing an upper bound on the probability of an event that happens rarely (thus, the probability is usually very small). The bound can be derived for any background noise.

Denote by $u(x)$ the step function, where $u(x) = 1$, if $x \geq 0$, and $u(x) = 0$, otherwise. Using the step function, the probability for the event that $X \geq x$ (which is usually called the tail probability) is given by

$$\begin{aligned}\Pr(X \geq x) &= \int_x^\infty f_X(\tau)\mathrm{d}\tau \\ &= \int_{-\infty}^\infty u(\tau - x) f_X(\tau)\mathrm{d}\tau \\ &= \mathcal{E}[u(X - x)],\end{aligned}$$

where $f_X(\tau)$ denotes the pdf of X. As shown in Fig. 2.6, since

$$u(x) \leq \mathrm{e}^x,$$

it can be shown that

$$\Pr(X \geq x) \leq \mathcal{E}[\mathrm{e}^{(X-x)}].$$

More generally,

$$\Pr(X \geq x) \leq \mathcal{E}[\mathrm{e}^{\lambda(X-x)}], \tag{2.21}$$

where $\lambda \geq 0$.

From (2.21), we can seek the tightest upper bound by minimizing the right-hand side with respect to λ. The resulting upper bound is called the Chernoff bound, which is given by

$$\Pr(X \geq x) \leq \min_{\lambda \geq 0} \mathrm{e}^{-\lambda x} \mathcal{E}[\mathrm{e}^{\lambda X}], \tag{2.22}$$

where $\mathcal{E}[\mathrm{e}^{sX}]$ is the moment generating function (mgf).

Example 2.3.1 Suppose that X is a Gaussian random variable with mean μ and variance σ^2. The mgf of X is given by

$$\mathcal{E}[e^{\lambda X}] = e^{\lambda\mu + \frac{1}{2}\lambda^2\sigma^2}.$$

The Chernoff bound becomes

$$\begin{aligned}\Pr(X \geq x) &\leq \min_{\lambda \geq 0} e^{-\lambda x} e^{\lambda\mu + \frac{1}{2}\lambda^2\sigma^2} \\ &= \min_{\lambda \geq 0} e^{\lambda(\mu - x) + \frac{1}{2}\lambda^2\sigma^2}.\end{aligned}$$

The solution of the minimization is found as

$$\begin{aligned}\lambda^* &= \arg\min_{\lambda \geq 0} e^{\lambda(\mu - x) + \frac{1}{2}\lambda^2\sigma^2} \\ &= \max\left\{0, \frac{x - \mu}{\sigma^2}\right\}.\end{aligned}$$

Provided that $s^* > 0$, the Chernoff bound is

$$\Pr(X \geq x) \leq e^{-\frac{(x-\mu)^2}{2\sigma^2}}. \tag{2.23}$$

For the error probability in (2.20), the Chernoff bound is given by

$$P_{\text{err}} \leq e^{-\frac{V^2}{2\sigma^2}} = e^{-\text{SNR}/2}.$$

We can have a useful bound from Example 2.3.1, which is often called the Chernoff bound:

$$\mathcal{Q}(x) \leq \exp\left(-\frac{x^2}{2}\right). \tag{2.24}$$

Of course, this is a special case of the Chernoff bound in (2.22).

2.4 Binary waveform signal detection

In communications over waveform channels, we transmit waveforms rather than discrete signals. For binary signaling, the received signal is given by

$$Y(t) = X(t) + N(t), \quad 0 \leq t < T, \tag{2.25}$$

where T denotes the signal duration and $N(t)$ is a white Gaussian random process with $\mathcal{E}[N(t)] = 0$ and $\mathcal{E}[N(t)N(\tau)] = \frac{N_0}{2}\delta(t - \tau)$. Here, $\delta(t)$ denotes the Dirac delta. The channel in (2.25) is called the additive white Gaussian noise (AWGN) channel. $X(t)$ is a binary waveform that is given by

$$X(t) = \begin{cases} s_0(t), & \text{under hypothesis } H_0; \\ s_1(t), & \text{under hypothesis } H_1. \end{cases}$$

For the signaling in (2.25), the transmission rate is $1/T$ bits per second.

2.4.1 Detection of waveform signals

We consider a heuristic approach to deal with waveform signal detection in this section. A formal approach will be discussed later with an expansion of waveforms. At the receiver, the decision is made with $Y(t)$, $0 \le t < T$. Denote by $y(t)$ a realization or observation of $Y(t)$. Suppose that we take L samples from $y(t)$. Let

$$\begin{aligned} y_l &= \int_{(l-1)\Delta}^{l\Delta} y(t)\mathrm{d}t; \\ s_{m,l} &= \int_{(l-1)\Delta}^{l\Delta} s_m(t)\mathrm{d}t; \\ n_l &= \int_{(l-1)\Delta}^{l\Delta} n(t)\mathrm{d}t, \end{aligned} \tag{2.26}$$

where $\Delta = T/L$ and $n(t)$ is a realization of $N(t)$. Then,

$$y_l = \begin{cases} s_{0,l} + n_l, & \text{under } H_0; \\ s_{1,l} + n_l, & \text{under } H_1. \end{cases}$$

Since $N(t)$ is a white process, the n_l's are independent. The mean of n_l is zero and the variance is given by

$$\begin{aligned} \sigma^2 &= \mathcal{E}[n_l^2] \\ &= \mathcal{E}\left[\left(\int_{(l-1)\Delta}^{l\Delta} N(t)\mathrm{d}t\right)^2\right] \\ &= \int_{(l-1)\Delta}^{l\Delta}\int_{(l-1)\Delta}^{l\Delta} \mathcal{E}[N(t)N(\tau)]\mathrm{d}t\mathrm{d}\tau \\ &= \int_{(l-1)\Delta}^{l\Delta}\int_{(l-1)\Delta}^{l\Delta} \frac{N_0}{2}\delta(t-\tau)\mathrm{d}t\mathrm{d}\tau \\ &= \frac{N_0\Delta}{2}. \end{aligned}$$

Let $\mathbf{y} = [y_1\ y_2\ \cdots\ y_L]^{\mathrm{T}}$. The LLR becomes

$$\mathrm{LLR}(\mathbf{y}) = \log \frac{f_0(\mathbf{y})}{f_1(\mathbf{y})},$$

where

$$f_m(\mathbf{y}) = \prod_{l=1}^{L} f(y_l|H_m).$$

It follows that

$$\begin{aligned}
\text{LLR}(\mathbf{y}) &= \sum_{l=1}^{L} \log \frac{f_0(y_l)}{f_1(y_l)} \\
&= \sum_{l=1}^{L} \log \left[\exp\left(-\frac{1}{N_0} \left((y_l - s_{0,l})^2 - (y_l - s_{1,l})^2 \right) \right) \right] \\
&= \frac{1}{N_0} \left(\sum_{l=1}^{L} (y_l - s_{1,l})^2 - (y_l - s_{0,l})^2 \right) \\
&= \frac{1}{N_0} \left(\sum_{l=1}^{L} 2y_l(s_{0,l} - s_{1,l}) + \sum_{l=1}^{L} (s_{1,l}^2 - s_{0,l}^2) \right) \\
&= \frac{1}{N_0} \left(2\mathbf{y}^{\mathrm{T}}(\mathbf{s}_0 - \mathbf{s}_1) - (\mathbf{s}_0^{\mathrm{T}}\mathbf{s}_0 - \mathbf{s}_1^{\mathrm{T}}\mathbf{s}_1) \right),
\end{aligned}$$

where $\mathbf{s}_m = [s_{m,1}\ s_{m,2} \ldots s_{m,L}]^{\mathrm{T}}$. Using the LLR, the MAP decision rule becomes

$$\mathbf{y}^{\mathrm{T}}(\mathbf{s}_0 - \mathbf{s}_1) \underset{H_1}{\overset{H_0}{\gtrless}} \sigma^2 \log\left(\frac{\Pr(H_1)}{\Pr(H_2)} \right) + \frac{1}{2}(\mathbf{s}_0^{\mathrm{T}}\mathbf{s}_0 - \mathbf{s}_1^{\mathrm{T}}\mathbf{s}_1).$$

For the LR-based decision rule, $\frac{\Pr(H_0)}{\Pr(H_1)}$ is replaced by the threshold τ. That is,

$$\mathbf{y}^{\mathrm{T}}(\mathbf{s}_0 - \mathbf{s}_1) \underset{H_1}{\overset{H_0}{\gtrless}} \sigma^2 \log \tau + \frac{1}{2}(\mathbf{s}_0^{\mathrm{T}}\mathbf{s}_0 - \mathbf{s}_1^{\mathrm{T}}\mathbf{s}_1).$$

2.4.2 Correlator detector and performance

In the sampling approach in (2.26), there could be signal information loss due to sampling operation if the number of samples during T seconds is small. To avoid any signal information loss, suppose that L is sufficiently large to approximate as

$$\mathbf{y}^{\mathrm{T}}\mathbf{s}_i \approx \frac{1}{T}\int_0^T y(t)s_m(t)\mathrm{d}t.$$

Then, the LR-based decision rule can be given by

$$\int_0^T y(t)(s_0(t) - s_1(t))\mathrm{d}t \underset{H_1}{\overset{H_0}{\gtrless}} \sigma^2 \log \tau + \frac{1}{2}\int_0^T \left(s_0^2(t) - s_1^2(t) \right) \mathrm{d}t.$$

Let

$$V_T = \sigma^2 \log \tau + \frac{1}{2}\int_0^T \left(s_0^2(t) - s_1^2(t) \right) \mathrm{d}t.$$

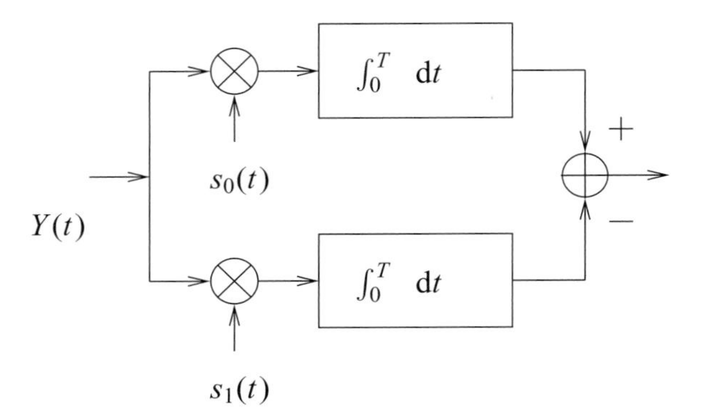

Figure 2.7 Correlator detector for binary signaling.

Then, the decision rule becomes

$$\int_0^T y(t)\,(s_0(t) - s_1(t))\,\mathrm{d}t \underset{H_1}{\overset{H_0}{\gtrless}} V_T .$$

This decision rule can be implemented as in Fig. 2.7, which is called the correlator detector.

For the performance analysis, consider the ML decision rule (i.e. $\tau = 1$ in the LR-based decision rule). We have $V_T = \frac{1}{2}\int_0^T \left(s_0^2(t) - s_1^2(t)\right) \mathrm{d}t$. Let

$$G = \int_0^T y(t)\,(s_0(t) - s_1(t))\,\mathrm{d}t - V_T .$$

Clearly,

$$\Pr(D_0|H_1) = \Pr(G > 0|H_1);$$
$$\Pr(D_1|H_0) = \Pr(G < 0|H_0).$$

To find the error probabilities, the random variable G has to be characterized. As $N(t)$ is assumed to be a Gaussian process, G is a Gaussian random variable. We note that the statistical properties of G depend on H_m as $Y(t) = s_m(t) + N(t)$ if H_m is true. Suppose that H_m is true. Then, to fully characterize G, we need to find the mean and variance as G is Gaussian. We have the mean as follows:

$$\begin{aligned}\mathcal{E}[G|H_m] &= \int_0^T \mathcal{E}[Y(t)|H_m]\,(s_0(t) - s_1(t))\,\mathrm{d}t - V_T \\ &= \int_0^T s_m(t)\,(s_0(t) - s_1(t))\,\mathrm{d}t - V_T .\end{aligned}$$

The variance is

$$\sigma_m^2 = \mathcal{E}[(G - \mathcal{E}[G|H_m])^2|H_m].$$

Let

$$E_s = \frac{1}{2}\int_0^T \left(s_0^2(t) + s_1^2(t)\right) dt$$

and

$$\rho = \frac{1}{E_s}\int_0^T s_0(t)s_1(t)dt.$$

Note that E_s is the average energy of the signals, $s_m(t)$, $m = 1, 2$, which are equally likely transmitted. Then,

$$\mathcal{E}[G|H_0] = E_s(1-\rho) \quad \text{and} \quad \mathcal{E}[G|H_1] = -E_s(1-\rho)$$

and

$$\sigma^2 = \sigma_0^2 = \sigma_1^2 = N_0 E_s(1-\rho).$$

Thus, the pdfs of G under H_0 and H_1 become

$$f_0(g) = \frac{1}{\sqrt{2\pi N_0 E_s(1-\rho)}} \exp\left(-\frac{(g - E_s(1-\rho))^2}{2N_0 E_s(1-\rho)}\right)$$

and

$$f_1(g) = \frac{1}{\sqrt{2\pi N_0 E_s(1-\rho)}} \exp\left(-\frac{(g + E_s(1-\rho))^2}{2N_0 E_s(1-\rho)}\right),$$

respectively. As in (2.20), we can find the error probability as follows:

$$P_{\text{err}} = \mathcal{Q}\left(\sqrt{\frac{E_s(1-\rho)}{N_0}}\right).$$

For a fixed signal energy, E_s, the error probability can be minimized when $\rho = -1$ and the corresponding minimum error probability is

$$P_{\text{err}} = \mathcal{Q}\left(\sqrt{\frac{2E_s}{N_0}}\right). \tag{2.27}$$

The resulting signals that minimize the error probability are antipodal signals:

$$s_0(t) = -s_1(t).$$

For an orthogonal signal set, we have

$$\rho = 0$$

and the corresponding error probability is

$$P_{\text{err}} = \mathcal{Q}\left(\sqrt{\frac{E_s}{N_0}}\right). \tag{2.28}$$

Consequently, from (2.27) and (2.28), we can see that there is a 3 dB gap (in SNR) between the antipodal and orthogonal signal sets.

2.5 *M*-ary signal detection

So far, we considered the binary signal detection, where $M = 2$. Now, suppose that there is a set of M waveforms, $\{s_1(t), s_2(t), \ldots, s_M(t)\}$, $0 \le t < T$, for M-ary communications. In this case, the transmission rate becomes $\frac{\log_2 M}{T}$ bits per seconds. Since the transmission rate increases with M, a large M would be preferable. However, in general, the detection performance becomes worse as M increases. In this section, we mainly focus on the detection problem with M-ary signals.

Under the mth hypothesis, it is assumed that the received signal is given by

$$Y(t) = s_m(t) + N(t), \ 0 \le t < T,$$

and the likelihood with L samples is given by

$$\begin{aligned} f_m(\mathbf{y}) &= \prod_{l=1}^{L} f_m(y_l) \\ &= \frac{1}{(\pi N_0)^{\frac{L}{2}}} \prod_{l=1}^{L} \exp\left(-\frac{(y_l - s_{m,l})^2}{N_0}\right). \end{aligned}$$

For convenience, we take the logarithm as follows:

$$\begin{aligned} \log f_m(\mathbf{y}) &= \log \frac{1}{(\pi N_0)^{\frac{L}{2}}} + \sum_{l=1}^{L} \log\left(\exp\left(-\frac{(y_l - s_{m,l})^2}{N_0}\right)\right) \\ &= \log \frac{1}{(\pi N_0)^{\frac{L}{2}}} - \frac{(y_l - s_{m,l})^2}{N_0}. \end{aligned}$$

Ignoring the common terms for all the hypotheses, the log-likelihood can be given by

$$\log f_m(\mathbf{y}) = \frac{1}{N_0}\left(\sum_{l=1}^{L} y_l s_{m,l} - \frac{1}{2}|s_{m,l}|^2\right).$$

Denote by $y(t)$ an observation of $Y(t)$. When L goes to infinity, we have

$$\log f_m(y(t)) = \frac{1}{N_0}\left(\int_0^T y(t)s_m(t)\mathrm{d}t - \frac{1}{2}\int_0^T s_m^2(t)\mathrm{d}t\right). \tag{2.29}$$

With the log-likelihood functions, the ML decision rule is given by

$$\text{Accept } H_m \text{ if } \log f_m(y(t)) \ge \log f_{m'}(y(t)), \ m' \in \{1, 2, \ldots, M\} \setminus \{m\}$$

or

$$\text{Accept } H_m \text{ if } \log \frac{f_m(y(t))}{f_{m'}(y(t))} \ge 0, \ m' \in \{1, 2, \ldots, M\} \setminus \{m\},$$

where $\setminus$ denotes the set difference ($A \setminus B = \{x \mid x \in A, x \notin B\}$). In (2.29), we note that $E_m = \int_0^T s_m^2(t)\mathrm{d}t$ is the signal energy of the mth signal, $s_m(t)$. In addition, $\int_0^T y(t)s_m(t)\mathrm{d}t$ is the correlation between $y(t)$ and $s_m(t)$. Thus, the ML decision rule can be implemented using a bank of the correlators as shown in Fig. 2.8.

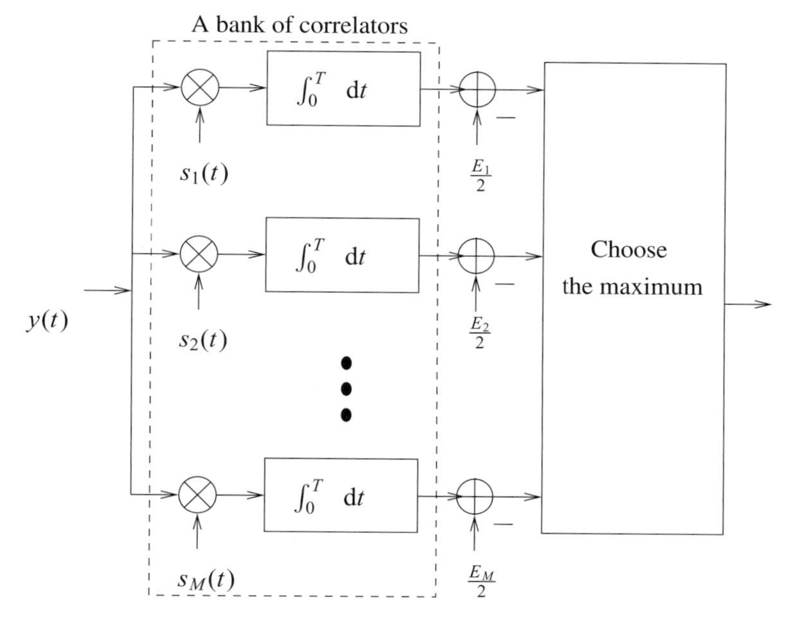

Figure 2.8 ML detector for M-ary signals using a bank of correlators.

The MAP decision rule can also be derived by modifying the ML decision rule above to take into account the a priori probability.

2.6 Signal detection in vector space

Although the decision rule can be built with continuous waveforms, it would be more convenient if the signals are in vector forms. In this section, we consider the signal detector in a vector space in which the signals are represented as vectors.

2.6.1 Karhunen–Loéve expansion

The Karhunen–Loéve (K–L) expansion is a way to represent a function by a weighted sum of basis functions. Suppose that there is a set of L orthonormal basis functions, $\{\phi_l(t)\}, l = 1, 2, \ldots, L; 0 \le t < T$. With $\{\phi_l(t)\}$, assume that the signal can be expressed as

$$s_m(t) = \sum_{l=1}^{L} s_{m,l}\phi_l(t), \ 0 \le t < T. \tag{2.30}$$

The expansion in (2.30) with coefficients, $s_{m,l}$, is called the K–L expansion. Using the orthonormality of the basis functions:

$$\int_0^T \phi_l(t)\phi_m^*(t)\mathrm{d}t = \delta_{l,m},$$

where

$$\delta_{l,m} = \begin{cases} 1, & \text{if } l = m; \\ 0, & \text{otherwise,} \end{cases}$$

we can show that

$$s_{m,l} = \int_0^T s_m(t)\phi_l^*(t)\mathrm{d}t.$$

As shown in (2.30), with $\{s_{m,l}\}$, we can reproduce $s_m(t)$.

Define a vector with $\{s_{m,l}\}$ as

$$\mathbf{s}_m = [s_{1,m}\ s_{2,m}\ \ldots\ s_{L,m}]^{\mathrm{T}}.$$

Then, we can show that $\mathbf{s}_m$ and $s_m(t)$ are equivalent. While $s_m(t)$ is a representation of the mth signal in a functional (or waveform) space, $\mathbf{s}_m$ is a representation in a vector space. It can be shown that the signal energy is invariant as follows:

$$\begin{aligned} E_m &= \int_0^T |s_m(t)|^2\mathrm{d}t \\ &= ||\mathbf{s}_m||^2. \end{aligned} \tag{2.31}$$

In addition, the distance is invariant:

$$\begin{aligned} d_{m,k} &= \sqrt{\int_0^T |s_m(t) - s_k(t)|^2\mathrm{d}t} \\ &= ||\mathbf{s}_m - \mathbf{s}_k||, \end{aligned} \tag{2.32}$$

where $d_{m,k}$ denotes the distance between the mth and kth signals.

Applying the K–L expansion, it can be shown that

$$\begin{aligned} r_l &= \int_0^T Y(t)\phi_l^*(t)\mathrm{d}t \\ &= s_{m,l} + n_\ell,\ l = 1, 2, \ldots, L, \end{aligned}$$

where $n_l = \int_0^T N(t)\phi_l^*(t)\mathrm{d}t$. Let

$$\begin{aligned} \mathbf{r} &= [r_1\ r_2\ \ldots\ r_L]^{\mathrm{T}} \\ &= \mathbf{s}_m + \mathbf{n}, \end{aligned} \tag{2.33}$$

where $\mathbf{n} = [n_1\ n_2\ \ldots\ n_L]^{\mathrm{T}}$. Note that the noise can be represented as

$$N(t) = \bar{N}(t) + \tilde{N}(t),\ 0 \le t < T,$$

where $\bar{N}(t) = \sum_{l=1}^L n_l\phi_l(t)$ and $\tilde{N}(t)$ is the noise component that cannot be represented by a weighted sum of basis functions:

$$\tilde{N}(t) = N(t) - \sum_{l=1}^L n_l\phi_l(t).$$

It is noteworthy that this noise component, $\tilde{N}(t)$, does not appear in $\mathbf{r}$ as shown in (2.33).

Since $N(t)$ is an additive white Gaussian noise process, n_l is also a white Gaussian random variable. The mean and variance are given by

$$\mathcal{E}[n_l] = \int_0^T \mathcal{E}[N(t)\phi_l^*(t)]\mathrm{d}t = 0$$

and

$$\begin{aligned}\mathcal{E}[|n_l|^2] &= \mathcal{E}\left[\left(\int_0^T N(t)\phi_l^*(t)\mathrm{d}t\right)\left(\int_0^T N(\tau)\phi_l^*(\tau)\mathrm{d}\tau\right)^*\right] \\ &= \int_0^T\int_0^T \mathcal{E}[N(t)N(\tau)]\phi_l^*(t)\phi_l(\tau)\mathrm{d}t\mathrm{d}\tau \\ &= \frac{N_0}{2},\end{aligned}$$

where $\int_0^T |\phi_l(t)|^2\mathrm{d}t = 1$ (since we assume the basis functions are orthonormal).

Example 2.6.1 For quadrature phase shift keying (QPSK), there are four signals as follows:

$$\begin{aligned} s_1(t) &= \sin(2\pi f_c t); \\ s_2(t) &= -\sin(2\pi f_c t); \\ s_3(t) &= \cos(2\pi f_c t); \\ s_4(t) &= -\cos(2\pi f_c t),\ 0 \le t < T, \end{aligned}$$

where T is an integer multiple of $1/f_c$ and f_c is the carrier frequency. Then, a set of the basis functions that can represent $s_m(t)$ is given by

$$\begin{aligned} \phi_1(t) &= \sqrt{\frac{2}{T}}\sin(2\pi f_c t); \\ \phi_2(t) &= \sqrt{\frac{2}{T}}\cos(2\pi f_c t). \end{aligned}$$

In Fig. 2.9, we show the four waveforms for QPSK with $f_c = T = 1$.

2.6.2 Signal detection in vector space

Using the K–L expansion, we can assume that the received signal can be represented by a vector form and given by

$$\mathbf{r} = \mathbf{s}_m + \mathbf{n},\ m = 0, 1, \ldots, M-1.$$

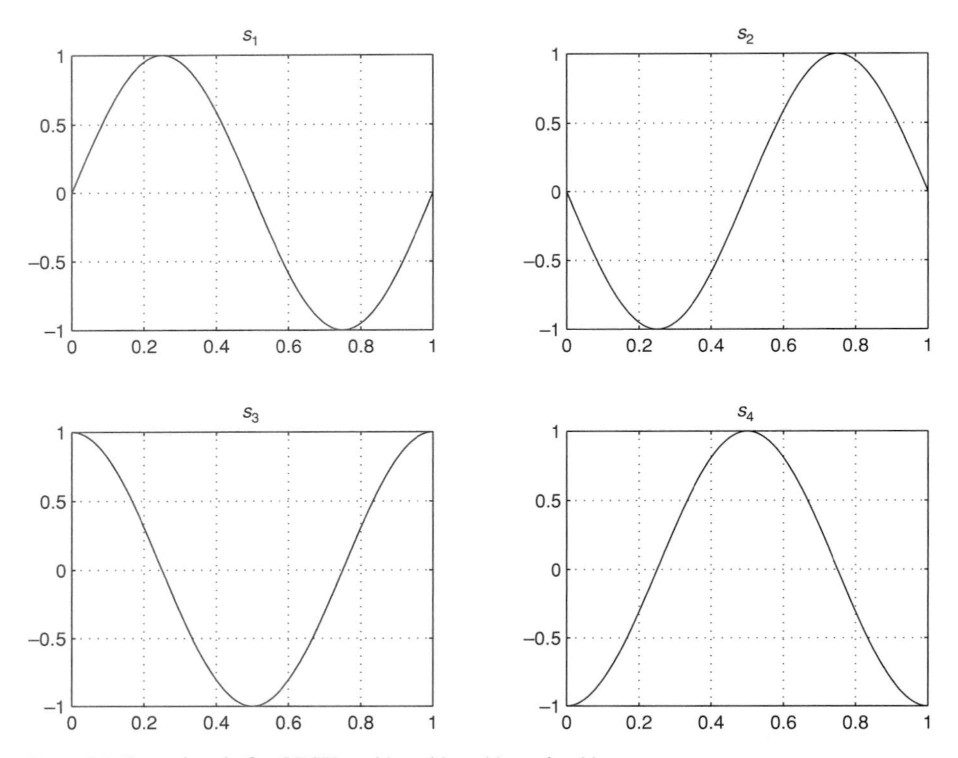

Figure 2.9 Four signals for QPSK; $s_1(t)$, $s_2(t)$, $s_3(t)$, and $s_4(t)$.

For a more general case, we now assume that $\mathbf{s}_m$ and $\mathbf{n}$ can be complex-valued vectors. In addition, it is assumed that $\mathbf{n}$ is a CSCG random vector with

$$\mathcal{E}[\mathbf{n}] = 0;$$
$$\mathcal{E}[\mathbf{n}\mathbf{n}^{\mathrm{H}}] = \mathbf{R}_{\mathbf{n}}.$$

For convenience, $\mathcal{CN}(\mathbf{m}, \mathbf{R}_{\mathbf{x}})$ denotes the pdf of a CSCG random vector with mean vector $\mathbf{m}$ and covariance matrix $\mathbf{R}_{\mathbf{x}}$.

For a given $\mathbf{r}$, the ML decision rule can be given by

$$\text{Accept } s_m(t) \text{ if } f_m(\mathbf{r}) \geq f_{m'}(\mathbf{r}), \text{ for } m' \neq m, \tag{2.34}$$

where $f_m(\mathbf{r})$ denotes the likelihood function of the mth hypothesis or $\mathbf{s}_m$. Since the noise is assumed to be a CSCG random vector, the likelihood function of $\mathbf{s}_m$ for a given $\mathbf{r}$ is given by

$$f(\mathbf{r}|\mathbf{s}_m) = \frac{1}{\pi^L \det(\mathbf{R}_{\mathbf{n}})} \exp\left(-(\mathbf{r} - \mathbf{s}_m)^{\mathrm{H}} \mathbf{R}_{\mathbf{n}}^{-1} (\mathbf{r} - \mathbf{s}_m)\right).$$

In addition, the log-likelihood function is given by

$$\log f_m(\mathbf{r}) = -(\mathbf{r} - \mathbf{s}_m)^{\mathrm{H}} \mathbf{R}_{\mathbf{n}}^{-1} (\mathbf{r} - \mathbf{s}_m) + \text{constant}$$

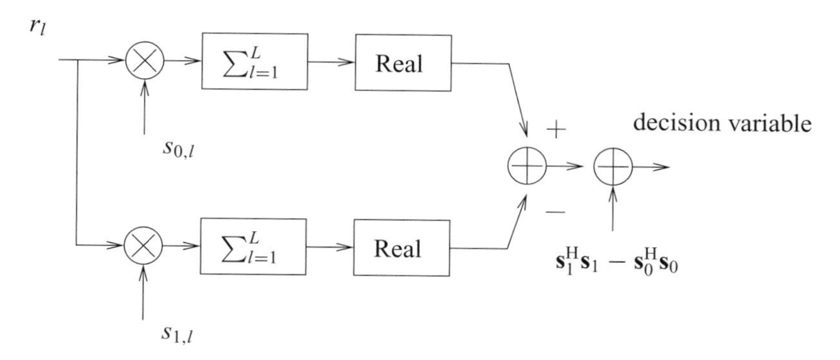

Figure 2.10 The ML detector for binary signals in the vector space.

or

$$\log f_m(\mathbf{r}) = 2\Re(\mathbf{s}_m^{\mathrm{H}}\mathbf{R}_{\mathbf{n}}^{-1}\mathbf{r}) - \mathbf{s}_m^{\mathrm{H}}\mathbf{R}_{\mathbf{n}}^{-1}\mathbf{s}_m + \text{constant}.$$

Thus, the ML decision rule can be reduced to

$$\text{Accept } s_m(t) \text{ if}$$
$$\Re(\mathbf{s}_m^{\mathrm{H}}\mathbf{R}_{\mathbf{n}}^{-1}\mathbf{r}) - \frac{\mathbf{s}_m^{\mathrm{H}}\mathbf{R}_{\mathbf{n}}^{-1}\mathbf{s}_m}{2} \geq \Re(\mathbf{s}_{m'}^{\mathrm{H}}\mathbf{R}_{\mathbf{n}}^{-1}\mathbf{r}) - \frac{\mathbf{s}_{m'}^{\mathrm{H}}\mathbf{R}_{\mathbf{n}}^{-1}\mathbf{s}_{m'}}{2}, \text{ for } m' \neq m. \tag{2.35}$$

Consider a binary signaling where $M = 2$. Then, the LLR becomes

$$\begin{aligned}\mathrm{LLR}(\mathbf{r}) &= \log\left(\frac{f_0(\mathbf{r})}{f_1(\mathbf{r})}\right) \\ &= \left((\mathbf{r}-\mathbf{s}_1)^{\mathrm{H}}\mathbf{R}_{\mathbf{n}}^{-1}(\mathbf{r}-\mathbf{s}_1) - (\mathbf{r}-\mathbf{s}_0)^{\mathrm{H}}\mathbf{R}_{\mathbf{n}}^{-1}(\mathbf{r}-\mathbf{s}_0)\right).\end{aligned}$$

As a special case, suppose that the n_l's are iid. In this case, the variances of the n_l's are the same. This implies $\mathbf{R}_{\mathbf{n}} = N_0\mathbf{I}$, $N_0 > 0$. The LLR becomes

$$\begin{aligned}\mathrm{LLR}(\mathbf{r}) &= \frac{1}{N_0}\left((\mathbf{s}_0^{\mathrm{H}}\mathbf{r} + \mathbf{r}^{\mathrm{H}}\mathbf{s}_0) - (\mathbf{s}_1^{\mathrm{H}}\mathbf{r} + \mathbf{r}^{\mathrm{H}}\mathbf{s}_1) + \mathbf{s}_1^{\mathrm{H}}\mathbf{s}_1 - \mathbf{s}_0^{\mathrm{H}}\mathbf{s}_0\right) \\ &= \frac{1}{N_0}\left(2\Re((\mathbf{s}_0-\mathbf{s}_1)^{\mathrm{H}}\mathbf{r}) + \mathbf{s}_1^{\mathrm{H}}\mathbf{s}_1 - \mathbf{s}_0^{\mathrm{H}}\mathbf{s}_0\right).\end{aligned}$$

This becomes the correlator detector with complex-valued signals as shown in Fig. 2.10.

2.6.3 Pairwise error probability and Bhattacharyya bound

While the error probability for a binary signaling can be easily found, the error probability for an M-ary signaling is not easy unless the signals are mutually orthogonal. However, an upper bound can be easily found as shown below.

Consider the symbol error probability of the ML decision rule. Suppose that the vector symbol $\mathbf{s}_m$ is sent. The error probability for the mth vector symbol is

$$P_m = \Pr(\mathbf{r} \in \Lambda_m^c \mid \mathbf{s}_m),$$

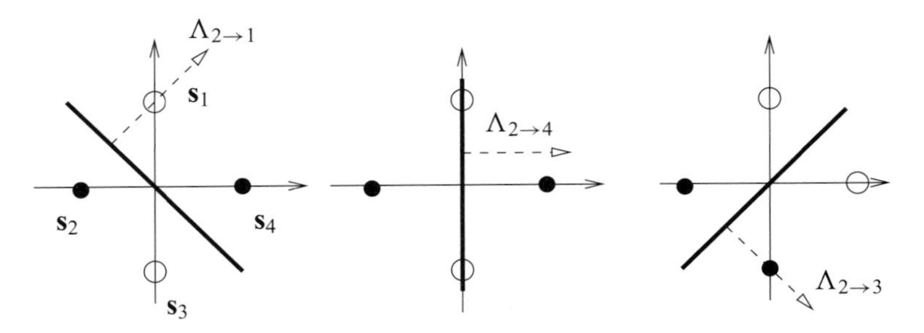

Figure 2.11 Illustration of $\Lambda_{m\to m'}$ to find the PEP's for a 4-ary signaling.

where Λ_m^c denotes the complementary set of Λ_m and Λ_m is the decision region for the mth signal defined as

$$\Lambda_m = \{\mathbf{r} \mid f_m(\mathbf{r}) \geq f_{m'}(\mathbf{r}),\ m' \neq m\}. \tag{2.36}$$

Here, $f_m(\mathbf{r}) = f(\mathbf{r}|\mathbf{s}_m)$. The average error probability of the ML decision rule is given by

$$\begin{aligned} P_{\mathrm{err}} &= \sum_{m=1}^{M} P_m \Pr(\mathbf{s}_m) \\ &= \sum_{m=1}^{M} \Pr(\mathbf{r} \in \Lambda_m^c \mid \mathbf{s}_m) \Pr(\mathbf{s}_m). \end{aligned}$$

In general, since it is not easy to compute P_m exactly, bounds are frequently used. In order to find an upper bound on the error probability, we can use the pairwise error probability (PEP) and the union bound. The PEP, $P(\mathbf{s}_m \to \mathbf{s}_{m'})$, is the error probability when $\mathbf{s}_m$ is sent and $\mathbf{s}_{m'}$ is the *only* alternative. Define

$$\Lambda_{m\to m'} = \{\mathbf{r} :\ f_{m'}(\mathbf{r}) \geq f_m(\mathbf{r})\}.$$

Then,

$$P(\mathbf{s}_m \to \mathbf{s}_{m'}) = \int_{\mathbf{r}\in\Lambda_{m\to m'}} f(\mathbf{r} \mid \mathbf{s}_m) d\mathbf{r}.$$

In Fig. 2.11, some examples for $\Lambda_{m\to m'}$ are illustrated for a 4-ary signaling.

It can be shown that

$$\begin{aligned} \Lambda_m^c &= \bigcup_{m'\neq m} \{\mathbf{r} :\ f(\mathbf{r} \mid \mathbf{s}_{m'}) \geq f(\mathbf{r} \mid \mathbf{s}_m)\} \\ &= \bigcup_{m'\neq m} \Lambda_{m\to m'}. \end{aligned}$$

Using the PEP and union bound, we can derive the following upper bound:

$$\begin{aligned} P_m &= \Pr(\mathbf{r} \in \Lambda_m^c \mid \mathbf{s}_m) \\ &\leq \sum_{m' \neq m} \Pr(\mathbf{r} \in \Lambda_{m \to m'} \mid \mathbf{s}_m) \\ &= \sum_{m' \neq m} P(\mathbf{s}_m \to \mathbf{s}_{m'}), \end{aligned}$$

where the inequality is due to the union bound. It is noteworthy that, in general, the PEP is not the error probability of the mth signal as shown below:

$$P(\mathbf{s}_m \to \mathbf{s}_{m'}) \neq \Pr(\text{decide } \mathbf{s}_{m'} \mid \mathbf{s}_m \text{ is sent}) = \Pr(\mathbf{r} \in \Lambda_{m'} \mid \mathbf{s}_m \text{ is sent}).$$

The PEP can be analytically obtained in most cases. Furthermore, bounds on the PEP or approximates are also available. Among them, we focus on the Bhattacharyya bound.

The PEP can be expressed as

$$P(\mathbf{s}_m \to \mathbf{s}_{m'}) = \int_{\mathbf{r}} \mathbf{1}(\mathbf{r} \in \Lambda_{m \to m'}) f(\mathbf{r} \mid \mathbf{s}_m) \mathrm{d}\mathbf{r},$$

where $\mathbf{1}(\cdot)$ is the indicator function that is given by

$$\mathbf{1}(\mathbf{r} \in \Lambda_{m \to m'}) = \begin{cases} 1, & \text{if } \mathbf{r} \in \Lambda_{m \to m'}; \\ 0, & \text{if } \mathbf{r} \notin \Lambda_{m \to m'} \end{cases}$$

The indicator function can be upper-bounded as

$$\mathbf{1}(\mathbf{r} \in \Lambda_{m \to m'}) = \begin{cases} 1 \leq \sqrt{\frac{f(\mathbf{r} \mid \mathbf{s}_{m'})}{f(\mathbf{r} \mid \mathbf{s}_m)}}, & \text{if } \mathbf{r} \in \Lambda_{m \to m'}; \\ 0 \leq \sqrt{\frac{f(\mathbf{r} \mid \mathbf{s}_{m'})}{f(\mathbf{r} \mid \mathbf{s}_m)}}, & \text{if } \mathbf{r} \notin \Lambda_{m \to m'}. \end{cases}$$

Using this upper bound, it can be shown that

$$\begin{aligned} P(\mathbf{s}_m \to \mathbf{s}_{m'}) &\leq \int_{\mathbf{r}} \sqrt{\frac{f(\mathbf{r} \mid \mathbf{s}_{m'})}{f(\mathbf{r} \mid \mathbf{s}_m)}} f(\mathbf{r} \mid \mathbf{s}_m) \mathrm{d}\mathbf{r} \\ &= \int_{\mathbf{r}} \sqrt{f(\mathbf{r} \mid \mathbf{s}_m) f(\mathbf{r} \mid \mathbf{s}_{m'})} \mathrm{d}\mathbf{r}. \end{aligned}$$

This is the Bhattacharyya bound on the PEP. Finally, we can have the following upper bound:

$$P_m \leq \sum_{m' \neq m} \int_{\mathbf{r}} \sqrt{f(\mathbf{r} \mid \mathbf{s}_m) f(\mathbf{r} \mid \mathbf{s}_{m'})} \mathrm{d}\mathbf{r}.$$

Example 2.6.2 If $\mathbf{n} \sim \mathcal{CN}(\mathbf{0}, \mathbf{R_n})$, we have

$$\sqrt{f(\mathbf{r} \mid \mathbf{s}_m) f(\mathbf{r} \mid \mathbf{s}_{m'})} = \frac{\mathrm{e}^{-\frac{1}{2}\left((\mathbf{r}-\mathbf{s}_m)^{\mathrm{H}}\mathbf{R}_{\mathbf{n}}^{-1}(\mathbf{r}-\mathbf{s}_m)+(\mathbf{r}-\mathbf{s}_{m'})^{\mathrm{H}}\mathbf{R}_{\mathbf{n}}^{-1}(\mathbf{r}-\mathbf{s}_{m'})\right)}}{\pi^L \det(\mathbf{R_n})}$$
$$= \frac{\mathrm{e}^{-(\mathbf{r}-\frac{1}{2}(\mathbf{s}_m+\mathbf{s}_{m'}))^{\mathrm{H}}\mathbf{R}_{\mathbf{n}}^{-1}(\mathbf{r}-\frac{1}{2}(\mathbf{s}_m+\mathbf{s}_{m'}))}}{\pi^L \det(\mathbf{R_n})}$$
$$\times \mathrm{e}^{-\frac{1}{4}(\mathbf{s}_m-\mathbf{s}_{m'})^{\mathrm{H}}\mathbf{R}_{\mathbf{n}}^{-1}(\mathbf{s}_m-\mathbf{s}_{m'})}.$$

The first term is the Gaussian pdf with mean $\frac{1}{2}(\mathbf{s}_m + \mathbf{s}_{m'})$ and covariance $\mathbf{R_n}$. Therefore, it can be shown that

$$\int_{\mathbf{r}} \sqrt{f(\mathbf{r} \mid \mathbf{s}_m) f(\mathbf{r} \mid \mathbf{s}_{m'})} \mathrm{d}\mathbf{r} = \mathrm{e}^{-\frac{1}{4}(\mathbf{s}_m-\mathbf{s}_{m'})^{\mathrm{H}}\mathbf{R}_{\mathbf{n}}^{-1}(\mathbf{s}_m-\mathbf{s}_{m'})}.$$

The Bhattacharyya bound becomes

$$P_m \le \sum_{m' \neq m} \mathrm{e}^{-\frac{1}{4}(\mathbf{s}_m-\mathbf{s}_{m'})^{\mathrm{H}}\mathbf{R}_{\mathbf{n}}^{-1}(\mathbf{s}_m-\mathbf{s}_{m'})}.$$

2.7 Signal detection with random parameters

In deriving decision rules for some applications, there could be the information bearing signal that may not be known to the receiver in advance. This leads to the signal detection problem with random parameters. In this section, we consider a general principle for the signal detection with random parameters and some examples.

2.7.1 LR-based detection with random parameters

Consider the following pair of hypotheses:

$$H_0 : X \sim f_0(x; \theta_0);$$
$$H_1 : X \sim f_1(x; \theta_1), \tag{2.37}$$

where $f_m(x; \theta_m)$ denotes the pdf of the observation under hypothesis m. Here, θ_m represents the parameter (vector) for the pdf of X under hypothesis m. In some cases, θ_m may not be known. However, if the pdf of θ_m is available, for a given $X = x$, the likelihood function can be given by

$$f_m(x) = \mathcal{E}_m[f_m(x; \theta_m)]$$
$$= \int f_m(x; \theta_m) h_m(\theta_m) \mathrm{d}\theta_m, \tag{2.38}$$

where $\mathcal{E}_m[\cdot]$ denotes the statistical expectation with respect to θ_m and $h_m(\theta_m)$ represents the pdf of θ_m. The LR-based decision rule with unknown parameters is now given by

$$\frac{f_0(x)}{f_1(x)} = \frac{\mathcal{E}_0[f_0(x;\theta_0)]}{\mathcal{E}_1[f_1(x;\theta_1)]} \underset{H_1}{\overset{H_0}{\gtrless}} \tau. \tag{2.39}$$

Another approach is based on the maximum likelihoods that are given by

$$f_m(x, \hat{\theta}_m) = \max_{\theta_m} f_m(x;\theta_m), \; m = 0, 1,$$

where

$$\hat{\theta}_m = \arg\max_{\theta_m} f_m(x;\theta_m), \; m = 0, 1.$$

Here, $\hat{\theta}_m$ is called the maximum likelihood estimate (MLE) (we will discuss the parameter estimation in Chapter 3). The generalized likelihood ratio (GLR) is given by

$$\mathrm{GLR}(x) = \frac{f_0(x;\hat{\theta}_0)}{f_1(x;\hat{\theta}_1)}. \tag{2.40}$$

The GLR-based decision rule that uses the likelihood ratio in (2.40) is called the generalized likelihood ratio test (GLRT).

2.7.2 Signals with random amplitude

Consider an example of detecting random signals. Suppose that the pair of hypotheses is given by

$$\begin{aligned} H_0 &: Y(t) = N(t); \\ H_1 &: Y(t) = A + N(t), \; 0 \le t < T, \end{aligned}$$

where A is unknown and considered to be a random variable whose pdf is denoted by $h(a)$. It is assumed that $N(t)$ is a white Gaussian random process with $\mathcal{E}[N(t)] = 0$ and $\mathcal{E}[N(t)N(\tau)] = \frac{N_0}{2}\delta(t-\tau)$.

For a given $A = a$ and observation $Y(t) = y(t)$, $0 \le t < T$, the likelihood functions become

$$f_0(y(t)) = C \exp\left(-\frac{1}{N_0}\int_0^T y^2(t)\mathrm{d}t\right);$$

$$f_1(y(t)|A = a) = C \exp\left(-\frac{1}{N_0}\left(\int_0^T y^2(t)\mathrm{d}t - 2a\int_0^T y(t)\mathrm{d}t + a^2T^2\right)\right),$$

where C is the normalization constant. As A is a random variable, the likelihood for H_1 becomes

$$f_1(y(t)) = \int_0^{\infty} f_1(y(t)|A = a)h(a)\mathrm{d}a$$
$$= C \exp\left(-\frac{1}{N_0}\int_0^T y^2(t)\mathrm{d}t\right)$$
$$\times \int_0^{\infty} \exp\left(\frac{1}{N_0}\left(2a\int_0^T y(t)\mathrm{d}t - a^2T^2\right)\right) h(a)\mathrm{d}a.$$

Then, the LR is given by

$$\mathrm{LR}(y(t)) = \frac{f_0(y(t))}{f_1(y(t))}$$
$$= \int_0^{\infty} \exp\left(-\frac{1}{N_0}\left(2a\int_0^T y(t)\mathrm{d}t - a^2T^2\right)\right) h(a)\mathrm{d}a$$
$$= \mathcal{E}\left[\exp\left(-\frac{1}{N_0}\left(2A\int_0^T y(t)\mathrm{d}t - A^2T^2\right)\right)\right],$$

where the expectation is carried out with respect to A.

2.7.3 Signals with random phase

Another example is to detect the signal with random phase. Consider the following pair of hypotheses:

$$H_0 : Y(t) = N(t);$$
$$H_1 : Y(t) = A\sin(2\pi f_c t + \Theta) + N(t),\ 0 \le t < T, \tag{2.41}$$

where A is a constant amplitude, $T \gg \frac{1}{f_c}$, Θ is the random phase, and f_c is the carrier frequency. Assume that $N(t)$ is a white Gaussian random process with $\mathcal{E}[N(t)] = 0$ and $\mathcal{E}[N(t)N(\tau)] = \frac{N_0}{2}\delta(t-\tau)$.

Let

$$s_0(t) = 0;$$
$$s_1(t) = A\sin(2\pi f_c t + \theta),\ 0 \le t < T.$$

For a given observation $Y(t) = y(t)$ and known phase, $\Theta = \theta$, the likelihood functions are given by

$$f_0(y(t)) = C\exp\left(-\frac{1}{N_0}\int_0^T y^2(t)\mathrm{d}t\right)$$

and

$$f_1(y(t)|\theta) = C\exp\left(-\frac{1}{N_0}\int_0^T \left(y^2(t) - A\sin(2\pi f_c t + \theta)\right)^2 \mathrm{d}t\right),$$

where C is the normalization constant.

Suppose that the pdf of the random phase is given by

$$f_\Theta(\theta) = \frac{1}{2\pi}, \quad 0 \le \theta < 2\pi,$$

This would be the worst-case distribution for the random phase. It follows that

$$\begin{aligned}
f_1(y(t)) &= \frac{1}{2\pi}\int_0^{2\pi} f_1(y(t)|\theta)\mathrm{d}\theta \\
&= \frac{C}{2\pi}\int_0^{2\pi} \exp\left(-\frac{1}{N_0}\left(\int_0^T (y(t) - A\sin(2\pi f_c t + \theta))^2 \mathrm{d}t\right)\right)\mathrm{d}\theta \\
&= \frac{C}{2\pi}\int_0^{2\pi} \exp(-\frac{1}{N_0}(\int_0^T (y^2(t) - 2y(t)A\sin(2\pi f_c t + \theta) \\
&\quad + A^2\sin^2(2\pi f_c t + \theta)dt))\mathrm{d}\theta \\
&= \frac{C}{2\pi}\exp\left(-\frac{1}{N_0}\int_0^T y^2(t)\mathrm{d}t\right) \\
&\quad \times \int_0^{2\pi} \exp\left(\frac{1}{N_0}\int_0^T 2y(t)A\sin(2\pi f_c t + \theta) - A^2\sin^2(2\pi f_c t + \theta)\mathrm{d}t\right)\mathrm{d}\theta.
\end{aligned}$$

For a sufficiently large T, we have

$$\int_0^T A^2\sin^2(2\pi f_c t + \theta)\mathrm{d}t \approx \frac{A^2 T}{2}.$$

It follows that

$$\begin{aligned}
f_1(y(t)) = f_0(y(t))&\left(\frac{1}{2\pi}\exp\left(-\frac{1}{N_0}\frac{A^2 T}{2}\right)\right) \\
&\times \int_0^{2\pi} \exp\left(\frac{1}{N_0}\int_0^T 2y(t)A\sin(2\pi f_c t + \theta)\mathrm{d}t\right)\mathrm{d}\theta.
\end{aligned}$$

Thus, the LLR becomes

$$\begin{aligned}
\mathrm{LLR}(y(t)) = &\left(\frac{1}{2\pi}\exp\left(-\frac{1}{N_0}\frac{A^2 T}{2}\right)\right)^{-1} \\
&\times \left(\int_0^{2\pi} \exp\left(\frac{2}{N_0}\int_0^T y(t)A\sin(2\pi f_c t + \theta)\mathrm{d}t\right)\mathrm{d}\theta\right)^{-1}.
\end{aligned}$$

Since

$$\sin(x \pm y) = \sin(x)\cos(y) \pm \cos(x)\sin(y),$$

it can be shown that

$$\int_0^T y(t)A\sin(2\pi f_c t+\theta)\mathrm{d}t = \cos(\theta)\int_0^T y(t)A\sin(2\pi f_c t)\mathrm{d}t + \sin(\theta)\int_0^T y(t)A\cos(2\pi f_c t)\mathrm{d}t.$$

Let

$$q\cos(\theta_0) = \int_0^T y(t)\sin(2\pi f_c t)\mathrm{d}t$$

and

$$q\sin(\theta_0) = \int_0^T y(t)\cos(2\pi f_c t)\mathrm{d}t,$$

where q will be found later. Then,

$$\int_0^T y(t)A\sin(2\pi f_c t+\theta)\mathrm{d}t = Aq\cos(\theta_0)\cos(\theta) + Aq\sin(\theta_0)\sin(\theta).$$

Since

$$\cos(x\pm y) = \cos(x)\cos(y)\mp\sin(x)\sin(y),$$

we have

$$\int_0^T y(t)A\sin(2\pi f_c t+\theta)\mathrm{d}t = Aq\cos(\theta-\theta_0).$$

Finally,

$$\mathrm{LLR}(y(t)) = \left[\frac{1}{2\pi}\exp\left(-\frac{1}{N_0}\frac{A^2T}{2}\right)\int_0^{2\pi}\exp\frac{2Aq\cos(\theta-\theta_0)}{N_0}\mathrm{d}\theta\right]^{-1}.$$

Note that the zeroth-order modified Bessel function is given by

$$I_0(x) = \frac{1}{2\pi}\int_0^{2\pi}\exp(x\cos(\theta))\mathrm{d}\theta.$$

It simplifies the LLR as

$$\mathrm{LLR}(y(t)) = \left[\exp\left(-\frac{A^2T}{2N_0}\right)I_0\left(\frac{2Aq}{N_0}\right)\right]^{-1}.$$

It is noteworthy that θ_0 is vanished and q can be found as

$$q = \sqrt{\left(\int_0^T y(t)\sin(2\pi f_c t)\mathrm{d}t\right)^2 + \left(\int_0^T y(t)\cos(2\pi f_c t)\mathrm{d}t\right)^2}.$$

For a given threshold, τ, the LR-based decision rule can be given by

$$\left[\exp\left(-\frac{A^2T}{4\sigma^2}\right)I_0\left(\frac{Aq}{\sigma^2}\right)\right]^{-1} \underset{H_1}{\overset{H_0}{\gtrless}} \tau. \tag{2.42}$$

2.7.4 Detection of random Gaussian vector

If the distribution of signal is known and Gaussian, the decision rule to test whether or not a signal is present can be easily obtained. In addition, its analysis is also straightforward.

Let $\mathbf{r}$ be the received signal vector of size $L \times 1$. Consider the following pair of hypotheses:

$$\begin{aligned} H_0 &: \mathbf{r} = \mathbf{n}; \\ H_1 &: \mathbf{r} = \mathbf{s} + \mathbf{n}, \end{aligned} \tag{2.43}$$

where $\mathbf{n} \sim \mathcal{CN}(0, \mathbf{R_n})$ is the background noise vector and $\mathbf{s} \sim \mathcal{CN}(0, \mathbf{R_s})$ is the unknown Gaussian signal vector. Under H_1, we can see that $\mathbf{r} \sim \mathcal{CN}(\mathbf{0}, \mathbf{R_s} + \mathbf{R_n})$. The LR becomes

$$\begin{aligned} \mathrm{LR}(\mathbf{r}) &= \frac{\frac{1}{\det(\mathbf{R_n})} e^{-\mathbf{r}^{\mathrm{H}} \mathbf{R_n}^{-1} \mathbf{r}}}{\frac{1}{\det(\mathbf{R_s}+\mathbf{R_n})} e^{-\mathbf{r}^{\mathrm{H}} (\mathbf{R_s}+\mathbf{R_n})^{-1} \mathbf{r}}} \\ &= \frac{\det(\mathbf{R_s} + \mathbf{R_n})}{\det(\mathbf{R_n})} e^{-\mathbf{r}^{\mathrm{H}} (\mathbf{R_n}^{-1} - (\mathbf{R_s}+\mathbf{R_n})^{-1}) \mathbf{r}}. \end{aligned} \tag{2.44}$$

If $\mathbf{R_s} = \sigma_s^2 \mathbf{I}$ and $\mathbf{R_n} = \sigma_n^2 \mathbf{I}$, the LLR becomes

$$\mathrm{LLR}(\mathbf{r}) = -\frac{\sigma_s^2}{\sigma_n^2(\sigma_s^2 + \sigma_n^2)} ||\mathbf{r}||^2 + \text{constant}. \tag{2.45}$$

Finally, the LR-based detection becomes

$$||\mathbf{r}||^2 \underset{H_0}{\overset{H_1}{\gtrless}} \tau. \tag{2.46}$$

Since $||\mathbf{r}||^2$ provides the energy of the received signal vector, this detector is called the energy detector.

The energy detector has been recently employed in cognitive radio systems to detect the presence of primary signals. Cognitive radio is a wireless communication system that operates over a frequency band that is licensed by primary wireless systems. If the primary wireless system does not transmit signals, a cognitive radio can be activated to transmit signals to associated receivers. As the cognitive radio does not have the license to access the frequency band, it has to stop its transmission immediately if the primary wireless system transmits signals. Therefore, the detection of primary signals is crucial for a cognitive radio not to interfere with the primary wireless system.

To find the FA and detection probabilities of the LR-based detection in (2.46), the pdf of $||\mathbf{r}||^2$ is required. Let

$$Z = ||\mathbf{r}||^2 = \sum_{l=1}^{L} |r_l|^2,$$

where r_l represents the lth element of $\mathbf{r}$. If $Z_l = |r_l|^2$ is a sum of two independent Gaussian random variables with mean zero and variance $\sigma^2/2$, the distribution of

$X_l = |r_l|^2$ is given by

$$f_X(x) = \frac{1}{\sigma^2} e^{-\frac{x}{\sigma^2}}, \ x \geq 0.$$

Using the mgf of X and inverse Laplace transform, the distribution of Z can be found as

$$f_Z(z) = \frac{1}{(L-1)!\sigma^{2N}} z^{L-1} e^{-\frac{z}{\sigma^2}}.$$

In addition, the cdf of Z is given by

$$\begin{aligned} F_Z(z) &= \Pr(Z \leq z) \\ &= 1 - e^{-\frac{z}{\sigma^2}} \sum_{k=0}^{L-1} \frac{\left(\frac{z}{\sigma^2}\right)^k}{k!}. \end{aligned} \tag{2.47}$$

Under H_0, σ^2 becomes σ_n^2 as there is no signal, while σ^2 becomes $\sigma_s^2 + \sigma_n^2$ under H_1.

Using (2.47), the FA probability can be found as

$$\begin{aligned} P_{\mathrm{FA}} &= \Pr\left(||\mathbf{r}||^2 > \tau \,|\, H_0\right) \\ &= e^{-\frac{\tau}{\sigma_n^2}} \sum_{k=0}^{L-1} \frac{\left(\frac{z}{\sigma_n^2}\right)^k}{k!}. \end{aligned}$$

The detection probability is given by

$$\begin{aligned} P_{\mathrm{D}} &= \Pr\left(||\mathbf{r}||^2 > \tau \,|\, H_1\right) \\ &= e^{-\frac{\tau}{\sigma_s^2+\sigma_n^2}} \sum_{k=0}^{L-1} \frac{\left(\frac{z}{\sigma_s^2+\sigma_n^2}\right)^k}{k!}. \end{aligned}$$

Figure 2.12 shows the ROCs for energy detector when $\sigma_n^2 = 1$ and $L = 4$. It is shown that the ROC is improved as the variance of the random signal $\mathbf{s}$ increases.

2.8 Summary and notes

In this chapter, various decision rules were presented with their applications to signal detection. To see the performance, error probabilities were derived. An important technique to find an upper bound on the error probability for M-ary signaling based on the PEP and union bound was also explained.

Signal detection theory is well established and a number of textbooks are available, e.g. (Whalen 1971) and (Kay 1998). A vector signal approach is employed to explain signal detection in (Scharf 1991). There are also a number of textbooks for signal

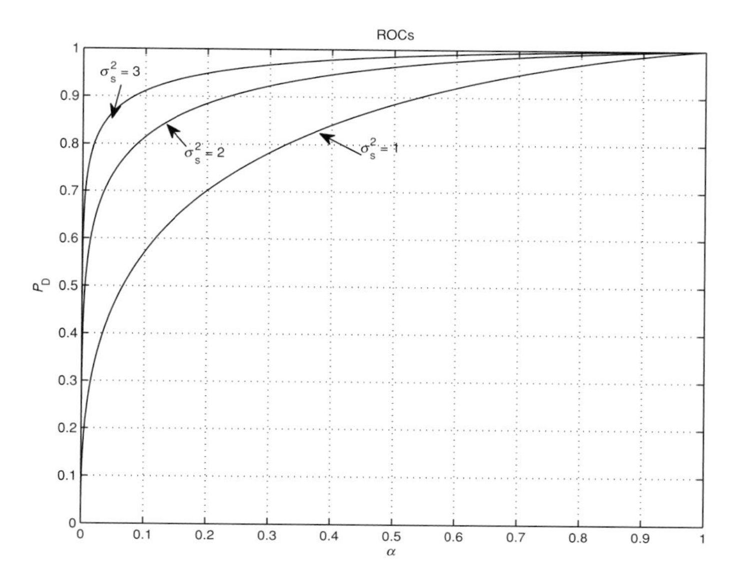

Figure 2.12 Receiver operating characteristics for energy detector when $\sigma_n^2 = 1$ and $L = 4$.

detection theory in the context of digital communications ((Viterbi & Omura 1979), (Wilson 1996), and (Proakis 1995)).

Problems

Problem 2.1 Derive (2.31).

Problem 2.2 Derive (2.32).

Problem 2.3 Suppose that there are two different kinds of dice, one is fair and the other is not fair. The probability distribution of unfair dice is given by

$$\Pr(x) = \frac{x}{21}, \; x \in \mathcal{X} = \{1, 2, \ldots, 6\}.$$

The observation, X, which is the number on the top of the chosen dice is available to decide whether or not the dice is fair. The pair of hypotheses is given by

$$H_0 : X \text{ is from a fair dice;}$$

$$H_1 : X \text{ is from an unfair dice.}$$

Assuming that a fair or unfair dice can be equally chosen, build Neyman–Pearson decision rule with $P_{\text{FA}} = 0.2$.

Problem 2.4 Derive the LR-based decision rule when the phase is known for the pair of hypotheses in (2.41). Draw the ROC curve for the LR-based decision rule in (2.42).

Problem 2.5 Consider the following signal alphabet:

$$\mathcal{S} = \{s_1, s_2, s_3, s_4\} = \{-3, -1, 1, 3\}.$$

The received signal is given by

$$Y = S + N,$$

where $S \in \mathcal{S}$ and N is the noise that has the following distribution (this distribution is a special case of the Laplace distribution):

$$f(n) = \frac{1}{2} \mathrm{e}^{-|n|}.$$

(i) Find all the PEPs according to the ML decision rule.
(ii) Find the upper bound on the average symbol error probability using the union bound and PEPs when the symbols are equally likely.

Problem 2.6 Consider the signal alphabet for M-ary amplitude shift keying (ASK) as follows:

$$\mathcal{S} = \{2m - 1 - M \mid m = 1, 2, \ldots, M\}.$$

The received signal is given by

$$r = s + n,$$

where $s \in \mathcal{S}$ is the transmitted signal and $n \sim \mathcal{N}(0, N_0/2)$ is the background noise. It is assumed that $s \in \mathcal{S}$ is equally likely.
(i) Show that the symbol energy is

$$\mathcal{E}[|s|^2] = \frac{M^2 - 1}{3}.$$

(ii) Show that the symbol error probability of M-ary ASK is given by

$$P_M = \frac{2(M-1)}{M} \mathcal{Q}\left(\sqrt{\frac{(6 \log_2 M) E_\mathrm{b}}{(M^2 - 1) N_0}}\right),$$

where E_b is the bit energy that is given by

$$E_\mathrm{b} = \frac{\mathcal{E}[|s|^2]}{\log_2 M}.$$

Problem 2.7 Suppose that $T = 1$ second. There are two basis functions, $\psi_1(t)$ and $\psi_2(t)$ for $0 \le t < T = 1$:

$$\psi_1(t) = 1, \quad 0 \le t < 1;$$

$$\psi_2(t) = \begin{cases} 1, & 0 \le t < 1/2; \\ -1, & 1/2 \le t < 1. \end{cases}$$

With the set of basis functions, $\{\psi_1(t), \psi_2(t)\}$, four waveforms for 4-ary signaling are given by

$$s_1(t) = 2a\psi_1(t);$$
$$s_2(t) = -2a\psi_1(t);$$
$$s_3(t) = a\psi_2(t);$$
$$s_4(t) = -a\psi_2(t),$$

where $a > 0$. The received signal is given by

$$Y(t) = s_m(t) + N(t),\ 0 \le t < 1,$$

where $N(t)$ is the white Gaussian noise with $\mathcal{E}[N(t)] = 0$ and $\mathcal{E}[N(t)N(t-\tau)] = \frac{N_0}{2}\delta(\tau)$.

(i) Find the symbol energy in terms of a and derive the SNR with N_0.
(ii) Find the pairwise error probabilities in terms of SNR.

Problem 2.8 Suppose that there are M signals, denoted by $\mathbf{s}_m$, $m = 1, 2, \ldots, M$, for M-ary communications. Assume that they are orthogonal, i.e.

$$\mathbf{s}_m^{\mathrm{T}}\mathbf{s}_n = \begin{cases} E_{\mathrm{s}}, & \text{if } m = n; \\ 0, & \text{if } m \ne n. \end{cases}$$

If $\mathbf{s}_m$ is transmitted, the received signal vector is given by

$$\mathbf{r} = \mathbf{s}_m + \mathbf{n},$$

where $\mathbf{n} \sim \mathcal{N}(\mathbf{0}, \frac{N_0}{2}\mathbf{I})$.

(i) Find the ML decision rule to detect the signal.
(ii) Show the symbol error probability is given by

$$P_{\mathrm{err}} = 1 - \int_{-\infty}^{\infty} \frac{1}{\sqrt{2\pi}} \exp\left(-\frac{z^2}{2}\right)\left(1 - \mathcal{Q}\left(z + \sqrt{\frac{2E_{\mathrm{s}}}{N_0}}\right)\right)^{M-1} \mathrm{d}z. \qquad (2.48)$$

Problem 2.9 Let $\mathbf{r}$ be the received signal vector of size $L \times 1$. Consider the following pair of hypotheses:

$$H_0 : \mathbf{r} = \mathbf{n};$$
$$H_1 : \mathbf{r} = \mathbf{s} + \mathbf{n}, \qquad (2.49)$$

where $\mathbf{n} \sim \mathcal{CN}(0, \sigma_n^2\mathbf{I})$ is the background noise vector and $\mathbf{s} \sim \mathcal{CN}(0, \sigma_s^2\mathbf{I})$ is the unknown Gaussian signal vector. We employ the energy detector to detect the presence of signals and the decision variable is given by

$$Z = \frac{||\mathbf{r}||^2}{L}.$$

(This energy detector is also considered in Subsection 2.7.4.)

(i) Verify that Z approaches a Gaussian random variable according to the central limit theorem (CLT) (Hint: use the Lyapunov condition in Subsection 1.4.2).

(ii) Assuming that Z is Gaussian and the LR-based decision rule is employed, show that the ROC is given by

$$P_{\mathrm{D}}(\alpha) = \mathcal{Q}\left(\frac{\mathcal{Q}^{-1}(\alpha)\sigma_n^2 - \sqrt{L}\sigma_s^2}{\sqrt{\sigma_s^4 + \sigma_n^4}}\right),$$

where $\alpha = P_{\mathrm{FA}}$ is the FA probability.

3 Fundamentals of estimation theory

In various statistical signal processing applications, we often assume that signals are generated from a system that can be characterized by a set of parameters. Therefore, in order to understand signals, it is desirable to know the values of parameters. Estimation theory provides statistical approaches to estimate unknown parameters' values from a set of measurements.

To carry out the estimation, we need a certain model (or system) of measurement signals. For example, the following linear model can be considered:

$$r = sh + n, \tag{3.1}$$

where r is the measurement signal, s is the data signal, h is the parameter to be estimated, and n is the measurement noise or modeling error. For example, for a given set of measurement signals, if we have the scatter diagram as shown in Fig. 3.1, the linear relation between measurement and input data in (3.1) can be adopted. The estimation problem associated with (3.1) is to estimate h from measurement r.

A general observation model can be considered as follows:

$$r = x(h) + n,$$

where $x(h)$ is a function of h, which can be a nonlinear function. For example, $x(h) = h^2$. Although it is possible to consider the estimation problem with a general model, we focus on linear models in this chapter as a number of estimation methods can be easily explained with linear models.

3.1 Least square and recursive least square estimation

The least square (LS) estimation method is a nonstatistical approach to find the values of the parameters of a certain observation model. We can find optimal values that minimize the difference between measurements and the model's outputs. In this section, we discuss LS estimation and its on-line version.

3.1.1 Least square estimation

Consider a generalized linear model from (3.1) with M unknown parameters, $h_1, h_2, \ldots, h_M$. Suppose that the lth measurement vector of size $N \times 1$ is a weighted

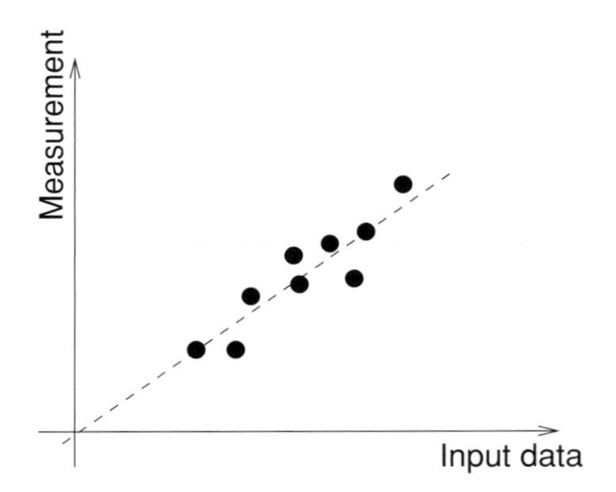

Figure 3.1 Scatter diagram for pairs of input data and measurement.

sum of the lth data vectors, $\mathbf{s}_{l,1}, \mathbf{s}_{l,2}, \ldots, \mathbf{s}_{l,M}$, and given by

$$\begin{aligned}\mathbf{r}_l &= \mathbf{s}_{l,1}h_1 + \mathbf{s}_{l,2}h_2 + \cdots + \mathbf{s}_{l,M}h_M + \mathbf{n}_l \\ &= \mathbf{S}_l\mathbf{h} + \mathbf{n}_l, \ l = 0, 1, \ldots, L-1,\end{aligned} \tag{3.2}$$

where L is the number of observations, $\mathbf{n}_l$ is the measurement noise vector of size $N \times 1$, and

$$\begin{aligned}\mathbf{S}_l &= [\mathbf{s}_{l,1}\ \mathbf{s}_{l,2}\ \cdots\ \mathbf{s}_{l,M}]; \\ \mathbf{h} &= [h_1\ h_2\ \ldots\ h_M]^{\mathrm{T}}.\end{aligned} \tag{3.3}$$

The model in (3.2) is linear with respect to the unknown parameters, $h_1, h_2, \ldots, h_M$. In general, the measurement noise, $\mathbf{n}_l$, is considered as a random vector and the data vectors, $\{\mathbf{s}_{l,m}\}$, are known.

Stacking the $\mathbf{r}_l$'s, we have the following linear model:

$$\begin{aligned}\mathbf{r} &= [\mathbf{r}_0^{\mathrm{T}}\ \mathbf{r}_1^{\mathrm{T}}\ \ldots\ \mathbf{r}_{L-1}^{\mathrm{T}}]^{\mathrm{T}} \\ &= \mathbf{S}\mathbf{h} + \mathbf{n},\end{aligned} \tag{3.4}$$

where

$$\mathbf{S} = \begin{bmatrix} \mathbf{S}_0 \\ \mathbf{S}_1 \\ \vdots \\ \mathbf{S}_{L-1} \end{bmatrix};$$

$$\mathbf{n} = [\mathbf{n}_0^{\mathrm{T}}\ \mathbf{n}_1^{\mathrm{T}}\ \ldots\ \mathbf{n}_{L-1}^{\mathrm{T}}]^{\mathrm{T}}.$$

The LS estimate of $\mathbf{h}$ is to minimize the squared difference between the measurement or observation, $\mathbf{r}$, and the model output without noise, $\mathbf{Sh}$. The LS estimate is given by

$$\hat{\mathbf{h}}_{\mathrm{ls}} = \arg\min_{\mathbf{h}} ||\mathbf{r} - \mathbf{S}\mathbf{h}||^2. \tag{3.5}$$

A generalization of LS estimation is weighted LS (WLS) estimation. The WLS estimate is given by

$$\hat{\mathbf{h}}_{\text{wls}} = \arg\min_{\mathbf{h}} ||\mathbf{r} - \mathbf{S}\mathbf{h}||_{\mathbf{W}}^2, \tag{3.6}$$

where $||\mathbf{x}||_{\mathbf{W}}^2 = \mathbf{x}^{\text{H}}\mathbf{W}\mathbf{x}$ and $\mathbf{W}$ is the weighting matrix that is symmetric and positive definite. It is easy to show that LS estimation is a special case of WLS estimation, where $\mathbf{W} = \mathbf{I}$.

We now consider the LS solution when $\mathbf{h}$ is a complex vector. Since $||\mathbf{r} - \mathbf{S}\mathbf{h}||_{\mathbf{W}}^2$ is quadratic with respect to $\mathbf{h}$, the LS estimate that minimizes $||\mathbf{r} - \mathbf{S}\mathbf{h}||_{\mathbf{W}}^2$ becomes the solution of the following equation:

$$\begin{aligned} 0 &= \frac{\partial}{\partial \mathbf{h}} ||\mathbf{r} - \mathbf{S}\mathbf{h}||_{\mathbf{W}}^2 \\ &= \frac{\partial}{\partial \mathbf{h}} (\mathbf{r} - \mathbf{S}\mathbf{h})^{\text{H}}\mathbf{W}(\mathbf{r} - \mathbf{S}\mathbf{h}), \end{aligned} \tag{3.7}$$

where $\partial/\partial\mathbf{x}$ denotes the complex gradient with respect to $\mathbf{x}$. Suppose that C is a real-valued function of (a complex-valued) $\mathbf{x}$. Then, the complex gradient of C is defined as

$$\frac{\partial C}{\partial \mathbf{x}} = \begin{bmatrix} \frac{\partial C}{\partial x_1} \\ \frac{\partial C}{\partial x_2} \\ \vdots \\ \frac{\partial C}{\partial x_M} \end{bmatrix},$$

where the complex derivative, $\partial C/\partial x$, is defined as

$$\frac{\partial C}{\partial x} = \frac{1}{2}\left(\frac{\partial C}{\partial \alpha} - j\frac{\partial C}{\partial \beta}\right). \tag{3.8}$$

Here, $x = \alpha + j\beta$, where α and β are the real and imaginary parts of x, respectively. The complex derivative can be generalized for a complex-valued function, C. The complex derivative is given by

$$\begin{aligned} \frac{\partial C}{\partial x} &= \frac{1}{2}\left(\frac{\partial}{\partial \alpha} - j\frac{\partial}{\partial \beta}\right)(\Re(C) + j\Im(C)) \\ &= \frac{1}{2}\left(\frac{\partial \Re(C)}{\partial \alpha} - j\frac{\partial \Re(C)}{\partial \beta}\right) + \frac{j}{2}\left(\frac{\partial \Im(C)}{\partial \alpha} - j\frac{\partial \Im(C)}{\partial \beta}\right). \end{aligned} \tag{3.9}$$

From (3.9), it follows that

$$\begin{aligned} \frac{\partial x}{\partial x} &= 1; \\ \frac{\partial x^*}{\partial x} &= 0. \end{aligned}$$

In addition, if $\mathbf{A} = \mathbf{A}^{\mathrm{H}}$,

$$\begin{aligned}\frac{\partial}{\partial \mathbf{x}} \mathbf{y}^{\mathrm{H}} \mathbf{x} &= \mathbf{y}^*; \\ \frac{\partial}{\partial \mathbf{x}} \mathbf{x}^{\mathrm{H}} \mathbf{y} &= \mathbf{0}; \\ \frac{\partial}{\partial \mathbf{x}} \mathbf{x}^{\mathrm{H}} \mathbf{A} \mathbf{x} &= 2(\mathbf{A}\mathbf{x})^*. \end{aligned} \tag{3.10}$$

Substituting (3.10) into (3.7), we can see that the LS estimate of $\mathbf{h}$ is the solution of the following equation:

$$\mathbf{S}^{\mathrm{H}} \mathbf{W} \mathbf{S} \mathbf{h} = \mathbf{S}^{\mathrm{H}} \mathbf{W} \mathbf{r}. \tag{3.11}$$

Consider the case that $\mathbf{x}$ is real. Since the real derivative is not a special case of the complex derivative in (3.8), the complex derivative has to be replaced by the real derivative that is defined as (for a complex-valued function C)

$$\frac{\partial C}{\partial x} = \frac{\partial \Re(C)}{\partial x} + j \frac{\partial \Im(C)}{\partial x}. \tag{3.12}$$

However, whether $\mathbf{h}$ is real or complex, the result in (3.11) is valid for both.

From (3.11), the WLS estimate, $\hat{\mathbf{h}}_{\mathrm{wls}}$, can be found as

$$\hat{\mathbf{h}}_{\mathrm{wls}} = (\mathbf{S}^{\mathrm{H}} \mathbf{W} \mathbf{S})^{-1} \mathbf{S}^{\mathrm{H}} \mathbf{W} \mathbf{r} \tag{3.13}$$

if $(\mathbf{S}^{\mathrm{H}} \mathbf{W} \mathbf{S})^{-1}$ exists or $\mathbf{S}^{\mathrm{H}} \mathbf{W} \mathbf{S}$ is full rank. A necessary condition that $(\mathbf{S}^{\mathrm{H}} \mathbf{W} \mathbf{S})^{-1}$ exists is that $NL \geq M$. If $(\mathbf{S}^{\mathrm{H}} \mathbf{W} \mathbf{S})^{-1}$ does not exist, a pseudo inverse of $\mathbf{S}^{\mathrm{H}} \mathbf{W} \mathbf{S}$, which is denoted by $(\mathbf{S}^{\mathrm{H}} \mathbf{W} \mathbf{S})^{\dagger}$, can be used. As a special case of the WLS, the LS estimate is given by

$$\hat{\mathbf{h}}_{\mathrm{ls}} = (\mathbf{S}^{\mathrm{H}} \mathbf{S})^{-1} \mathbf{S}^{\mathrm{H}} \mathbf{r}. \tag{3.14}$$

Note that if all the quantities are real-valued, (3.13) is reduced to

$$\hat{\mathbf{h}}_{\mathrm{wls}} = (\mathbf{S}^{\mathrm{T}} \mathbf{W} \mathbf{S})^{-1} \mathbf{S}^{\mathrm{T}} \mathbf{W} \mathbf{r}. \tag{3.15}$$

It is also possible to avoid the complex derivative by transforming complex-valued vectors and matrices into real-valued ones as follows:

$$\begin{aligned}\mathbf{x} &\to \begin{bmatrix} \Re(\mathbf{x}) \\ \Im(\mathbf{x}) \end{bmatrix}; \\ \mathbf{S} &\to \begin{bmatrix} \Re(\mathbf{S}) & -\Im(\mathbf{S}) \\ \Im(\mathbf{S}) & \Re(\mathbf{S}) \end{bmatrix}. \end{aligned}$$

In (3.1) we consider a linear model. This linear model can be extended to include an (unknown) offset. In this case, the measurement signal is given by

$$\begin{aligned} r_l &= s_l h + a + n_l \\ &= [s_l \; 1] \begin{bmatrix} h_1 \\ h_2 \end{bmatrix} + n_l, \; l = 0, 1, \ldots, L-1, \end{aligned} \tag{3.16}$$

where $h_1 = h$ and $h_2 = a$. The linear model in (3.2) can also be further extended to include more offset terms as follows:

$$\begin{aligned}\mathbf{r}_l &= \mathbf{s}_{l,1}h_1 + \mathbf{s}_{l,2}h_2 + \cdots + \mathbf{s}_{l,M}h_M + \mathbf{a} + \mathbf{n}_l \\ &= [\mathbf{S}_l\ \mathbf{I}]\begin{bmatrix}\mathbf{h}\\ \mathbf{a}\end{bmatrix} + \mathbf{n}_l,\ l = 0, 1, \ldots, L-1,\end{aligned} \tag{3.17}$$

where $\mathbf{a}$ is a vector that consists of unknown offset parameters.

Example 3.1.1 Suppose that the output signal from a circuit is given by

$$s(t) = 1 - \mathrm{e}^{-at},\ a > 0,\ t \geq 0,$$

where a is the parameter to be estimated. We can take a sample of $s(t)$ at $t = \epsilon_l$ during the lth measurement and it is known that $a\epsilon_l \ll 1$. The lth measurement signal is given by

$$r_l = s(\epsilon_l) + n_l,\ l = 0, 1, \ldots, L-1,$$

where n_l is the lth measurement noise. Although $s(\epsilon_l)$ is a nonlinear function of a, if $a\epsilon_l$ is sufficiently small, we can have the following approximation:

$$\begin{aligned}s(\epsilon_l) &= 1 - \mathrm{e}^{-a\epsilon_l} \\ &= a\epsilon_l + O(a^2\epsilon_l^2).\end{aligned}$$

Ignoring the higher-order terms, we have

$$r_l = a\epsilon_l + n_l,\ l = 0, 1, \ldots, L-1.$$

Letting $\mathbf{s} = [\epsilon_0\ \epsilon_1\ \ldots \epsilon_{L-1}]^{\mathrm{T}}$ and $\mathbf{r} = [r_0\ r_1\ \ldots r_{L-1}]^{\mathrm{T}}$, the LS estimate becomes

$$\begin{aligned}\hat{a} &= (\mathbf{s}^{\mathrm{H}}\mathbf{s})^{-1}\mathbf{s}^{\mathrm{H}}\mathbf{r} \\ &= \frac{\sum_{l=0}^{L-1}\epsilon_l r_l}{\sum_{l=0}^{L-1}\epsilon_l^2}.\end{aligned}$$

3.1.2 Recursive least square approach

The recursive LS (RLS) algorithm is an on-line algorithm to perform LS estimation. Consider the following linear model:

$$r_l = \mathbf{s}_l^{\mathrm{H}}\mathbf{h} + n_l,\ l = 0, 1, \ldots, \tag{3.18}$$

where $\mathbf{s}_l$ is the lth data vector of size $M \times 1$. We assume that the measurement r_l is available at time l. Define an accumulated or sum squared error (SSE) as

$$\mathrm{SSE}_l = \sum_{k=0}^{l} |r_k - \mathbf{s}_k^{\mathrm{H}}\mathbf{h}|^2. \tag{3.19}$$

Suppose that it is necessary to find the LS estimate of $\mathbf{h}$ that minimizes SSE_l for each time l. Based on (3.14), the LS estimate can be found. However, this approach requires a matrix inversion for every l and the resulting total computational complexity can be too high in some applications. To avoid this high computational complexity, the RLS algorithm can be employed. The main idea of the RLS algorithm is to update the LS estimate from the previous one recursively when a new measurement is available. To derive the RLS, we need to introduce the forgetting factor, λ. From (3.19), the exponential weighted SSE up to time l is given by

$$\text{EWSSE}_l = \sum_{k=0}^{l} \lambda^{l-k} |r_k - \mathbf{s}_k^{\text{H}} \mathbf{h}|^2, \tag{3.20}$$

where $0 < \lambda < 1$. A recent squared error has more weight than a past squared error since $0 < \lambda < 1$. This weight difference increases as λ decreases. From (3.20), we can show that

$$\text{EWSSE}_l = \lambda \times \text{EWSSE}_{l-1} + |r_l - \mathbf{s}_l^T \mathbf{h}|^2.$$

The LS estimate that minimizes EWSSE_l, denoted by $\mathbf{h}_l$, satisfies the following equation:

$$\begin{aligned}
0 &= \nabla_{\mathbf{h}} \text{EWSSE}_l \\
&= \nabla_{\mathbf{h}} \left(\lambda \times \text{EWSSE}_{l-1} + |r_l - \mathbf{s}_l^{\text{H}} \mathbf{h}|^2 \right) \\
&= \left(\lambda \left(\sum_{k=0}^{l-1} \lambda^{l-1-k} \mathbf{s}_k \mathbf{s}_k^{\text{H}} \right) + \mathbf{s}_l \mathbf{s}_l^{\text{H}} \right) \mathbf{h}^{(l)} \\
&\quad - \lambda \left(\sum_{k=0}^{l-1} \lambda^{l-1-k} \mathbf{s}_k r_k \right) - \mathbf{s}_l r_l.
\end{aligned} \tag{3.21}$$

Define

$$\mathbf{\Sigma}_l = \sum_{k=0}^{l} \lambda^{l-k} \mathbf{s}_k \mathbf{s}_k^{\text{H}}$$

and

$$\mathbf{c}_l = \sum_{k=0}^{l} \lambda^{l-k} \mathbf{s}_k r_k.$$

We can readily show that

$$\mathbf{\Sigma}_l = \lambda \mathbf{\Sigma}_{l-1} + \mathbf{s}_l \mathbf{s}_l^{\text{H}} \tag{3.22}$$

and

$$\mathbf{c}_l = \lambda \mathbf{c}_{l-1} + \mathbf{s}_l r_l. \tag{3.23}$$

From these, (3.21) can be rewritten as

$$\left(\lambda \mathbf{\Sigma}_{l-1} + \mathbf{s}_l \mathbf{s}_l^{\text{H}} \right) \mathbf{h}^{(l)} = (\lambda \mathbf{c}_{l-1} + \mathbf{s}_l r_l). \tag{3.24}$$

Using the following matrix inversion lemma or Woodbury's identity:

$$(\mathbf{A} + \gamma^2 \mathbf{u}\mathbf{u}^{\mathrm{H}})^{-1} = \mathbf{A}^{-1} - \frac{\gamma^2}{1 + \gamma^2 \mathbf{u}^{\mathrm{H}} \mathbf{A}^{-1} \mathbf{u}} \mathbf{A}^{-1} \mathbf{u}\mathbf{u}^{\mathrm{H}} \mathbf{A}^{-1}, \tag{3.25}$$

where $\mathbf{A}$ is a full-rank Hermitian matrix, we have

$$\begin{aligned} \mathbf{h}^{(l)} &= \mathbf{\Sigma}_l^{-1} \mathbf{c}_l \\ &= \left(\lambda \mathbf{\Sigma}_{l-1} + \mathbf{s}_l \mathbf{s}_l^{\mathrm{H}}\right)^{-1} (\lambda \mathbf{c}_{l-1} + \mathbf{s}_l r_l) \\ &= \mathbf{\Sigma}_{l-1}^{-1} \mathbf{c}_{l-1} - \lambda \mathbf{\Omega}_l \mathbf{c}_{l-1} + \lambda^{-1} \mathbf{\Sigma}_{l-1}^{-1} \mathbf{s}_l r_l - \mathbf{\Omega}_l \mathbf{s}_l r_l, \end{aligned} \tag{3.26}$$

where

$$\mathbf{\Omega}_l = \frac{\lambda^{-2}}{1 + \lambda^{-1} \mathbf{s}_l^{\mathrm{H}} \mathbf{\Sigma}_{l-1}^{-1} \mathbf{s}_l} \mathbf{\Sigma}_{l-1}^{-1} \mathbf{s}_l \mathbf{s}_l^{\mathrm{H}} \mathbf{\Sigma}_{l-1}^{-1}.$$

For further simplification, we define

$$\begin{aligned} \beta_l &= \frac{\lambda^{-1}}{1 + \lambda^{-1} \mathbf{s}_l^{\mathrm{H}} \mathbf{\Sigma}_{l-1}^{-1} \mathbf{s}_l}; \\ \mathbf{m}_l &= \beta_l \mathbf{\Sigma}_{l-1}^{-1} \mathbf{s}_l. \end{aligned} \tag{3.27}$$

Then, $\mathbf{\Omega}_l$ is rewritten as

$$\mathbf{\Omega}_l = (\lambda \beta_l)^{-1} \mathbf{m}_l \mathbf{m}_l^{\mathrm{H}}.$$

Substituting this into (3.26), we have

$$\begin{aligned} \mathbf{h}^{(l)} &= \mathbf{h}^{(l-1)} - \mathbf{m}_l \mathbf{s}_l^{\mathrm{H}} \mathbf{\Sigma}_{l-1}^{-1} \mathbf{c}_{l-1} + (\lambda \beta_l)^{-1} \mathbf{m}_l r_l - (\lambda \beta_l)^{-1} \mathbf{m}_l \mathbf{m}_l^{\mathrm{H}} \mathbf{s}_l r_l \\ &= \mathbf{h}^{(l-1)} - \mathbf{m}_l \mathbf{s}_l^{\mathrm{H}} \mathbf{h}^{(l-1)} + (\lambda \beta_l)^{-1} \mathbf{m}_l (1 - \mathbf{m}_l^{\mathrm{H}} \mathbf{s}_l) r_l \\ &= \mathbf{h}^{(l-1)} + \mathbf{m}_l \left((\lambda \beta_l)^{-1} (1 - \mathbf{m}_l^{\mathrm{H}} \mathbf{s}_l) r_l - \mathbf{s}_l^{\mathrm{H}} \mathbf{h}^{(l-1)} \right). \end{aligned} \tag{3.28}$$

Using the definitions of β_l and $\mathbf{m}_l$, we can show that

$$1 - \mathbf{m}_l^{\mathrm{H}} \mathbf{s}_l = 1 - \frac{\lambda^{-1} \mathbf{s}_l^{\mathrm{H}} \mathbf{\Sigma}_{l-1}^{-1} \mathbf{s}_l}{1 + \lambda^{-1} \mathbf{s}_l^{\mathrm{H}} \mathbf{\Sigma}_{l-1}^{-1} \mathbf{s}_l} = \lambda \beta_l.$$

This finally simplifies the recursion for the RLS algorithm as

$$\mathbf{h}^{(l)} = \mathbf{h}^{(l-1)} + \mathbf{m}_l \left(r_l - \mathbf{s}_l^{\mathrm{H}} \mathbf{h}^{(l-1)} \right). \tag{3.29}$$

In the RLS algorithm, we need to compute the inverse of $\mathbf{\Sigma}_l$. This matrix inversion requires the complexity of an order of $O(M^3)$. To avoid a matrix inversion for every available measurement, r_l, we can recursively update $\mathbf{\Sigma}_l^{-1}$ from the previous one, $\mathbf{\Sigma}_{l-1}^{-1}$. To this end, let $\mathbf{\Phi}_l = \mathbf{\Sigma}_l^{-1}$. Then, applying (3.25) to (3.22), we have the following recursion:

$$\begin{aligned} \mathbf{\Phi}_l &= \lambda^{-1} \mathbf{\Phi}_{l-1} - \mathbf{\Omega}_l \\ &= \lambda^{-1} \left(\mathbf{\Phi}_{l-1} - \mathbf{m}_l \mathbf{s}_l^{\mathrm{H}} \mathbf{\Phi}_{l-1} \right). \end{aligned} \tag{3.30}$$

This implies that we can obtain the matrix inversion of $\boldsymbol{\Sigma}_l$ recursively with the complexity of an order of $O(M^2)$.

Finally, the RLS algorithm can be summarized as follows:

$$\begin{aligned}
\mathbf{h}^{(l)} &= \mathbf{h}^{(l-1)} + \beta_l \boldsymbol{\Phi}_{l-1} \mathbf{s}_l \left(r_l - \mathbf{s}_l^{\mathrm{H}} \mathbf{h}^{(l-1)} \right); \\
\beta_l &= \frac{\lambda^{-1}}{1 + \lambda^{-1} \mathbf{s}_l^{\mathrm{H}} \boldsymbol{\Phi}_{l-1} \mathbf{s}_l}; \\
\mathbf{m}_l &= \beta_l \boldsymbol{\Phi}_{l-1} \mathbf{s}_l; \\
\boldsymbol{\Phi}_l &= \lambda^{-1} \left(\boldsymbol{\Phi}_{l-1} - \mathbf{m}_l \mathbf{s}_l^{\mathrm{H}} \boldsymbol{\Phi}_{l-1} \right).
\end{aligned} \tag{3.31}$$

The initial matrix, $\boldsymbol{\Phi}_0$, can be given by

$$\boldsymbol{\Phi}_0 = \phi_0 \mathbf{I},$$

where $\phi_0 > 0$.

3.2 Minimum variance unbiased and best linear unbiased estimation

LS estimation is not a statistical approach. Thus, LS estimation can be used without any statistical knowledge of signal and noise. If statistical properties of signal and noise are available, we can derive statistical approaches for the parameter estimation.

Suppose that there is a measurement $\mathbf{x}$. We assume that $\mathbf{x}$ is a realization of a random vector whose pdf is denoted by $f_{\mathbf{h}}(\mathbf{x})$, where $\mathbf{h}$ is the parameter vector to be estimated. For example, consider a Gaussian random variable X whose pdf is given by

$$f(x) = \frac{1}{\sqrt{2\pi\sigma^2}} \mathrm{e}^{-\frac{1}{2\sigma^2}(x-\mu)^2},$$

where μ and σ^2 are the mean and variance, respectively. If μ is the parameter to be estimated, while σ^2 is known, the pdf of X can be written as $f_\mu(x)$.

In the statistical estimation, an estimator of $\mathbf{h}$ is a function of realizations of $\mathbf{x}$ and usually denoted by $\hat{\mathbf{h}}(\mathbf{x})$. An estimator is called *unbiased* if

$$\mathcal{E}[\hat{\mathbf{h}}(\mathbf{x})] = \mathbf{h}. \tag{3.32}$$

An estimator becomes minimum variance unbiased (MVUB) if

$$\mathcal{E}[(\hat{\mathbf{h}}(\mathbf{x}) - \mathbf{h})^{\mathrm{H}}(\hat{\mathbf{h}}(\mathbf{x}) - \mathbf{h})] \le \mathcal{E}[(\mathbf{d}(\mathbf{x}) - \mathbf{h})^{\mathrm{H}}(\mathbf{d}(\mathbf{x}) - \mathbf{h})], \tag{3.33}$$

for any other unbiased estimator $\mathbf{d}(\mathbf{x})$. In both (3.32) and (3.33), the expectation is carried out with respect to $\mathbf{x}$. Unfortunately, the existence of the MVUB estimator is not always guaranteed. To verify the existence of the MVUB estimator, we need to understand the notion of sufficient statistics.

A *statistic* $\mathbf{t}(\mathbf{x})$ is a vector function of measurement $\mathbf{x}$. If the conditional pdf of $\mathbf{x}$ given $\mathbf{t}(\mathbf{x})$ does not depend on the parameter vector $\mathbf{h}$, $\mathbf{t}(\mathbf{x})$ is called a *sufficient statistic* for the parameter vector $\mathbf{h}$. Let $f_{\mathbf{h}}(\mathbf{x})$ and $f(\mathbf{x}|\mathbf{t})$ denote the pdf of $\mathbf{x}$ and the conditional pdf of

$\mathbf{x}$ given $\mathbf{t} = \mathbf{t}(\mathbf{x})$, respectively. Noting that

$$f_{\mathbf{h}}(\mathbf{x}) = \mathcal{E}[f(\mathbf{x}|\mathbf{t})] = \int f(\mathbf{x}|\mathbf{t}) g_{\mathbf{h}}(\mathbf{t}) d\mathbf{t}, \quad (3.34)$$

where the expectation is carried out with respect to $\mathbf{t}$ and $g_{\mathbf{h}}(\mathbf{t})$ denotes the pdf of $\mathbf{t}$, we can see that if $\mathbf{t}(\mathbf{x})$ is a sufficient statistic, $f(\mathbf{x}|\mathbf{t})$ should not be a function of $\mathbf{h}$. Thus, the parameter vector $\mathbf{h}$ is related to $f_{\mathbf{h}}(\mathbf{x})$ only through $g_{\mathbf{h}}(\mathbf{t})$. In other words, there is no loss of information to estimate $\mathbf{h}$ with $\mathbf{t}(\mathbf{x})$.

Theorem 3.2.1 *The statistic* $\mathbf{t}(\mathbf{x})$ *is sufficient for* $\mathbf{h}$ *if and only if there exists a factorization of* $f_{\mathbf{h}}(\mathbf{x})$ *as follows:*

$$f_{\mathbf{h}}(\mathbf{x}) = \phi_{\mathbf{h}}(\mathbf{t}(\mathbf{x})) v(\mathbf{x}),$$

where $\phi_{\mathbf{h}}(\mathbf{t}(\mathbf{x})) \geq 0$ *and* $v(\mathbf{x}) \geq 0$ *is a function of* $\mathbf{x}$, *not* $\mathbf{h}$.

Example 3.2.1 Consider independent and identically distributed (iid) real-valued Gaussian random variables with unknown mean μ. Assuming that the variance is 1, the pdf of X_l is given by

$$f(x_l) = \frac{1}{\sqrt{2\pi}} e^{-(x_l-\mu)^2/2}, \ l = 0, 1, \ldots, L-1.$$

The sample mean is given by

$$t(\mathbf{x}) = \frac{1}{L} \sum_{l=0}^{L-1} X_l. \quad (3.35)$$

Since the joint pdf of the X_l's is

$$\begin{aligned} f(\mathbf{x}) &= \frac{1}{(2\pi)^{L/2}} \exp\left(-\sum_{l=0}^{L-1} (x_l - \mu)^2/2\right) \\ &= \frac{1}{(2\pi)^{L/2}} \exp\left(-\frac{1}{2}\left(\sum_{l=0}^{L-1} x_l^2 - 2\mu \sum_{l=0}^{L-1} x_l + L\mu^2\right)\right) \\ &= \frac{1}{(2\pi)^{L/2}} \exp\left(-\frac{1}{2}\sum_{l=0}^{L-1} x_l^2\right) \exp\left(\mu L t(\mathbf{x}) - \frac{L}{2}\mu^2\right), \end{aligned}$$

we can see that the statistic $t(\mathbf{x})$ is sufficient for μ.

Once we have a sufficient statistic, we can find an MVUB estimate via the following Rao–Blackwell theorem.

Theorem 3.2.2 *Suppose that there exists a sufficient statistic* $\mathbf{t}(\mathbf{x})$ *for* $\mathbf{h}$. *If there exists an unbiased estimate* $\hat{\mathbf{h}}'(\mathbf{x})$, *then* $\hat{\mathbf{h}}(\mathbf{t}) = \mathcal{E}[\hat{\mathbf{h}}'(\mathbf{x})|\mathbf{t}]$ *is an MVUB estimate of* $\mathbf{h}$ *and it is unique.*

Proof: In this proof, we first consider a special case, where $\mathbf{h} = h$ is a scalar. In this case, we can easily prove the theorem using Jensen's inequality.

We can show that $\hat{h}(\mathbf{t})$ is unbiased using the law of total expectation.

$$\begin{aligned}\mathcal{E}[\hat{h}(\mathbf{t})] &= \mathcal{E}_{\mathbf{t}}[\mathcal{E}_{\mathbf{x}}[\hat{h}'(\mathbf{x})|\mathbf{t}]] \\ &= \mathcal{E}[\hat{h}'(\mathbf{x})] \\ &= h.\end{aligned}$$

Since $\text{Var}(X) = \mathcal{E}[X^2] - \mathcal{E}^2[X]$ and $\hat{h}(\mathbf{t})$ and $\hat{h}'(\mathbf{x})$ are unbiased, in order to prove $\text{Var}(\hat{h}(\mathbf{t})) \leq \text{Var}(\hat{h}'(\mathbf{x}))$, it is sufficient to show

$$\mathcal{E}[(\hat{h}(\mathbf{t}))^2] \leq \mathcal{E}[(\hat{h}'(\mathbf{x}))^2]. \tag{3.36}$$

If the inequality in (3.36) holds, we can conclude that $\hat{\mathbf{h}}(\mathbf{t})$ is MVUB. Using Jensen's inequality, we can show that

$$\begin{aligned}\mathcal{E}[(\hat{h}(\mathbf{t}))^2] &= \mathcal{E}[(\mathcal{E}[\hat{h}'(\mathbf{x})|\mathbf{t}])^2] \\ &\leq \mathcal{E}\left[\mathcal{E}\left[\left(\hat{h}'(\mathbf{x})\right)^2 |\mathbf{t}\right]\right] \\ &= \mathcal{E}[(\hat{h}'(\mathbf{x}))^2].\end{aligned}$$

This completes the proof for a scalar parameter.

Consider a general case, where $\mathbf{h}$ is a vector. It can be easily shown that $\hat{\mathbf{h}}(\mathbf{t})$ is unbiased using the law of total expectation. The variance of $\hat{\mathbf{h}}'(\mathbf{x})$ is given by

$$\begin{aligned}\mathcal{E}[(\hat{\mathbf{h}}'(\mathbf{x}) - \mathbf{h})^{\mathrm{H}}(\hat{\mathbf{h}}'(\mathbf{x}) - \mathbf{h})] &= \mathcal{E}[(\hat{\mathbf{h}}'(\mathbf{x}) - \hat{\mathbf{h}}(\mathbf{t}) + \hat{\mathbf{h}}(\mathbf{t}) - \mathbf{h})^{\mathrm{H}} \\ &\quad \times (\hat{\mathbf{h}}'(\mathbf{x}) - \hat{\mathbf{h}}(\mathbf{t}) + \hat{\mathbf{h}}(\mathbf{t}) - \mathbf{h})] \\ &= \mathcal{E}[(\hat{\mathbf{h}}'(\mathbf{x}) - \hat{\mathbf{h}}(\mathbf{t})^{\mathrm{H}}(\hat{\mathbf{h}}'(\mathbf{x}) - \hat{\mathbf{h}}(\mathbf{t})] \\ &\quad + \mathcal{E}[(\hat{\mathbf{h}}(\mathbf{t}) - \mathbf{h})^{\mathrm{H}}(\hat{\mathbf{h}}(\mathbf{t}) - \mathbf{h})] \\ &\quad + 2\Re(\mathcal{E}[(\hat{\mathbf{h}}'(\mathbf{x}) - \hat{\mathbf{h}}(\mathbf{t}))^{\mathrm{H}}(\hat{\mathbf{h}}(\mathbf{t}) - \mathbf{h})]).\end{aligned} \tag{3.37}$$

In (3.37) the first term is nonnegative and the second term is the variance of $\hat{\mathbf{h}}(\mathbf{t})$. For the third term, we can show that

$$\mathcal{E}[(\hat{\mathbf{h}}'(\mathbf{x}) - \hat{\mathbf{h}}(\mathbf{t}))^{\mathrm{H}}(\hat{\mathbf{h}}(\mathbf{t}) - \mathbf{h})] = \mathcal{E}_{\mathbf{t}}[\mathcal{E}_{\mathbf{x}}[(\hat{\mathbf{h}}'(\mathbf{x}) - \hat{\mathbf{h}}(\mathbf{t}))^{\mathrm{H}}(\hat{\mathbf{h}}(\mathbf{t}) - \mathbf{h})\big|\mathbf{t}]].$$

Since $\mathcal{E}[\hat{\mathbf{h}}'(\mathbf{x})|\mathbf{t}] = \hat{\mathbf{h}}(\mathbf{t})$ by definition, we can show that

$$\begin{aligned}\mathcal{E}_{\mathbf{x}}[(\hat{\mathbf{h}}'(\mathbf{x}) - \hat{\mathbf{h}}(\mathbf{t}))^{\mathrm{H}}(\hat{\mathbf{h}}(\mathbf{t}) - \mathbf{h})\big|\mathbf{t}] &= (\hat{\mathbf{h}}(\mathbf{t}) - \hat{\mathbf{h}}(\mathbf{t}))^{\mathrm{H}}(\hat{\mathbf{h}}(\mathbf{t}) - \mathbf{h}) \\ &= 0.\end{aligned}$$

From this, we can show that

$$\text{Var}(\hat{\mathbf{h}}(\mathbf{t})) \leq \text{Var}(\hat{\mathbf{h}}'(\mathbf{x})).$$

That is, $\hat{\mathbf{h}}(\mathbf{t})$ is MVUB. For a more complete treatment of the Rao–Backwell theorem (including the uniqueness), see (Porat 1994). □

Although the Rao–Backwell theorem can produce MVUB estimates, they are found only when sufficient statistics are available. On the other hand, compared with the MVUB estimator, a linear estimator, which is a linear function of measurement, $\mathbf{x}$, can be easily derived without using the notion of sufficient statistics. The best linear unbiased (BLU) estimate of $\mathbf{h}$ is a linear estimate and given by

$$\hat{\mathbf{h}}_{\text{blu}} = \mathbf{L}_{\text{blu}}\mathbf{x},$$

where $\mathbf{L}_{\text{blu}}$ is a matrix that satisfies the following conditions:

$$\begin{aligned} &\mathcal{E}[\hat{\mathbf{h}}_{\text{blu}}] = \mathbf{h}; \\ &\mathcal{E}[(\mathbf{h} - \hat{\mathbf{h}}_{\text{blu}})^{\text{H}}(\mathbf{h} - \hat{\mathbf{h}}_{\text{blu}})] \le \mathcal{E}[(\mathbf{h} - \mathbf{Lr})^{\text{H}}(\mathbf{h} - \mathbf{Lr})], \text{ for any } \mathbf{L}. \end{aligned} \tag{3.38}$$

Due to the first condition in (3.38), the estimate becomes unbiased. The variance of the estimate is minimized among all linear estimates due to the second condition.

The BLU estimate of $\mathbf{h}$ can be derived for the following linear model:

$$\mathbf{r} = \mathbf{Sh} + \mathbf{n}, \tag{3.39}$$

where the sizes of $\mathbf{r}$, $\mathbf{S}$, $\mathbf{h}$, and $\mathbf{n}$ are $N \times 1$, $N \times M$, $M \times 1$, and $N \times 1$, respectively. We assume that the mean vector and covariance matrix of $\mathbf{n}$ are $\mathcal{E}[\mathbf{n}] = \mathbf{0}$ and $\mathcal{E}[\mathbf{nn}^{\text{H}}] = \mathbf{R}_{\mathbf{n}}$, respectively. It is noteworthy that we do not need to know the pdf of $\mathbf{n}$ to derive the BLU estimate. Due to the first condition in (3.38), we should have

$$\mathbf{L}_{\text{blu}}\mathbf{S} = \mathbf{I}_M. \tag{3.40}$$

From this, it is clear that $\mathbf{S}$ should be full-rank and $L \ge M$. Let $\mathbf{u}_k$ denote the unit vector that is given by

$$[\mathbf{u}_m]_k = \begin{cases} 1, & \text{if } k = m; \\ 0, & \text{otherwise,} \end{cases}$$

and let

$$\mathbf{L}_{\text{blu}} = \begin{bmatrix} \mathbf{a}_1^{\text{H}} \\ \mathbf{a}_2^{\text{H}} \\ \vdots \\ \mathbf{a}_M^{\text{H}} \end{bmatrix}.$$

Noting that

$$\mathcal{E}[||\mathbf{h} - \hat{\mathbf{h}}||^2] = \sum_{m=1}^{M} \mathcal{E}[|h_m - \mathbf{a}_m^{\text{H}}\mathbf{r}|^2],$$

and the first condition in (3.40) becomes

$$\mathbf{S}^{\text{H}}\mathbf{a}_m = \mathbf{u}_m, \ m = 1, 2, \dots, M, \tag{3.41}$$

a minimization can be considered to derive the mth row vector of $\mathbf{L}_{\text{blu}}$, $\mathbf{a}_m^{\text{H}}$, as follows:

$$\min_{\mathbf{a}_m} C(\mathbf{a}_m, \mathbf{d}_m)$$

where

$$\begin{aligned} C(\mathbf{a}_m, \mathbf{d}_m) &= \mathcal{E}[|h_m - \mathbf{a}_m^{\text{H}}\mathbf{r}|^2] - \mathbf{d}_m^{\text{H}}(\mathbf{S}^{\text{H}}\mathbf{a}_m - \mathbf{u}_m) \\ &= \mathcal{E}[|(h_m - \mathbf{a}_m^{\text{H}}\mathbf{S}\mathbf{h}) - \mathbf{a}_m^{\text{H}}\mathbf{n}|^2] - \mathbf{d}_m^{\text{H}}(\mathbf{S}^{\text{H}}\mathbf{a}_m - \mathbf{u}_m). \end{aligned} \tag{3.42}$$

Here, $\mathbf{d}_m$ is the Lagrangian multiplier vector for the constraint in (3.41). A necessary condition for minimizing $C(\mathbf{a}_m, \mathbf{d}_m)$ is

$$\begin{aligned} \mathbf{0} &= \frac{\partial C(\mathbf{a}_m, \mathbf{d}_m)}{\partial \mathbf{a}_m} \\ &= \mathbf{R}_{\mathbf{n}}\mathbf{a}_m - \mathcal{E}[\mathbf{n}(h_m - \mathbf{a}_m^{\text{H}}\mathbf{S}\mathbf{h})^*] - \mathbf{S}\mathbf{d}_m \\ &= \mathbf{R}_{\mathbf{n}}\mathbf{a}_m - \mathbf{S}\mathbf{d}_m. \end{aligned}$$

It follows that

$$\mathbf{a}_m = \mathbf{R}_{\mathbf{n}}^{-1}\mathbf{S}\mathbf{d}_m. \tag{3.43}$$

Using the first condition, we can determine $\mathbf{d}_m$ as follows:

$$\mathbf{d}_m = (\mathbf{S}^{\text{H}}\mathbf{R}_{\mathbf{n}}^{-1}\mathbf{S})^{-1}\mathbf{u}_m. \tag{3.44}$$

Substituting (3.44) into (3.43), we have

$$\mathbf{a}_m = \mathbf{R}_{\mathbf{n}}^{-1}\mathbf{S}(\mathbf{S}^{\text{H}}\mathbf{R}_{\mathbf{n}}^{-1}\mathbf{S})^{-1}\mathbf{u}_m$$

or

$$\begin{aligned} [\mathbf{a}_1\ \mathbf{a}_2\ \ldots\ \mathbf{a}_M] &= \mathbf{R}_{\mathbf{n}}^{-1}\mathbf{S}(\mathbf{S}^{\text{H}}\mathbf{R}_{\mathbf{n}}^{-1}\mathbf{S})^{-1}[\mathbf{u}_1\ \mathbf{u}_2\ \ldots\ \mathbf{u}_M] \\ &= \mathbf{R}_{\mathbf{n}}^{-1}\mathbf{S}(\mathbf{S}^{\text{H}}\mathbf{R}_{\mathbf{n}}^{-1}\mathbf{S})^{-1}. \end{aligned}$$

Finally, we can show that

$$\mathbf{L}_{\text{blu}} = (\mathbf{S}^{\text{H}}\mathbf{R}_{\mathbf{n}}^{-1}\mathbf{S})^{-1}\mathbf{S}^{\text{H}}\mathbf{R}_{\mathbf{n}}^{-1} \tag{3.45}$$

and the BLU estimate of $\mathbf{h}$ is given by

$$\hat{\mathbf{h}}_{\text{blu}} = (\mathbf{S}^{\text{H}}\mathbf{R}_{\mathbf{n}}^{-1}\mathbf{S})^{-1}\mathbf{S}^{\text{H}}\mathbf{R}_{\mathbf{n}}^{-1}\mathbf{r}. \tag{3.46}$$

In order to see the error variance, consider the following estimation error:

$$\begin{aligned} \tilde{\mathbf{h}} &= \mathbf{h} - \hat{\mathbf{h}}_{\text{blu}} \\ &= \mathbf{h} - (\mathbf{S}^{\text{H}}\mathbf{R}_{\mathbf{n}}^{-1}\mathbf{S})^{-1}\mathbf{S}^{\text{H}}\mathbf{R}_{\mathbf{n}}^{-1}(\mathbf{S}\mathbf{h} + \mathbf{n}) \\ &= (\mathbf{S}^{\text{H}}\mathbf{R}_{\mathbf{n}}^{-1}\mathbf{S})^{-1}\mathbf{S}^{\text{H}}\mathbf{R}_{\mathbf{n}}^{-1}\mathbf{n} \\ &= \mathbf{L}_{\text{blu}}\mathbf{n}. \end{aligned} \tag{3.47}$$

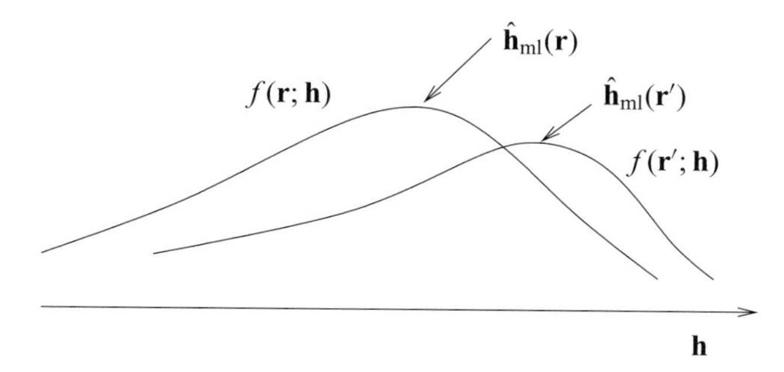

Figure 3.2 ML estimates and likelihood functions are dependent on realizations.

The covariance matrix of the BLU estimate can be easily found as

$$\begin{aligned}\mathcal{E}[(\mathbf{h}-\hat{\mathbf{h}}_{\rm blu})(\mathbf{h}-\hat{\mathbf{h}}_{\rm blu})^{\rm H}] &= \mathbf{L}_{\rm blu}\mathbf{R}_{\mathbf{n}}\mathbf{L}_{\rm blu}^{\rm H} \\ &= (\mathbf{S}^{\rm H}\mathbf{R}_{\mathbf{n}}^{-1}\mathbf{S})^{-1}.\end{aligned} \tag{3.48}$$

Theorem 3.2.3 *The BLU estimate is a special case of the WLS estimate in (3.13) when* $\mathbf{W}=\mathbf{R}_{\mathbf{n}}^{-1}$.

3.3 Maximum likelihood estimation

Like the BLU and WLS estimation, the maximum likelihood (ML) estimation is widely used to estimate unknown parameters. However, the ML estimation requires more statistical information of measurement than the BLU estimation.

Suppose that the pdf of a random vector $\mathbf{r}$ is given by $f_{\mathbf{h}}(\mathbf{r})=f(\mathbf{r};\mathbf{h})$, where $\mathbf{h}$ is the parameter vector to be estimated. For a given observation or realization of $\mathbf{r}$, $f(\mathbf{r};\mathbf{h})$ becomes a function of $\mathbf{h}$ (while $\mathbf{r}$ is serving as known coefficients), called the likelihood function (of $\mathbf{h}$). The ML estimate of $\mathbf{h}$ is the vector $\mathbf{h}$ that maximizes the likelihood function as follows:

$$\hat{\mathbf{h}}_{\rm ml} = \arg\max_{\mathbf{h}} f(\mathbf{r};\mathbf{h}). \tag{3.49}$$

In most cases, numerical approaches can be used to find the optimal solution of $\mathbf{h}$ in (3.49) that maximizes the likelihood function. Constrained optimization techniques can also be involved if there are constraints on the parameter set of $\mathbf{h}$. The log-likelihood, $\log f(\mathbf{r};\mathbf{h})$ can replace the likelihood $f(\mathbf{r};\mathbf{h})$. Note that since the logarithm function is strictly monotone, the logarithm does not affect the maximization and the maximization of $f(\mathbf{r};\mathbf{h})$ is identical to that of $\log f(\mathbf{r};\mathbf{h})$.

As shown in Fig. 3.2, the likelihood function can be different from one realization of $\mathbf{r}$ to another. Thus, $\hat{\mathbf{h}}_{\rm ml}$ is a function of realization or observation $\mathbf{r}$. To emphasize this, $\hat{\mathbf{h}}_{\rm ml}(\mathbf{r})$ can replace $\hat{\mathbf{h}}_{\rm ml}$.

Example 3.3.1 Consider a binary source that produces independent binary symbols, $X_l \in \{0, 1\}, l = 0, 1, \ldots, L-1$. The probability that $X_l = 0$ is θ. The probability mass function of $\{X_l\}$ is given by

$$f(x_0, x_1, \ldots, x_{L-1}) = \prod_{l=0}^{L-1} \theta^{x_l}(1-\theta)^{1-x_l}.$$

From a set of realizations or measurements, suppose that the probability θ is to be estimated. Let $\{x_l\}$ denote the set of realizations. Then, the likelihood function of θ is given by

$$\begin{aligned} f(\{x_l\}|\theta) &= \prod_{l=0}^{L-1} \theta^{x_l}(1-\theta)^{1-x_l} \\ &= \theta^t(1-\theta)^{L-t}, \end{aligned} \tag{3.50}$$

where

$$t = \sum_{l=0}^{L-1} x_l.$$

The ML estimate of θ that maximizes the likelihood function becomes

$$\begin{aligned} \hat{\theta}_{\text{ml}} &= \arg\max_{\theta} \theta^t(1-\theta)^{L-t} \\ &= \frac{t}{L}. \end{aligned}$$

We can also show that $\hat{\theta}_{\text{ml}}$ is unbiased. In addition, from (3.50) we can see that t is a sufficient statistic for θ. This implies that $\hat{\theta}_{\text{ml}} = t/L$ is MVUB.

It is noteworthy that the ML estimate does not need to be unbiased as shown in the following example.

Example 3.3.2 Consider the following observations:

$$r_l = \sqrt{\gamma} s_l + n_l, \; l = 0, 1, \ldots, L-1, \tag{3.51}$$

where γ is the parameter to be estimated and s_l is the lth known signal. The observation, r_l, can be considered as the received signal over a memoryless channel of channel gain $\sqrt{\gamma}$ with background noise, n_l. We assume that $n_l \sim \mathcal{CN}(0, \sigma^2)$ and n_l is independent of each other. The ML estimate of γ is given by

$$\begin{aligned} \hat{\gamma} &= \arg\max_{\gamma} \sum_{l=0}^{L-1} |r_l - \sqrt{\gamma} s_l|^2 \\ &= \left| \frac{\sum_{l=0}^{L-1} r_l s_l^*}{\sum_{l=0}^{L-1} |s_l|^2} \right|^2. \end{aligned} \tag{3.52}$$

Substituting $r_l = \sqrt{\gamma} s_l + n_l$ into (3.52), we have

$$\hat{\gamma} = |\sqrt{\gamma} + n|^2,$$

where $n = \frac{\sum_{l=0}^{L-1} n_l s_l^*}{\sum_{l=0}^{L-1} |s_l|^2}$. It can be easily shown that $n \sim \mathcal{CN}\left(0, \frac{\sigma^2}{\sum_{l=0}^{L-1} |s_l|^2}\right)$. Since the mean of $\hat{\gamma}$ is

$$\mathcal{E}[\hat{\gamma}] = \gamma + \frac{\sigma^2}{\sum_{l=0}^{L-1} |s_l|^2} > \gamma,$$

we can see that the ML estimate of γ, $\hat{\gamma}$, is biased. However, if $\sum_{l=0}^{L-1} |s_l|^2$ approaches infinity as $L \to \infty$ and $\sigma^2 < \infty$, $\mathcal{E}[\hat{\gamma}]$ approaches γ. This shows that $\hat{\gamma}$ becomes unbiased as $L \to \infty$. If an estimate becomes unbiased as the number of samples approaches infinity, it is called asymptotically unbiased.

We now focus on the ML estimation for the linear model in (3.39). Suppose that the pdf of $\mathbf{n}$ in (3.39) is known and denoted by $f(\mathbf{n})$. The likelihood function of $\mathbf{h}$ is given by

$$f(\mathbf{n}) = f(\mathbf{r} - \mathbf{S}\mathbf{h}). \tag{3.53}$$

Thus, the ML estimate of $\mathbf{h}$ is given by

$$\hat{\mathbf{h}}_{\mathrm{ml}} = \arg\max_{\mathbf{h}} f(\mathbf{r} - \mathbf{S}\mathbf{h}). \tag{3.54}$$

For example, from (3.39), suppose that $\mathbf{n}$ is a zero-mean CSCG random vector. The pdf of $\mathbf{n}$ is given by

$$f(\mathbf{n}) = \frac{1}{\det(\pi \mathbf{R})} \exp\left(-\mathbf{n}^{\mathrm{H}} \mathbf{R}^{-1} \mathbf{n}\right), \tag{3.55}$$

where $\mathbf{R} = \mathcal{E}[\mathbf{n}\mathbf{n}^{\mathrm{H}}]$ is the covariance matrix of $\mathbf{n}$. Then, the ML estimate of $\mathbf{h}$ in (3.39) becomes

$$\begin{aligned}
\hat{\mathbf{h}}_{\mathrm{ml}} &= \arg\max_{\mathbf{h}} f(\mathbf{r} - \mathbf{S}\mathbf{h}) \\
&= \arg\max_{\mathbf{h}} \frac{1}{\det(\pi \mathbf{R})} \exp\left(-(\mathbf{r} - \mathbf{S}\mathbf{h})^{\mathrm{H}} \mathbf{R}^{-1} (\mathbf{r} - \mathbf{S}\mathbf{h})\right) \\
&= \arg\min_{\mathbf{h}} (\mathbf{r} - \mathbf{S}\mathbf{h})^{\mathrm{H}} \mathbf{R}^{-1} (\mathbf{r} - \mathbf{S}\mathbf{h}) \\
&= \arg\min_{\mathbf{h}} ||\mathbf{r} - \mathbf{S}\mathbf{h}||^2_{\mathbf{R}^{-1}} \\
&= (\mathbf{S}^{\mathrm{H}} \mathbf{R}_{\mathbf{n}}^{-1} \mathbf{S})^{-1} \mathbf{S}^{\mathrm{H}} \mathbf{R}_{\mathbf{n}}^{-1} \mathbf{r}.
\end{aligned} \tag{3.56}$$

This shows that the ML estimate is identical to the BLU estimate if the background noise is Gaussian. In addition, if $\mathbf{W} = \mathbf{R}^{-1}$, it can be shown that

$$\hat{\mathbf{h}}_{\mathrm{ml}} = \hat{\mathbf{h}}_{\mathrm{wls}}.$$

The ML estimate has the following important properties (under certain conditions; see (Porat 1994)):

(i) If $\hat{\mathbf{h}}_{\rm ml}(\mathbf{r})$ is the ML estimate of $\mathbf{h}$, then $g(\hat{\mathbf{h}}_{\rm ml}(\mathbf{r}))$ is the ML estimate of $g(\mathbf{h})$.
(ii) The ML estimate is an efficient estimate if there exists an efficient estimate (the meaning of *efficient* is given in the next section).
(iii) The ML estimate is asymptotically normal.

3.4 Cramer–Rao bound

We consider a lower bound of the error covariance matrix of unbiased estimate in this section. If an unbiased estimator can achieve this bound, this estimate is considered to be a good estimator (or *efficient*).

We first consider a special case where $\mathbf{h} = h$ (parameter) and $\mathbf{r} = R$ (measurement or observation) are scalars. Assume that the pdf of R, denoted by $f_h(r) = f(r; h)$, is differentiable. Taking R as a random variable, define the following random variable:

$$\begin{aligned} V &= \frac{\partial}{\partial h} \log f(R; h) \\ &= \frac{\frac{\partial}{\partial h} f(R; h)}{f(R; h)}, \end{aligned} \tag{3.57}$$

where V is called the score. The mean of the score V is given by

$$\begin{aligned} \mathcal{E}[V] &= \int \left(\frac{\frac{\partial}{\partial h} f(r; h)}{f(r; h)} \right) f(r; h) \mathrm{d}r \\ &= \int \frac{\partial}{\partial h} f(r; h) \mathrm{d}r \\ &= \frac{\partial}{\partial h} \int f(r; h) \mathrm{d}r \\ &= \frac{\partial}{\partial h} 1 \\ &= 0. \end{aligned} \tag{3.58}$$

Therefore, the variance of V is

$$\begin{aligned} \mathrm{Var}(V) &= \mathcal{E}[V^2] \\ &= \mathcal{E}\left[\left(\frac{\partial}{\partial h} \log f(R; h) \right)^2 \right]. \end{aligned} \tag{3.59}$$

The variance of the score, V, is called the *Fisher information* and denoted by $J(h)$. Consider any unbiased estimate of h, say $X(R)$. With $X(R)$ and V, using the Cauchy–Schwarz inequality, we can have the following inequality:

$$(\mathcal{E}[(V - \mathcal{E}[V])(X - \mathcal{E}[X])])^2 \le \mathcal{E}[(V - \mathcal{E}[V])^2]\mathcal{E}[(X - \mathcal{E}[X])^2]. \tag{3.60}$$

Since $\mathcal{E}[V] = 0$ and $X = X(R)$ is unbiased, it can be shown that

$$\begin{aligned}
\mathcal{E}[(V - \mathcal{E}[V])(X - \mathcal{E}[X])] &= \mathcal{E}[VX] \\
&= \int \frac{\frac{\partial}{\partial h} f(r;h)}{f(r;h)} X(r) f(r;h) \mathrm{d}r \\
&= \int X(r) \frac{\partial}{\partial h} f(r;h) \mathrm{d}r \\
&= \frac{\partial}{\partial h} \int X(r) f(r;h) \mathrm{d}r \\
&= \frac{\partial}{\partial h} h \\
&= 1.
\end{aligned}$$

Thus, the term on the left-hand side in (3.60) becomes 1 and (3.60) is rewritten as

$$1 \leq J(h)\mathrm{Var}(X)$$

or

$$\mathrm{Var}(X(R)) \geq \frac{1}{J(h)} = \frac{1}{\mathcal{E}\left[\left(\frac{\partial}{\partial h} \log f(R;h)\right)^2\right]}. \tag{3.61}$$

This inequality is called the *Cramer–Rao* bound (CRB) or the information inequality.

We can also derive a general case as follows. If $\mathbf{h}$ is a vector, the *Fisher's information matrix* $\mathbf{J}(\mathbf{h})$ is defined as

$$[\mathbf{J}(\mathbf{h})]_{n,k} = \mathcal{E}\left[\frac{\partial \log f(\mathbf{r};\mathbf{h})}{\partial h_n} \frac{\partial \log f(\mathbf{r};\mathbf{h})}{\partial h_k}\right], \tag{3.62}$$

where h_n denotes the nth element of $\mathbf{h}$ and $f_{\mathbf{h}}(\mathbf{r}) = f(\mathbf{r};\mathbf{h})$ represents the pdf of $\mathbf{r}$.

Theorem 3.4.1 *An unbiased estimate has the following inequality:*

$$\mathcal{E}[(\hat{\mathbf{h}}(\mathbf{r}) - \mathbf{h})(\hat{\mathbf{h}}(\mathbf{r}) - \mathbf{h})^{\mathrm{T}}] \geq \mathbf{J}^{-1}(\mathbf{h}),$$

where $\mathbf{J}^{-1}(\mathbf{h})$ *is the CRB.*

Proof: To derive the CRB, we need to use the following inequality for a covariance matrix:

$$\begin{aligned}
\mathcal{E}[(\mathbf{x}_1 - \mathbf{C}_{12}\mathbf{C}_{22}^{-1}\mathbf{x}_2)(\mathbf{x}_1 - \mathbf{C}_{12}\mathbf{C}_{22}^{-1}\mathbf{x}_2)^{\mathrm{T}}] &= \mathbf{C}_{11} - \mathbf{C}_{12}\mathbf{C}_{22}^{-1}\mathbf{C}_{12}^{\mathrm{T}} \\
&\geq \mathbf{0},
\end{aligned}$$

where $\mathbf{x}_1$ and $\mathbf{x}_2$ are zero-mean random vectors and $\mathbf{C}_{nk} = \mathcal{E}[\mathbf{x}_n\mathbf{x}_k^{\mathrm{T}}]$. The inequality is valid since a covariance matrix is positive semi-definite. In addition, the equality holds if

$$\mathbf{x}_1 = \mathbf{C}_{12}\mathbf{C}_{22}^{-1}\mathbf{x}_2. \tag{3.63}$$

Let $\mathbf{x}_1 = \hat{\mathbf{h}}(\mathbf{r}) - \mathbf{h}$ and $\mathbf{x}_2 = \frac{\partial \log f(\mathbf{r};\mathbf{h})}{\partial \mathbf{h}}$. Then, we have

$$\begin{aligned} \mathbf{C}_{11} &= \mathcal{E}[(\hat{\mathbf{h}}(\mathbf{r}) - \mathbf{h})(\hat{\mathbf{h}}(\mathbf{r}) - \mathbf{h})^{\mathrm{T}}] \\ &\geq \mathbf{C}_{12}\mathbf{C}_{22}^{-1}\mathbf{C}_{12}^{\mathrm{T}}. \end{aligned}$$

It can be shown that

$$\begin{aligned} \mathbf{C}_{12} &= \mathcal{E}\left[(\hat{\mathbf{h}}(\mathbf{r}) - \mathbf{h})\left(\frac{\partial \log f(\mathbf{r};\mathbf{h})}{\partial \mathbf{h}}\right)^{\mathrm{T}}\right] \\ &= \mathcal{E}\left[\hat{\mathbf{h}}(\mathbf{r})\left(\frac{\partial \log f(\mathbf{r};\mathbf{h})}{\partial \mathbf{h}}\right)^{\mathrm{T}}\right] - \mathcal{E}\left[\mathbf{h}\left(\frac{\partial \log f(\mathbf{r};\mathbf{h})}{\partial \mathbf{h}}\right)^{\mathrm{T}}\right]. \end{aligned}$$

Furthermore, we can show that

$$\begin{aligned} \mathcal{E}\left[\hat{\mathbf{h}}(\mathbf{r})\left(\frac{\partial \log f(\mathbf{r};\mathbf{h})}{\partial \mathbf{h}}\right)^{\mathrm{T}}\right] &= \int \hat{\mathbf{h}}(\mathbf{r})\left(\frac{\partial f(\mathbf{r};\mathbf{h})}{\partial \mathbf{h}}\right)^{\mathrm{T}} \mathrm{d}\mathbf{r} \\ &= \left[\frac{\partial}{\partial h_k}\underbrace{\int \hat{h}_n(\mathbf{r}) f(\mathbf{r};\mathbf{h})\mathrm{d}\mathbf{r}}_{=h_n}\right] \\ &= \mathbf{I} \end{aligned}$$

and

$$\begin{aligned} \mathcal{E}\left[\mathbf{h}\left(\frac{\partial \log f(\mathbf{r};\mathbf{h})}{\partial \mathbf{h}}\right)^{\mathrm{T}}\right] &= \mathbf{h}\int \left(\frac{\partial f(\mathbf{r};\mathbf{h})}{\partial \mathbf{h}}\right)^{\mathrm{T}} \mathrm{d}\mathbf{r} \\ &= \mathbf{h}\left(\frac{\partial}{\partial \mathbf{h}}\int f(\mathbf{r};\mathbf{h})\mathrm{d}\mathbf{r}\right)^{\mathrm{T}} \\ &= 0, \end{aligned}$$

where $\hat{h}_n(\mathbf{r})$ denotes the nth element of $\hat{\mathbf{h}}(\mathbf{r})$. From above, we have

$$\mathbf{C}_{12} = \mathbf{I}.$$

Since $\mathbf{C}_{22} = \mathbf{J}(\mathbf{h})$, we conclude that

$$\mathcal{E}[(\hat{\mathbf{h}}(\mathbf{r}) - \mathbf{h})(\hat{\mathbf{h}}(\mathbf{r}) - \mathbf{h})^{\mathrm{T}}] \geq \mathbf{C}_{22}^{-1} = \mathbf{J}^{-1}(\mathbf{h}).$$

This completes the proof. □

Theorem 3.4.2 *We can show that*

$$[\mathbf{J}(\mathbf{h})]_{n,k} = -\mathcal{E}\left[\frac{\partial^2 \log f(\mathbf{r};\mathbf{h})}{\partial h_n \partial h_k}\right]. \tag{3.64}$$

Proof: To show (3.64), we can demonstrate that

$$\frac{\partial^2 \log f(\mathbf{r};\mathbf{h})}{\partial h_n \partial h_k} = -\frac{1}{f^2(\mathbf{r};\mathbf{h})}\frac{\partial f(\mathbf{r};\mathbf{h})}{\partial h_n}\frac{\partial f(\mathbf{r};\mathbf{h})}{\partial h_k} + \frac{1}{f(\mathbf{r};\mathbf{h})}\frac{\partial^2 f(\mathbf{r};\mathbf{h})}{\partial h_n \partial h_k}.$$

Since

$$\frac{\partial \log f(\mathbf{r};\mathbf{h})}{\partial h_n}\frac{\partial \log f(\mathbf{r};\mathbf{h})}{\partial h_k} = \frac{1}{f^2(\mathbf{r};\mathbf{h})}\frac{\partial f(\mathbf{r};\mathbf{h})}{\partial h_n}\frac{\partial f(\mathbf{r};\mathbf{h})}{\partial h_k}$$

and

$$\begin{aligned}\mathcal{E}\left[\frac{1}{f(\mathbf{r};\mathbf{h})}\frac{\partial^2 f(\mathbf{r};\mathbf{h})}{\partial h_n \partial h_k}\right] &= \int \frac{\partial^2 f(\mathbf{r};\mathbf{h})}{\partial h_n \partial h_k}\mathrm{d}\mathbf{r} \\ &= \frac{\partial^2}{\partial h_n \partial h_k}\underbrace{\int f(\mathbf{r};\mathbf{h})\mathrm{d}\mathbf{r}}_{=1} \\ &= 0,\end{aligned}$$

(3.64) is true. □

An unbiased estimate $\hat{\mathbf{h}}(\mathbf{r})$ is said to be *efficient* if it achieves the CRB. From (3.63), an efficient estimate can be written as

$$\hat{\mathbf{h}}(\mathbf{r}) = \mathbf{h} + \mathbf{J}^{-1}(\mathbf{h})\frac{\partial \log f(\mathbf{r};\mathbf{h})}{\partial \mathbf{h}}.$$

For more discussions and mathematical details of the CRB, the reader is referred to (Porat 1994).

Theorem 3.4.3 *Let $\mathbf{r}_l$ be iid random vectors with pdf $f(\mathbf{r}_l;\mathbf{h})$. If $\mathbf{h}$ is estimated from $\mathbf{r}_0, \mathbf{r}_1, \ldots, \mathbf{r}_{L-1}$, the CRB is given by*

$$\mathbf{J}_L^{-1}(\mathbf{h}) = \frac{1}{L}\mathbf{J}^{-1}(\mathbf{h}).$$

That is, the CRB decreases linearly with L.

Example 3.4.1 A group of distributions is called an *exponential family* if the distribution has the following form:

$$f_\theta(x) = C(\theta)\mathrm{e}^{g(\theta)T(x)}u(x),$$

where $C(\theta)$ and $g(\theta)$ are functions of θ, which is the parameter to be estimated, and $T(x)$ and $u(x)$ are functions of x. This group of distributions has the CRB that depends only on $g(\theta)$ and the variance of $T(X)$. To see this, we can find the Fisher's information as follows:

$$J(\theta) = \mathcal{E}\left[\left(\frac{\partial \log f_\theta(X)}{\partial \theta}\right)^2\right]. \tag{3.65}$$

Note that $C(\theta)$ is the normalization factor that can be given by

$$C(\theta) = \frac{1}{\int \mathrm{e}^{g(\theta)T(x)}u(x)\mathrm{d}x}.$$

We can show that

$$\begin{aligned}
\frac{\partial \log f_\theta(x)}{\partial \theta} &= \frac{\partial}{\partial \theta}\left(g(\theta)T(x) + \log u(x) - \log \int e^{g(\theta)T(x)} u(x)\mathrm{d}x\right) \\
&= g'(\theta)T(x) - \frac{\int g'(\theta)T(x)e^{g(\theta)T(x)}u(x)\mathrm{d}x}{\int e^{g(\theta)T(x)}u(x)\mathrm{d}x} \\
&= g'(\theta)T(x) - \frac{\int g'(\theta)T(x)C(\theta)e^{g(\theta)T(x)}u(x)\mathrm{d}x}{\underbrace{\int C(\theta)e^{g(\theta)T(x)}u(x)\mathrm{d}x}_{=1}} \\
&= g'(\theta)T(x) - \mathcal{E}[g'(\theta)T(X)] \\
&= g'(\theta)\left(T(x) - \mathcal{E}[T(X)]\right).
\end{aligned} \tag{3.66}$$

Substituting (3.66) into (3.65), we have

$$J(\theta) = \left(g'(\theta)\right)^2 \mathrm{Var}(T(X)). \tag{3.67}$$

This shows that the CRB depends on $g(\theta)$ and the variance of $T(X)$.

Example 3.4.2 Let X_l be iid and $X_l \sim f_\mu(x) = \mathcal{N}(\mu, \sigma^2)$, where σ^2 is known. The ML estimate of μ is $\bar{X} = \sum_{l=0}^{L-1} X_l$ and its variance is σ^2/L. We can also show that

$$\begin{aligned}
J(\mu) &= \mathcal{E}\left[\frac{\partial \log f_\mu(\bar{X})}{\partial \mu}\frac{\partial \log f_\mu(\bar{X})}{\partial \mu}\right] \\
&= \mathcal{E}\left[\frac{(\bar{X}-\mu)^2}{(\sigma^2/L)^2}\right] \\
&= \frac{\sigma^2}{L}.
\end{aligned}$$

Thus, we can see that $\bar{X}$ is unbiased and efficient.

Example 3.4.3 If $\mathbf{n}$ is Gaussian, the ML estimate of $\mathbf{h}$ for the linear model in (3.39) becomes the BLU estimate as shown in (3.56). Using (3.64), it can be shown that the Fisher's information matrix is given by

$$\mathbf{J}(\mathbf{h}) = \mathbf{S}^{\mathrm{H}}\mathbf{R}_{\mathbf{n}}^{-1}\mathbf{S}.$$

This Fisher's information matrix is identical to the covariance matrix of the BLU or ML estimate in (3.48) (thus, the ML or BLU estimate is efficient in this case).

3.5 MAP estimation

Suppose that the parameter vector, $\mathbf{h}$, is now a random vector. Note that in the ML estimation, the pdf of $\mathbf{r}$ is denoted by $f_{\mathbf{h}}(\mathbf{r}) = f(\mathbf{r};\mathbf{h})$, where $\mathbf{h}$ is not assumed to be a random vector. Once we assume $\mathbf{h}$ is a random vector, $f(\mathbf{r};\mathbf{h})$ becomes the pdf of $\mathbf{r}$ conditioned on $\mathbf{h}$. Thus, it would be proper to replace $f(\mathbf{r};\mathbf{h})$ with $f(\mathbf{r}|\mathbf{h})$, which is a conditional pdf.

If the a priori pdf of the parameter vector $\mathbf{h}$ is available, the maximum a posteriori probability (MAP) estimation can be formulated as follows:

$$\hat{\mathbf{h}}_{\rm map} = \arg\max_{\mathbf{h}} f(\mathbf{h}|\mathbf{r}), \tag{3.68}$$

where

$$f(\mathbf{h}|\mathbf{r}) = \frac{f(\mathbf{r}|\mathbf{h})f(\mathbf{h})}{f(\mathbf{r})}$$

and $f(\mathbf{h})$ is the a priori pdf of $\mathbf{h}$. If $\mathbf{h}$ is uniformly distributed over the parameter space, $f(\mathbf{h})$ is a constant. In this case, the MAP estimation reduces to the ML estimation:

$$\begin{aligned}\hat{\mathbf{h}}_{\rm map} &= \arg\max_{\mathbf{h}} f(\mathbf{h}|\mathbf{r}) \\ &= \arg\max_{\mathbf{h}} f(\mathbf{r}|\mathbf{h}).\end{aligned}$$

Example 3.5.1 Consider the linear model in (3.39). Assume that $\mathbf{S}$ is known, $\mathbf{n} \sim \mathcal{CN}(\mathbf{0}, \mathbf{R}_{\mathbf{n}})$, and $\mathbf{h} \sim \mathcal{CN}(\bar{\mathbf{h}}, \mathbf{R}_{\mathbf{h}})$. Noting that $\mathbf{r}$ is a Gaussian random vector conditioned on $\mathbf{h}$, the a posteriori pdf of $\mathbf{h}$ is given by

$$\begin{aligned}f(\mathbf{h}|\mathbf{r}) &\propto f(\mathbf{r}|\mathbf{h})f(\mathbf{h}) \\ &\propto e^{-(\mathbf{r}-\mathbf{S}\mathbf{h})^{\rm H}\mathbf{R}_{\mathbf{n}}^{-1}(\mathbf{r}-\mathbf{S}\mathbf{h})} e^{-(\mathbf{h}-\bar{\mathbf{h}})^{\rm H}\mathbf{R}_{\mathbf{h}}^{-1}(\mathbf{h}-\bar{\mathbf{h}})} \\ &\propto e^{-\mathbf{h}^{\rm H}(\mathbf{S}^{\rm H}\mathbf{R}_{\mathbf{n}}^{-1}\mathbf{S}+\mathbf{R}_{\mathbf{h}}^{-1})\mathbf{h}+2\Re\{(\mathbf{r}^{\rm H}\mathbf{R}_{\mathbf{n}}^{-1}\mathbf{S}+\bar{\mathbf{h}}^{\rm H}\mathbf{R}_{\mathbf{h}}^{-1})\mathbf{h}\}}.\end{aligned} \tag{3.69}$$

Thus, the MAP estimate of $\mathbf{h}$ is given by

$$\begin{aligned}\hat{\mathbf{h}}_{\rm map} &= \arg\max_{\mathbf{h}} \mathbf{h}^{\rm H}(\mathbf{S}^{\rm H}\mathbf{R}_{\mathbf{n}}^{-1}\mathbf{S} + \mathbf{R}_{\mathbf{h}}^{-1})\mathbf{h} + 2\Re\{(\mathbf{r}^{\rm H}\mathbf{R}_{\mathbf{n}}^{-1}\mathbf{S} + \bar{\mathbf{h}}^{\rm H}\mathbf{R}_{\mathbf{h}}^{-1})\mathbf{h}\} \\ &= (\mathbf{S}^{\rm H}\mathbf{R}_{\mathbf{n}}^{-1}\mathbf{S} + \mathbf{R}_{\mathbf{h}}^{-1})^{-1}(\mathbf{S}^{\rm H}\mathbf{R}_{\mathbf{n}}^{-1}\mathbf{r} + \mathbf{R}_{\mathbf{h}}^{-1}\bar{\mathbf{h}}).\end{aligned} \tag{3.70}$$

Although the mean of $\hat{\mathbf{h}}_{\rm map}$ is identical to $\bar{\mathbf{h}}$:

$$\mathcal{E}[\hat{\mathbf{h}}_{\rm map}] = (\mathbf{S}^{\rm H}\mathbf{R}_{\mathbf{n}}^{-1}\mathbf{S} + \mathbf{R}_{\mathbf{h}}^{-1})^{-1}(\mathbf{S}^{\rm H}\mathbf{R}_{\mathbf{n}}^{-1}\mathbf{S} + \mathbf{R}_{\mathbf{h}}^{-1})\bar{\mathbf{h}} = \bar{\mathbf{h}},$$

the MAP estimate is not unbiased. To see whether biased or not, we need to see the conditional mean, $\mathcal{E}[\hat{\mathbf{h}}|\mathbf{h}]$. Let $\bar{\mathbf{h}} = \mathbf{0}$ and consider the conditional mean as follows:

$$\mathcal{E}[\hat{\mathbf{h}}_{\rm map}|\,\mathbf{h}] = (\mathbf{S}^{\rm H}\mathbf{R}_{\mathbf{n}}^{-1}\mathbf{S} + \mathbf{R}_{\mathbf{h}}^{-1})^{-1}\mathbf{S}^{\rm H}\mathbf{R}_{\mathbf{n}}^{-1}\mathbf{S}\mathbf{h}. \tag{3.71}$$

Thus, $\mathcal{E}[\hat{\mathbf{h}}_{\mathrm{map}} |\, \mathbf{h}] \neq \mathbf{h}$ unless $\mathbf{R}_{\mathbf{h}}$ is zero. Note that the MAP estimate can also be given by

$$\mathbf{h}_{\mathrm{map}} = \mathbf{L}_{\mathrm{map}} \mathbf{r}, \tag{3.72}$$

where

$$\mathbf{L}_{\mathrm{map}} = (\mathbf{S}^{\mathrm{H}} \mathbf{R}_{\mathbf{n}}^{-1} \mathbf{S} + \mathbf{R}_{\mathbf{h}}^{-1})^{-1} \mathbf{S}^{\mathrm{H}} \mathbf{R}_{\mathbf{n}}^{-1}. \tag{3.73}$$

This shows that the MAP estimator becomes linear in this example.

Example 3.5.2 Consider Example 3.3.2 again. Now we assume that γ is exponentially distributed and its pdf (exponential pdf) is given by:

$$f(\gamma) = \frac{1}{\bar{\gamma}} \mathrm{e}^{-\frac{\gamma}{\bar{\gamma}}}, \ \gamma \geq 0, \tag{3.74}$$

where $\bar{\gamma}$ is the mean of γ. The MAP estimate of γ is given by

$$\begin{aligned}
\hat{\gamma} &= \arg\max_{\gamma} \left(\prod_{l=0}^{L-1} \mathrm{e}^{-\frac{1}{\sigma^2} |r_l - \sqrt{\gamma} s_l|^2} \right) \mathrm{e}^{-\frac{\gamma}{\bar{\gamma}}} \\
&= \arg\min_{\gamma} \sum_{l=0}^{L-1} |r_l - \sqrt{\gamma} s_l|^2 + \frac{\sigma^2}{\bar{\gamma}} \gamma \\
&= \left| \frac{\sum_{l=0}^{L-1} r_l s_l^*}{\sum_{l=0}^{L-1} |s_l|^2 + \frac{\sigma^2}{\bar{\gamma}}} \right|^2 .
\end{aligned}$$

3.6 MMSE estimation

The minimum mean squared error (MMSE) estimation is widely used for various estimation problems as it provides reasonably good performance with relatively less statistical information of measurements. In this section, we focus on the linear MMSE estimation, which is a special case of the MMSE estimation where a linear estimator is employed. This linear MMSE estimation is simply referred to as the MMSE estimation, while the MMSE estimation without the linear estimator constraint is referred to as the ideal MMSE estimation in this section.

Using the notion of the orthogonality principle, we will derive the MMSE estimator and discuss its properties. Note that as in the MAP estimation, we also assume that the parameter vector is random in the MMSE estimation.

3.6.1 MSE cost with linear estimator

Suppose that there are two *zero-mean* random vectors, $\mathbf{h}$ and $\mathbf{r}$. A linear transform of $\mathbf{r}$ can be considered to estimate $\mathbf{h}$. An estimate of $\mathbf{h}$ from a linear estimator, $\mathbf{L}$, is given by

$$\hat{\mathbf{h}} = \mathbf{L}\mathbf{r}.$$

To quantify the difference between $\mathbf{h}$ and its estimate, $\hat{\mathbf{h}}$, the MSE cost can be considered as follows:

$$\text{MSE}(\mathbf{L}) = \mathcal{E}[||\mathbf{h} - \hat{\mathbf{h}}||^2],$$

where the MSE cost becomes a function of $\mathbf{L}$ and the expectation is carried out with respect to $\mathbf{h}$ and $\mathbf{r}$. The linear MMSE or simply MMSE estimator can be found as

$$\begin{aligned}\mathbf{L}_{\text{mmse}} &= \arg\min_{\mathbf{L}} \text{MSE}(\mathbf{L}) \\ &= \arg\min_{\mathbf{L}} \mathcal{E}[||\mathbf{h} - \mathbf{L}\mathbf{r}||^2].\end{aligned}$$

The minimum can be achieved by setting the first-order derivative of the MSE with respect to $\mathbf{L}$ equal to zero, because the MSE is a quadratic function of $\mathbf{L}$. Since

$$\frac{\partial}{\partial L_{n,k}} \mathcal{E}\left[\sum_p |h_p - \sum_m L_{p,m} y_m|^2\right] = 2\mathcal{E}\left[(h_n - \sum_m L_{n,m} y_m) y_k^*\right], \tag{3.75}$$

the condition to obtain the minimum can be derived as

$$\mathcal{E}[(\mathbf{h} - \mathbf{L}\mathbf{r})\mathbf{r}^{\text{H}}] = \mathbf{0}. \tag{3.76}$$

Let $\mathbf{e} = \mathbf{h} - \mathbf{L}\mathbf{r}$ be the error vector. Then, (3.76) implies that the error vector should be uncorrelated with the observation vector $\mathbf{r}$ to achieve the minimum. This is the orthogonality principle for the MMSE estimation.

From (3.76), the MMSE estimator is given by

$$\begin{aligned}\mathbf{L}_{\text{mmse}} &= \mathcal{E}[\mathbf{h}\mathbf{r}^{\text{H}}](\mathcal{E}[\mathbf{r}\mathbf{r}^{\text{H}}])^{-1} \\ &= \mathbf{R}_{\mathbf{hr}}\mathbf{R}_{\mathbf{r}}^{-1},\end{aligned}$$

where $\mathbf{R}_{\mathbf{hr}} = \mathcal{E}[\mathbf{h}\mathbf{r}^{\text{H}}]$ and $\mathbf{R}_{\mathbf{r}} = \mathcal{E}[\mathbf{r}\mathbf{r}^{\text{H}}]$. The MMSE covariance matrix, $\mathbf{R}_{\text{mmse}}$, can be found as

$$\begin{aligned}\mathbf{R}_{\text{mmse}} &= \mathcal{E}[(\mathbf{h} - \mathbf{L}_{\text{mmse}}\mathbf{r})(\mathbf{h} - \mathbf{L}_{\text{mmse}}\mathbf{r})^{\text{H}}] \\ &= \mathcal{E}[(\mathbf{h} - \mathbf{L}_{\text{mmse}}\mathbf{r})\mathbf{h}^{\text{H}}] \\ &= \mathbf{R}_{\mathbf{h}} - \mathbf{L}_{\text{mmse}}\mathbf{R}_{\mathbf{rh}} \\ &= \mathbf{R}_{\mathbf{h}} - \mathbf{R}_{\mathbf{hr}}\mathbf{R}_{\mathbf{r}}^{-1}\mathbf{R}_{\mathbf{rh}},\end{aligned}$$

where $\mathbf{R}_{\mathbf{h}} = \mathcal{E}[\mathbf{h}\mathbf{h}^{\text{H}}]$ and $\mathbf{R}_{\mathbf{rh}} = \mathcal{E}[\mathbf{r}\mathbf{h}^{\text{H}}]$.

Example 3.6.1 Consider the linear model in (3.39). Assume that $\mathbf{h}$ and $\mathbf{n}$ are uncorrelated and $\mathcal{E}[\mathbf{h}] = \mathbf{0}$ and $\mathcal{E}[\mathbf{n}] = \mathbf{0}$. Then,

$$\mathbf{R}_{\mathbf{rh}} = \mathbf{S}\mathbf{R}_{\mathbf{h}};$$
$$\mathbf{R}_{\mathbf{r}} = \mathbf{S}\mathbf{R}_{\mathbf{h}}\mathbf{S}^{\mathrm{H}} + \mathbf{R}_{\mathbf{n}},$$

where $\mathbf{R}_{\mathbf{h}}$ and $\mathbf{R}_{\mathbf{n}}$ denote the covariance matrices of $\mathbf{h}$ and $\mathbf{n}$, respectively. It follows that

$$\mathbf{L}_{\mathrm{mmse}} = \mathbf{R}_{\mathbf{h}}\mathbf{S}^{\mathrm{H}}(\mathbf{S}\mathbf{R}_{\mathbf{h}}\mathbf{S}^{\mathrm{H}} + \mathbf{R}_{\mathbf{n}})^{-1}. \tag{3.77}$$

In this particular case, we can show that the MAP estimator in (3.73) is identical to the MMSE estimator since $\mathbf{L}_{\mathrm{mmse}}$ in (3.77) is identical to $\mathbf{L}_{\mathrm{map}}$.

It is necessary to extend the MMSE estimation for nonzero-mean random vectors. If $\mathcal{E}[\mathbf{h}] = \bar{\mathbf{h}} \neq \mathbf{0}$ and $\mathcal{E}[\mathbf{r}] = \bar{\mathbf{r}} \neq \mathbf{0}$, the MMSE estimation problem is given by

$$\{\mathbf{L}_{\mathrm{mmse}}, \mathbf{c}_{\mathrm{mmse}}\} = \arg\min_{\mathbf{L},\mathbf{c}} \mathcal{E}[||\mathbf{h} - \mathbf{L}\mathbf{r} - \mathbf{c}||^2],$$

where $\mathbf{c}$ is a vector. The vector $\mathbf{c}$ is included to deal with nonzero-mean vectors. Let $\tilde{\mathbf{h}} = \mathbf{h} - \bar{\mathbf{h}}$ and $\tilde{\mathbf{r}} = \mathbf{r} - \bar{\mathbf{r}}$. Then, the MSE is rewritten as

$$\mathcal{E}[||\tilde{\mathbf{h}} - \mathbf{L}\tilde{\mathbf{r}} + \bar{\mathbf{h}} - \mathbf{L}\bar{\mathbf{r}} - \mathbf{c}||^2].$$

Let $\mathbf{u} = \bar{\mathbf{h}} - \mathbf{L}\bar{\mathbf{r}} - \mathbf{c}$. Then, the MMSE estimation problem is rewritten as

$$\{\mathbf{L}_{\mathrm{mmse}}, \mathbf{u}_{\mathrm{mmse}}\} = \arg\min_{\mathbf{L},\mathbf{u}} \mathcal{E}[||\mathbf{h} - \mathbf{L}\mathbf{r} + \mathbf{u}||^2].$$

The optimal vector $\mathbf{u}_{\mathrm{mmse}}$ is zero to minimize the MSE and it implies that

$$\mathbf{c}_{\mathrm{mmse}} = \bar{\mathbf{h}} - \mathbf{L}\bar{\mathbf{r}}.$$

The optimal estimator $\mathbf{L}_{\mathrm{mmse}}$ is given by

$$\begin{aligned}\mathbf{L}_{\mathrm{mmse}} &= \mathcal{E}[\tilde{\mathbf{h}}\tilde{\mathbf{r}}^{\mathrm{H}}](\mathcal{E}[\tilde{\mathbf{r}}\tilde{\mathbf{r}}^{\mathrm{H}}])^{-1} \\ &= \mathbf{R}_{\mathbf{hr}}\mathbf{R}_{\mathbf{r}}^{-1},\end{aligned}$$

where $\mathbf{R}_{\mathbf{hr}}$ and $\mathbf{R}_{\mathbf{r}}$ are now $\mathbf{R}_{\mathbf{hr}} = \mathcal{E}[\tilde{\mathbf{h}}\tilde{\mathbf{r}}^{\mathrm{H}}]$ and $\mathbf{R}_{\mathbf{r}} = \mathcal{E}[\tilde{\mathbf{r}}\tilde{\mathbf{r}}^{\mathrm{H}}]$. With nonzero-mean vectors, the MMSE estimate of $\mathbf{h}$ is given by

$$\begin{aligned}\hat{\mathbf{h}}_{\mathrm{mmse}} &= \mathbf{L}_{\mathrm{mmse}}\mathbf{r} + \mathbf{c}_{\mathrm{mmse}} \\ &= \mathbf{L}_{\mathrm{mmse}}\mathbf{r} + \bar{\mathbf{h}} - \mathbf{L}_{\mathrm{mmse}}\bar{\mathbf{r}} \\ &= \bar{\mathbf{h}} + \mathbf{L}_{\mathrm{mmse}}(\mathbf{r} - \bar{\mathbf{r}}).\end{aligned}$$

3.6.2 Conditional mean and Gaussian vectors

Previously, we considered the MMSE estimation with linear estimators. If there is no constraint on estimators, the ideal MMSE estimate of $\mathbf{h}$ given $\mathbf{r}$ can be found as

$$\hat{\mathbf{h}}_{\mathrm{mmse}*} = \arg\min_{\hat{\mathbf{h}}} \mathcal{E}[||\mathbf{h} - \hat{\mathbf{h}}||^2|\mathbf{r}].$$

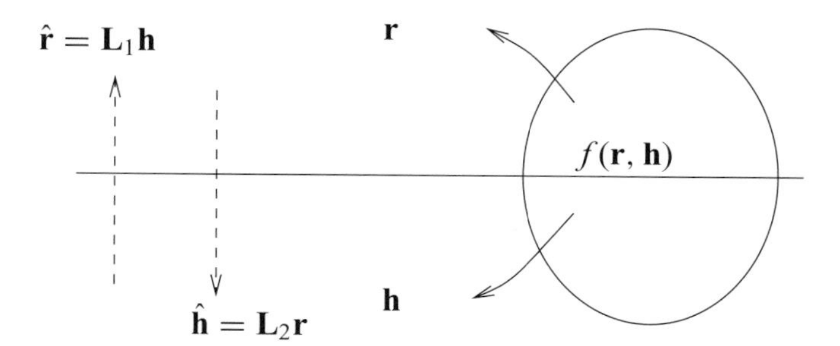

Figure 3.3 Realizations of jointly Gaussian random vectors and MMSE estimates.

In general, it is not necessarily true that $\hat{\mathbf{h}}_{\text{mmse}*} = \hat{\mathbf{h}}_{\text{mmse}}$. Taking the derivative with respect to $\hat{\mathbf{h}}$ and setting it equal to zero, we have

$$\begin{aligned} 0 &= \mathcal{E}[(\mathbf{h} - \hat{\mathbf{h}})|\mathbf{r}] \\ \Rightarrow \hat{\mathbf{h}}_{\text{mmse}*} &= \mathcal{E}[\mathbf{h}|\mathbf{r}]. \end{aligned}$$

This shows that the conditional mean vector of $\mathbf{h}$ given $\mathbf{r}$ is the ideal MMSE estimate of $\mathbf{h}$. In general, the ideal MMSE estimate of $\mathbf{h}$ given $\mathbf{r}$ is different from the linear MMSE estimate $\hat{\mathbf{h}}_{\text{mmse}}$. However, if $\mathbf{r}$ and $\mathbf{h}$ are jointly Gaussian random vectors, we have $\hat{\mathbf{h}}_{\text{mmse}*} = \hat{\mathbf{h}}_{\text{mmse}}$, because

$$\begin{aligned} \mathcal{E}[\mathbf{h}|\mathbf{r}] &= \hat{\mathbf{h}}_{\text{mmse}} \\ &= \bar{\mathbf{h}} + \mathbf{L}_{\text{mmse}}(\mathbf{r} - \bar{\mathbf{r}}) \\ &= \bar{\mathbf{h}} + \mathbf{R}_{\mathbf{hr}}\mathbf{R}_{\mathbf{r}}^{-1}(\mathbf{r} - \bar{\mathbf{r}}). \end{aligned} \tag{3.78}$$

From this, we can also find the conditional mean of $\mathbf{r}$ as

$$\begin{aligned} \mathcal{E}[\mathbf{r}|\mathbf{h}] &= \hat{\mathbf{r}}_{\text{mmse}} \\ &= \bar{\mathbf{r}} + \mathbf{R}_{\mathbf{rh}}\mathbf{R}_{\mathbf{h}}^{-1}(\mathbf{h} - \bar{\mathbf{h}}). \end{aligned} \tag{3.79}$$

As shown in Fig. 3.3, if $\mathbf{r}$ and $\mathbf{h}$ are zero-mean and jointly Gaussian, the ideal MMSE estimation of each random vector can be found by a linear function of a realization of the other random vector.

3.6.3 Geometrical interpretation

The linear MMSE estimation can be explained from a geometric point of view. Consider the projection of a vector onto a subspace as shown in Fig. 3.4. To find the best vector that approximates $\mathbf{h}$ from the subspace spanned by $\mathbf{r}_1$ and $\mathbf{r}_2$, denoted by $\mathcal{R}$, we can consider the squared norm of error as follows:

$$||\mathbf{h} - (a_1\mathbf{r}_1 + a_2\mathbf{r}_2)||^2,$$

where a_1 and a_2 are the coefficients to be determined. For convenience, we assume real-valued vectors in this subsection.

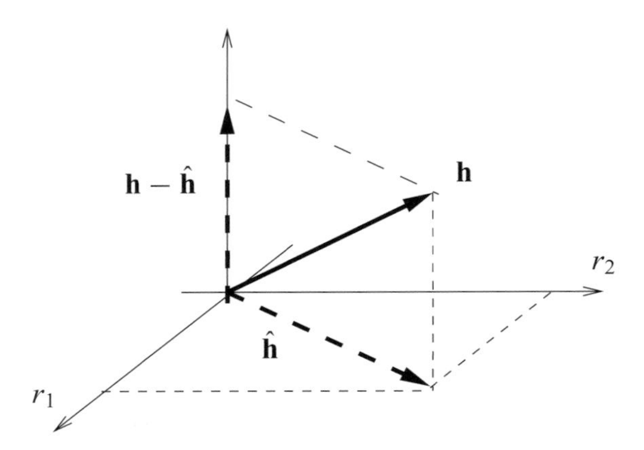

Figure 3.4 Projection in vector space.

Let $\hat{\mathbf{h}} = a_1\mathbf{r}_1 + a_2\mathbf{r}_2$. Obviously, $\hat{\mathbf{h}}$ is in $\mathcal{R}$. If $\hat{\mathbf{h}}$ minimizes the squared norm of error, the error vector, $\mathbf{h} - \hat{\mathbf{h}}$, is orthogonal to $\mathcal{R}$ as illustrated in Fig. 3.4. From this, the coefficients a_1 and a_2 can be obtained from the following set of equations:

$$< \mathbf{r}_1, \mathbf{h} - (a_1\mathbf{r}_1 + a_2\mathbf{r}_2) > = 0$$
$$< \mathbf{r}_2, \mathbf{h} - (a_1\mathbf{r}_1 + a_2\mathbf{r}_2) > = 0,$$

where $< \mathbf{x}, \mathbf{y} >= \mathbf{x}^{\mathrm{T}}\mathbf{y}$.

A generalization with an L-dimensional subspace can be considered using the following squared norm of error:

$$||\mathbf{h} - \sum_{l=0}^{L-1} \mathbf{r}_l a_l||^2 = ||\mathbf{h} - \mathbf{R}\mathbf{a}||^2,$$

where $\mathbf{R} = [\mathbf{r}_0 \ \mathbf{r}_1 \ \cdots \ \mathbf{r}_{L-1}]$, $\mathbf{a} = [a_0 \ a_1 \ \cdots \ a_{L-1}]^{\mathrm{T}}$, and $\hat{\mathbf{h}} = \mathbf{R}\mathbf{a}$. If $\hat{\mathbf{h}}$ minimizes the squared norm of error, we have

$$\mathbf{R}^{\mathrm{T}}(\mathbf{h} - \mathbf{R}\mathbf{a}) = \mathbf{0}.$$

This implies that the minimum squared error (MIN-SE) solution of $\mathbf{a}$ is given by

$$\mathbf{a}_{\text{min-se}} = (\mathbf{R}^{\mathrm{T}}\mathbf{R})^{-1}\mathbf{R}^{\mathrm{T}}\mathbf{h} \tag{3.80}$$

if $(\mathbf{R}^{\mathrm{T}}\mathbf{R})^{-1}$ exists.

In the MMSE estimation, we have the same result. Consider the following MSE cost:

$$\text{MSE}(\mathbf{a}) = \mathcal{E}\left[|h - \sum_{l=0}^{L-1} a_l r_l|^2\right],$$

where $\mathbf{a} = [a_0\ a_1\ \cdots\ a_{L-1}]^{\mathrm{T}}$. The means of h and r_l are assumed to be zero. We now deal with random variables as vectors and define new norm and inner product as follows:

$$||\mathbf{x}||^2 \leftrightarrow \mathcal{E}[|X|^2];$$
$$< \mathbf{x}, \mathbf{y} > \leftrightarrow \mathcal{E}[XY].$$

Thus, if we say $(h - \sum_{l=0}^{L-1} a_l r_l)$ and r_k are orthogonal, this means their inner product is zero, i.e.

$$(h - \sum_{l=0}^{L-1} a_l r_l) \perp r_k \leftrightarrow \mathcal{E}[(h - \sum_{l=0}^{L-1} a_l r_l) r_k] = 0.$$

Let $\mathbf{r} = [r_0\ r_1\ \ldots\ r_{L-1}]^{\mathrm{T}}$. The vector $\mathbf{a}$ that minimizes the MSE cost or squared norm of error can be obtained by solving the following equation:

$$\mathcal{E}[(h - \sum_{l=0}^{L-1} a_l r_l)\mathbf{r}] = 0. \tag{3.81}$$

Let $\mathbf{R}_{\mathbf{r}} = \mathcal{E}[\mathbf{r}\mathbf{r}^{\mathrm{T}}]$ and $\mathbf{c} = \mathcal{E}[\mathbf{r}h]$. Then, from (3.81), the MMSE solution of $\mathbf{a}$ is given by

$$\mathbf{a}_{\text{mmse}} = \mathbf{R}_{\mathbf{r}}^{-1}\mathbf{c}.$$

For comparison with (3.80), we can show the following equivalent relations:

$$\begin{array}{rcl} \text{(MMSE with random variables)} & & \text{(MIN-SE in vector space)} \\ [\mathbf{R}_{\mathbf{r}}]_{l,k} = E[r_l r_k] & \leftrightarrow & < \mathbf{r}_l, \mathbf{r}_k >= \mathbf{r}_l^{\mathrm{T}}\mathbf{r}_k \\ [\mathbf{c}]_l = E[r_l h] & \leftrightarrow & < \mathbf{r}_l, \mathbf{h} >= \mathbf{r}_l^{\mathrm{T}}\mathbf{h}. \end{array}$$

3.7 Example: channel estimation

Suppose that a data sequence is transmitted over a dispersive channel. The received signal is given by

$$r_l = \sum_{p=0}^{P-1} h_p s_{l-p} + n_l,\ l = 0, 1, \ldots, L-1, \tag{3.82}$$

where s_l is the lth data symbol, $\{h_p\}$ is the channel impulse response, and n_l is the background noise. Here, P denotes the length of the channel impulse response and we assume that n_l is iid and $n_l \sim \mathcal{N}(0, \sigma^2)$. Stacking the r_l's, we have

$$\begin{aligned} \mathbf{r} &= [r_0\ r_1\ \cdots\ r_{L-1}]^{\mathrm{T}} \\ &= \underbrace{\begin{bmatrix} s_0 & s_{-1} & \cdots & s_{-P+1} \\ s_1 & s_0 & \cdots & s_{-P+2} \\ \vdots & \vdots & \ddots & \vdots \\ s_{L-1} & s_{L-2} & \cdots & s_{L-P} \end{bmatrix}}_{=\mathbf{S}} \mathbf{h} + \mathbf{n}, \end{aligned} \tag{3.83}$$

where $\mathbf{n} = [n_0 \; n_1 \; \cdots \; n_{L-1}]^{\mathrm{T}}$ and $\mathbf{h} = [h_0 \; h_1 \; \cdots \; h_{P-1}]^{\mathrm{T}}$. We now have a linear model as in (3.83) and the parameter vector to be estimated is $\mathbf{h}$, which is the channel impulse response. For the channel estimation, in general, it is assumed that $\{s_l\}$ is known and this data sequence is often called the pilot signal. Since $\mathbf{R}_{\mathbf{n}} = \sigma^2 \mathbf{I}$, the BLU or ML estimate of $\mathbf{h}$ becomes

$$\hat{\mathbf{h}}_{\mathrm{ml}} = (\mathbf{S}^{\mathrm{H}}\mathbf{S})^{-1}\mathbf{S}^{\mathrm{H}}\mathbf{r}.$$

As shown in Example 3.4.3, this ML estimate is unbiased and efficient. The covariance matrix is

$$\mathcal{E}[(\hat{\mathbf{h}}_{\mathrm{ml}} - \mathbf{h})(\hat{\mathbf{h}}_{\mathrm{ml}} - \mathbf{h})^{\mathrm{H}}] = \sigma^2 \left(\mathbf{S}^{\mathrm{H}}\mathbf{S}\right)^{-1}.$$

This shows that an undesirable result can happen if the same symbol is repeatedly transmitted, because $\mathbf{S}$ can be linearly dependent and $\left(\mathbf{S}^{\mathrm{H}}\mathbf{S}\right)^{-1}$ does not exist. To avoid this problem, the pilot signal, $\{s_l\}$, should be well designed.

Example 3.7.1 Suppose that $P = 2$ and $L = 4$. The pilot signal is given by

$$\{s_0, s_1, s_2, s_3\} = \{1, -1, 1, -1\},$$

while $s_l = 0$ for $l < 0$. If the noise variance is unity (i.e. $\sigma^2 = 1$), the covariance matrix of the ML estimate of $\mathbf{h}$ becomes

$$\begin{aligned}\mathcal{E}[(\hat{\mathbf{h}}_{\mathrm{ml}} - \mathbf{h})(\hat{\mathbf{h}}_{\mathrm{ml}} - \mathbf{h})^{\mathrm{H}}] &= (\mathbf{S}^{\mathrm{H}}\mathbf{S})^{-1} \\ &= \begin{bmatrix} 1 & 1 \\ 1 & \frac{4}{3} \end{bmatrix}\end{aligned}$$

and the error variance is $\mathcal{E}[(\hat{\mathbf{h}}_{\mathrm{ml}} - \mathbf{h})^{\mathrm{H}}(\hat{\mathbf{h}}_{\mathrm{ml}} - \mathbf{h})] = 7/3$.

If a different pilot signal is used, the error variance can be reduced. Consider

$$\{s_0, s_1, s_2, s_3\} = \{1, -1, -1, 1\}.$$

Then, we can show that $\mathcal{E}[(\hat{\mathbf{h}}_{\mathrm{ml}} - \mathbf{h})^{\mathrm{H}}(\hat{\mathbf{h}}_{\mathrm{ml}} - \mathbf{h})] = 7/11$. This shows that the design of the pilot signal is important in minimizing the error variance of the estimated channel.

If the statistical properties of the channel impulse response are known, the MAP estimation is available. If $\mathbf{h}$ is assumed to be Gaussian and $\bar{\mathbf{h}} = \mathcal{E}[\mathbf{h}]$ and $\mathbf{R}_{\mathbf{h}} = \mathcal{E}[(\mathbf{h} - \bar{\mathbf{h}})(\mathbf{h} - \bar{\mathbf{h}})^{\mathrm{H}}]$, the MAP estimate is given in (3.70).

3.8 Summary and notes

We presented various approaches for the parameter estimation, including the ML, MAP, and MMSE estimation methods. A linear model was mainly considered to derive estimation methods.

There are a number of books on estimation theory from different point of views. For signal processing and communications, we can recommend (Kay 1993), (Porat 1994), and (Scharf 1991). There are also books that deal with statistical estimation theory from a more theoretical point of view, e.g (Lehmann 1983) and (Cox & Hinkley 1974).

It is noteworthy that we do not deal with dynamic models (which are essential for Kalman filtering) in this chapter. The reader is referred to (Kay 1993), (Anderson & Moore 1979), and (Sage & Melsa 1971) for a detailed account of the estimation with dynamic models.

Problems

Problem 3.1 Suppose that $X_l, l = 0, 1, \ldots, L-1$, are iid and its pdf is $\sim \mathcal{N}(\mu, \sigma^2)$.
(i) Find the ML estimate of σ^2 when μ is known.
(ii) Find the MAP estimate of $\theta = \sigma^2$ if prior density is

$$f(\theta) = \begin{cases} \alpha e^{-\alpha\theta}, & \theta \geq 0; \\ 0, & \theta < 0. \end{cases}$$

Problem 3.2 Suppose that the X_l's are L iid random variables with the following pdf:

$$f(x) = \frac{1}{\theta} e^{-\frac{x}{\theta}}, \; x \geq 0,$$

where $\theta > 0$. Denote by X_l observations.
(i) Find the ML estimate of θ.
(ii) Show that the ML estimate is also MVU.
(iii) Find the CRB and show that the ML estimate is efficient.
(iv) Show that the following estimator for θ:

$$\hat{\theta} = \frac{1}{L+1} \sum_{l=1}^{L} X_l$$

has smaller MSE than the ML estimate.

Problem 3.3 Suppose that the X_l's are L iid random variables with the following pdf:

$$f(x) = \alpha e^{-\alpha x}, \; x \geq 0,$$

where $\alpha > 0$.
(i) Find the ML estimate.
(ii) Is the ML estimate MVU (find the mean of the ML estimate)?

Problem 3.4 Consider the following signals:

$$r_l = \sqrt{\gamma} s_l + n_l, \; l = 0, 1, \ldots, L-1, \tag{3.84}$$

where γ is the parameter to be estimated and s_l is the lth known signal. The signal, r_l, can be considered as the received signal over a memoryless channel of channel gain,

$\sqrt{\gamma}$ with background noise, n_l. We assume that $n_l \sim \mathcal{CN}(0, \sigma^2)$ and n_l is independent of each other.

(i) Let $a = \sqrt{\gamma}$. Find the ML estimate of a from $\{r_0, r_1, \ldots, r_{L-1}\}$.

(ii) Show that the ML estimate of γ is $\hat{a}^2$, where $\hat{a}$ is the ML estimate of a.

Problem 3.5 Show that $\mathbf{L}_{\text{map}}$ in (3.73):

$$\mathbf{L}_{\text{map}} = (\mathbf{S}^{\text{H}}\mathbf{R}_{\mathbf{n}}^{-1}\mathbf{S} + \mathbf{R}_{\mathbf{h}}^{-1})^{-1}\mathbf{S}^{\text{H}}\mathbf{R}_{\mathbf{n}}^{-1},$$

is identical to $\mathbf{L}_{\text{mmse}}$ in (3.77):

$$\mathbf{L}_{\text{mmse}} = \mathbf{R}_{\mathbf{h}}\mathbf{S}^{\text{H}}(\mathbf{S}\mathbf{R}_{\mathbf{h}}\mathbf{S}^{\text{H}} + \mathbf{R}_{\mathbf{n}})^{-1}.$$

Problem 3.6 Derive (3.78) and (3.79).

Problem 3.7 Suppose that the L observed signal vectors of $N \times 1$ can be written as

$$\mathbf{r}_l = \mathbf{H}\mathbf{s}_l + \mathbf{n}_l, \; l = 1, 2, \ldots, L, \tag{3.85}$$

where $\mathbf{H}$ is a constant matrix of size $N \times K$ to be estimated. We assume that $\mathbf{s}_l$ is known and $\mathbf{n}_l \sim \mathcal{N}(0, \sigma^2)$ is iid.

(i) If $L \geq K$, find the LS estimate of $\mathbf{H}$.

(ii) Suppose that $\mathbf{H}$ is a random matrix which is independent of $\mathbf{n}_l$. Let $\mathbf{a}_p^{\text{H}}$ denote the pth row vector of $\mathbf{H}$, $p = 1, 2, \ldots, N$. If $\mathcal{E}[\mathbf{a}_p\mathbf{a}_q^{\text{H}}] = \sigma_a^2\mathbf{I}\delta_{p,q}$ and $\mathcal{E}[\mathbf{a}_p] = \mathbf{0}$ for all p, find the MMSE estimate of $\mathbf{H}$.

Problem 3.8 Suppose that the output of a certain linear system is given by

$$\begin{aligned} y_l &= a_1 y_{l-1} + a_2 y_{l-2} + \cdots + a_P y_{l-P} + u_l \\ &= \sum_{p=1}^{P} a_l y_{l-p} + u_l, \end{aligned} \tag{3.86}$$

where y_l and u_l denote the lth (real-valued) output and input of the system, respectively, and $\{a_p\}$ is the set of unknown (real-valued) parameters. It is assumed that u_l is an uncorrelated process with zero mean and variance σ_u^2. The output process is called an autoregressive (AR) process and denoted by AR(P), where P is the order of the process.

(i) Find $\{a_p\}$ in terms of the auto-correlations of y_l (i.e. $\mathcal{E}[y_l y_{l-p}]$) and the variance of u_l.

(ii) Assume that $u_l \sim \mathcal{N}(0, \sigma_u^2)$. Derive the ML estimate of $\mathbf{a} = [a_1 \; a_2 \; \ldots \; a_P]^{\text{T}}$ if $y_{-P}, y_{-P+1}, \ldots, y_L$ are available, where $L \geq P$, while u_l is not observable.

4 Optimal combining: single-signal

For a better signal reception, it is often desirable to use multiple sensors or antennas at a receiver. (Note that we will assume that receive antenna and sensor are interchangeable and they are considered as a device that can receive signals through a certain channel medium. For convenience, however, we prefer antenna throughout the book with wireless communication applications in mind.) To extract a signal of interest, multiple signals received by multiple antennas are to be properly combined. For signal combining, we need to take into account the desired signal's (statistical or deterministic) properties as well as statistical properties of background noise.

Although there are various signal combining techniques, we focus on linear combining techniques in this chapter, because they can be relatively easily implemented and their analysis is tractable. In addition, only second-order moments of a desired signal and noise are usually required to find a linear combiner under the MMSE criterion.

4.1 Signals in space

Suppose that there are N sensors or antennas to receive a signal of interest generated from a source, which can be a radio signal or a voice. In general, the signal is received through a certain channel medium with channel attenuation or distortion and corrupted by noise. Since multiple observations of a signal are available using multiple sensors or antennas, the signal can be seen as a vector in a vector space as illustrated in Fig. 4.1. As N increases, the dimension of the vector space increases and a better subspace to extract the desired signal can be found.

Let s_l denote the transmitted signal at discrete time l. The signals received by N antennas at time l are given by

$$\begin{aligned} r_{1,l} &= h_{1,l}s_l + n_{1,l}; \\ r_{2,l} &= h_{2,l}s_l + n_{2,l}; \\ &\vdots \\ r_{N,l} &= h_{N,l}s_l + n_{N,l}, \end{aligned} \tag{4.1}$$

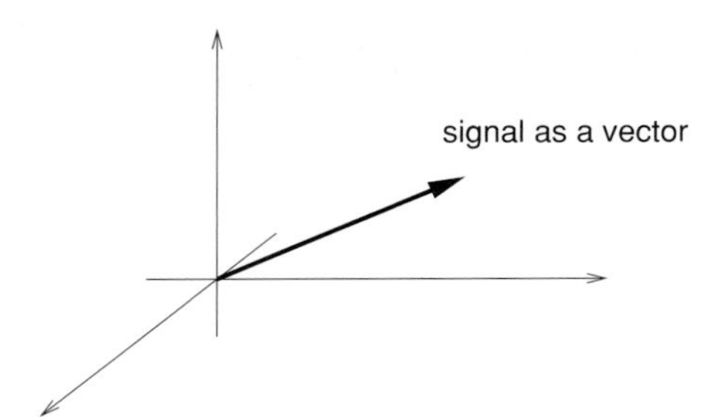

Figure 4.1 A signal as a vector in a vector space.

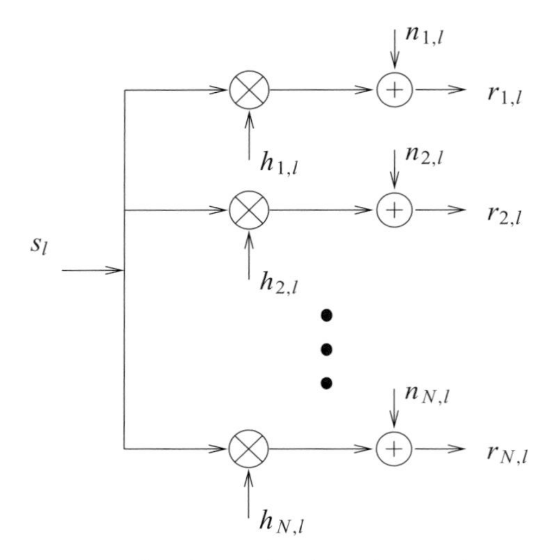

Figure 4.2 Signal reception through multiple sensors or antennas.

where $h_{q,l}$ is the channel gain or attenuation and $n_{q,l}$ is the noise of the qth received signal. Stacking the r_l's, we have

$$\begin{aligned}\mathbf{r}_l &= [r_{1,l}\; r_{2,l}\; \ldots\; r_{N,l}]^{\mathrm{T}} \\ &= \mathbf{h}_l s_l + \mathbf{n}_l,\; l = 0, 1, \ldots, \infty, \end{aligned} \tag{4.2}$$

where $\mathbf{h}_l = [h_{1,l}\; h_{2,l}\; \ldots\; h_{N,l}]^{\mathrm{T}}$ is the channel gain vector or simply channel vector for the desired signal s_l and $\mathbf{n}_l = [n_{1,l}\; n_{2,l}\; \ldots\; n_{N,l}]^{\mathrm{T}}$ is the noise vector. The channel vector $\mathbf{h}_l$ is characterized by the channel and plays a crucial role in signal combining. Figure 4.2 shows a channel model for multi-antenna reception. Since multiple copies of the desired signal are available, a good estimate of the desired signal can be obtained by combining multiple observations.

An estimate of s_l obtained by linear combining of $\mathbf{r}_l$ is given by

$$\hat{s}_l = \sum_{q=1}^{N} w_{q,l}^* r_{q,l} \\ = \mathbf{w}_l^{\mathrm{H}} \mathbf{r}_l, \qquad (4.3)$$

where $\mathbf{w}_l = [w_{1,l}\ w_{2,l}\ \ldots\ w_{N,l}]^{\mathrm{T}}$ denotes the linear combining vector. We mainly focus on linear combining in this chapter as it is easy to implement and its performance analysis is mathematically tractable. Furthermore, as shown in Chapter 3, the ideal MMSE combining becomes the linear MMSE combining if signals and noise are Gaussian.

Linear combining is a general notion that can be applied without multiple antennas or sensors in some applications. An example is given as follows.

Example 4.1.1 Suppose that the received signal is a weighted sum of delayed versions of a signal, denoted by s_l, as follows:

$$r_l = h_0 s_l + h_1 s_{l-1} + h_2 s_{l-2} + n_l,$$

where h_p denotes the channel gain of the pth delayed signal, s_{l-p}, $p = 0, 1, 2$, and n_l is the background noise. If a channel has memory, then the received signal is usually a weighted sum of delayed versions of the transmitted signal and this channel is called an inter-symbol interference (ISI) channel. If s_l is considered as the desired signal, the delayed signals in r_l, i.e. s_{l-1} and s_{l-2}, can be treated as interfering signals.

The desired signal, s_l, can be found in r_l as well as r_{l+1} and r_{l+2} as follows:

$$r_l = h_0 s_l + h_1 s_{l-1} + h_2 s_{l-2} + n_l;$$

$$r_{l+1} = h_0 s_{l+1} + h_1 s_l + h_2 s_{l-1} + n_{l+1};$$

$$r_{l+2} = h_0 s_{l+2} + h_1 s_{l+1} + h_2 s_l + n_{l+2}.$$

Let $\mathbf{r}_l = [r_l\ r_{l+1}\ r_{l+2}]^{\mathrm{T}}$. It can be shown that

$$\mathbf{r}_l = \mathbf{h} s_l + \mathbf{u}_l + \mathbf{n}_l,$$

where $\mathbf{h} = [h_0\ h_1\ h_2]^{\mathrm{T}}$, $\mathbf{n}_l = [n_l\ n_{l+1}\ n_{l+2}]^{\mathrm{T}}$, and

$$\mathbf{u}_l = \begin{bmatrix} h_2 & h_1 & 0 & 0 \\ 0 & h_2 & h_0 & 0 \\ 0 & 0 & h_1 & h_0 \end{bmatrix} \begin{bmatrix} s_{l-2} \\ s_{l-1} \\ s_{l+1} \\ s_{l+2} \end{bmatrix}.$$

Using linear combining, as in (4.3), s_l can be estimated as follows:

$$\hat{s}_l = \mathbf{w}^{\mathrm{H}} \mathbf{r}_l \\ = w_1^* r_l + w_2^* r_{l+1} + w_3^* r_{l+2}.$$

This linear combining is called linear equalization (Proakis 1995).

In wireless communications, there are techniques that provide multiple copies of the signal at a receiver to improve the reliability of transmission. For example, as mentioned above, multiple receive antennas can be used. Alternatively, signals can be repeatedly transmitted. If each signal can be transmitted through a different channel and received under a different condition, the SNR of each signal or the signal's strength ($h_{q,l}$ in (4.1)) is usually different. As mentioned earlier, the received signal can be seen as a vector in a vector space and the component in each dimension can be different. Although the signal strength can be weak in one dimension, the overall signal strength can be strong and increases as the number of dimensions increases.

For convenience, the vector space where the received signal becomes a vector is called the space domain and the gain generated by having multiple dimensions (or multiple copies of signals) is called the spatial diversity gain. In this chapter, we discuss various combining techniques with emphasis on the MMSE combining and applications in wireless communications.

4.2 MMSE combining with known channel vector

The MMSE approach to estimate a random signal is well established since Wiener (Wiener 1949). The combining vector can be decided to minimize the MSE between the desired signal, s_l, and the output of a linear combiner. To perform this linear MMSE combining, we need to know the channel vector $\mathbf{h}_l$ and statistical properties of s_l and $\mathbf{n}_l$. For convenience, we will omit time index l. Without loss of generality, we assume that $\mathcal{E}[s] = 0$ and $\mathcal{E}[\mathbf{n}] = \mathbf{0}$. In addition, the covariance matrix of $\mathbf{n}$, $\mathcal{E}[\mathbf{n}\mathbf{n}^{\mathrm{H}}] = \mathbf{R}_{\mathbf{n}}$, and the variance of s, $\mathcal{E}[|s|^2] = \sigma_s^2$, are assumed to be known. Throughout this chapter, it is assumed that $\mathbf{R}_{\mathbf{n}}$ is full-rank.

The MSE between s and its estimate, which is the output of a linear combiner, is given by

$$\mathcal{E}[|s - \hat{s}|^2] = \mathcal{E}[|s - \mathbf{w}^{\mathrm{H}}\mathbf{r}|^2].$$

It can be easily shown that the MSE is a function of $\mathbf{w}$. Using the orthogonality principle, the optimal combining vector can be found as

$$\begin{aligned} \mathbf{w}_{\mathrm{mmse}} &= \arg\min_{\mathbf{w}} \mathcal{E}[|s - \mathbf{w}^{\mathrm{H}}\mathbf{r}|^2] \\ &= \mathbf{R}_{\mathbf{r}}^{-1}\mathbf{c}, \end{aligned} \tag{4.4}$$

where $\mathbf{R}_{\mathbf{r}} = \mathcal{E}[\mathbf{r}\mathbf{r}^{\mathrm{H}}]$ is the covariance matrix of $\mathbf{r}$ and $\mathbf{c} = \mathcal{E}[\mathbf{r}s^*]$ is the correlation vector between $\mathbf{r}$ and s^*. If s and $\mathbf{n}$ are uncorrelated, we have

$$\begin{aligned} \mathbf{R}_{\mathbf{r}} &= \mathbf{h}\mathbf{h}^{\mathrm{H}}\sigma_s^2 + \mathbf{R}_{\mathbf{n}}; \\ \mathbf{c} &= \mathbf{h}\sigma_s^2. \end{aligned} \tag{4.5}$$

Then, the optimal MMSE combining vector becomes

$$\mathbf{w}_{\mathrm{mmse}} = \left(\mathbf{h}\mathbf{h}^{\mathrm{H}}\sigma_s^2 + \mathbf{R}_{\mathbf{n}}\right)^{-1}\mathbf{h}\sigma_s^2. \tag{4.6}$$

Using Woodbury's identity, it can be shown that

$$\mathbf{w}_{\text{mmse}} = \alpha \mathbf{R}_{\mathbf{n}}^{-1} \mathbf{h}, \tag{4.7}$$

where

$$\alpha = \frac{\sigma_s^2}{1 + \sigma_s^2 \mathbf{h}^{\mathrm{H}} \mathbf{R}_{\mathbf{n}}^{-1} \mathbf{h}}.$$

In addition, the MMSE becomes

$$\begin{aligned}
\epsilon_{\text{mmse}}^2 &= \mathcal{E}[|s - \mathbf{w}_{\text{mmse}}^{\mathrm{H}} \mathbf{r}|^2] \\
&= \sigma_s^2 \left(1 - \mathbf{h}^{\mathrm{H}} \left(\mathbf{h}\mathbf{h}^{\mathrm{H}} \sigma_s^2 + \mathbf{R}_{\mathbf{n}}\right)^{-1} \mathbf{h} \sigma_s^2\right) \\
&= \frac{\sigma_s^2}{1 + \sigma_s^2 \mathbf{h}^{\mathrm{H}} \mathbf{R}_{\mathbf{n}}^{-1} \mathbf{h}}.
\end{aligned} \tag{4.8}$$

As $\sigma_s^2 \to \infty$, we have

$$\lim_{\sigma_s^2 \to \infty} \epsilon_{\text{mmse}}^2 = \frac{1}{\mathbf{h}^{\mathrm{H}} \mathbf{R}_{\mathbf{n}}^{-1} \mathbf{h}}. \tag{4.9}$$

This implies that the asymptotic MMSE depends on the channel vector and noise covariance matrix and cannot approach zero if $\mathbf{h}^{\mathrm{H}} \mathbf{R}_{\mathbf{n}}^{-1} \mathbf{h}$ is finite although $\sigma_s^2 \to \infty$.

Note that we can also show that $\alpha = \epsilon_{\text{mmse}}^2$,

$$\mathbf{w}_{\text{mmse}} = \epsilon_{\text{mmse}}^2 \mathbf{R}_{\mathbf{n}}^{-1} \mathbf{h}, \tag{4.10}$$

and

$$\begin{aligned}
\hat{s}_{\text{mmse}} &= \epsilon_{\text{mmse}}^2 \mathbf{h}^{\mathrm{H}} \mathbf{R}_{\mathbf{n}}^{-1} \mathbf{r} \\
&= \mu s + u,
\end{aligned} \tag{4.11}$$

where

$$\begin{aligned}
\mu &= \epsilon_{\text{mmse}}^2 \mathbf{h}^{\mathrm{H}} \mathbf{R}_{\mathbf{n}}^{-1} \mathbf{h}; \\
\mathcal{E}[|u|^2] &= \epsilon_{\text{mmse}}^4 \mathbf{h}^{\mathrm{H}} \mathbf{R}_{\mathbf{n}}^{-1} \mathbf{h}.
\end{aligned} \tag{4.12}$$

From this, we can easily show that the SNR becomes

$$\begin{aligned}
\text{SNR} &= \frac{\mathcal{E}[|\mu s|^2]}{\mathcal{E}[|u|^2]} \\
&= \sigma_s^2 \mathbf{h}^{\mathrm{H}} \mathbf{R}_{\mathbf{n}}^{-1} \mathbf{h}.
\end{aligned} \tag{4.13}$$

Interestingly, the SNR can also be given by

$$\text{SNR} = \frac{\sigma_s^2}{\lim_{\sigma_s^2 \to \infty} \epsilon_{\text{mmse}}^2}.$$

As discussed in Chapter 3, in general, the linear MMSE estimate is different from the ideal MMSE estimate that is the conditional mean unless $\mathbf{r}$ and s are jointly Gaussian. That is,

$$\hat{s}_{\text{mmse}} = \mathbf{w}_{\text{mmse}}^{\text{H}} \mathbf{r} \neq \hat{s}_{\text{mmse}*} = \mathcal{E}[s \,|\, \mathbf{r}].$$

In the following example, we show the relationship between the linear MMSE and ideal MMSE for binary signals.

Example 4.2.1 The ideal MMSE estimate of s is the conditional mean that is given by

$$\hat{s}_{\text{mmse}*} = \mathcal{E}[s \,|\, \mathbf{r}]. \tag{4.14}$$

Suppose that $s \in \{-1, +1\}$ and equally likely and $\mathbf{n} \sim \mathcal{CN}(0, \mathbf{R}_{\mathbf{n}})$. Since

$$\Pr(s \,|\, \mathbf{r}) = \frac{e^{\beta s}}{e^{\beta} + e^{-\beta}},$$

where $\beta = \Re(\mathbf{r}^{\text{H}} \mathbf{R}_{\mathbf{n}}^{-1} \mathbf{h})$, we have

$$\begin{aligned} \hat{s}_{\text{mmse}*} &= \Pr(s = 1 \,|\, \mathbf{r}) - \Pr(s = -1 \,|\, \mathbf{r}) \\ &= \frac{e^{\beta} - e^{-\beta}}{e^{\beta} + e^{-\beta}} \\ &= \tanh(\beta). \end{aligned} \tag{4.15}$$

Note that since $\beta = \frac{\Re(\hat{s}_{\text{mmse}})}{\epsilon_{\text{mmse}}^2}$,

$$\hat{s}_{\text{mmse}*} = \tanh\left(\frac{\Re(\hat{s}_{\text{mmse}})}{\epsilon_{\text{mmse}}^2}\right).$$

Figure 4.3 shows the relationship between the linear MMSE estimate and ideal MMSE estimate when $s \in \{-1, +1\}$ is binary signal.

4.3 Information-theoretical optimality of MMSE combining

Certainly, MMSE combining is optimal under the MSE criterion. We can also show that MMSE combining is information-theoretically optimal as this combining does not lose any information of signal under a certain condition.

Recall the received signal vector:

$$\mathbf{r} = \mathbf{h}s + \mathbf{n}. \tag{4.16}$$

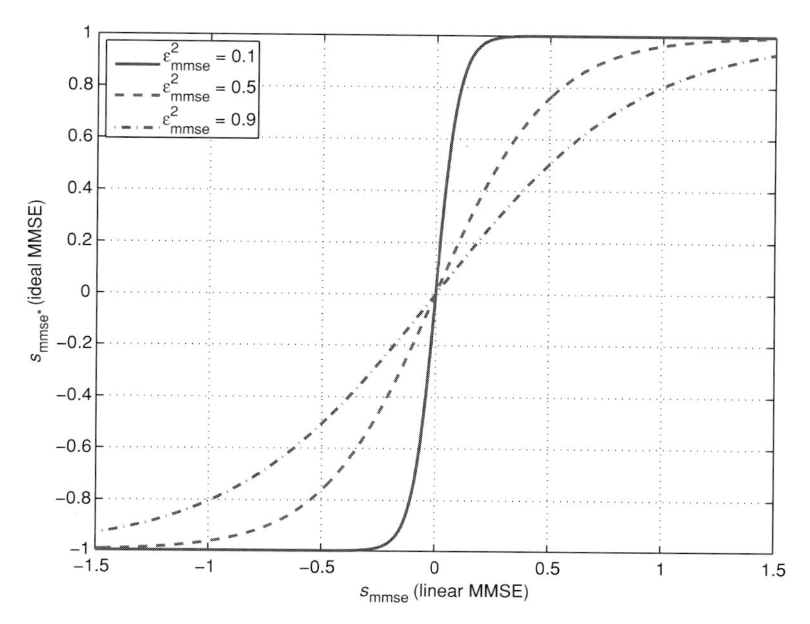

Figure 4.3 Relationship between the linear MMSE estimate and ideal MMSE estimate when $s \in \{-1, +1\}$ is binary signal.

We assume that s and $\mathbf{n}$ are zero-mean independent CSCG random variable and vector, respectively. The mutual information between $\mathbf{r}$ and s becomes

$$\begin{aligned} \mathsf{I}(\mathbf{r}; s) &= \mathsf{h}(\mathbf{r}) - \mathsf{h}(\mathbf{r}|s) \\ &= \log_2(\det(\pi e \mathbf{R_r})) - \log_2(\det(\pi e \mathbf{R_n})) \\ &= \log_2 \det \mathbf{R_r} \mathbf{R_n}^{-1}, \end{aligned} \tag{4.17}$$

where $\mathsf{h}(\cdot)$ denotes differential entropy (see Appendix 2 for a brief overview of information theory or (Cover & Thomas 1991) for a detailed account). Suppose that

$$\mathbf{R_n} = N_0 \mathbf{I},$$

i.e. the noise vector is spatially white and each element of $\mathbf{n}$ has the same variance, N_0. Then, it follows that

$$\begin{aligned} \mathbf{R_r} &= \sigma_s^2 \mathbf{h}\mathbf{h}^{\mathrm{H}} + N_0 \mathbf{I}; \\ \mathbf{R_r}\mathbf{R_n}^{-1} &= \frac{\sigma_s^2}{N_0} \mathbf{h}\mathbf{h}^{\mathrm{H}} + \mathbf{I}. \end{aligned} \tag{4.18}$$

Substituting (4.18) into (4.17), the mutual information is given by

$$\mathsf{I}(\mathbf{r}; s) = \log_2 \left(1 + \frac{\sigma_s^2}{N_0} ||\mathbf{h}||^2 \right). \tag{4.19}$$

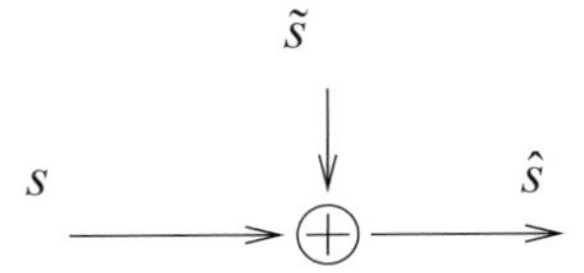

Figure 4.4 Equivalent channel model after MMSE combining.

We now consider the mutual information between s and its MMSE estimate which is the output of the MMSE combiner. The output of the MMSE combiner is

$$\hat{s} = \mathbf{w}_{\text{mmse}}^{\text{H}} \mathbf{r}, \tag{4.20}$$

where $\mathbf{w}_{\text{mmse}} = \mathbf{R}_{\mathbf{r}}^{-1} \mathbf{h} \sigma_s^2$. Let $\tilde{s} = \hat{s} - s$. Due to the orthogonality principle of the MMSE combining, $\tilde{s}$ and $\mathbf{r}$ are uncorrelated (or independent as $\mathbf{r}$ and s are Gaussian). This implies that $\tilde{s}$ and $\hat{s}$ are uncorrelated. The variance of the estimation error becomes

$$\begin{aligned} \sigma_{\tilde{s}}^2 &= \mathcal{E}[|\tilde{s}|^2] \\ &= \mathcal{E}[(s - \mathbf{w}_{\text{mmse}}^{\text{H}} \mathbf{r}) s^*] \\ &= \sigma_s^2 (1 - \sigma_s^2 \mathbf{h}^{\text{H}} \mathbf{R}_{\mathbf{r}}^{-1} \mathbf{h}) \\ &= \frac{\sigma_s^2}{\frac{\sigma_s^2}{N_0} ||\mathbf{h}||^2 + 1}. \end{aligned} \tag{4.21}$$

Noting that

$$\hat{s} = s + \tilde{s}, \tag{4.22}$$

the output of the MMSE combiner, $\hat{s}$, can be considered as the channel output, while s and $\tilde{s}$ are the channel input and the background noise, respectively. The equivalent channel model after the MMSE combining is shown in Fig. 4.4. Then, the mutual information between $\hat{s}$ and s is given by

$$\begin{aligned} \mathsf{I}(\hat{s}; s) &= \mathsf{h}(s) - \mathsf{h}(s | \hat{s}) \\ &= \mathsf{h}(s) - \mathsf{h}(\tilde{s}), \end{aligned} \tag{4.23}$$

because

$$\begin{aligned} \mathsf{h}(s | \hat{s}) &= \mathsf{h}(\hat{s} - \tilde{s} | \hat{s}) \\ &= \mathsf{h}(\tilde{s}). \end{aligned}$$

The last equality results from the fact that $\tilde{s}$ and $\hat{s} = \mathbf{w}_{\text{mmse}}^{\text{H}} \mathbf{r}$ are uncorrelated and independent under the assumption that s and $\mathbf{n}$ are Gaussian. As s and $\tilde{s}$ are Gaussian, it can be shown that

$$\begin{aligned} \mathsf{h}(s) &= \log_2(\pi \mathrm{e} \sigma_s^2); \\ \mathsf{h}(\tilde{s}) &= \log_2(\pi \mathrm{e} \sigma_{\tilde{s}}^2). \end{aligned} \tag{4.24}$$

Finally, it follows that

$$\begin{aligned}\mathsf{I}(\hat{s}; s) &= \mathsf{I}(\mathbf{r}; s) \\ &= \log_2\left(1 + \frac{\sigma_s^2}{N_0}||\mathbf{h}||^2\right).\end{aligned} \tag{4.25}$$

From (4.25), we can see that the MMSE combining is information-lossless in terms of the mutual information if s and $\mathbf{n}$ are Gaussian.

The same result is also valid when the noise is not spatially white, i.e. $\mathbf{R_n} \neq N_0\mathbf{I}$.

Theorem 4.3.1 *Suppose that s and $\mathbf{n}$ are independent CSCG random variable and vector, respectively. The MMSE combining is information-lossless in terms of the mutual information since*

$$\begin{aligned}\mathsf{I}(\hat{s}; s) &= \mathsf{I}(\mathbf{r}; s) \\ &= \log_2\left(1 + \sigma_s^2\mathbf{h}^{\mathrm{H}}\mathbf{R}_{\mathbf{n}}^{-1}\mathbf{h}\right).\end{aligned} \tag{4.26}$$

Proof: From (4.8), the mutual information between $\hat{s}$ and s is given by

$$\begin{aligned}\mathsf{I}(\hat{s}; s) &= \mathsf{h}(s) - \mathsf{h}(\tilde{s}) \\ &= \log_2\left(1 + \sigma_s^2\mathbf{h}^{\mathrm{H}}\mathbf{R}_{\mathbf{n}}^{-1}\mathbf{h}\right).\end{aligned}$$

From (4.17), we can show that

$$\begin{aligned}\mathsf{I}(\mathbf{r}; s) &= \log_2 \det \mathbf{R_r}\mathbf{R}_{\mathbf{n}}^{-1} \\ &= \log_2 \det\left(\mathbf{I} + \sigma_s^2\mathbf{h}\mathbf{h}^{\mathrm{H}}\mathbf{R}_{\mathbf{n}}^{-1}\right).\end{aligned} \tag{4.27}$$

Since $\sigma_s^2\mathbf{h}\mathbf{h}^{\mathrm{H}}\mathbf{R}_{\mathbf{n}}^{-1}$ is rank 1, we have

$$\det\left(\mathbf{I} + \sigma_s^2\mathbf{h}\mathbf{h}^{\mathrm{H}}\mathbf{R}_{\mathbf{n}}^{-1}\right) = 1 + \sigma_s^2\mathbf{h}^{\mathrm{H}}\mathbf{R}_{\mathbf{n}}^{-1}\mathbf{h}.$$

Then, it follows

$$\mathsf{I}(\mathbf{r}; s) = \log_2\left(1 + \sigma_s^2\mathbf{h}^{\mathrm{H}}\mathbf{R}_{\mathbf{n}}^{-1}\mathbf{h}\right).$$

This completes the proof. □

As the MMSE combining is information-lossless, we can consider that the detection with MMSE combined signals is an information-lossless approach.

There is also another interesting relationship between the mutual information and MMSE. For the case of spatially correlated noise, define the SNR as follows:

$$\gamma = \sigma_s^2\mathbf{h}^{\mathrm{H}}\mathbf{R}_{\mathbf{n}}^{-1}\mathbf{h}.$$

Then, the normalized MMSE is given by

$$\begin{aligned}\bar{\epsilon}^2(\gamma) &= \frac{\epsilon_{\mathrm{mmse}}^2}{\sigma_s^2} \\ &= \frac{1}{1+\gamma}.\end{aligned} \tag{4.28}$$

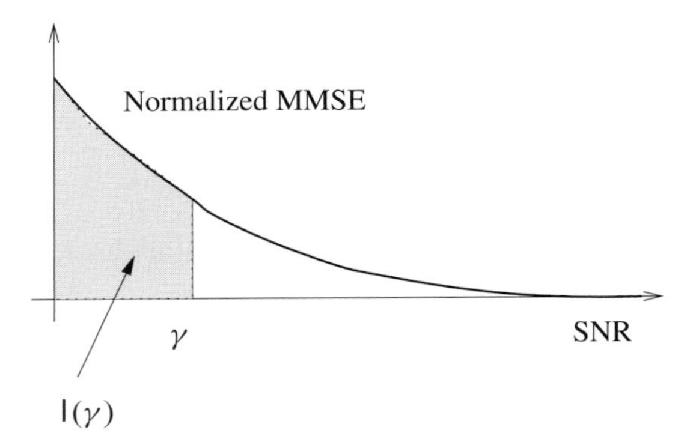

Figure 4.5 Relationship between the MMSE and mutual information.

Letting $\mathsf{I}(\gamma) = \mathsf{I}(\mathbf{r}; s) = \log_2(1 + \gamma)$, it can be shown that

$$\frac{\mathrm{d}}{\mathrm{d}\gamma}\mathsf{I}(\gamma) = \bar{\epsilon}^2(\gamma) \log_2 \mathrm{e}. \tag{4.29}$$

Alternatively, we have

$$\mathsf{I}(\gamma) = \log_2 \mathrm{e} \int_0^{\gamma} \bar{\epsilon}^2(z)\mathrm{d}z. \tag{4.30}$$

As shown in Fig. 4.5, the mutual information can be expressed as an area of the normalized MSE curve. The relationship in (4.29) is found in (Guo, Shamai & Verdu 2005) and it is shown that this relationship is valid for any distribution of s (as long as the background noise is Gaussian).[1]

4.4 Relation with other combining techniques

In this section, we discuss the relationship between the MMSE combining and other well-known combining techniques. Again, we will omit time index l for convenience.

4.4.1 MSNR combining

Consider the linear combining that maximizes the SNR. With the combining vector, $\mathbf{w}$, the combined signal becomes

$$\begin{aligned} \hat{s} &= \mathbf{w}^{\mathrm{H}}\mathbf{r} \\ &= \mathbf{w}^{\mathrm{H}}\mathbf{h}s + \mathbf{w}^{\mathrm{H}}\mathbf{n}. \end{aligned}$$

[1] If s is not Gaussian, the linear MMSE estimate has to be replaced by the ideal MMSE estimate, which is the conditional mean, and the corresponding MMSE should replace the MMSE obtained by the linear estimator.

Since the first and second terms on the right-hand side are the signal and noise terms, respectively, the SNR is given by

$$\begin{aligned}\text{SNR} &= \frac{\mathcal{E}[|\mathbf{w}^{\mathrm{H}}\mathbf{h}s|^2]}{\mathcal{E}[|\mathbf{w}^{\mathrm{H}}\mathbf{n}|^2]} \\ &= \frac{|\mathbf{w}^{\mathrm{H}}\mathbf{h}|^2\sigma_s^2}{\mathbf{w}^{\mathrm{H}}\mathbf{R}_{\mathbf{n}}\mathbf{w}}. \end{aligned} \tag{4.31}$$

Since $\mathbf{R}_{\mathbf{n}}$ is positive definite as a full-rank covariance matrix, we can factorize it as $\mathbf{R}_{\mathbf{n}} = \mathbf{R}_{\mathbf{n}}^{\mathrm{H}/2}\mathbf{R}_{\mathbf{n}}^{1/2}$, where $\mathbf{R}_{\mathbf{n}}^{\mathrm{H}/2} = (\mathbf{R}_{\mathbf{n}}^{1/2})^{\mathrm{H}}$. Then, the SNR can be upper-bounded as

$$\begin{aligned}\text{SNR} &= \frac{|(\mathbf{R}_{\mathbf{n}}^{1/2}\mathbf{w})^{\mathrm{H}}(\mathbf{R}_{\mathbf{n}}^{-\mathrm{H}/2}\mathbf{h})|^2\sigma_s^2}{(\mathbf{R}_{\mathbf{n}}^{1/2}\mathbf{w})^{\mathrm{H}}(\mathbf{R}_{\mathbf{n}}^{1/2}\mathbf{w})} \\ &\le \frac{||\mathbf{R}_{\mathbf{n}}^{1/2}\mathbf{w}||^2\ ||\mathbf{R}_{\mathbf{n}}^{-\mathrm{H}/2}\mathbf{h}||^2\sigma_s^2}{(\mathbf{R}_{\mathbf{n}}^{1/2}\mathbf{w})^{\mathrm{H}}(\mathbf{R}_{\mathbf{n}}^{1/2}\mathbf{w})} \\ &= ||\mathbf{R}_{\mathbf{n}}^{-\mathrm{H}/2}\mathbf{h}||^2\sigma_s^2 \\ &= \mathbf{h}^{\mathrm{H}}\mathbf{R}_{\mathbf{n}}^{-1}\mathbf{h}\sigma_s^2. \end{aligned} \tag{4.32}$$

The inequality in (4.32) is due to the Cauchy–Schwarz inequality and the equality holds if and only if $\mathbf{R}_{\mathbf{n}}^{1/2}\mathbf{w} = \alpha\mathbf{R}_{\mathbf{n}}^{-\mathrm{H}/2}\mathbf{h}$, where α is a non-zero constant. For the equality to hold, let

$$\begin{aligned}\mathbf{w} &= \alpha\mathbf{R}_{\mathbf{n}}^{-1/2}\mathbf{R}_{\mathbf{n}}^{-\mathrm{H}/2}\mathbf{h} \\ &= \alpha(\mathbf{R}_{\mathbf{n}}^{\mathrm{H}/2}\mathbf{R}_{\mathbf{n}}^{1/2})^{-1}\mathbf{h} \\ &= \alpha\mathbf{R}_{\mathbf{n}}^{-1}\mathbf{h}. \end{aligned}$$

From this, the maximum SNR (MSNR) combining vector is given by

$$\mathbf{w}_{\text{msnr}} = \alpha\mathbf{R}_{\mathbf{n}}^{-1}\mathbf{h}. \tag{4.33}$$

From (4.33) and (4.7), we can see that the MMSE combining is an MSNR combining.

4.4.2 ML combining

Taking s as a parameter to be estimated, ML estimation can be considered. Consider the ML estimation of s when $\mathbf{n}$ in (4.16) is a zero-mean CSCG random vector, $\mathbf{n} \sim \mathcal{CN}(\mathbf{0}, \mathbf{R}_{\mathbf{n}})$. For given $\mathbf{r}$ in (4.16), the likelihood function of s is given by

$$f(\mathbf{r}|s) = \frac{1}{\det(\pi\mathbf{R}_{\mathbf{n}})}\exp\left(-(\mathbf{r}-\mathbf{h}s)^{\mathrm{H}}\mathbf{R}_{\mathbf{n}}^{-1}(\mathbf{r}-\mathbf{h}s)\right). \tag{4.34}$$

The ML estimate of s becomes

$$\begin{aligned}\hat{s}_{\mathrm{ml}} &= \arg\max_{s} f(\mathbf{r}|s) \\ &= \arg\min_{s} (\mathbf{r} - \mathbf{h}s)^{\mathrm{H}} \mathbf{R}_{\mathbf{n}}^{-1} (\mathbf{r} - \mathbf{h}s) \\ &= \frac{(\mathbf{R}_{\mathbf{n}}^{-1}\mathbf{h})^{\mathrm{H}}\mathbf{r}}{\mathbf{h}^{\mathrm{H}}\mathbf{R}_{\mathbf{n}}^{-1}\mathbf{h}}. \end{aligned} \tag{4.35}$$

Letting

$$\begin{aligned}\mathbf{w}_{\mathrm{ml}} &= \frac{1}{\mathbf{h}^{\mathrm{H}}\mathbf{R}_{\mathbf{n}}^{-1}\mathbf{h}} \mathbf{R}_{\mathbf{n}}^{-1}\mathbf{h} \\ &\propto \mathbf{R}_{\mathbf{n}}^{-1}\mathbf{h}, \end{aligned} \tag{4.36}$$

it follows that

$$\hat{s}_{\mathrm{ml}} = \mathbf{w}_{\mathrm{ml}}^{\mathrm{H}}\mathbf{r}.$$

We can see that the ML estimate is obtained by a linear combining operation, called the ML combining, with the weighting vector, $\mathbf{w}_{\mathrm{ml}}$. From (4.36), we can observe that the ML combining is also an MSNR combining.

As shown above, the ML estimate can be expressed as the output of a linear combiner if $\mathbf{n}$ is Gaussian. However, the ML estimate cannot be expressed as a linear combination of observations in other cases. We present such an example below.

Example 4.4.1 Suppose that the n_q's are iid with the following distribution:

$$f(n_q) = \frac{1}{2} \mathrm{e}^{-|n_q|},$$

which is called the standard double-exponential distribution. In this case, the ML estimate of s is given by

$$\begin{aligned}\hat{s}_{\mathrm{ml}} &= \arg\max_{s} \prod_{q=1}^{N} \frac{1}{2} \mathrm{e}^{-|r_q - h_q s|} \\ &= \arg\min_{s} \sum_{q=1}^{N} |r_q - h_q s|. \end{aligned} \tag{4.37}$$

If $N = 2$ and $h_q \neq 0$, $q = 1, 2$, the ML estimate is given by

$$\hat{s}_{\mathrm{ml}} = \begin{cases} \frac{r_1}{h_1}, & \text{if } |r_2 - \frac{r_1}{h_1} h_2| \leq |r_1 - \frac{r_2}{h_2} h_1|; \\ \frac{r_2}{h_2}, & \text{if } |r_2 - \frac{r_1}{h_1} h_2| > |r_1 - \frac{r_2}{h_2} h_1|. \end{cases} \tag{4.38}$$

Clearly, the ML estimate cannot be expressed by a linear combination of r_1 and r_2 in this case.

Table 4.1 Conditions for MSNR combining.

Combining methods	Conditions for MSNR combining
Linear MMSE combining	no condition is required
ML combining	$\mathbf{n}$ is Gaussian
MAP combining	s and $\mathbf{n}$ are Gaussian

Example 4.4.2 The MAP estimate of s can be found as follows:

$$\hat{s}_{\text{map}} = \arg\max_s f(\mathbf{r}|s)g(s),$$

where $g(s)$ denotes the distribution of s. If $s \sim \mathcal{CN}(0, \sigma_s^2)$, it can be shown that

$$\begin{aligned}\hat{s}_{\text{map}} &= \frac{\mathbf{h}^{\text{H}}\mathbf{R}_{\mathbf{n}}^{-1}\mathbf{r}}{\frac{1}{\sigma_s^2} + \mathbf{h}^{\text{H}}\mathbf{R}_{\mathbf{n}}^{-1}\mathbf{h}} \\ &= \mathbf{w}_{\text{mmse}}^{\text{H}}\mathbf{r} \\ &= \hat{s}_{\text{mmse}}.\end{aligned} \tag{4.39}$$

Thus, MAP combining is identical to MMSE combining if s is Gaussian.

In Table 4.1, we show the required conditions for each combining method to be MSNR combining. Indeed, most combining techniques reduce to MSNR combining under Gaussian assumptions.

4.4.3 Maximal ratio and equal gain combining

In wireless communications, the channel gain between a transmitter and receiver depends on a number of factors, including the distance between them. While the distance decides the average channel gain, the instantaneous channel gain can be fluctuated over time due to multipaths and the motion of the transmitter or receiver or both. In addition, any moving obstacles between them can contribute to this time-varying channel gain. This results in fading channels in wireless communications. Under a fading channel environment, the SNR can be modeled as a random variable. Since the receiver cannot detect signals if the SNR is lower than a certain threshold, it is desirable to minimize the probability that the SNR is lower than a certain threshold, which is called the outage probability. For the robust signal detection to minimize the outage probability, diversity techniques can be used with multiple receive antennas. The resulting diversity gain is called the spatial diversity gain.

Maximal ratio combining (MRC) and equal gain combining (EGC) are two popular linear combining techniques to improve the performance over fading channels ((Schwartz, Bennet & Stein 1966), (Lee 1982)). Indeed, MRC can be considered as a special case of MSNR combining. In deriving the MRC, it is assumed that the noise terms in (4.1) are uncorrelated and have the same variance.[2] That is, $\mathcal{E}[\mathbf{n}\mathbf{n}^{\mathrm{H}}] = N_0\mathbf{I}$. In this case, from (4.31), the SNR is given by

$$\mathrm{SNR} = \frac{|\mathbf{w}^{\mathrm{H}}\mathbf{h}|^2\sigma_s^2}{N_0||\mathbf{w}||^2}.$$

Using the Cauchy–Schwartz inequality, we can find that the combining vector that maximizes the SNR is $\mathbf{w} = \alpha\mathbf{h}$, which is a special case of the MSNR combining vector in (4.33). The linear combining with $\mathbf{w} = \alpha\mathbf{h}$ is called the MRC. The SNR after the MRC becomes

$$\begin{aligned}\mathrm{SNR}_{\mathrm{mrc}} &= \frac{||\mathbf{h}||^2\sigma_s^2}{N_0} \\ &= \sum_{q=1}^{N} \frac{|h_q|^2\sigma_s^2}{N_0}.\end{aligned} \tag{4.40}$$

The EGC is another linear combining technique, but not optimal in any sense. Let $h_q = |h_q|\mathrm{e}^{j\phi_q}$, where $|h_q|$ and ϕ_q are the amplitude and phase of h_q, respectively. The combining vector for the EGC is given by

$$\mathbf{w} = \alpha[\mathrm{e}^{j\phi_1}\ \mathrm{e}^{j\phi_2}\ \ldots\ \mathrm{e}^{j\phi_N}]^{\mathrm{T}},$$

where α is a normalization constant. While the gains or amplitudes of the weights are all the same in EGC, the phase distortion caused by channels is to be compensated for coherent combining (MRC is also coherent combining as the phase distortion is compensated in combining). Assuming that α is real and positive, the output after EGC is given by

$$\begin{aligned}\hat{s} &= \alpha \sum_{q=1}^{N} \mathrm{e}^{-j\phi_q} r_q \\ &= \alpha \sum_{q=1}^{N} \mathrm{e}^{-j\phi_q} \left(|h_q|\mathrm{e}^{j\phi_q} s + n_q\right) \\ &= \alpha \left(\sum_{q=1}^{N} |h_q| s + \sum_{q=1}^{N} \mathrm{e}^{-j\phi_q} n_q \right).\end{aligned} \tag{4.41}$$

Thus, the signals are coherently combined. The resulting SNR is given by

$$\mathrm{SNR}_{\mathrm{egc}} = \frac{\left|\sum_{q=1}^{N} |h_q|\right|^2 \sigma_s^2}{N N_0}.$$

The performance is worse than that of MRC. However, in terms of implementation, EGC would be easier than MRC as EGC only needs to compensate the phase distortion. For a detailed account of the properties of EGC, the reader is referred to (Schwartz *et al.* 1966) and (Goldsmith 2005).

[2] As the noise is assumed to be Gaussian, this also means that the n_q's are iid.

Example 4.4.3 Suppose that $N = 4$ and $\mathbf{h}$ is given by

$$\mathbf{h} = [2 \; -2 \; j \; -j]^{\mathrm{T}}.$$

Assuming that $\sigma_s^2/N_0 = 1$, we can find the SNR for the following three combining techniques: (i) MRC, (ii) EGC, (iii) dumb combining without any information of $\mathbf{h}$, where $\mathbf{w} = [1 \; 1 \; 1 \; 1]^{\mathrm{T}}$. We have

$$\mathrm{SNR}_{\mathrm{mrc}} = ||\mathbf{h}||^2 = 10;$$

$$\mathrm{SNR}_{\mathrm{egc}} = \frac{|6|^2}{4} = 9.$$

For dumb combining with $\mathbf{w} = [1 \; 1 \; 1 \; 1]^{\mathrm{T}}$, the SNR becomes 0, because $\mathbf{w}^{\mathrm{H}}\mathbf{h} = 0$. This shows that the phase information of channel, $\mathbf{h}$, is important in signal combining.

4.4.4 Generalized selection diversity combining

Selection diversity (SD) is also a popular technique to exploit the spatial diversity in wireless communications. While all the received signals are combined to maximize the SNR in MRC, only the strongest signal is chosen among N received signals in SD. Thus, there is no need to combine N signals in the SD and the implementation cost would be lower as only one signal among N signals is processed. The resulting SNR is given by

$$\gamma_{\mathrm{sd}} = \max\{\gamma_1, \gamma_2, \ldots, \gamma_N\},$$

where $\gamma_q = |h_q|^2\sigma_s^2/N_0$.

To improve the performance of SD, a generalization of SD is proposed (Kong & Milstein 1999). Generalized SD combining (GSDC) is to combine M signals out of N received signals. If $M = N$, this approach becomes MRC (for the best performance), while it becomes SD if $M = 1$ (for the simplest implementation). Thus, without losing a full diversity gain (the definition of diversity gain is discussed later), GSDC enjoys the tradeoff between the performance and implementation complexity. If the M signals are chosen under the MSNR criterion, the resulting SNR becomes

$$\gamma_{\mathrm{gsdc}} = \sum_{q=1}^{M} \gamma_{(q)},$$

where $\gamma_{(q)}$ denotes the qth highest SNR among the γ_q's.

4.5 MMSE combining with unknown channel vector

So far, we have assumed that the channel vector, $\mathbf{h}$, is known for signal combining. In some applications, however, $\mathbf{h}$ may not be known and can be considered as a random vector. For example, if the channel vector is to be estimated, we can assume that the (estimated) channel vector is random due to estimation error. In this section, we discuss an extension of the MMSE combining approach for a random channel vector.

Denote by $\mathcal{E}[\mathbf{h}_l] = \bar{\mathbf{h}}_l$ and $\mathcal{E}[(\mathbf{h}_l - \bar{\mathbf{h}}_l)(\mathbf{h}_l - \bar{\mathbf{h}}_l)^{\mathrm{H}}] = \mathbf{C}_l$ the mean vector and covariance matrix of the lth channel vector, $\mathbf{h}_l$, respectively, and assume that they are known. If $\mathbf{h}_l$ is to be estimated, the mean of $\mathbf{h}_l$ can be replaced by its estimate, while the covariance matrix, $\mathbf{C}_l$, is replaced by the estimation error covariance matrix. For convenience, let

$$\mathbf{R}_l = \mathcal{E}[\mathbf{h}_l \mathbf{h}_l^{\mathrm{H}}] = \mathbf{C}_l + \bar{\mathbf{h}}_l \bar{\mathbf{h}}_l^{\mathrm{H}}.$$

In this case, the MMSE combining vector is given by

$$\begin{aligned} \mathbf{w}_{\mathrm{mmse},l} &= \mathcal{E}[|s_l - \mathbf{w}_l^{\mathrm{H}} \mathbf{r}_l|^2] \\ &= (\mathbf{R}_l \sigma_s^2 + \mathbf{R}_{\mathbf{n}})^{-1} \bar{\mathbf{h}}_l \sigma_s^2. \end{aligned} \tag{4.42}$$

If $\bar{\mathbf{h}}_l = \mathbf{0}$, the MMSE combining vector becomes $\mathbf{0}$. Unfortunately, this means that MMSE combining fails. Thus, to avoid this problem, the MMSE combining approach has to be modified (the modification can be different depending on applications). We consider an example for the modification of MMSE combining as follows.

Suppose that the channel vector is given by $\mathbf{h}_l = \mathrm{e}^{j\phi_l}\mathbf{h}$, where ϕ_l is a random phase and $\mathbf{h}$ is a non-zero constant vector. If ϕ_l is uniformly distributed, $\bar{\mathbf{h}}_l = \mathbf{0}$ although $\mathbf{h}$ is not zero. This is the case where the amplitude of the channel gain remains constant, but the common phase is time-varying. If the phase change is slow (i.e., $\phi_l \simeq \phi_{l+1}$), robust signaling techniques, including differential phase shift keying (DPSK), can be used so that the receiver can detect signals without knowing or estimating phase. Consider the received signal that is given by

$$\begin{aligned} \mathbf{r}_l &= \mathrm{e}^{j\phi_l} \mathbf{h} s_l + \mathbf{n}_l \\ &= \mathbf{h} \mathrm{e}^{j\phi_l} s_l + \mathbf{n}_l. \end{aligned} \tag{4.43}$$

Taking $x_l = \mathrm{e}^{j\phi_l} s_l$ as a new desired signal, the MMSE combining vector is given by

$$\begin{aligned} \mathbf{w}_{\mathrm{mmse},l} &= \mathbf{w}_{\mathrm{mmse}} \\ &= \arg\min_{\mathbf{w}} \mathcal{E}[|\mathrm{e}^{j\phi_l} s_l - \mathbf{w}^{\mathrm{H}} \mathbf{r}_l|^2] \\ &= (\mathbf{h}\mathbf{h}^{\mathrm{H}} \sigma_s^2 + \mathbf{R}_{\mathbf{n}})^{-1} \mathbf{h} \sigma_s^2. \end{aligned} \tag{4.44}$$

It is noteworthy that the MMSE combining vector is time-invariant. The output of the MMSE combiner is given by

$$\begin{aligned} \hat{x}_l &= \mathbf{w}_{\mathrm{mmse}}^{\mathrm{H}} \mathbf{r} \\ &= \mathbf{w}_{\mathrm{mmse}}^{\mathrm{H}} \mathbf{h} x_l + \mathbf{w}_{\mathrm{mmse}}^{\mathrm{H}} \mathbf{n}_l. \end{aligned} \tag{4.45}$$

Now, the detection can be carried out with $\hat{x}_l$.

Example 4.5.1 Suppose that s_l is encoded in the DPSK modulation format as $s_l = \mathrm{e}^{j\pi d_l}$, where d_l is given by

$$d_l = d_l \oplus b_l.$$

Here, $\oplus$ denotes modulo-2 addition and $b_l \in \{0, 1\}$ is the data symbol. From the combined signal in (4.45), the data symbol can be detected by exploiting the phase difference:

$$\begin{aligned} u_l &= \hat{x}_l \hat{x}_{l-1}^* \\ &= |A|^2 \mathrm{e}^{j(\phi_l - \phi_{l-1}) + \pi(d_l - d_{l-1})} + A x_l z_{l-1}^* + A^* x_{l-1}^* z_l + z_l z_{l-1}^*, \end{aligned}$$

where $A = \mathbf{w}_{\mathrm{mmse}}^{\mathrm{H}} \mathbf{h}$ and $z_l = \mathbf{w}_{\mathrm{mmse}}^{\mathrm{H}} \mathbf{n}_l$. Thus, if $|\phi_l - \phi_{l-1}| \ll \pi$, the phase difference, $\pi(d_l - d_{l-1})$, can be used to detect the signal b_l. Note that MMSE combining can mitigate the background noise, but cannot suppress the phase noise $\phi_l - \phi_{l-1}$. Thus, although the SNR after combining can be improved, the performance is limited by the phase noise.

In general, if the random channel vector can be decomposed or factorized into two different terms – one is random and the other is constant or known – the MMSE combining approach is applicable by modifying the desired signal to include the random term.

4.6 Performance of optimal combining under fading environment

In the context of wireless communications, diversity combining techniques including MRC or SD are used to exploit the spatial diversity gain. It is assumed that each received signal (i.e. r_q in (4.1)) experiences a different channel environment and the resulting gain, h_q, is different for each signal. If the receive antennas are spaced far apart, we can expect that the h_q's are statistically independent. Thus, even though some channel gains are low, the probability that the other channel gains are high could be high. Consequently, the probability that the total SNR is low decreases with the number of antennas and a more reliable communication quality is expected.

The performance measure (SNR or MSE) depends on $\mathbf{h}$ as it is a function of $\mathbf{h}$. To characterize an average performance with respect to different realizations of $\mathbf{h}$, we now assume that $\mathbf{h}$ is a random vector. The distribution of $\mathbf{h}$ depends on spatial channel environments. To maximize the spatial diversity gain, it is desirable to have spatially uncorrelated channel gains.

4.6.1 Performance of MRC

In wireless communications, the outage probability is an important performance measure. The outage probability is the probability that the SNR is lower than a certain threshold SNR:

$$P_{\text{out}}(\Gamma) = \Pr(\gamma \leq \Gamma), \tag{4.46}$$

where Γ is the threshold SNR. If the SNR, γ, is higher than Γ, we assume that a reliable communication quality is available. Otherwise, signals are weak and they are not detectable or decodable (i.e. outage happens). Therefore, a low outage probability is desirable.

As shown in (4.46), the outage probability is a function of the threshold SNR. As the threshold SNR increases, the outage probability also increases. The threshold SNR can be considered the required SNR to keep a certain reliable communication quality. Therefore, the threshold SNR depends on the modulation scheme, code, and other parameters.

The SNR of the MRC in (4.40) is a sum of N random variables. In wireless communications, it is often assumed that h_q is a CSCG random variable. Then, we can show that $|h_q|$ is Rayleigh distributed and $|h_q|^2$ becomes a chi-square random variable with two degrees of freedom or exponential random variable. Let $\gamma_q = |h_q|^2\sigma_s^2/N_0$. If $\mathcal{E}[|h_q|^2] = \sigma_h^2$, the pdf of γ_q is given by

$$f(\gamma_q) = \frac{1}{\bar{\gamma}_q} \mathrm{e}^{-\frac{\gamma_q}{\bar{\gamma}_q}}, \tag{4.47}$$

where $\bar{\gamma}_q = \mathcal{E}[\gamma_q] = \sigma_h^2\sigma_s^2/N_0$. The moment generating function (mgf) of γ_q is given by

$$\begin{aligned}\mathcal{E}[\mathrm{e}^{s\gamma_q}] &= \int_0^\infty \mathrm{e}^{s\gamma_q} \frac{1}{\bar{\gamma}_q} \mathrm{e}^{-\frac{\gamma_q}{\bar{\gamma}_q}} \, \mathrm{d}\gamma_q \\ &= \frac{1}{1 - s\bar{\gamma}_q}.\end{aligned}$$

Since $\mathcal{E}[\mathrm{e}^{-s\gamma_q}]$ can be seen as the Laplace transform of the pdf of γ_q, once the mgf is available, the pdf can be found using the inverse Laplace transform.

If the γ_q's are assumed to be independent, the mgf of $\gamma = \sum_{q=1}^N \gamma_q$ becomes

$$\begin{aligned}\mathcal{E}[\mathrm{e}^{s\gamma}] &= \mathcal{E}\left[\prod_{q=1}^N \mathrm{e}^{s\gamma_q}\right] \\ &= \prod_{q=1}^N \frac{1}{1 - s\bar{\gamma}_q}.\end{aligned} \tag{4.48}$$

If we assume that $\bar{\gamma}_q = \bar{\gamma}, q = 1, 2, \ldots, N$, from (4.48), it can be shown that

$$\mathcal{E}[\mathrm{e}^{-s\gamma}] = \frac{1}{(1 + s\bar{\gamma})^N}.$$

Using the inverse Laplace transform, the pdf of γ can be found and given by

$$f(\gamma) = \frac{1}{(N-1)!\bar{\gamma}^N} \gamma^{N-1} \mathrm{e}^{-\frac{\gamma}{\bar{\gamma}}}. \tag{4.49}$$

This is a special case of the Gamma distribution. The pdf of the Gamma distribution is given by

$$f(x; k, \theta) = \frac{1}{\Gamma(k)\theta^k} x^{k-1} e^{-\frac{x}{\theta}}, \; x \geq 0, \; \theta > 0, \tag{4.50}$$

which is often denoted by $\Gamma(k, \theta)$. If $k = n/2$ and $\theta = 2$, the Gamma distribution becomes the chi-square distribution[3] with n degrees of freedom.

From (4.49), we can find the outage probability with the threshold SNR, Γ, as follows:

$$\begin{aligned} P_{\text{out}}(\Gamma) &= \int_0^{\Gamma} \frac{1}{(N-1)!\bar{\gamma}^N} \gamma^{N-1} e^{-\frac{\gamma}{\bar{\gamma}}} d\gamma \\ &= \frac{1}{(N-1)!\bar{\gamma}^N} \int_0^{\Gamma} \gamma^{N-1} e^{-\frac{\gamma}{\bar{\gamma}}} d\gamma. \end{aligned} \tag{4.51}$$

In order to have a closed-form expression, we can use the incomplete gamma function. The incomplete gamma function is defined as

$$\gamma(a, x) = \int_0^x t^{a-1} e^{-t} dt. \tag{4.52}$$

If a is an (positive) integer, $\gamma(a, x)$ has the following closed-form expression:

$$\gamma(a, x) = \Gamma(a) - \Gamma(a) e^{-x} \sum_{k=0}^{a-1} \frac{x^k}{k!}, \tag{4.53}$$

where $\Gamma(a)$ is the gamma function which is $\Gamma(a) = (a-1)!$ for an integer a. Using (4.53), the outage probability in (4.51) can be expressed as

$$P_{\text{out}}(\Gamma) = 1 - e^{-\frac{\Gamma}{\bar{\gamma}}} \sum_{k=0}^{N-1} \frac{(\Gamma/\bar{\gamma})^k}{k!}. \tag{4.54}$$

It is noteworthy that the outage probability is a cdf. Since

$$e^x = \sum_{k=0}^{\infty} \frac{x^k}{k!},$$

it can be shown that

$$\lim_{N \to \infty} \sum_{k=0}^{N-1} \frac{(\Gamma/\bar{\gamma})^k}{k!} = e^{\frac{\Gamma}{\bar{\gamma}}}.$$

Using this, we can show that

$$\begin{aligned} \lim_{N \to \infty} P_{\text{out}}(\Gamma) &= \lim_{N \to \infty} 1 - e^{-\frac{\Gamma}{\bar{\gamma}}} \sum_{k=0}^{N-1} \frac{(\Gamma/\bar{\gamma})^k}{k!} \\ &= 1 - e^{-\frac{\Gamma}{\bar{\gamma}}} e^{\frac{\Gamma}{\bar{\gamma}}} \\ &= 0. \end{aligned} \tag{4.55}$$

[3] Throughout this book, if k is an integer, the Gamma distribution is also called the chi-square distribution.

This shows that if the dimension of the signal space or the number of receive antennas approaches infinity, an absolute reliability of communications over fading channels, where the outage probability becomes zero for any threshold SNR, can be achieved.

The outage probability can also be expressed as

$$\begin{aligned} P_{\text{out}}(\Gamma) &= 1 - \mathrm{e}^{-\frac{\Gamma}{\bar{\gamma}}}\left(\mathrm{e}^{\frac{\Gamma}{\bar{\gamma}}} - \sum_{k=N}^{\infty}\frac{(\Gamma/\bar{\gamma})^k}{k!}\right) \\ &= \mathrm{e}^{-\frac{\Gamma}{\bar{\gamma}}}\sum_{k=N}^{\infty}\frac{(\Gamma/\bar{\gamma})^k}{k!} \\ &= \mathrm{e}^{-\Gamma/\bar{\gamma}}\left(\frac{(\Gamma/\bar{\gamma})^N}{N!} + O\left((\Gamma/\bar{\gamma})^{N+1}\right)\right). \end{aligned} \tag{4.56}$$

Thus, if $\Gamma \ll \bar{\gamma}$,

$$P_{\text{out}}(\Gamma) \simeq \frac{1}{N!}\left(\frac{\Gamma}{\bar{\gamma}}\right)^N. \tag{4.57}$$

It is also possible to derive this approximate outage probability by another (heuristic) approach. If $\gamma \ll \Gamma$ in (4.49), the pdf is approximated by

$$f(\gamma) \simeq \frac{1}{(N-1)!\bar{\gamma}^N}\gamma^{N-1}. \tag{4.58}$$

Then, the outage probability is approximated as

$$\begin{aligned} P_{\text{out}}(\Gamma) &\simeq \int_0^{\Gamma}\frac{1}{(N-1)!\bar{\gamma}^N}\gamma^{N-1}\mathrm{d}\gamma \\ &= \frac{1}{N!}\left(\frac{\Gamma}{\bar{\gamma}}\right)^N. \end{aligned}$$

Thus, for a large $\bar{\gamma}$, we have

$$P_{\text{out}}(\Gamma) \propto \bar{\gamma}^{-N}.$$

This shows that the outage probability decreases with $\bar{\gamma}$ more rapidly as N increases. This is confirmed by the outage probability curves in Fig. 4.6, where the outage probability for various values of N is shown when the SNR threshold, Γ, is $= 10$ dB. In this figure, the mean SNR on the x-axis is $\bar{\gamma} = \sigma_h^2\sigma_s^2/N_0$.

It is shown in Fig. 4.6 that the negative slope of the outage probability in logarithm scales is proportional to N. As N increases, the slope becomes steeper. This slope is related to the diversity gain or diversity order that is defined as

$$\text{Diversity gain} = -\lim_{\bar{\gamma}\to\infty}\frac{\log_2 \Pr(\gamma < \Gamma)}{\log_2 \bar{\gamma}}. \tag{4.59}$$

From the approximate outage probability in (4.57), we can show that the diversity gain of MRC is N. If we let $d =$ diversity gain, (4.59) implies

$$\Pr(\gamma < \Gamma) \doteq \bar{\gamma}^{-d},\ \bar{\gamma} \gg 1,$$

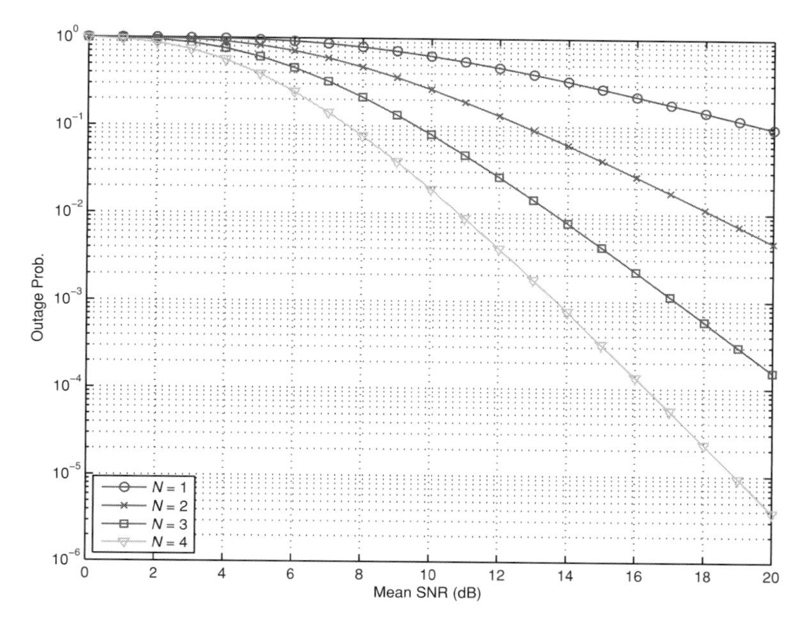

Figure 4.6 Outage probability of MRC when $\Gamma = 10$ dB.

where $\doteq$ denotes the *exponential equality*: if

$$\lim_{\bar{\gamma}\to\infty} \frac{\log_2 g(\bar{\gamma})}{\log_2 \bar{\gamma}} = b,$$

we write $g(\bar{\gamma}) \doteq \bar{\gamma}^b$. This definition of diversity gain is based on (Zheng & Tse 2003), in which the diversity gain is defined with the information outage probability.

The information outage probability is the probability that the channel capacity is less than the transmission rate, R:

$$P_{\text{out}}(R) = \Pr(\log_2(1+\gamma) < R). \tag{4.60}$$

Note that the channel capacity $\log_2(1+\gamma)$ is a random variable if γ is random. Thus, for a fixed transmission rate, the signal can be successfully decoded if the channel capacity is greater than the rate. On the other hand, the signal cannot be decoded if the channel capacity is lower than the rate. We can show that

$$P_{\text{out}}(R) = \Pr(\gamma < 2^R - 1).$$

Therefore, if $\Gamma = 2^R - 1$, the information outage probability is identical to the outage probability in (4.46). For a fixed R or Γ, we can show that the diversity order in (4.59) is N.

4.6.2 Performance of SD

The outage probability of SD can be obtained using order statistics. The probability that γ_{sd} is less than the threshold SNR, Γ, is identical to the probability that all the γ_q's are

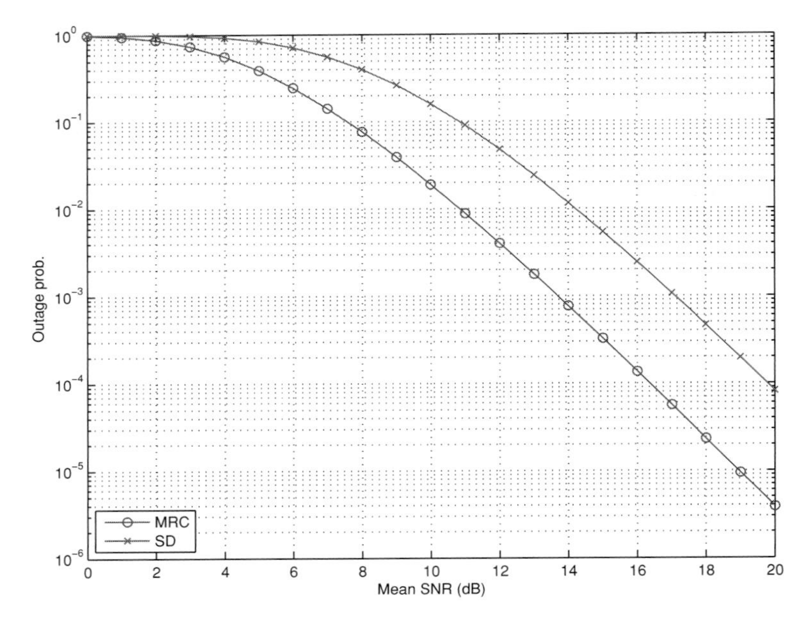

Figure 4.7 Outage probability of MRC and SD when $\Gamma = 10$ dB when $N = 4$.

less than Γ. That is,

$$\begin{aligned} P_{\text{out}}(\Gamma) &= \Pr(\gamma_{\text{sd}} < \Gamma) \\ &= \Pr(\max_q \gamma_q < \Gamma) \\ &= \Pr(\gamma_1 < \Gamma, \gamma_2 < \Gamma, \ldots, \gamma_N < \Gamma) \\ &= \prod_{q=1}^{N} \Pr(\gamma_q < \Gamma), \end{aligned} \tag{4.61}$$

where the last equality is valid if the γ_q's are independent. If the pdf of γ_q is given in (4.47), from (4.61), the outage probability of SD becomes

$$P_{\text{out}}(\Gamma) = \left(1 - e^{-\frac{\Gamma}{\bar{\gamma}}}\right)^N. \tag{4.62}$$

Again, if $\bar{\gamma} \gg \Gamma$, we have

$$P_{\text{out}}(\Gamma) \simeq \left(\frac{\Gamma}{\bar{\gamma}}\right)^N. \tag{4.63}$$

This shows that the diversity gain of SD is N, which is the same as that of MRC. However, compared with the equivalent approximate outage probability of MRC in (4.58), we can see that the MRC gain exceeds that for the SD by a factor of $(N!)^{1/N}$.

Figure 4.7 shows the outage probabilities of MRC and SD for comparison purposes. It is shown that the diversity order is the same (the slopes of the two outage probabilities are the same), but the performance of MRC is better than that of SD.

4.6.3 Performance over spatially correlated channels

If the h_q's are independent, the diversity order becomes N, which is the number of receive antennas. This is the case where the channel vector is spatially uncorrelated. If the channel vector is correlated, the diversity order can be less than N. To see this case, let us assume that $\mathbf{h}$ is a spatially correlated random vector. In particular, we assume that $\mathbf{h} \sim \mathcal{CN}(0, \mathbf{R_h})$.

Consider the eigen-decomposition of $\mathbf{R_h}$ as follows:

$$\mathbf{R_h} = \sum_{n=1}^{N} \lambda_n \mathbf{a}_n \mathbf{a}_n^{\mathrm{H}}, \tag{4.64}$$

where λ_n represents the nth-largest eigenvalue and $\mathbf{a}_n$ denotes its corresponding eigenvector. When $\mathbf{h}$ is spatially correlated, some eigenvalues could be zero. Thus, we assume that there are Q ($\leq N$) non-zero eigenvalues. In this case, $\mathbf{h}$ can be represented as a weighted sum of the eigenvectors corresponding to non-zero eigenvalues:

$$\begin{aligned} \mathbf{h} &= \sum_{q=1}^{Q} \mathbf{a}_q c_q \\ &= \mathbf{Ac}, \end{aligned} \tag{4.65}$$

where $\mathbf{A} = [\mathbf{a}_1\ \mathbf{a}_2\ \ldots\ \mathbf{a}_Q]$ and $\mathbf{c} = [c_1\ c_2\ \ldots\ c_Q]^{\mathrm{T}}$. As $\mathbf{h}$ is Gaussian, $\mathbf{c}$ is also Gaussian and

$$\mathcal{E}[\mathbf{cc}^{\mathrm{H}}] = \mathrm{Diag}(\lambda_1, \lambda_2, \ldots, \lambda_Q).$$

This implies that the c_q's are independent and $c_q \sim \mathcal{CN}(0, \lambda_q)$, $q = 1, 2, \ldots, Q$.

From (4.65), the SNR of MRC is given by

$$\begin{aligned} \gamma &= \sum_{q=1}^{N} \frac{|h_q|^2 \sigma_s^2}{N_0} \\ &= \frac{\sigma_s^2}{N_0} ||\mathbf{h}||^2 \\ &= \frac{\sigma_s^2}{N_0} ||\mathbf{c}||^2 \\ &= \sum_{q=1}^{Q} \gamma_q, \end{aligned} \tag{4.66}$$

where γ_q is re-defined as

$$\gamma_q = \frac{\sigma_s^2 |c_q|^2}{N_0}.$$

Using the mgf of γ_q, the pdf of γ can be found. The pdf of γ_q is

$$f(\gamma_q) = \frac{1}{\bar{\gamma}_q} \mathrm{e}^{-\frac{\gamma_q}{\bar{\gamma}_q}},$$

where $\bar{\gamma}_q = \mathcal{E}[\gamma_q] = \sigma_s^2 \lambda_q / N_0$. Thus,

$$\mathcal{E}[\mathrm{e}^{-s\gamma}] = \prod_{q=1}^{Q} \frac{1}{1 + s\bar{\gamma}_q}. \quad (4.67)$$

In (Proakis 1995), the distribution is derived as

$$f(\gamma) = \sum_{q=1}^{Q} \frac{\pi_q}{\bar{\gamma}_q} \mathrm{e}^{-\frac{\gamma}{\bar{\gamma}_q}},$$

where

$$\pi_q = \prod_{k=1, k \neq q}^{Q} \frac{\bar{\gamma}_q}{\bar{\gamma}_q - \bar{\gamma}_k}.$$

The outage probability becomes

$$P_{\text{out}}(\Gamma) = \sum_{q=1}^{Q} \pi_q \left(1 - \mathrm{e}^{-\frac{\Gamma}{\bar{\gamma}_q}}\right). \quad (4.68)$$

We can show that the diversity gain is Q. Thus, when $\mathbf{h}$ is spatially correlated, the diversity order is decided by the number of non-zero eigenvalues rather than the number of receive antennas.

Example 4.6.1 Suppose that $Q = 1$. In this case, $\mathbf{h} = \mathbf{a}_1 c_1$ (this is the case where the channel vector is perfectly correlated). Then, the pdf becomes

$$f(\gamma) = \frac{1}{\bar{\gamma}_1} \mathrm{e}^{-\frac{\gamma}{\bar{\gamma}_1}}$$

and the outage probability is given by

$$P_{\text{out}}(\Gamma) = 1 - \mathrm{e}^{-\frac{\Gamma}{\bar{\gamma}_1}}.$$

For a large $\bar{\gamma}_1$,

$$P_{\text{out}}(\Gamma) \simeq \frac{\Gamma}{\bar{\gamma}_1},$$

This shows that the diversity order becomes 1 if $Q = 1$ or $\mathbf{h}$ is perfectly correlated. Figure 4.8 shows the outage probabilities when $\mathbf{h}$ is correlated and uncorrelated. For fair comparison, the mean SNR is set to be the same for both the cases (i.e., $\mathcal{E}[\gamma] = \mathcal{E}[||\mathbf{h}||^2]\sigma_s^2/N_0$ is the same for both the cases). It is shown that the MRC for spatially uncorrelated channels can provide a much lower outage probability for a sufficiently high average SNR. However, as the average SNR decreases, the MRC for spatially correlated channels provides lower outage probability. This behavior results from a large variation of the SNR when $\mathbf{h}$ is spatially correlated.

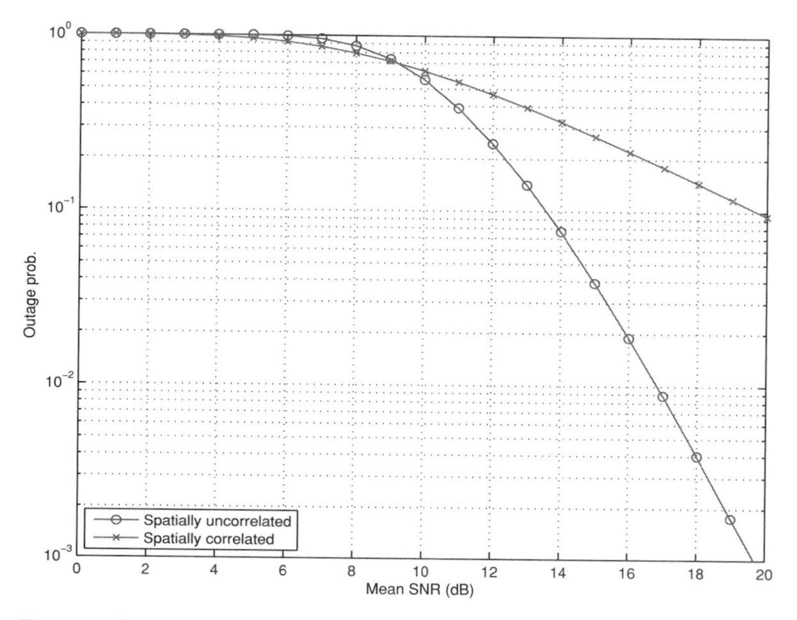

Figure 4.8 Outage probabilities of MRC when $\mathbf{h}$ is correlated and uncorrelated (threshold SNR, $\Gamma = 16$ dB).

4.7 Summary and notes

We presented various linear signal combining techniques in this chapter when multiple observations are available. It was shown that the linear MMSE combiner is optimal in various cases (in particular, when signals and noise are jointly Gaussian). Performance issues were discussed when signal combiners are used for a receiver equipped with multiple receive antennas in wireless communications. As key performance measures over fading channels, the outage probability and diversity gain have been defined and discussed.

Antenna diversity techniques are crucial in wireless communications. In (Schwartz *et al.* 1966) and (Lee 1982), various approaches including MRC and EGC are discussed with their performance analysis. For a detailed account of general diversity techniques and performance analysis, the reader is referred to (Goldsmith 2005) and (Tse & Viswanath 2005).

Problems

Problem 4.1 Show that the linear MMSE estimate, $\hat{s}_{\text{mmse}} = \mathbf{w}_{\text{mmse}}^{\text{H}}\mathbf{r}$ is biased.

Problem 4.2 In MRC, it is usually assumed that the noise is spatially white, i.e., $\mathcal{E}[\mathbf{n}\mathbf{n}^{\text{H}}] = N_0\mathbf{I}$. If the noise is not spatially white, we can first apply noise-whitening filtering to the received signal vector and then apply MRC. The noise-whitening filtering to $\mathbf{r}$ results in the following signal vector:

$$\mathbf{z} = \mathbf{G}^{-1}\mathbf{r},$$

where $\mathbf{G}$ is the Cholesky factor of $\mathbf{R_n}$ when $\mathbf{R_n} = \mathbf{G}\mathbf{G}^{\mathrm{H}}$. Then, MRC is applied to $\mathbf{z}$ to estimate s. Show that the resulting combining is identical to MSNR combining in (4.33).

Problem 4.3 Derive (4.42).

Problem 4.4 Suppose that

$$\mathbf{h} = \bar{\mathbf{h}} + \sqrt{\alpha}\mathbf{e},$$

where $\alpha > 0$ is a known constant and $\mathbf{e} \sim \mathcal{CN}(0, \mathbf{I})$ is an unknown random vector, while $\bar{\mathbf{h}}$ is known. Derive the linear MMSE combining vector with known $\bar{\mathbf{h}}$. In addition, find the SNR after MMSE combining with $\bar{\mathbf{h}}$ and compare this SNR with the average SNR after MMSE combining when $\mathbf{h}$ is perfectly known.

Problem 4.5 We would like to design a receiver with diversity combining techniques. The target outage probability is 0.1. It is known that the average SNR for each antenna is $\bar{\gamma} = 10$ dB. Find the smallest number of receive antennas to meet the target outage probability when MRC is used.

Problem 4.6 Repeat Problem 4.5 when the SD is used.

Problem 4.7 Consider the received signal that is given by

$$\mathbf{r} = \mathbf{h}s + \mathbf{n},$$

where $\mathbf{n} \sim \mathcal{CN}(0, N_0\mathbf{I})$ and $s \in \{-A, A\}$. Assume that s is equally likely (i.e. $\Pr(s = -A) = \Pr(s = A) = \frac{1}{2}$).

(i) Suppose that the ML detector is employed to detect s from $\mathbf{r}$. For a given $\mathbf{h}$, find the error probability, denoted by $P_{\text{err}}(\mathbf{h})$.

(ii) Assume that $\mathbf{h} \sim \mathcal{CN}(0, \mathbf{I})$ and let $\bar{\gamma} = A^2/N_0$. Show that ML combining is equivalent to MRC and find the outage probability with the threshold SNR, Γ.

(iii) Under the same assumption in (ii), show that the average error probability is upper-bounded as

$$\begin{aligned}\bar{P}_{\text{err}} &= \mathcal{E}[P_{\text{err}}(\mathbf{h})] \\ &\le e^{-\Gamma}(1 - P_{\text{out}}(\Gamma)) + P_{\text{out}}(\Gamma),\end{aligned} \tag{4.69}$$

where $P_{\text{out}}(\Gamma)$ denotes the outage probability with the threshold SNR, Γ. For a fixed Γ, show that the average error probability is given by

$$\bar{P}_{\text{err}} \doteq \bar{\gamma}^{-N}, \; \bar{\gamma} \gg 1.$$

Problem 4.8 Suppose that we employ a re-transmission scheme to improve the reliability of transmission over fading channels. It is assumed that a codeword is divided into multiple sub-codewords and each sub-codeword is transmitted over an independent fading channel. This re-transmission scheme is called an incremental redundancy re-transmission diversity scheme. If there are N sub-codewords, the total mutual information becomes

$$\mathsf{I}_N = \sum_{n=1}^{N} \log_2(1 + \gamma_n),$$

where γ_n is the channel SNR for the nth re-transmission. Let $\bar{R}$ denote the transmission rate for each sub-codeword. The throughput of the re-transmission scheme with N re-transmissions can be defined as

$$\begin{aligned} T_N(\bar{R}) &= N\bar{R} \times (1 - P_{\text{out}}) + 0 \times P_{\text{out}} \\ &= N\bar{R}\,\Pr(\mathsf{I}_N \geq N\bar{R}), \end{aligned} \tag{4.70}$$

where P_{out} is the outage probability after N re-transmissions. Show that the normalized throughput, which is defined as $\frac{1}{N}T_N(\bar{R})$, becomes $\mathcal{E}[\log_2(1+\gamma_n)]$ as $N \to \infty$ if γ_n is iid using the strong law of large numbers (SLLN). That is,

$$\max_{\bar{R}} \lim_{N\to\infty} \frac{1}{N} T_N(\bar{R}) = \mathcal{E}[\log_2(1+\gamma_n)].$$

Problem 4.9 We consider again Problem 4.8 with a different re-transmission scheme. It is assumed that the same codeword is repeatedly transmitted over an independent fading channel. This scheme is called a re-transmission diversity scheme. Show that the normalized throughput is now given by

$$\max_{\bar{R}} \lim_{N\to\infty} \frac{1}{N} T_N(\bar{R}) = \log_2(1 + \mathcal{E}[\gamma_n]).$$

Note that using Jensen's inequality, it can be shown that the throughput of the incremental redundancy re-transmission diversity scheme is higher than that of the re-transmission diversity scheme.

5 Array signal processing and smart antenna

Array signal processing or array processing has diverse applications including radar and sonar systems. An array of multiple sensors can provide spatial selectivity in receiving signals. In general, the more sensors, the better spatial selectivity can be achieved. Exploiting this spatial selectivity, we can have a clearer signal or less performance degradation due to interfering signals. An important application of array processing in cellular communication systems is smart antennas. The base stations equipped with antenna arrays for spatial selectivity can be considered as smart antenna systems. In cellular systems, since the performance is degraded by the inter-cell interference (which is the signal transmitted from adjacent cells where the same frequency band is used for communications), it is important to mitigate inter-cell interference. Smart antennas can mitigate inter-cell interference using the spatial selectivity and improve the performance of cellular systems.

In this chapter, we focus on signal combining and related techniques for array processing and discuss how array processing can be applied to smart antenna systems.

5.1 Antenna arrays

An antenna array is an array of antenna elements that allows spatial processing, which is also called array processing. For array processing, we need to take into account array configuration and spatial characteristics of signals. Array processing can be extended in both space and time domains. In this case, spatial and temporal processing is to be jointly performed. Array processing can be considered for both signal reception and transmission. Throughout this chapter, we mainly focus on antenna arrays for signal reception.

5.1.1 Plane wave model

Consider a receiver system equipped with an antenna array. We assume free space propagation, which is an ideal condition and not applicable to practical situations (e.g., terrestrial wireless communications). However, it is helpful in understanding the role of antenna arrays in terms of spatial selectivity.

As shown in Fig. 5.1, suppose that there are point sources that transmit signals uniformly to all directions (i.e. omni-directional transmission) and we are interested

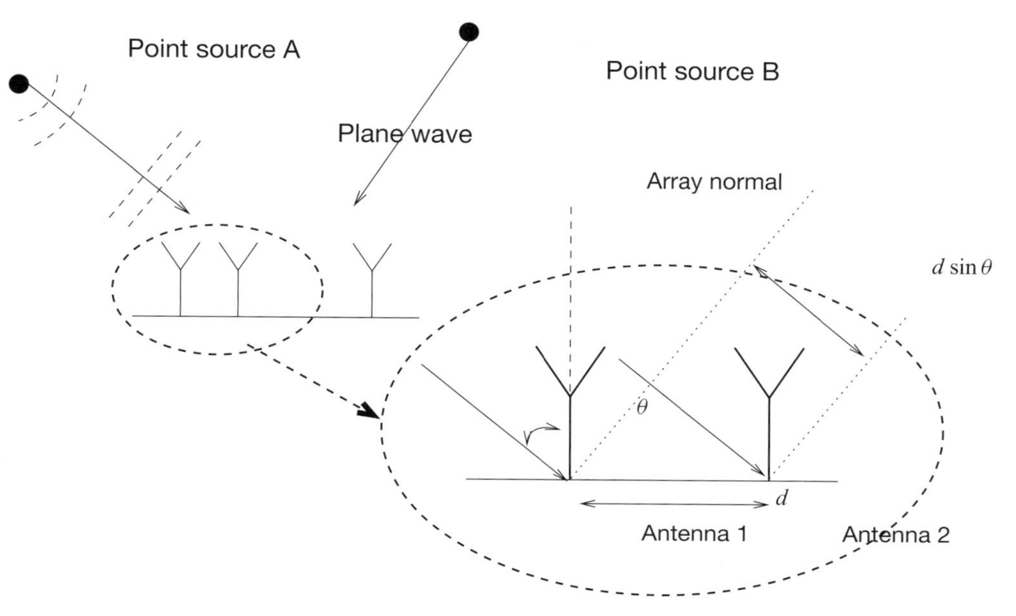

Figure 5.1 Array reception model (a passive array receiving signals from point sources).

in a particular signal source, say source A. If the distance between the signal source (i.e., transmitter) and antenna array is sufficiently long, the propagated signal from an omni-directional signal source becomes a plane wave as shown in Fig. 5.1. In general, the plane wave assumption simplifies the mathematical modeling for impinging signals on an antenna array.

Using the plane wave model, we can define the angle of arrival (AoA) of a signal. Consider an array consisting of two antenna elements. The plane wave from a signal source arrives at antenna elements in the array at different times. Denote by $r(t)$ the received signal at the reference antenna element, which is antenna 1, without any background noise. The received signals are given by

$$r_1(t) = r(t);$$
$$r_2(t) = r(t - \tau),$$

where $r_q(t)$ is the received signal at antenna q and τ is the inter-arrival time difference (simply delay). According to the example shown in Fig. 5.1, the signal travels longer to arrive at antenna 2 and the inter-arrival time difference between the two antenna elements, say τ, can be obtained from the following relation:

$$d \sin\theta = c\tau,$$

where c is the propagation speed (i.e. the speed of light for radio signal transmission), d is the antenna spacing, and θ is the AoA of the impinging signal on the array with respect to the array normal. In other words, the delay becomes

$$\tau = \frac{d}{c} \sin \theta.$$

An inter-arrival time difference can result in a phase difference between the received signals at different antenna elements when the transmitted signal is modulated by a carrier of high frequency. Suppose that

$$r(t) = s(t) \cos(2\pi f_c t + \phi(t)),$$

where f_c and $\phi(t)$ are the carrier frequency and (time-variant) phase of the carrier, respectively, and $s(t)$ is a baseband signal. Then, the received signal with a time delay of τ is given by

$$r(t - \tau) = s(t - \tau) \cos(2\pi f_c(t - \tau) + \phi(t - \tau)).$$

If τ is sufficiently small, we can have

$$\begin{aligned} s(t - \tau) &\simeq s(t); \\ \phi(t - \tau) &\simeq \phi(t). \end{aligned}$$

Using these approximations, $r(t - \tau)$ is approximated as

$$r(t - \tau) \simeq s(t) \cos(2\pi f_c t + \phi(t) - 2\pi f_c \tau), \tag{5.1}$$

where we can see that a phase change of $-2\pi f_c \tau$ is introduced by a time delay of τ. Using complex envelope signal notations, let

$$\tilde{r}(t) = a(t) \mathrm{e}^{j 2\pi f_c t},$$

where $a(t) = s(t) e^{j\phi(t)}$ is the complex baseband signal. It is readily shown that $r(t) = \Re(\tilde{r}(t))$ and

$$\tilde{r}(t - \tau) \simeq \tilde{r}(t) \mathrm{e}^{-j 2\pi f_c \tau}.$$

This clearly shows that a time delay becomes a phase change when the signal is modulated by a carrier of high frequency. Note that although τ is small, $f_c \tau$ may not be small as the carrier frequency, f_c, is high.

Example 5.1.1 To consider the approximation in (5.1), suppose that $s(t) = A \cos(2\pi f_m t)$, where $A > 0$ is the amplitude and f_m is the maximum frequency of the baseband signal. Furthermore, assume that $f_c = 1\mathrm{GHz}$ and $f_m = 1\mathrm{MHz}$. Since the inter-arrival time delay is bounded as

$$|\tau| = \left| \frac{d}{c} \sin \theta \right| \leq \tau_{\max} = \frac{d}{c},$$

if $d = 1$ meter, with the maximum delay, we have

$$f_c \tau_{max} = \frac{10}{3};$$
$$f_m \tau_{max} = \frac{1}{3} \times 10^{-2}$$

and

$$s(t - \tau_{max}) = A\cos(2\pi f_m t)\cos\left(2\pi \frac{10^{-2}}{3}\right) + A\sin(2\pi f_m t)\sin\left(2\pi \frac{10^{-2}}{3}\right).$$

Since $\cos\left(2\pi \frac{10^{-2}}{3}\right) \simeq 1$ and $\sin\left(2\pi \frac{10^{-2}}{3}\right) \simeq 0.02$, we have

$$s(t - \tau_{max}) \simeq A\cos(2\pi f_m t) = s(t).$$

From this, we can see that if $f_m \tau_{max} \ll 1$, the approximation in (5.1) is valid.

5.1.2 Array response vector

In the previous subsection, we only consider two antenna elements. It is straightforward to consider more than two antenna elements. Assume that there are N antennas in an array. Then, the received signal has a vector form and is given by

$$\begin{aligned}\mathbf{r}(t) &= \begin{bmatrix} r_1(t) \\ r_2(t) \\ \vdots \\ r_N(t) \end{bmatrix} \\ &= \begin{bmatrix} r(t) \\ r(t-\tau_1) \\ \vdots \\ r(t-\tau_{N-1}) \end{bmatrix} \\ &\simeq \mathbf{a}r(t), \end{aligned} \tag{5.2}$$

where τ_q is the inter-arrival time difference between the reference antenna (i.e. the first antenna) and the $(q+1)$th antenna and

$$\mathbf{a} = \begin{bmatrix} 1 \\ \exp(-j2\pi f_c \tau_1) \\ \vdots \\ \exp(-j2\pi f_c \tau_{N-1}) \end{bmatrix}.$$

Here, $\mathbf{a}$ is called the array response vector (ARV), which is a function of the AoA, θ, and depends on array configuration.

Example 5.1.2 **Uniform linear array (ULA)**: Suppose that the antenna elements in an array are uniformly linearly located with spacing d. Then, for a given AoA θ, the inter-arrival time difference at the $(q+1)$th antenna is given by

$$\tau_q = \frac{qd}{c}\sin\theta,\ q = 0, 1, \ldots, N-1.$$

The ARV becomes

$$\mathbf{a} = \mathbf{a}(\theta) = [1\ \mathrm{e}^{-j2\pi\frac{d}{\lambda}\sin\theta}\ \cdots\ \mathrm{e}^{-j2\pi(N-1)\frac{d}{\lambda}\sin\theta}]^{\mathrm{T}}, \tag{5.3}$$

where $\lambda = c/f_{\mathrm{c}}$ is the wavelength of the carrier.

The ULA with half-wavelength spacing is called the *standard* ULA. In this case, the ARV becomes

$$\mathbf{a}(\theta) = [1\ \mathrm{e}^{-j\pi\sin\theta}\ \cdots\ \mathrm{e}^{-j\pi(N-1)\sin\theta}]^{\mathrm{T}}.$$

It can be readily shown that the ARV and AoA has a one-to-one mapping when $-\pi/2 \le \theta < \pi/2$. It is also noteworthy that since $\sin\theta = \sin(\pi-\theta)$, the standard ULA has an ambiguity (i.e. the ARVs are the same for the AoAs, θ and $\pi-\theta$) and this ambiguity exists in other arrays.

The normalized directional gain pattern or beam pattern of an array is defined as

$$B(\theta) = \frac{1}{N}\mathbf{1}^{\mathrm{T}}\mathbf{a}(\theta). \tag{5.4}$$

For a ULA, the normalized beam pattern becomes

$$\begin{aligned} B(\theta) &= \frac{1}{N}\sum_{l=0}^{N-1}\mathrm{e}^{-jl2\pi\frac{d}{\lambda}\sin\theta} \\ &= \frac{\mathrm{e}^{-j(N-1)2\pi\frac{d}{\lambda}u}}{N}\frac{\sin\left(N\frac{d}{\lambda}\pi u\right)}{\sin\left(\frac{d}{\lambda}\pi u\right)}, \end{aligned} \tag{5.5}$$

where $u = \sin\theta$. Note that the normalized beam pattern is uniquely decided only when $-\lambda/2d \le u < \lambda/2d$ or

$$-\sin^{-1}\left(\frac{\lambda}{2d}\right) \le \theta < \sin^{-1}\left(\frac{\lambda}{2d}\right). \tag{5.6}$$

This range of AoA is called the visible region. For the standard ULA, the normalized beam pattern is given by

$$B(\theta) = \frac{\mathrm{e}^{-j(N-1)\pi u}}{N}\frac{\sin\left(N\frac{\pi}{2}u\right)}{\sin\left(\frac{\pi}{2}u\right)}.$$

The visible region for the standard ULA is $-\pi/2 \le \theta < \pi/2$.

Noting that the beam pattern is defined by the normalized inner product of the two ARVs: one is the ARV of AoA 0° and the other ARV of AoA θ, a generalized beam

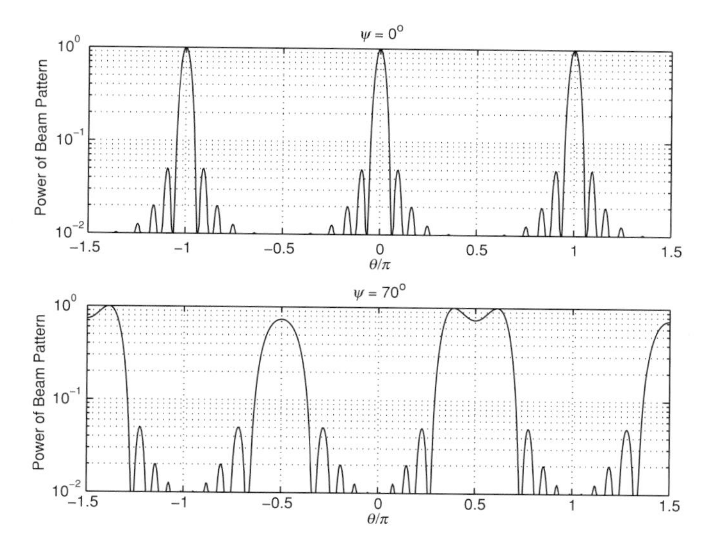

Figure 5.2 Powers of generalized beam patterns of the standard ULA of $N = 10$: (a) $\psi = 0°$; (b) $\psi = 70°$.

pattern of an array can be defined as

$$B(\psi : \theta) = \frac{1}{N} \mathbf{a}^{\mathrm{H}}(\psi)\mathbf{a}(\theta), \tag{5.7}$$

which is the normalized inner product of the two ARVs: one is the ARV of AoA ψ and the other ARV of AoA θ. Thus, $B(0° : \theta) = B(\theta)$. For a ULA, the generalized beam pattern becomes

$$B(\psi : \theta) = \frac{\mathrm{e}^{-j(N-1)2\frac{d}{\lambda}\pi(u(\theta)-u(\psi))}}{N} \frac{\sin\left(N\frac{d}{\lambda}\pi(u(\theta) - u(\psi))\right)}{\sin\left(\frac{d}{\lambda}\pi(u(\theta) - u(\psi))\right)},$$

where $u(\theta) - u(\psi) = \sin\theta - \sin\psi$. The generalized beam pattern can be considered as the array gain of the impinging signal with AoA ψ at look-direction θ. The array gain is usually maximized when the look-direction is identical to the AoA of the impinging signal, i.e. $\theta = \psi$.

Figure 5.2 shows the power of beam pattern, $|B(\psi : \theta)|^2$, of the standard ULA of $N = 10$ with a fixed ψ. It is shown that the beam pattern has a peak when $\theta = \psi$. However, the beamwidth varies depending on ψ.

5.2 AoA estimation

In array processing, it is often required to estimate the AoAs of signals. As a parametric estimation problem, there are various statistical approaches for the AoA estimation, including the maximum likelihood (ML) approach. In this section, we present well-known approaches.

5.2.1 ML approaches

Suppose that there are K independent (narrowband) signals impinging on an array of N antennas. The received signal vector is given by

$$\mathbf{r}(t) = \sum_{k=1}^{K} \mathbf{a}(\theta_k) s_k(t) + \mathbf{n}(t), \tag{5.8}$$

where $s_k(t)$ and θ_k are the kth signal and its AoA, respectively, and $\mathbf{n}(t)$ is the background noise. Suppose that the signals are sampled after filtering. Denote by $s_{k,l}$ the lth sampled signal of $s_k(t)$. In addition, let $n_{q,l}$ denote the lth noise sample of the qth antenna element. We assume that the noise is spatially and temporally white. Furthermore, we assume that $\mathbf{n}_l = [n_{1,l}\ n_{2,l}\ \dots\ n_{N,l}]^{\mathrm{T}}$ is a CSCG random vector, i.e. $\mathbf{n}_l \sim \mathcal{CN}(0, N_0\mathbf{I})$. Let $\mathbf{s}_l = [s_{1,l}\ s_{2,l}\ \dots\ s_{K,l}]^{\mathrm{T}}$. The received signal vector after sampling is given by

$$\mathbf{r}_l = \sum_{k=1}^{K} \mathbf{a}(\theta_k) s_{k,l} + \mathbf{n}_l, \tag{5.9}$$

Provided that $\{\mathbf{s}_l\}$ is unknown, but deterministic, the likelihood function of $\boldsymbol{\theta} = [\theta_1\ \theta_2\ \dots\ \theta_K]^{\mathrm{T}}$ and $\{\mathbf{s}_l\}$ is given by

$$\begin{aligned} f(\{\mathbf{r}_l\}|\boldsymbol{\theta}, \{\mathbf{s}_l\}) &= \prod_{l=0}^{L-1} f(\mathbf{r}_l|\boldsymbol{\theta}, \mathbf{s}_l) \\ &= \prod_{l=0}^{L-1} \frac{1}{(\pi N_0)^N} \exp\left(-\frac{1}{N_0}||\mathbf{r}_l - \sum_{k=1}^{K} \mathbf{a}(\theta_k) s_{k,l}||^2\right) \\ &= \frac{1}{(\pi N_0)^{NL}} \exp\left(-\frac{1}{N_0}\sum_{l=0}^{L-1} ||\mathbf{r}_l - \mathbf{A}(\boldsymbol{\theta})\mathbf{s}_l||^2\right), \end{aligned} \tag{5.10}$$

where

$$\mathbf{A}(\boldsymbol{\theta}) = [\mathbf{a}(\theta_1)\ \mathbf{a}(\theta_2)\ \dots\ \mathbf{a}(\theta_K)].$$

Let

$$\begin{aligned} \mathbf{R} &= [\mathbf{r}_0\ \mathbf{r}_1\ \dots\ \mathbf{r}_{L-1}]; \\ \mathbf{S} &= [\mathbf{s}_0\ \mathbf{s}_1\ \dots\ \mathbf{s}_{L-1}], \end{aligned}$$

The joint ML estimates of $\boldsymbol{\theta}$ and $\mathbf{S}$ are given by

$$\begin{aligned} \{\hat{\boldsymbol{\theta}}_{\mathrm{cml}}, \hat{\mathbf{S}}_{\mathrm{cml}}\} &= \arg\max_{\boldsymbol{\theta}, \mathrm{s}} f(\{\mathbf{r}_l\}|\boldsymbol{\theta}, \mathbf{S}) \\ &= \arg\min_{\boldsymbol{\theta}, \mathrm{s}} V(\boldsymbol{\theta}, \mathbf{S}), \end{aligned} \tag{5.11}$$

where

$$V(\boldsymbol{\theta}, \mathbf{S}) = \sum_{l=0}^{L-1} ||\mathbf{r}_l - \mathbf{A}(\boldsymbol{\theta})\mathbf{s}_l||^2$$
$$= ||\mathbf{R} - \mathbf{A}(\boldsymbol{\theta})\mathbf{S}||_{\mathrm{F}}^2. \tag{5.12}$$

The ML estimate of $\boldsymbol{\theta}$, denoted by $\hat{\boldsymbol{\theta}}_{\mathrm{cml}}$, is called the conditional (or deterministic) ML (CML) estimate, because it is assumed that $\mathbf{S}$ is given and deterministic.

The minimization in (5.11) can be divided into two steps as follows:

$$\{\hat{\boldsymbol{\theta}}_{\mathrm{cml}}, \hat{\mathbf{S}}_{\mathrm{cml}}\} = \arg\min_{\boldsymbol{\theta}} \left\{ \min_{\mathbf{S}} ||\mathbf{R} - \mathbf{A}(\boldsymbol{\theta})\mathbf{S}||_{\mathrm{F}}^2 \right\}. \tag{5.13}$$

From (5.12), we can show that

$$\min_{\mathbf{S}} ||\mathbf{R} - \mathbf{A}(\boldsymbol{\theta})\mathbf{S}||_{\mathrm{F}}^2 = \sum_{l=0}^{L-1} \min_{\mathbf{s}_l} ||\mathbf{r}_l - \mathbf{A}(\boldsymbol{\theta})\mathbf{s}_l||^2.$$

If $K \le N$, we have

$$\hat{\mathbf{s}}_l = \min_{\mathbf{s}_l} ||\mathbf{r}_l - \mathbf{A}(\boldsymbol{\theta})\mathbf{s}_l||^2$$
$$= \left(\mathbf{A}^{\mathrm{H}}(\boldsymbol{\theta})\mathbf{A}(\boldsymbol{\theta})\right)^{-1} \mathbf{A}^{\mathrm{H}}(\boldsymbol{\theta})\mathbf{r}_l$$

or

$$\hat{\mathbf{S}}_{\mathrm{cml}} = \left(\mathbf{A}^{\mathrm{H}}(\boldsymbol{\theta})\mathbf{A}(\boldsymbol{\theta})\right)^{-1} \mathbf{A}^{\mathrm{H}}(\boldsymbol{\theta})\mathbf{R}. \tag{5.14}$$

Substituting (5.14) into (5.13), it follows that

$$\hat{\boldsymbol{\theta}}_{\mathrm{cml}} = \arg\min_{\boldsymbol{\theta}} V(\boldsymbol{\theta}, \hat{\mathbf{S}}_{\mathrm{cml}})$$
$$= \arg\min_{\boldsymbol{\theta}} ||\mathbf{R} - \mathbf{A}(\boldsymbol{\theta})\hat{\mathbf{S}}_{\mathrm{cml}}||_{\mathrm{F}}^2$$
$$= \arg\min_{\boldsymbol{\theta}} ||\mathbf{R} - \mathbf{A}(\boldsymbol{\theta})\left(\mathbf{A}^{\mathrm{H}}(\boldsymbol{\theta})\mathbf{A}(\boldsymbol{\theta})\right)^{-1} \mathbf{A}^{\mathrm{H}}(\boldsymbol{\theta})\mathbf{R}||_{\mathrm{F}}^2$$
$$= \arg\min_{\boldsymbol{\theta}} ||(\mathbf{I} - \mathbf{P}(\boldsymbol{\theta}))\mathbf{R}||_{\mathrm{F}}^2, \tag{5.15}$$

where $\mathbf{P}(\boldsymbol{\theta})$ is the projection matrix onto the subspace spanned by $\mathbf{A}(\boldsymbol{\theta})$, i.e.

$$\mathbf{P}(\boldsymbol{\theta}) = \mathbf{A}(\boldsymbol{\theta})\left(\mathbf{A}^{\mathrm{H}}(\boldsymbol{\theta})\mathbf{A}(\boldsymbol{\theta})\right)^{-1} \mathbf{A}^{\mathrm{H}}(\boldsymbol{\theta}).$$

Note that if $K = N$ and $\mathbf{A}(\boldsymbol{\theta})$ is full-rank, $\mathbf{P}(\boldsymbol{\theta}) = \mathbf{I}$ and the CML approach becomes unavailable. Thus, it is necessary that $K < N$. Define the orthogonal projection matrix onto the subspace spanned by $\mathbf{A}(\boldsymbol{\theta})$ as

$$\mathbf{P}^{\perp}(\boldsymbol{\theta}) = \mathbf{I} - \mathbf{P}(\boldsymbol{\theta}).$$

Let $\boldsymbol{\theta}_{\mathrm{o}}$ denote the true AoA vector. For convenience, let

$$V_{\mathrm{cml}}(\boldsymbol{\theta}) = V(\boldsymbol{\theta}, \hat{\mathbf{S}}_{\mathrm{cml}}).$$

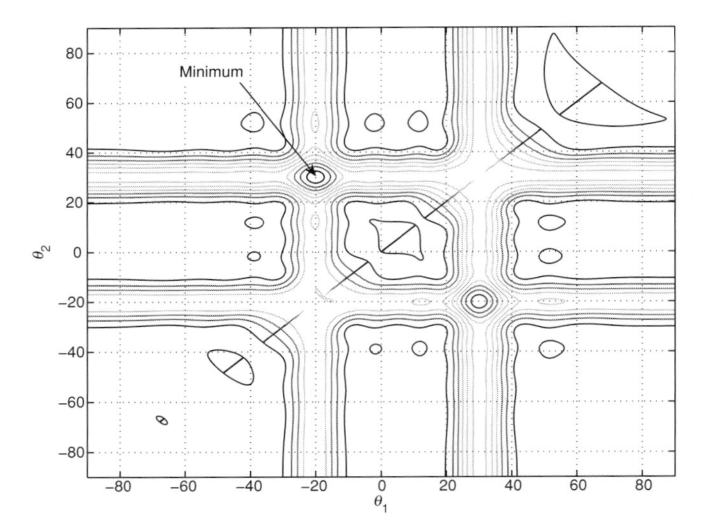

Figure 5.3 Contour of a CML cost function, $V_{\mathrm{cml}}(\boldsymbol{\theta})$ when $K = 2$ with $\theta_1 = -20^\circ$ and $\theta_2 = 30^\circ$. The two minima exist at $\boldsymbol{\theta} = [-20^\circ \; 30^\circ]^{\mathrm{T}}$ and $[30^\circ \; -20^\circ]^{\mathrm{T}}$ and it is shown that $V_{\mathrm{cml}}(\boldsymbol{\theta})$ is symmetric.

The CML estimate of $\boldsymbol{\theta}$ can be given by

$$\hat{\boldsymbol{\theta}}_{\mathrm{cml}} = \arg \min_{\boldsymbol{\theta}} V_{\mathrm{cml}}(\boldsymbol{\theta}). \tag{5.16}$$

If $\mathbf{n}_l = 0$ for all l, we can easily verify that

$$\begin{aligned} V_{\mathrm{cml}}(\boldsymbol{\theta}_{\mathrm{o}}) &= ||\mathbf{P}^{\perp}(\boldsymbol{\theta}_{\mathrm{o}})\mathbf{R}||_{\mathrm{F}}^2 \\ &= ||\mathbf{P}^{\perp}(\boldsymbol{\theta}_{\mathrm{o}})\mathbf{A}(\boldsymbol{\theta}_{\mathrm{o}})\mathbf{S}||_{\mathrm{F}}^2 \\ &= 0. \end{aligned}$$

This implies that $\boldsymbol{\theta}_{\mathrm{o}}$ is a solution that minimizes $V_{\mathrm{cml}}(\boldsymbol{\theta})$ (≥ 0) in the absence of background noise.

In general, since $V_{\mathrm{cml}}(\boldsymbol{\theta})$ is a nonlinear function of $\boldsymbol{\theta}$, a closed-form solution is not available to the minimization in (5.16). Thus, numerical optimization techniques are required to find a solution.

Figure 5.3 shows an illustration of the cost function of the CML approach, $V_{\mathrm{cml}}(\boldsymbol{\theta})$ when $K = 2$ with $\theta_1 = -20^\circ$ and $\theta_2 = 30^\circ$. A standard ULA of $N = 10$ is used. We assume that $\mathbf{s}_l \sim \mathcal{CN}(0, \mathbf{I})$ with $L = 100$. It is shown in Fig. 5.3 that the two minima exist at $\boldsymbol{\theta} = [-20^\circ \; 30^\circ]^{\mathrm{T}}$ and $[30^\circ \; -20^\circ]^{\mathrm{T}}$. Since the subspaces spanned by $\mathbf{A} = [\mathbf{a}(\theta_1) \; \mathbf{a}(\theta_2)]$ and $\mathbf{A}' = [\mathbf{a}(\theta_2) \; \mathbf{a}(\theta_1)]$ are identical, we can have the two minima, which results in the ambiguity as shown in Fig. 5.3.

There is another ML approach, called unconditional (or stochastic) ML (UML) approach, to estimate AoAs. Since we are only interested in the AoA estimation, we can assume that the $\mathbf{s}_l$'s are random vectors and independent of the $\mathbf{n}_l$'s. In particular, assume that $\mathbf{s}_l \sim \mathcal{CN}(0, \mathbf{R}_\mathbf{s})$. Then, we can show that $\mathbf{r}_l$ is a CSCG random vector with

$$\begin{aligned}\mathcal{E}[\mathbf{r}_l] &= \mathbf{0};\\ \mathcal{E}[\mathbf{r}_l\mathbf{r}_l^{\mathrm{H}}] &= \mathbf{R}_\mathbf{r} = \mathbf{A}(\boldsymbol{\theta})\mathbf{R}_\mathbf{s}\mathbf{A}^{\mathrm{H}}(\boldsymbol{\theta}) + N_0\mathbf{I}.\end{aligned} \tag{5.17}$$

From this and with an additional assumption that the $\mathbf{s}_l$'s are independent, the ML estimate is given by

$$\begin{aligned}\hat{\boldsymbol{\theta}}_{\mathrm{uml}} &= \arg\max_{\boldsymbol{\theta}} f(\{\mathbf{r}_l\}|\boldsymbol{\theta})\\ &= \arg\max_{\boldsymbol{\theta}} \prod_{l=0}^{L-1} f(\mathbf{r}_l|\boldsymbol{\theta})\\ &= \arg\max_{\boldsymbol{\theta}} \prod_{l=0}^{L-1} \frac{1}{\pi^N \det(\mathbf{R}_\mathbf{r})} \exp\left(-\mathbf{r}_l^{\mathrm{H}}\mathbf{R}_\mathbf{r}^{-1}\mathbf{r}_l\right)\\ &= \arg\max_{\boldsymbol{\theta}} \frac{1}{\pi^{NL} \det^L(\mathbf{R}_\mathbf{r})} \exp\left(-\sum_{l=0}^{L-1} \mathbf{r}_l^{\mathrm{H}}\mathbf{R}_\mathbf{r}^{-1}\mathbf{r}_l\right).\end{aligned} \tag{5.18}$$

Note that $\mathbf{R}_\mathbf{r}$ is a function of $\boldsymbol{\theta}$. The resulting ML estimate is called the UML estimate. Taking logarithms, we can also show that

$$\begin{aligned}\hat{\boldsymbol{\theta}}_{\mathrm{uml}} &= \arg\min_{\boldsymbol{\theta}} \sum_{l=0}^{L-1} \mathbf{r}_l^{\mathrm{H}}\mathbf{R}_\mathbf{r}^{-1}\mathbf{r}_l + L\log\det(\mathbf{R}_\mathbf{r})\\ &= \arg\min_{\boldsymbol{\theta}} \operatorname{tr}(\mathbf{R}_\mathbf{r}^{-1}\hat{\mathbf{R}}_\mathbf{r}) + \log\det(\mathbf{R}_\mathbf{r}),\end{aligned} \tag{5.19}$$

where $\hat{\mathbf{R}}_\mathbf{r}$ is the sample covariance matrix of $\mathbf{r}_l$:

$$\hat{\mathbf{R}}_\mathbf{r} = \frac{1}{L}\sum_{l=0}^{L-1} \mathbf{r}_l\mathbf{r}_l^{\mathrm{H}}.$$

Since $\mathbf{R}_\mathbf{r}$ is a function of $\boldsymbol{\theta}$ as shown in (5.17), if we define the UML cost function as

$$V_{\mathrm{uml}}(\boldsymbol{\theta}) = \operatorname{tr}(\mathbf{R}_\mathbf{r}^{-1}\hat{\mathbf{R}}_\mathbf{r}) + \log\det(\mathbf{R}_\mathbf{r}), \tag{5.20}$$

we have

$$\hat{\boldsymbol{\theta}}_{\mathrm{uml}} = \arg\min_{\boldsymbol{\theta}} V_{\mathrm{uml}}(\boldsymbol{\theta}).$$

We consider the UML estimates of AoAs under the same environment as in Fig. 5.3. An illustration of the contour of the cost function of the UML approach, $V_{\mathrm{uml}}(\boldsymbol{\theta})$, is shown in Fig. 5.4. Again, since $\mathbf{R}_\mathbf{r}$ is the same for $\mathbf{A} = [\mathbf{a}(\theta_1)\ \mathbf{a}(\theta_2)]$ and $\mathbf{A}' = [\mathbf{a}(\theta_2)\ \mathbf{a}(\theta_1)]$, we can have the two minima at $\boldsymbol{\theta} = [-20^\circ\ 30^\circ]^{\mathrm{T}}$ and $[30^\circ\ -20^\circ]^{\mathrm{T}}$.

While an optimal performance can be achieved using the ML approaches, their complexity can be prohibitively high for a large K. This high computational complexity results from joint minimization.

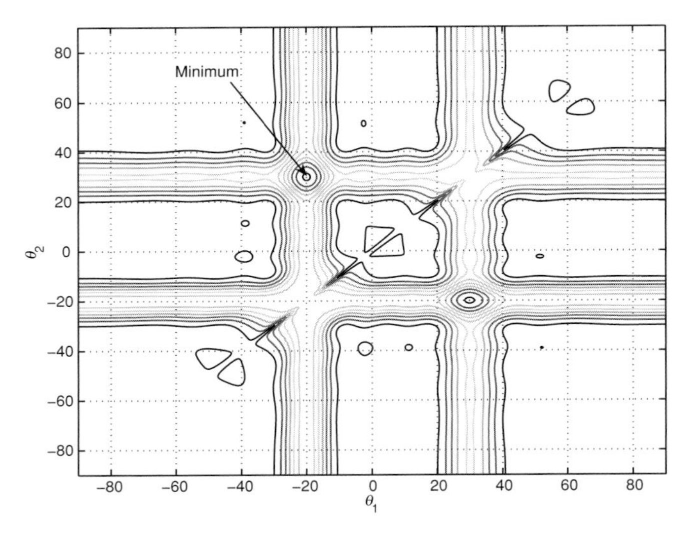

Figure 5.4 Contour of the UML cost function, $V_{\rm uml}(\boldsymbol{\theta})$ when $K = 2$ with $\theta_1 = -20°$ and $\theta_2 = 30°$. The two minima exist at $\boldsymbol{\theta} = [-20° \ 30°]^{\rm T}$ and $[30° \ -20°]^{\rm T}$ and it is shown that $V_{\rm cml}(\boldsymbol{\theta})$ is symmetric.

5.2.2 Subspace approaches

The subspace method was originally invented for the AoA estimation in (Schmidt 1986). As it can provide an excellent performance with low complexity, it has become popular and has been applied to a number of applications as a parametric estimation method.

In order to derive subspace approaches for the AoA estimation, we assume that the $\mathbf{s}_l$'s are random and independent of $\mathbf{n}_l$. In particular, it is assumed that $\mathcal{E}[\mathbf{s}_l] = \mathbf{0}$ and $\mathcal{E}[\mathbf{s}_l \mathbf{s}_l^{\rm H}] = \mathbf{R}_{\mathbf{s}}$. Furthermore, assume that the rank of $\mathbf{R}_{\mathbf{s}}$ is K (i.e. full-rank) and $K < N$. The covariance matrix of $\mathbf{r}_l$ becomes

$$\mathbf{R}_{\mathbf{r}} = \mathbf{A}(\boldsymbol{\theta})\mathbf{R}_{\mathbf{s}}\mathbf{A}^{\rm H}(\boldsymbol{\theta}) + N_0\mathbf{I}. \tag{5.21}$$

Consider the eigendecomposition of $\mathbf{R}_{\mathbf{r}}$:

$$\begin{aligned}\mathbf{R}_{\mathbf{r}} &= \sum_{q=1}^{N} \lambda_q \mathbf{e}_q \mathbf{e}_q^{\rm H} \\ &= \mathbf{E}\boldsymbol{\Lambda}\mathbf{E}^{\rm H},\end{aligned} \tag{5.22}$$

where $\mathbf{E} = [\mathbf{e}_1 \ \mathbf{e}_2 \ \ldots \ \mathbf{e}_N]^{\rm T}$ and $\boldsymbol{\Lambda} = {\rm Diag}(\lambda_1, \lambda_2, \ldots, \lambda_N)$. Here, λ_q and $\mathbf{e}_q$ are the qth eigenvalue and its associated eigenvector of $\mathbf{R}_{\mathbf{r}}$, respectively. In addition, we assume that

$$\lambda_1 \geq \lambda_2 \geq \ldots \geq \lambda_N.$$

Since $\mathbf{R}_{\mathbf{s}}$ is a $K \times K$ matrix and $K < N$, the rank of $\mathbf{A}(\boldsymbol{\theta})\mathbf{R}_{\mathbf{s}}\mathbf{A}^{\rm H}(\boldsymbol{\theta})$ is less than or equal to K. This implies that the $(N - K)$ smallest eigenvalues of $\mathbf{R}_{\mathbf{r}}$ are N_0, i.e.,

$$\lambda_{K+1} = \lambda_{K+2} = \cdots = \lambda_N = N_0.$$

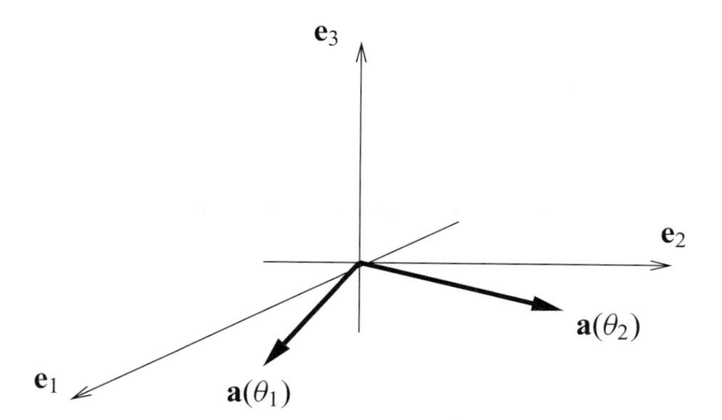

Figure 5.5 Subspaces generated by the ARVs of two AoAs, θ_1 and θ_2.

Thus,

$$\mathbf{R_r e}_q = \lambda_q \mathbf{e}_q$$
$$= N_0 \mathbf{e}_q, \ q = K+1, K+2, \ldots, N. \quad (5.23)$$

Substituting (5.21) into (5.23), we have

$$(\mathbf{A}(\boldsymbol{\theta})\mathbf{R_s}\mathbf{A}^{\mathrm{H}}(\boldsymbol{\theta}) + N_0\mathbf{I})\mathbf{e}_q = N_0\mathbf{e}_q, \ q = K+1, K+2, \ldots, N. \quad (5.24)$$

This implies that $\mathbf{A}^{\mathrm{H}}(\boldsymbol{\theta})\mathbf{e}_q = \mathbf{0}$ for $q \in \{K+1, K+2, \ldots, N\}$. Consequently,

$$\mathrm{Range}(\mathbf{A}(\boldsymbol{\theta})) \perp \mathrm{Span}\{\mathbf{e}_{K+1}, \mathbf{e}_{K+2}, \ldots, \mathbf{e}_N\}. \quad (5.25)$$

Furthermore,

$$\mathrm{Range}(\mathbf{A}(\boldsymbol{\theta})) = \mathrm{Span}\{\mathbf{e}_1, \mathbf{e}_2, \ldots, \mathbf{e}_K\}. \quad (5.26)$$

For convenience, the subspace spanned by $\{\mathbf{e}_{K+1}, \mathbf{e}_{K+2}, \ldots, \mathbf{e}_N\}$ is called the noise subspace, while the subspace spanned by $\{\mathbf{e}_1, \mathbf{e}_2, \ldots, \mathbf{e}_K\}$ is called the signal subspace. Clearly, they are orthogonal to each other. Using (5.25), we can show that

$$\mathbf{E}_{\mathrm{N}}\mathbf{a}^{\mathrm{H}}(\theta_k) = \mathbf{0}, \ k = 1, 2, \ldots, K,$$

where

$$\mathbf{E}_{\mathrm{N}} = [\mathbf{e}_{K+1} \ \mathbf{e}_{K+2} \ \ldots \ \mathbf{e}_N].$$

For convenience, we also define

$$\mathbf{E}_{\mathrm{S}} = [\mathbf{e}_1 \ \mathbf{e}_2 \ \ldots \ \mathbf{e}_K].$$

Now, we can have

$$\text{Signal subspace} = \mathrm{Range}(\mathbf{E}_{\mathrm{S}});$$
$$\text{Noise subspace} = \mathrm{Range}(\mathbf{E}_{\mathrm{N}}). \quad (5.27)$$

In Fig. 5.5, the subspaces are illustrated when $K = 2$ and $N = 3$. The signal subspace is generated by $\{\mathbf{a}(\theta_1), \mathbf{a}(\theta_2)\}$. The orthonormal basis vectors for the signal subspace are denoted by $\{\mathbf{e}_1, \mathbf{e}_2\}$. The noise subspace is spanned by $\mathbf{e}_3$, which is orthogonal to $\mathbf{e}_1$ and $\mathbf{e}_2$. Thus, we can see that $\mathbf{a}(\theta_k)$, $k = 1, 2$, is orthogonal to $\mathbf{e}_3$ or the noise subspace.

To perform the AoA estimation using the subspaces, define a cost function as follows:

$$\begin{aligned} f_{\text{music}}(\theta) &= \mathbf{a}^{\text{H}}(\theta)\mathbf{E}_{\text{N}}\mathbf{E}_{\text{N}}^{\text{H}}\mathbf{a}(\theta) \\ &= ||\mathbf{E}_{\text{N}}^{\text{H}}\mathbf{a}(\theta)||^2. \end{aligned} \tag{5.28}$$

From (5.25), we expect that $f_{\text{music}}(\theta) = 0$ if $\theta = \theta_k$, $k = 1, 2, \ldots, K$, and $f_{\text{music}}(\theta) > 0$ otherwise. The approach to find the AoAs using the cost function in (5.28) is called the multiple signal classification (MUSIC) approach (Schmidt 1986) (see also (Bienvenu & Kopp 1983)). In addition, $f_{\text{music}}(\theta)$, which is a function of θ, is called the MUSIC spectrum. The MUSIC approach is an example of the subspace method.

In the subspace method, we have a subspace that depends on the parameters to be estimated. This subspace is called the signal subspace. The orthogonal subspace, called the noise subspace, is generally utilized to estimate the parameters. Since the projection of any vector that lies in the signal subspace onto the noise subspace becomes zero, the parameter estimation can be performed using a parameter vector that lies in the signal subspace if the parameter is true. In the context of the MUSIC approach, the parameter vector is $\mathbf{a}(\theta)$ and the signal subspace is the subspace spanned by the column vectors of $\mathbf{E}_{\text{S}}$. It can be shown that

$$\mathbf{a}(\theta) \in \text{Range}(\mathbf{E}_{\text{S}}), \ \theta \in \{\theta_1, \theta_2, \ldots, \theta_K\}.$$

The projection onto the noise subspace is given by

$$\mathbf{a}_{\text{N}}(\theta) = \mathbf{E}_{\text{N}}\mathbf{E}_{\text{N}}^{\text{H}}\mathbf{a}(\theta).$$

It can be readily verified that

$$f_{\text{music}}(\theta) = ||\mathbf{a}_{\text{N}}(\theta)||^2.$$

If the true covariance matrix, $\mathbf{R}_{\mathbf{r}}$, is available, the MUSIC spectrum, $f_{\text{music}}(\theta)$, has K zeros when $\theta \in \{\theta_1, \theta_2, \ldots, \theta_K\}$. However, if a sample covariance matrix is used, which is a realistic case, the eigenvectors could be perturbed from the true eigenvectors. Let $\hat{\lambda}_n$ denote the nth largest eigenvalue of the sample covariance matrix $\hat{\mathbf{R}}_{\mathbf{r}}$ and denote by $\hat{\mathbf{e}}_n$ the eigenvector corresponding to $\hat{\lambda}_n$. With the estimated eigenvectors, the MUSIC cost function is given by

$$\hat{f}_{\text{music}}(\theta) = \mathbf{a}^{\text{H}}(\theta)\hat{\mathbf{E}}_{\text{N}}\hat{\mathbf{E}}_{\text{N}}^{\text{H}}\mathbf{a}(\theta), \tag{5.29}$$

where

$$\hat{\mathbf{E}}_{\text{N}} = [\hat{\mathbf{e}}_{K+1} \ \hat{\mathbf{e}}_{K+2} \ \ldots \ \hat{\mathbf{e}}_N].$$

In this case, the MUSIC approach can estimate the AoAs by finding the K minima of $\hat{f}_{\text{music}}(\theta)$ for $\theta \in \Theta$, where Θ is the set of all the possible AoAs (e.g. $\Theta = [-\pi/2, \pi/2)$ for a standard ULA).

As shown above, the MUSIC algorithm does not require a joint minimization, its complexity is much lower than that of the ML approaches. In particular, for a large K, the MUSIC algorithm becomes more computationally efficient than the ML approaches.

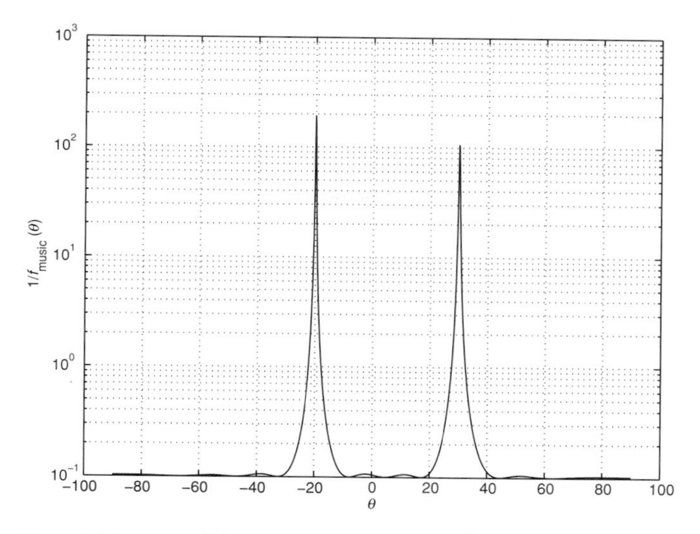

Figure 5.6 Inverse of the MUSIC spectrum, $1/\hat{f}_{\mathrm{music}}(\theta)$, from a sample covariance matrix when $K = 2$ with $\theta_1 = -20^\circ$ and $\theta_2 = 30^\circ$. We can observe the two maxima of the inverse MUSIC spectrum at $\theta = -20^\circ$ and 30°.

Figure 5.6 shows the inverse of the MUSIC spectrum, $1/\hat{f}_{\mathrm{music}}(\theta)$, from a sample covariance matrix for $\theta \in [-90^\circ, 90^\circ]$ when $K = 2$ with $\theta_1 = -20^\circ$ and $\theta_2 = 30^\circ$. A standard ULA of $N = 10$ is used. We have $\mathbf{s}_l \sim \mathcal{CN}(0, \mathbf{I})$ with $L = 100$. There are two maxima at $\theta = -20^\circ$ and 30° which correspond to the true AoAs.

There are other subspace methods including the minimum-norm algorithm (Kumaresan & Tufts 1983). The minimum-norm algorithm uses a single non-zero vector in the noise subspace. Let $\mathbf{d} \in \mathrm{Range}(\mathbf{E}_{\mathrm{N}})$ and $\mathbf{d} \neq \mathbf{0}$. This implies that

$$\mathbf{a}^{\mathrm{H}}(\theta)\mathbf{d} = 0, \ \theta \in \{\theta_1, \theta_2, \ldots, \theta_K\}$$

and

$$\mathbf{E}_{\mathrm{S}}^{\mathrm{H}}\mathbf{d} = \mathbf{0}. \tag{5.30}$$

Using $\mathbf{d}$, we can define the following cost function:

$$f_{\mathbf{d}}(\theta) = \mathbf{a}^{\mathrm{H}}(\theta)\mathbf{d}\mathbf{d}^{\mathrm{H}}\mathbf{a}(\theta). \tag{5.31}$$

The minimum-norm algorithm uses the vector $\mathbf{d}$ that has the minimum norm subject to the constraint in (5.30).

In order to find the minimum norm vector of $\mathbf{d}$, let

$$\mathbf{E}_{\mathrm{S}}^{\mathrm{H}} = [\mathbf{g}\ \mathbf{X}], \tag{5.32}$$

where $\mathbf{g}$ denotes the first column vector $\mathbf{E}_{\mathrm{S}}^{\mathrm{H}}$ and $\mathbf{X}$ is a submatrix of $\mathbf{E}_{\mathrm{S}}^{\mathrm{H}}$ obtained after removing $\mathbf{g}$. We assume that the first element of $\mathbf{d}$ is unity. That is,

$$\mathbf{d} = \begin{bmatrix} 1 \\ \bar{\mathbf{d}} \end{bmatrix}.$$

This condition is necessary to make sure that $\mathbf{d}$ is not a zero vector in the minimization of the norm of $\mathbf{d}$. Then, it follows that

$$\mathbf{E}_S^H \mathbf{d} = \mathbf{g} + \mathbf{X}\bar{\mathbf{d}} = \mathbf{0}. \tag{5.33}$$

The minimization of the norm of $\mathbf{d}$ (with the first element of $\mathbf{d}$ unity) is equivalent to the minimization of the norm of $\bar{\mathbf{d}}$. This minimization can be formulated as follows:

$$\begin{aligned} \bar{\mathbf{d}}_{\text{min-norm}} &= \arg\min_{\bar{\mathbf{d}}} ||\bar{\mathbf{d}}||^2 \\ &\text{subject to } \mathbf{g} + \mathbf{X}\bar{\mathbf{d}} = \mathbf{0}. \end{aligned} \tag{5.34}$$

Using the Lagrangian multiplier, we can find the solution as follows:

$$\bar{\mathbf{d}}_{\text{min-norm}} = -\mathbf{X}^H(\mathbf{X}\mathbf{X}^H)^{-1}\mathbf{g}. \tag{5.35}$$

Noting that

$$\begin{aligned} \mathbf{I} &= \mathbf{E}_S^H \mathbf{E}_S \\ &= \mathbf{g}\mathbf{g}^H + \mathbf{X}\mathbf{X}^H, \end{aligned} \tag{5.36}$$

$\bar{\mathbf{d}}_{\text{min-norm}}$ becomes

$$\begin{aligned} \bar{\mathbf{d}}_{\text{min-norm}} &= -\mathbf{X}^H(\mathbf{I} - \mathbf{g}\mathbf{g}^H)^{-1}\mathbf{g} \\ &= -\mathbf{X}^H\left(\mathbf{I} + \frac{1}{1-||\mathbf{g}||^2}\mathbf{g}\mathbf{g}^H\right)\mathbf{g} \\ &= -\frac{1}{1-||\mathbf{g}||^2}\mathbf{X}^H\mathbf{g}. \end{aligned} \tag{5.37}$$

Finally, we have

$$\begin{aligned} \mathbf{d}_{\text{min-norm}} &= \begin{bmatrix} 1 \\ \bar{\mathbf{d}}_{\text{min-norm}} \end{bmatrix} \\ &= \begin{bmatrix} 1 \\ -\frac{1}{1-||\mathbf{g}||^2}\mathbf{X}^H\mathbf{g} \end{bmatrix}. \end{aligned} \tag{5.38}$$

The minimum-norm vector, $\mathbf{d}_{\text{min-norm}}$, can also be expressed by $\mathbf{E}_N$. Let

$$\mathbf{E}_N^H = [\mathbf{c}\ \mathbf{Y}]. \tag{5.39}$$

Using (5.32) and (5.39), we can show that

$$\begin{aligned} \mathbf{I} &= [\mathbf{E}_S\ \mathbf{E}_N]\begin{bmatrix} \mathbf{E}_S^H \\ \mathbf{E}_N^H \end{bmatrix} \\ &= \begin{bmatrix} \mathbf{g}^H & \mathbf{c}^H \\ \mathbf{X}^H & \mathbf{Y}^H \end{bmatrix}\begin{bmatrix} \mathbf{g} & \mathbf{X} \\ \mathbf{c} & \mathbf{Y} \end{bmatrix} \\ &= \begin{bmatrix} \underbrace{||\mathbf{g}||^2 + ||\mathbf{c}||^2}_{=1} & \mathbf{g}^H\mathbf{X} + \mathbf{c}^H\mathbf{Y} \\ \underbrace{\mathbf{X}^H\mathbf{g} + \mathbf{Y}^H\mathbf{c}}_{=\mathbf{0}} & \mathbf{X}^H\mathbf{X} + \mathbf{Y}^H\mathbf{Y} \end{bmatrix}. \end{aligned} \tag{5.40}$$

From this, it follows that

$$\begin{aligned} 1 - ||\mathbf{g}||^2 &= ||\mathbf{c}||^2; \\ -\mathbf{X}^{\mathrm{H}}\mathbf{g} &= \mathbf{Y}^{\mathrm{H}}\mathbf{c}. \end{aligned} \tag{5.41}$$

Substituting (5.41) into (5.38), we have

$$\begin{aligned} \mathbf{d}_{\text{min-norm}} &= \begin{bmatrix} 1 \\ \frac{1}{||\mathbf{c}||^2}\mathbf{Y}^{\mathrm{H}}\mathbf{c} \end{bmatrix} \\ &= \frac{1}{||\mathbf{c}||^2} \begin{bmatrix} \mathbf{c}^{\mathrm{H}} \\ \mathbf{Y}^{\mathrm{H}} \end{bmatrix} \mathbf{c} \\ &= \mathbf{E}_{\mathrm{N}} \frac{\mathbf{c}}{||\mathbf{c}||^2}. \end{aligned} \tag{5.42}$$

This confirms that $\mathbf{d}_{\text{min-norm}} \in \text{Range}(\mathbf{E}_{\mathrm{N}})$. The minimum-norm cost function is given by

$$\begin{aligned} f_{\text{min-norm}}(\theta) &= \mathbf{a}^{\mathrm{H}}(\theta)\mathbf{d}_{\text{min-norm}}\mathbf{d}^{\mathrm{H}}_{\text{min-norm}}\mathbf{a}(\theta) \\ &= \mathbf{a}^{\mathrm{H}}(\theta)\mathbf{E}_{\mathrm{N}}\mathbf{W}_{\text{min-norm}}\mathbf{E}^{\mathrm{H}}_{\mathrm{N}}\mathbf{a}(\theta), \end{aligned} \tag{5.43}$$

where

$$\mathbf{W}_{\text{min-norm}} = \frac{1}{||\mathbf{c}||^4}\mathbf{c}\mathbf{c}^{\mathrm{H}}. \tag{5.44}$$

The min-norm cost function or spectrum is denoted by $\hat{f}_{\text{min-norm}}(\theta)$ if a sample covariance matrix is used to estimate the eigenvectors.

Figure 5.7 shows the inverses of the MUSIC and minimum-norm spectra, $1/\hat{f}_{\text{music}}(\theta)$ and $1/\hat{f}_{\text{min-norm}}(\theta)$, respectively, from a sample covariance matrix for $\theta \in [-40^\circ, 40^\circ]$ when $K = 2$ with $\theta_1 = -5^\circ$ and $\theta_2 = 5^\circ$. A standard ULA of $N = 10$ is used. We have $\mathbf{s}_l \sim \mathcal{CN}(0, \mathbf{I})$ with $L = 100$. There are two maxima at $\theta = -5^\circ$ and 5° which correspond to the true AoAs.

The probability of resolution is a performance criterion for AoA estimation methods when the AoAs of two signal sources are sufficiently close to each other. A better resolution performance is expected with a higher probability of resolution. In (Kaveh & Barabell 1986), (Lee & Wengrovitz 1991), and (Zhang 1995), the probability of resolution is studied with slightly different definitions. In (Lee & Wengrovitz 1991), when $K = 2$, the probability of resolution of the MUSIC algorithm is defined as

$$P_{\text{res}} = \Pr\left(\hat{f}_{\text{music}}(\theta_{\mathrm{m}}) > \max\{\hat{f}_{\text{music}}(\theta_1) + \hat{f}_{\text{music}}(\theta_2)\}\right), \tag{5.45}$$

where $\theta_{\mathrm{m}} = \frac{1}{2}(\theta_1 + \theta_2)$. A slightly different definition of the probability of resolution is used in (Zhang 1995) as follows:

$$P_{\text{res}} = \Pr\left(\hat{f}_{\text{music}}(\theta_{\mathrm{m}}) > \frac{1}{2}\left(\hat{f}_{\text{music}}(\theta_1) + \hat{f}_{\text{music}}(\theta_2)\right)\right). \tag{5.46}$$

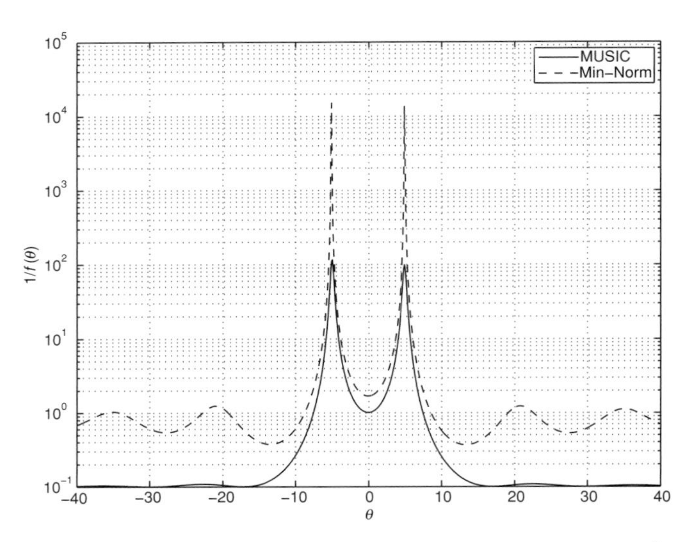

Figure 5.7 Inverses of the MUSIC and minimum-norm spectra, $1/\hat{f}_{\text{music}}(\theta)$ and $1/\hat{f}_{\text{min-norm}}(\theta)$, respectively, from a sample covariance matrix when $K = 2$ with $\theta_1 = -5^\circ$ and $\theta_2 = 5^\circ$. We can observe the two peaks at $\theta = -5^\circ$ and 5° in both inverse-spectra.

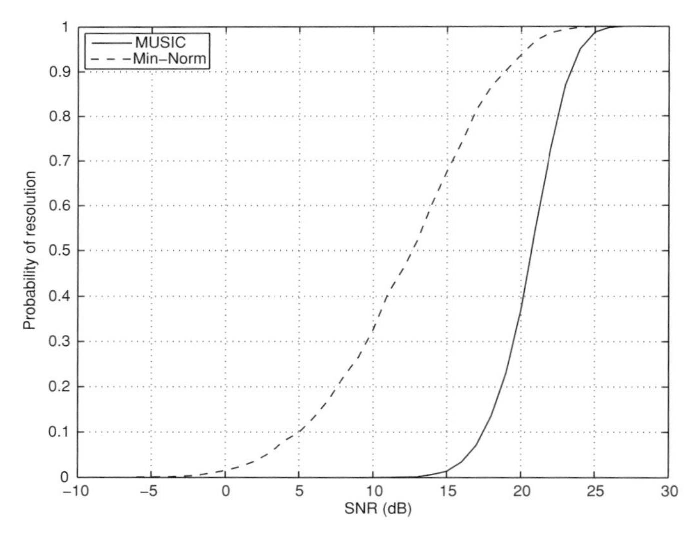

Figure 5.8 Probability of resolution of the MUSIC and minimum-norm algorithms, when $\theta_1 = -1^\circ$ and $\theta_2 = 1^\circ$.

Figure 5.8 shows the probability of resolution of the MUSIC and minimum-norm algorithm based on the definition in (5.45). We assume that there are two uncorrelated equal-power Gaussian sources (i.e., $\mathbf{s}_l \sim \mathcal{CN}(0, \mathbf{I})$) at $\theta_1 = -1^\circ$ and $\theta_2 = 1^\circ$ (i.e., $K = 2$). A standard ULA of $N = 10$ is used and a sample covariance matrix is obtained with $L = 100$. The result shows that the minimum-norm algorithm provides a better resolution capability than the MUSIC algorithm.

Based on the expression in (5.43), we can generalize the MUSIC cost function, which is called the weighted MUSIC cost function, as follows:

$$f_{\text{g-music}}(\theta) = \mathbf{a}^{\mathrm{H}}(\theta)\mathbf{E}_{\mathrm{N}}\mathbf{W}\mathbf{E}_{\mathrm{N}}^{\mathrm{H}}\mathbf{a}(\theta), \tag{5.47}$$

where $\mathbf{W}$ is symmetric and positive semi-definite. Clearly, if $\mathbf{W} = \mathbf{I}$, we can have the original MUSIC cost function. For the minimum-norm cost function, $\mathbf{W}$ becomes

$$\mathbf{W} = \frac{1}{||\mathbf{c}||^4}\mathbf{c}\mathbf{c}^{\mathrm{H}}.$$

5.2.3 Specialized algorithms for ULA

For a ULA, we can derive computationally efficient algorithms to estimate AoAs using root-finding algorithms for polynomials.

A polynomial representation can be adopted for the ARV of a ULA as follows:

$$\mathbf{a}(z) = \begin{bmatrix} 1 \; z^{-1} \; \dots \; z^{-N+1} \end{bmatrix}^{\mathrm{T}},$$

where $z = e^{j\psi}$ and ψ represents a phase shift. For a standard ULA, $\psi = \pi \sin\theta$. Then, the MUSIC cost function becomes a polynomial of z as follows:

$$\begin{aligned} f_{\text{music}}(z) &= \mathbf{a}^{\mathrm{H}}(z)\mathbf{E}_{\mathrm{N}}\mathbf{E}_{\mathrm{N}}^{\mathrm{H}}\mathbf{a}(z) \\ &= \mathbf{a}^{\mathrm{T}}(z^{-1})\mathbf{E}_{\mathrm{N}}\mathbf{E}_{\mathrm{N}}^{\mathrm{H}}\mathbf{a}(z). \end{aligned} \tag{5.48}$$

Using any root finding algorithm, we will be able to obtain the roots of $f_{\text{music}}(z)$. If $\theta = \theta_k$ for any $k \in \{1, 2, \dots, K\}$, we have

$$f_{\text{music}}(z)\big|_{z=\exp(j\psi_k)} = 0,$$

where ψ_k is the phase shift corresponding to the AoA θ_k. This implies that K roots are located on the unit circle. In other words, from the K roots on the unit circle, the AoAs can be estimated. Let $z_k = \exp(j\psi_k)$. Then, for a standard ULA, we have

$$\begin{aligned} \theta_k &= \sin^{-1}\frac{\psi_k}{\pi} \\ &= \sin^{-1}\frac{\angle z_k}{\pi}, \quad k = 1, 2, \dots, K. \end{aligned}$$

In practice, however, since we use a sample covariance matrix, the MUSIC cost function becomes

$$f_{\text{music}}(z) = \mathbf{a}^{\mathrm{T}}(z^{-1})\hat{\mathbf{E}}_{\mathrm{N}}\hat{\mathbf{E}}_{\mathrm{N}}^{\mathrm{H}}\mathbf{a}(z). \tag{5.49}$$

In this case, we need to choose the K roots that are inside the unit circle and closest to the unit circle. If these roots are denoted by $\hat{z}_k$, $k = 1, 2, \dots, K$, the AoA estimates for a standard ULA is given by

$$\theta_k = \sin^{-1}\frac{\angle \hat{z}_k}{\pi}, \quad k = 1, 2, \dots, K.$$

In a certain parametric estimation problem, if a parameter vector can be represented as a polynomial form with a parameter to be estimated, the subspace approach can be applied. An example is given below.

Example 5.2.1 Suppose that $\mathbf{a}(\rho)$ is given by

$$\mathbf{a}(\rho) = [1 \ \rho \ \rho^2]^{\mathrm{T}},$$

where $\rho \in [0, 1]$ is the parameter to be estimated. The received signal is a superposition of two signals which is given by

$$\mathbf{r}_l = \mathbf{a}(\rho_1)s_{1,l} + \mathbf{a}(\rho_2)s_{2,l} + \mathbf{n}_l,$$

where the $s_{k,l}$'s are zero-mean uncorrelated random signals and $\mathbf{n}_l$ is the background noise vector with $\mathcal{E}[\mathbf{n}_l] = \mathbf{0}$ and $\mathcal{E}[\mathbf{n}_l\mathbf{n}_l^{\mathrm{H}}] = \mathbf{I}$. If $\mathcal{E}[|s_{k,l}|^2] = 1$, the covariance matrix of $\mathbf{r}_l$ becomes

$$\begin{aligned}\mathbf{R}_{\mathbf{r}} &= \mathbf{a}(\rho_1)\mathbf{a}^{\mathrm{H}}(\rho_1) + \mathbf{a}(\rho_2)\mathbf{a}^{\mathrm{H}}(\rho_2) + \mathbf{I} \\ &= \begin{bmatrix} 3 & \rho_1 + \rho_2 & \rho_1^2 + \rho_2^2 \\ \rho_1 + \rho_2 & \rho_1^2 + \rho_2^2 + 1 & \rho_1^3 + \rho_2^3 \\ \rho_1^2 + \rho_2^2 & \rho_1^3 + \rho_2^3 & \rho_1^4 + \rho_2^4 + 1 \end{bmatrix}.\end{aligned}$$

If $\rho_1 = 1/2$ and $\rho_2 = 1/4$, we have

$$\mathbf{e}_3 = [-0.0995 \ 0.5970 \ -0.7960]^{\mathrm{T}}.$$

The estimates of ρ_1 and ρ_2 can be obtained using the following MUSIC spectrum

$$f_{\text{music}}(\rho) = |\mathbf{e}_3^{\mathrm{H}}\mathbf{a}(\rho)|^2.$$

Alternatively, noting that $\mathbf{e}_3^{\mathrm{H}}\mathbf{a}(\rho)$ is a polynomial of ρ, they can also be obtained by finding the roots of $g(\rho) = \mathbf{e}_3^{\mathrm{H}}\mathbf{a}(\rho) = -0.7960\rho^2 + 0.5970\rho - 0.0995$.

5.3 Beamforming methods

With known ARV or AoA of the desired signal, a beam can be formed to improve the signal reception quality or SNR using a linear combiner, which is called the beamformer, with the received signals from an array. The beamforming operation is depicted in Fig. 5.9.

Beamforming can be considered as spatial filtering since it can provide spatial selectivity (Veen & Buckley 1988). Using the power of the normalized beam pattern, the

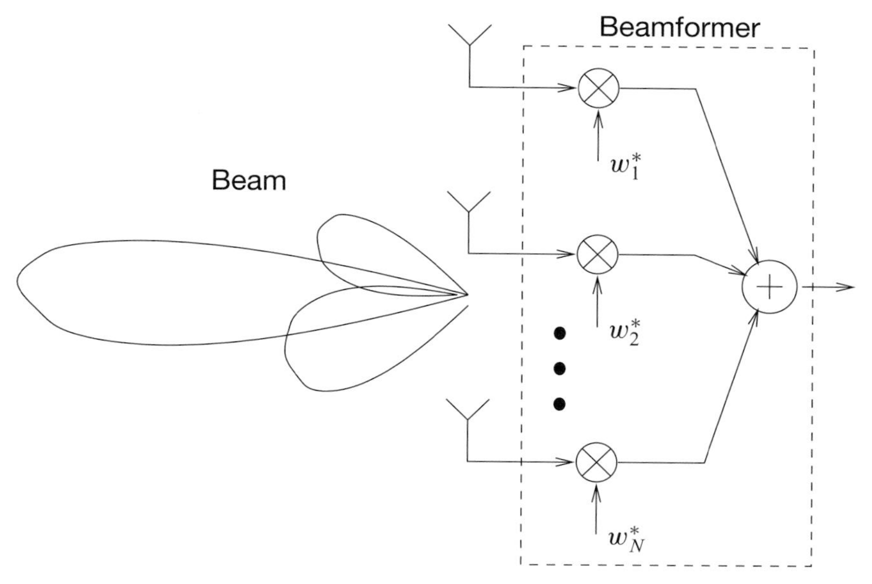

Figure 5.9 Beamforming with N receive antennas.

spatial selectivity of the beamforming vector, denoted by $\mathbf{w}$, can be shown as follows:

$$\begin{aligned} V(\theta; \mathbf{w}) &= \left| \frac{1}{\sqrt{N}||\mathbf{w}||} \mathbf{a}^{\mathrm{H}}(\theta)\mathbf{w} \right|^2 \\ &= \frac{1}{N||\mathbf{w}||^2} \mathbf{a}^{\mathrm{H}}(\theta)\mathbf{w}\mathbf{w}^{\mathrm{H}}\mathbf{a}(\theta). \end{aligned} \tag{5.50}$$

If the impact of interfering signals is not strong, it is generally desirable that the beamforming vector is determined to maximize the gain, $V(\theta; \mathbf{w})$, if θ is the AoA of the desired signal. On the other hand, if there are strong interfering signals, the beamforming gain has to be minimized at the AoA of the strong interfering signals. To derive a beamforming vector for a general case, there could be statistical and deterministic approaches. In this section, we focus on statistical approaches as the signals are characterized by statistical properties.

5.3.1 MMSE beamforming

Consider the case that there exist other signals, which are interfering signals, and background noise. In a free space propagation, the received (baseband) signal at an antenna array (the carrier $e^{j2\pi f_c t}$ is suppressed) is given by

$$\mathbf{r}(t) = \mathbf{a}s(t) + \mathbf{i}(t) + \mathbf{n}(t),$$

where $\mathbf{a} = \mathbf{a}(\theta)$ is the ARV of AOA θ, $s(t)$ is the desired (baseband) signal, $\mathbf{i}(t)$ is the interfering signal, and $\mathbf{n}(t)$ is the background noise vector. The sampled signal is also

given by

$$\mathbf{r}_l = \mathbf{a}s_l + \mathbf{i}_l + \mathbf{n}_l, \ l = 0, 1, \ldots \tag{5.51}$$

The vector $\mathbf{a}$ is often called the spatial signature vector, which could be a generalization of ARV for a multipath fading channel later. A beamformer is applied to $\mathbf{r}_l$ to estimate s_l.

For a given beamforming vector $\mathbf{w}$, the output of the beamformer is given by

$$\begin{aligned} \hat{s}_l &= \mathbf{w}^{\mathrm{H}}\mathbf{r}_l \\ &= \mathbf{w}^{\mathrm{H}}\mathbf{a}s_l + \mathbf{w}^{\mathrm{H}}(\mathbf{i}_l + \mathbf{n}_l). \end{aligned} \tag{5.52}$$

Using the minimum mean square error (MMSE) criterion, the beamforming vector $\mathbf{w}$ that minimizes the mean square error (MSE) can be found. The error is given by

$$e_l = s_l - \mathbf{w}^{\mathrm{H}}\mathbf{r}_l.$$

For convenience, assume that s_l, $\mathbf{i}_l$, and $\mathbf{n}_l$ have zero means, i.e., $\mathcal{E}[s_l] = 0$, $\mathcal{E}[\mathbf{i}_l] = \mathcal{E}[\mathbf{n}_l] = \mathbf{0}$. The MSE is given by

$$\begin{aligned} \epsilon^2 &= \mathcal{E}[|e_l|^2] \\ &= \mathcal{E}[|s_l|^2] - \left(\mathbf{w}^{\mathrm{H}}\mathcal{E}[\mathbf{r}_l s_l^*] + \left(\mathcal{E}[\mathbf{r}_l s_l^*]\right)^{\mathrm{H}}\mathbf{w}\right) + \mathbf{w}^{\mathrm{H}}\mathcal{E}[\mathbf{r}_l\mathbf{r}_l^{\mathrm{H}}]\mathbf{w}. \end{aligned} \tag{5.53}$$

Let $\mathbf{R}_{\mathbf{r}} = \mathcal{E}[\mathbf{r}_l\mathbf{r}_l^{\mathrm{H}}]$ and $\mathbf{r}_{\mathbf{r},s} = \mathcal{E}[\mathbf{r}_l s_l^*]$. Since the MSE is a quadratic function of $\mathbf{w}$, the optimal solution of $\mathbf{w}$ that minimizes the MSE can be obtained by taking the derivative with respect to $\mathbf{w}$:

$$\frac{\partial}{\partial \mathbf{w}}\epsilon^2 = -2\mathbf{r}_{\mathbf{r},s} + 2\mathbf{R}_{\mathbf{r}}\mathbf{w}.$$

Since the optimal weight vector makes the derivative zero, we have

$$\left.\frac{\partial}{\partial \mathbf{w}}\epsilon^2\right|_{\mathbf{w}=\mathbf{w}_{\mathrm{mmse}}} = 0,$$

where $\mathbf{w}_{\mathrm{mmse}}$ denotes the MMSE solution (also called the Wiener solution). It follows that

$$\mathbf{w}_{\mathrm{mmse}} = \mathbf{R}_{\mathbf{r}}^{-1}\mathbf{r}_{\mathbf{r},s}. \tag{5.54}$$

The MMSE becomes

$$\epsilon^2_{\min} = E[|s_l|^2] - \mathbf{r}_{\mathbf{r},s}^{\mathrm{H}}\mathbf{R}_{\mathbf{r}}^{-1}\mathbf{r}_{\mathbf{r},s}.$$

It is noteworthy that we need to have second-order moments of $\mathbf{r}_l$ and s_l.

If s_l, $\mathbf{i}_l$, and $\mathbf{n}_l$ are mutually uncorrelated, we have

$$\begin{aligned} \mathcal{E}[\mathbf{r}_l s_l^*] &= \mathbf{a}\sigma_s^2; \\ \mathcal{E}[\mathbf{r}_l\mathbf{r}_l^{\mathrm{H}}] &= \mathbf{a}\mathbf{a}^{\mathrm{H}}\sigma_s^2 + \mathbf{R}_{\mathbf{i}} + \mathbf{R}_{\mathbf{n}}, \end{aligned}$$

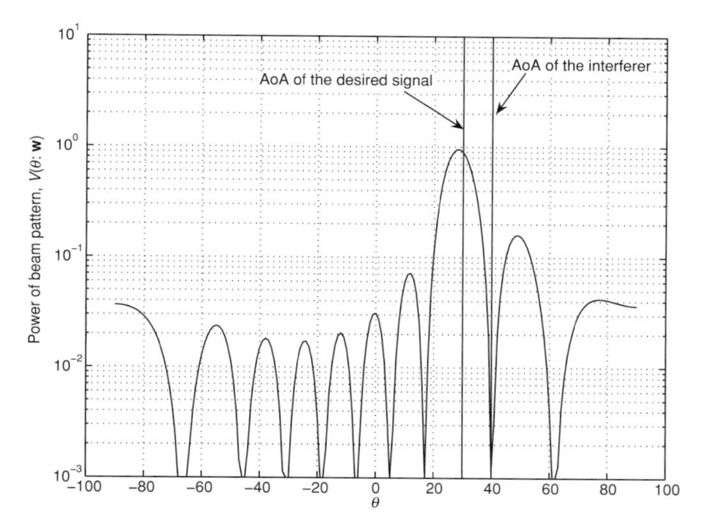

Figure 5.10 Power of the normalized beam pattern for the MMSE beamformer when the AoAs of the desired signal and interferer are 30° and 40°, respectively.

where $\sigma_s^2 = \mathcal{E}[|s_l|^2]$, $\mathbf{R_i} = \mathcal{E}[\mathbf{i}_l \mathbf{i}_l^{\mathrm{H}}]$, and $\mathbf{R_n} = \mathcal{E}[\mathbf{n}_l \mathbf{n}_l^{\mathrm{H}}]$. Using the results in Chapter 4, we can show that the MMSE beamforming vector becomes

$$\mathbf{w}_{\mathrm{mmse}} = \epsilon_{\mathrm{min}}^2 \left(\mathbf{R_i} + \mathbf{R_n}\right)^{-1} \mathbf{a}, \tag{5.55}$$

where the MMSE is given by

$$\epsilon_{\mathrm{min}}^2 = \frac{\sigma_s^2}{1 + \sigma_s^2 \mathbf{a}^{\mathrm{H}}(\mathbf{R_i} + \mathbf{R_n})^{-1}\mathbf{a}}. \tag{5.56}$$

Example 5.3.1 Suppose that the standard ULA of $N = 10$ is used. The AoA of the desired signal is 30°, i.e. $\mathbf{a} = \mathbf{a}(\theta) = \mathbf{a}(30°)$ and there is an interferer whose AoA is 40°. In particular, we have

$$\mathbf{i}_l = \mathbf{a}(40°)x_l,$$

where x_l represents the interfering signal. We assume that $\mathcal{E}[s_l] = \mathcal{E}[x_l] = 0$ and $\mathcal{E}[|s_l|^2] = \mathcal{E}[|x_l|^2] = 1$. Furthermore, $\mathbf{n}_l \sim \mathcal{CN}(0, \mathbf{I})$. Then, we have

$$\mathbf{w}_{\mathrm{mmse}} = \epsilon_{\mathrm{min}}^2 \left(\mathbf{a}(40°)\mathbf{a}^{\mathrm{H}}(40°) + \mathbf{I}\right)^{-1} \mathbf{a}(30°).$$

Figure 5.10 shows the power of the normalized beam pattern for the MMSE beamformer. It is shown that a null is formed to the AoA of the interferer. Note that the beamforming gain is not maximized at $\theta = 30°$, which is the AoA of the desired signal. The MMSE beamformer does not necessarily maximize the beamforming gain at the AoA of the desired signal, while it is crucial to mitigate the impact of interfering signals by forming nulls to minimize the MSE as shown in Fig. 5.10.

5.3.2 MSINR beamforming

The beamforming vector can be derived to maximize the signal-to-interference-plus-noise ratio (SINR). From (5.52), the SINR is given by

$$\mathrm{SINR} = \frac{\mathcal{E}[|\mathbf{w}^{\mathrm{H}}\mathbf{a}s_l|^2]}{\mathcal{E}[|\mathbf{w}^{\mathrm{H}}\mathbf{u}_l|^2]}, \tag{5.57}$$

where $\mathbf{u}_l = \mathbf{i}_l + \mathbf{n}_l$. Let $\mathbf{R}_{\mathbf{u}} = E[\mathbf{u}_l\mathbf{u}_l^{\mathrm{H}}]$. Then, the SINR becomes

$$\mathrm{SINR} = \frac{\mathbf{w}^{\mathrm{H}}\mathbf{a}\mathbf{a}^{\mathrm{H}}\mathbf{w}}{\mathbf{w}^{\mathrm{H}}\mathbf{R}_{\mathbf{u}}\mathbf{w}}\sigma_s^2. \tag{5.58}$$

Using the Cauchy–Schwarz inequality, we have

$$\mathrm{SINR} \le \sigma_s^2\mathbf{a}^{\mathrm{H}}\mathbf{R}_{\mathbf{u}}^{-1}\mathbf{a}, \tag{5.59}$$

where the term on the right-hand side is the maximum SINR. In (5.59), the equality is achieved if

$$\mathbf{w}_{\mathrm{msinr}} = \kappa\mathbf{R}_{\mathbf{u}}^{-1}\mathbf{a}, \tag{5.60}$$

where κ is any non-zero complex scalar. This beamforming is the maximum SINR beamforming. The MMSE beamforming vector also a MSINR beamforming vector as shown in Chapter 4.

5.3.3 MVDR beamforming

The minimum variance distortionless response (MVDR) criterion is widely used for beamforming. There are two different approaches for MVDR beamforming. In the first approach, we attempt to minimize the noise variance after beamforming. In the second approach, the variance of the beamforming output itself is to be minimized. In both approaches, the same constraint for distortionless response is employed.

To derive the first approach for the MVDR beamforming, consider the beamformer's output from (5.52):

$$\hat{s}_l = \mathbf{w}^{\mathrm{H}}\mathbf{a}s_l + \mathbf{w}^{\mathrm{H}}\mathbf{u}_l. \tag{5.61}$$

The variance of the interference-plus-noise is given by

$$\mathcal{E}[|\mathbf{w}^{\mathrm{H}}\mathbf{u}_l|^2] = \mathbf{w}^{\mathrm{H}}\mathbf{R}_{\mathbf{u}}\mathbf{w}. \tag{5.62}$$

To avoid any distortion to the desired signal in minimizing the variance of the interference-plus-noise, the following constraint can be employed:

$$\mathbf{w}^{\mathrm{H}}\mathbf{a} = 1. \tag{5.63}$$

Once this constraint is imposed, the beamformer's output becomes $\hat{s}_l = s_l$ in the absence of interference and noise. From the above, we can formulate the MVDR optimization problem as follows:

$$\begin{aligned} \mathbf{w}_{\mathrm{mvdr}} &= \arg\min_{\mathbf{w}} \mathbf{w}^{\mathrm{H}}\mathbf{R}_{\mathbf{u}}\mathbf{w} \\ &\text{subject to } \mathbf{w}^{\mathrm{H}}\mathbf{a} = 1. \end{aligned} \tag{5.64}$$

The solution of (5.64) can be found by using the Lagrange multiplier. Let λ be the Lagrange multiplier. Then, the constrained optimization in (5.64) becomes the following unconstrained optimization problem:

$$\mathbf{w}(\lambda) = \arg\min_{\mathbf{w}} \left\{ \mathbf{w}^{\mathrm{H}}\mathbf{R}_{\mathbf{u}}\mathbf{w} + \lambda(1 - \mathbf{w}^{\mathrm{H}}\mathbf{a}) + \lambda^*(1 - \mathbf{a}^{\mathrm{H}}\mathbf{w}) \right\}. \tag{5.65}$$

It follows that

$$\mathbf{w}(\lambda) = \lambda \mathbf{R}_{\mathbf{u}}^{-1}\mathbf{a}.$$

Since $\mathbf{w}^{\mathrm{H}}\mathbf{a} = 1$, λ becomes

$$\lambda = \frac{1}{\mathbf{a}^{\mathrm{H}}\mathbf{R}_{\mathbf{u}}^{-1}\mathbf{a}}.$$

The resulting beamforming vector is given by

$$\mathbf{w}_{\mathrm{mvdr}} = \frac{1}{\mathbf{a}^{\mathrm{H}}\mathbf{R}_{\mathbf{u}}^{-1}\mathbf{a}}\mathbf{R}_{\mathbf{u}}^{-1}\mathbf{a}. \tag{5.66}$$

In the second approach for MVDR beamforming, we attempt to minimize the variance of the beamformer's output. Thus, the MVDR optimization problem is given by

$$\begin{aligned} \mathbf{w}_{\mathrm{mvdr}} &= \arg\min\nolimits_{\mathbf{w}} \mathcal{E}[|\mathbf{w}^{\mathrm{H}}\mathbf{r}_l|^2] \\ &\text{subject to } \mathbf{w}^{\mathrm{H}}\mathbf{a} = 1. \end{aligned} \tag{5.67}$$

The optimal solution is given by

$$\mathbf{w}_{\mathrm{mvdr}} = \frac{1}{\mathbf{a}^{\mathrm{H}}\mathbf{R}_{\mathbf{r}}^{-1}\mathbf{a}}\mathbf{R}_{\mathbf{r}}^{-1}\mathbf{a}. \tag{5.68}$$

In addition, the estimate of s_l from the MVDR beamformer is given by

$$\begin{aligned} \hat{s}_{\mathrm{mvdr},l} &= \mathbf{w}_{\mathrm{mvdr}}^{\mathrm{H}}\mathbf{r}_l \\ &= s_l + \frac{\mathbf{a}^{\mathrm{H}}\mathbf{R}_{\mathbf{r}}^{-1}\mathbf{u}_l}{\mathbf{a}^{\mathrm{H}}\mathbf{R}_{\mathbf{r}}^{-1}\mathbf{a}}. \end{aligned} \tag{5.69}$$

It can also be shown that if $\mathbf{u}_l$ and s_l are uncorrelated with each other, the MVDR beamforming vector becomes an MSINR beamforming vector. Furthermore, using Woodbury's identity, we can prove that the MVDR beamforming vectors from the two different approaches are the same by showing

$$\frac{1}{\mathbf{a}^{\mathrm{H}}\mathbf{R}_{\mathbf{u}}^{-1}\mathbf{a}}\mathbf{R}_{\mathbf{u}}^{-1}\mathbf{a} = \frac{1}{\mathbf{a}^{\mathrm{H}}\mathbf{R}_{\mathbf{r}}^{-1}\mathbf{a}}\mathbf{R}_{\mathbf{r}}^{-1}\mathbf{a}.$$

Since the covariance matrices are to be estimated in real implementation, the second approach becomes preferable where the sample covariance matrix of $\mathbf{r}_l$, $\hat{\mathbf{R}}_{\mathbf{r}} = \frac{1}{L}\sum_{l=0}^{L-1}\mathbf{r}_l\mathbf{r}_l^{\mathrm{H}}$, can be used to replace $\mathbf{R}_{\mathbf{r}}$.

5.3.4 Generalized MVDR beamformer

A generalization of the MVDR beamformer can be devised by modifying the constraint as follows:

$$\mathbf{Cw} = \mathbf{c},$$

where $\mathbf{C}$ is the constraint matrix and $\mathbf{c}$ is the constraining vector. This generalization is called linearly constrained minimum variance (LCMV) beamforming and becomes useful when the spatial information of interfering signals is available. For example, suppose that the interfering signal is given by

$$\mathbf{i}_l = \mathbf{u}x_l,$$

where $\mathbf{u}$ is the spatial signature vector of the (scalar) interfering signal, x_l. Then, the constraint becomes

$$\underbrace{\begin{bmatrix} \mathbf{a}^{\mathrm{H}} \\ \mathbf{u}^{\mathrm{H}} \end{bmatrix}}_{\triangleq \mathbf{C}} \mathbf{w} = \underbrace{\begin{bmatrix} 1 \\ 0 \end{bmatrix}}_{\triangleq \mathbf{c}}$$

to suppress the interfering signal without any distortion of the desired signal.

We can show that the LCMV beamforming problem is given by

$$\begin{gathered} \min_{\mathbf{w}} \mathbf{w}^{\mathrm{H}} \mathbf{R}_{\mathbf{r}} \mathbf{w} \\ \text{subject to } \mathbf{Cw} = \mathbf{c}. \end{gathered}$$

Using the Lagrange multiplier, the solution becomes

$$\mathbf{w}_{\mathrm{lcmv}} = \mathbf{R}_{\mathbf{r}}^{-1} \mathbf{C}^{\mathrm{H}} (\mathbf{C} \mathbf{R}_{\mathbf{r}}^{-1} \mathbf{C}^{\mathrm{H}})^{-1} \mathbf{c}. \tag{5.70}$$

An example from multiuser communications can be considered for the LCMV beamforming problem. Suppose that there are K users and the received signal sampled at the lth symbol time is

$$\mathbf{r}_l = \sum_{k=1}^{K} \mathbf{a}(\theta_k) s_{k,l} + \mathbf{n}_l, \tag{5.71}$$

where $s_{k,l}$ and θ_k are the kth user's signal and its AoA, respectively. If the signal from user 1 is the desired signal, the constraint is given by

$$\underbrace{\begin{bmatrix} \mathbf{a}^{\mathrm{H}}(\theta_1) \\ \mathbf{a}^{\mathrm{H}}(\theta_2) \\ \vdots \\ \mathbf{a}^{\mathrm{H}}(\theta_K) \end{bmatrix}}_{\triangleq \mathbf{C}} \mathbf{w} = \underbrace{\begin{bmatrix} 1 \\ 0 \\ \vdots \\ 0 \end{bmatrix}}_{\triangleq \mathbf{c}}.$$

The constraint becomes feasible if $N \geq K$. Using this LCMV beamforming, the other users' signals are suppressed. If a joint detection/decoding is available, the signals to be

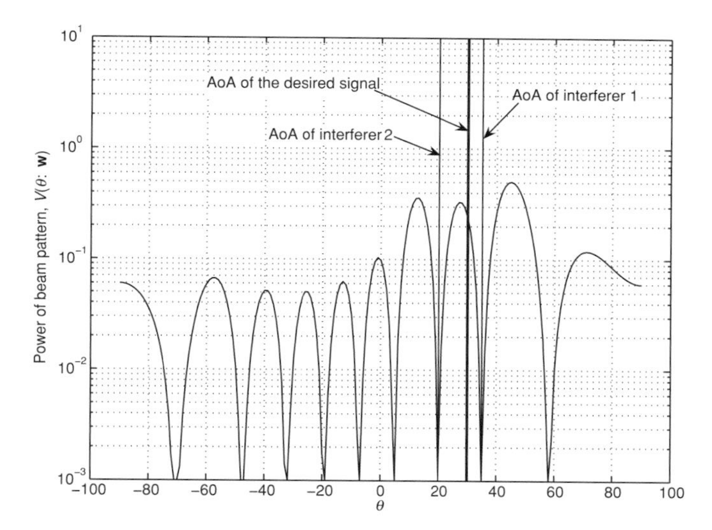

Figure 5.11 Power of normalized beam pattern for the MVDR beamformer when the AoA of the desired signal is 30° and the AoAs of interferers are 20° and 35°.

jointly detected/decoded should not be suppressed. In this case, the constraint has to be modified.

Example 5.3.2 Suppose that the standard ULA of $N = 10$ is used. The AoA of the desired signal is 30° and there are two interferers whose AoAs are 35° and 20°. We assume that the variances of the desired signal and interferers are all unity. Furthermore, $\mathbf{n}_l \sim \mathcal{CN}(0, \mathbf{I})$. Figure 5.11 shows the power of beam pattern for the MVDR beamformer. There are the nulls at 20° and 35° to suppress the interferers. As the AoAs of the interferers are close to the AoA of the desired signal, the beamforming gain at the AoA of the desired signal, 30°, is not the highest.

5.4 Adaptive beamforming

For beamforming in a practical system, there are various issues to be addressed. For example, the AoAs of signals are to be estimated. We also need to estimate second-order statistics in most beamforming methods. Furthermore, in the case that the array geometry has been changed, array calibration is necessary to update the ARV. Among various implementation issues, in this section, we only focus on the approaches to build beamforming vectors from actual signals. The reader is referred to (Van Trees 2002), (Monzingo & Miller 1980), and (Johnson & Dudgeon 1993) for a detailed account of implementation issues and other array signal processing techniques.

In order to find beamforming vectors, second-order statistics of the signals are required for MMSE, MSINR, or MVDR beamforming. This implies that the beamforming can be carried out in two steps: (i) estimating second-order statistics; and (ii) finding beamforming vectors using estimated second-order statistics. In general, this approach is called the sample matrix inversion (SMI) method or direct matrix inversion (DMI) method (Monzingo & Miller 1980), (Van Trees 2002) since the inverse of a sample covariance matrix is required. However, there are alternative approaches to find beamforming vectors including adaptive algorithms. In this subsection, we present well-known adaptive algorithms for beamforming.

5.4.1 LMS algorithm

Since MMSE beamforming requires a matrix inversion, the computational complexity can be high if the size of antenna array (or the number of antenna elements) is large. Without matrix inversion, fortunately, it is possible to find the MMSE solution using iterative techniques by noting that the MSE cost function is convex.

Suppose that we have a cost function of a vector $\mathbf{w}$, denoted by $C(\mathbf{w})$. It is desirable to find the optimal vector that minimizes the cost function. The optimal vector that minimizes the cost function is given by

$$\mathbf{w}_{\mathrm{o}} = \arg \min_{\mathbf{w}} C(\mathbf{w}). \tag{5.72}$$

In the steepest descent (SD) algorithm, we can recursively search for a better vector that has a smaller cost than the previous one. Let $\mathbf{w}_0$ denote the initial vector, which could be any vector. The cost becomes $C(\mathbf{w}_0)$. Now, we need to find a new vector, denoted by $\mathbf{w}_1$, that has a smaller cost than $C(\mathbf{w}_0)$, i.e. $C(\mathbf{w}_1) < C(\mathbf{w}_0)$. If $C(\mathbf{w})$ is a smooth function and its derivative or gradient exists, a new vector obtained by moving from $\mathbf{w}_0$ to the direction of the negative gradient of $C(\mathbf{w})$ can have a smaller cost. That is, the new vector can be found as

$$\mathbf{w}_1 = \mathbf{w}_0 - \mu \left. \frac{\partial C(\mathbf{w})}{\partial \mathbf{w}} \right|_{\mathbf{w}=\mathbf{w}_0},$$

where $\left. \frac{\partial C(\mathbf{w})}{\partial \mathbf{w}} \right|_{\mathbf{w}=\mathbf{w}_0}$ denotes the gradient vector at $\mathbf{w} = \mathbf{w}_0$ and μ is a positive constant, which is called the step size or adaptation gain. To make sure that $C(\mathbf{w}_1) < C(\mathbf{w}_0)$, μ needs to be sufficiently small. Based on this observation, the SD algorithm can be formulated as

$$\mathbf{w}_{l+1} = \mathbf{w}_l - \mu \left. \frac{\partial C(\mathbf{w})}{\partial \mathbf{w}} \right|_{\mathbf{w}=\mathbf{w}_l}. \tag{5.73}$$

We can easily verify that if $\mathbf{w}_l$ becomes the optimal vector, i.e. $\mathbf{w}_l = \mathbf{w}_{\mathrm{o}}$, there is no more updating, i.e. $\mathbf{w}_{l+1} = \mathbf{w}_l = \mathbf{w}_{\mathrm{o}}$, since the gradient is zero at $\mathbf{w} = \mathbf{w}_{\mathrm{o}}$:

$$\left. \frac{\partial C(\mathbf{w})}{\partial \mathbf{w}} \right|_{\mathbf{w}=\mathbf{w}_{\mathrm{o}}} = \mathbf{0}.$$

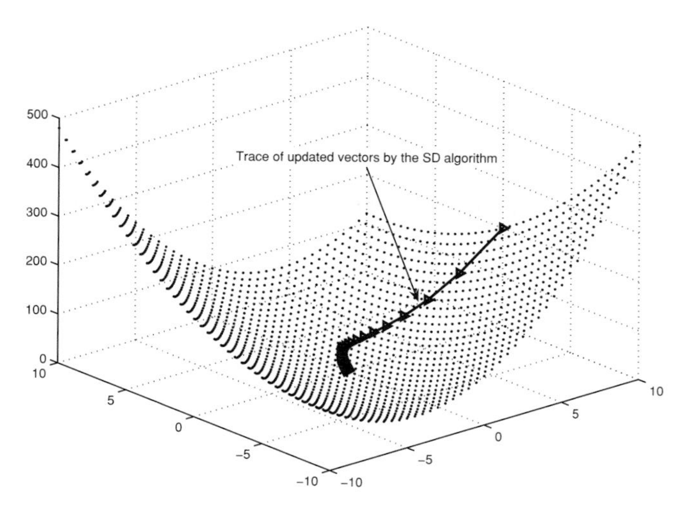

Figure 5.12 Surface of a cost function and trace of the updated vectors by the SD algorithm.

As shown above, the SD algorithm can find a local minimum where the gradient is zero. Thus, if the cost function is convex, the SD algorithm can find the global minimum.

In general, the convergence properties of the SD algorithm depend on the cost function and step size. For a detailed account, the reader is referred to (Solo & Kong 1995), (Choi 2006).

In Fig. 5.12, an illustration of the trace of the updated vectors from the SD algorithm is shown when $C(\mathbf{w}) = \mathbf{w}^{\mathrm{T}}\mathbf{A}\mathbf{w}$, where

$$\mathbf{A} = \begin{bmatrix} 2 & -0.9 \\ -0.9 & 1 \end{bmatrix}.$$

The initial vector is $\mathbf{w}_0 = [10\ 0]^{\mathrm{T}}$ and the step size is 0.1. It is shown that the SD algorithm can find the optimal vector, $\mathbf{w}_{\mathrm{o}} = [0\ 0]^{\mathrm{T}}$.

To find the MMSE beamforming vector, the SD algorithm can be applied. As shown in (5.53), the cost function is a quadratic function. In this case, the SD algorithm can converge to the optimal solution if μ is sufficiently small. The SD algorithm to find the MMSE beamforming vector is given by

$$\mathbf{w}_{l+1} = \mathbf{w}_l - \mu \frac{\partial}{\partial \mathbf{w}} \epsilon^2 \bigg|_{\mathbf{w}=\mathbf{w}_l}$$

or

$$\mathbf{w}_{l+1} = \mathbf{w}_l - 2\mu \left(\mathbf{R}_{\mathbf{r}} \mathbf{w}_l - \mathbf{r}_{\mathbf{r},s} \right). \tag{5.74}$$

From (5.74), we note that the second-order statistics, $\mathbf{R}_{\mathbf{r}}$ and $\mathbf{r}_{\mathbf{r},s}$, are needed to run the SD algorithm. As they are not available, we may replace them with their estimates from L samples. Alternatively, an on-line algorithm can be devised by updating the weight vector when a new sample is available.

From (5.74), it is straightforward to derive an on-line algorithm, which is the least mean square (LMS) algorithm. Noting

$$\begin{aligned}
\frac{\partial}{\partial \mathbf{w}}\epsilon^2 &= \frac{\partial}{\partial \mathbf{w}}\mathcal{E}[|e_l|^2] \\
&= \frac{\partial}{\partial \mathbf{w}}\mathcal{E}[|s_l - \mathbf{w}^{\mathrm{H}}\mathbf{r}_l|^2] \\
&= \mathcal{E}\left[\frac{\partial}{\partial \mathbf{w}}|s_l - \mathbf{w}^{\mathrm{H}}\mathbf{r}_l|^2\right] \\
&= \mathcal{E}[-2e_l\mathbf{r}_l],
\end{aligned}$$

an estimate of $\frac{\partial}{\partial \mathbf{w}}\epsilon^2$ is given by

$$\begin{aligned}
\widehat{\frac{\partial}{\partial \mathbf{w}}\epsilon^2} &= \mathcal{E}\widehat{[-2e_l\mathbf{r}_l]} \\
&= -2e_l\mathbf{r}_l.
\end{aligned} \tag{5.75}$$

With the estimate $\frac{\partial}{\partial \mathbf{w}}\epsilon^2$, the SD algorithm can be modified as

$$\begin{aligned}
\mathbf{w}_{l+1} &= \mathbf{w}_l - \mu\widehat{\frac{\partial}{\partial \mathbf{w}}\epsilon^2}\Bigg|_{\mathbf{w}=\mathbf{w}_l} \\
&= \mathbf{w}_l + 2\mu\xi_l^*\mathbf{r}_l \\
&= \mathbf{w}_l - 2\mu(\mathbf{w}_l^{\mathrm{H}}\mathbf{r}_l - s_l)^*\mathbf{r}_l,
\end{aligned} \tag{5.76}$$

where

$$\begin{aligned}
\xi_l &= e_l\Big|_{\mathbf{w}=\mathbf{w}_l} \\
&= s_l - \mathbf{w}_l^{\mathrm{H}}\mathbf{r}_l.
\end{aligned}$$

As shown in (5.76), the transmitted signal, s_l, should be available to run the LMS algorithm at the receiver. This implies that the transmitter needs to transmit a certain known sequence first so that the receiver can run the LMS algorithm for MMSE beamforming.

Although the LMS algorithm is derived from the SD algorithm, the LMS algorithm can be considered as an algorithm for stochastic approximation (Albert & Gardner 1966). Note that a sequence of the weight vectors generated from the LMS algorithm is random, while that from the SD algorithm is deterministic. Thus, the convergence analysis of the LMS algorithm is different from that of the SD algorithm and is more complicated. For a detailed account, the reader is referred to (Solo & Kong 1995).

Using the RLS algorithm in Chapter 3 (Subsection 3.1.2), we can also build an adaptive algorithm to find the weight vector for MMSE beamforming. With the exponential weighted sum squared error (EWSSE) which is given by

$$\mathrm{EWSSE}_l = \sum_{k=0}^{l}\lambda^{l-k}|s_k - \mathbf{w}^{\mathrm{H}}\mathbf{r}_l|^2, \tag{5.77}$$

where λ is the forgetting factor, the RLS algorithm can be readily obtained.

5.4.2 Frost's adaptive algorithm

Frost's adaptive algorithm (Frost 1972) is an on-line approach for the MVDR beamforming. Frost's adaptive algorithm is a stochastic version of the SD algorithm for the MVDR problem (in this view, we can see that the LMS algorithm is a stochastic version of the SD algorithm for the MMSE problem). The SD algorithm can be devised for the MVDR problem as

$$\mathbf{w}_{l+1} = \mathbf{w}_l - \mu \frac{\partial}{\partial \mathbf{w}} \left(\mathbf{w}^{\mathrm{H}} \mathbf{R}_{\mathbf{r}} \mathbf{w} + \lambda(1 - \mathbf{a}^{\mathrm{H}} \mathbf{w}) \right) \big|_{\mathbf{w}=\mathbf{w}_l},$$

where λ is the Lagrange multiplier for the constraint $\mathbf{a}^{\mathrm{H}}\mathbf{w} = 1$. It follows that

$$\mathbf{w}_{l+1} = \mathbf{w}_l - \mu \left(\mathbf{R}_{\mathbf{r}} \mathbf{w}_l - \lambda \mathbf{a} \right). \tag{5.78}$$

The Lagrange multiplier has to be decided to satisfy the following constraint:

$$\mathbf{a}^{\mathrm{H}} \mathbf{w}_{l+1} = 1. \tag{5.79}$$

In (5.78), the covariance matrix of the received signal vector, $\mathbf{R}_{\mathbf{r}}$ can be replaced by its estimate $\mathbf{r}_l \mathbf{r}_l^{\mathrm{H}}$. Then, Frost's adaptive algorithm is given by

$$\mathbf{w}_{l+1} = \mathbf{a}_c + \mathbf{P}_{\mathbf{a}}^{\perp} \left(\mathbf{r}_l - \mu z_l^* \mathbf{r}_l \right), \tag{5.80}$$

where $\mathbf{a}_c = \frac{1}{||\mathbf{a}||^2} \mathbf{a}$, $\mathbf{P}_{\mathbf{a}}^{\perp} = \mathbf{I} - \mathbf{a}_c \mathbf{a}^{\mathrm{H}} = \mathbf{I} - \frac{1}{||\mathbf{a}||^2} \mathbf{a}\mathbf{a}^{\mathrm{H}}$, and $z_l = \mathbf{w}_l^{\mathrm{H}} \mathbf{r}_l$.

5.5 Smart antenna systems

Antenna arrays can be employed in wireless communication systems to provide spatial selectivity, which becomes crucial, in particular, in a multiuser system. Wireless communication systems with antenna arrays are called smart antenna systems (Winters 1998), (Paulraj & Ng 1998), and (Liberti & Rappaport 1999). Depending on how antenna arrays are utilized, there are different smart antenna systems. In this section, we consider some examples of smart antenna systems and related processing.

5.5.1 Interference suppression in cellular systems

In cellular systems, the users in each cell have dedicated channels which are generally orthogonal. Therefore, there is no intra-cell interference. However, since the same channels can be used in other cells, there could be inter-cell interference. To mitigate inter-cell interference, beamforming techniques can be employed for a base station equipped with an antenna array as shown in Fig. 5.13. If the base station can estimate the AoAs of the desired signal and interferers, a beam can be constructed to suppress the interfering signals as illustrated in Example 5.3.2.

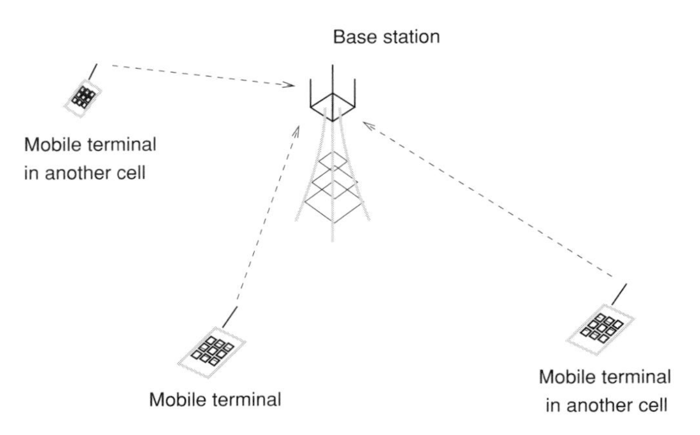

Figure 5.13 Beamforming at a base station to mitigate inter-cell interference.

5.5.2 Space division multiple access

Space division multiple access (SDMA) is a multiple access approach that uses antenna arrays. Suppose that K users attempt to transmit their signals through a common channel simultaneously to a receiver equipped with N receive antennas. In cellular communications, this is the case for uplink channels (from mobile users to a base station).

The received signal can be modeled as in (5.9) if a free space propagation is assumed:

$$\mathbf{r}_l = \mathbf{A}\mathbf{s}_l + \mathbf{n}_l, \tag{5.81}$$

where $\mathbf{A} = [\mathbf{a}(\theta_1)\ \mathbf{a}(\theta_2)\ \ldots\ \mathbf{a}(\theta_K)]$. If $K \leq N$ and $\mathbf{A}$ is full-rank, there exists a pseudo-inverse of $\mathbf{A}$ as

$$\mathbf{A}^\dagger = (\mathbf{A}^{\mathrm{H}}\mathbf{A})^{-1}\mathbf{A}^{\mathrm{H}}.$$

Using $\mathbf{A}^\dagger$ (for the ZF approach), the interference can be removed as follows:

$$\begin{aligned}\mathbf{A}^\dagger\mathbf{r}_l &= \mathbf{A}^\dagger(\mathbf{A}\mathbf{s}_l + \mathbf{n}_l)\\ &= \mathbf{s}_l + \mathbf{A}^\dagger\mathbf{n}_l.\end{aligned} \tag{5.82}$$

This implies that there can be up to $K = N$ users who transmit signals through a common channel and those signals can be detected without interference when the receiver is equipped with N receive antennas. Since the multiple signals can be separated in the space domain (created by an antenna array), the multiple signal transmission to an antenna array is called SDMA.

In separating multiple signals in (5.82), we do not take into account the background noise. The background noise can be enhanced by $\mathbf{A}^\dagger$. For example, if two AoAs are close to each other, the corresponding two ARVs are almost the same and $\mathbf{A}$ becomes rank deficient. In this case, the elements of $\mathbf{A}^\dagger$ become larger. To avoid this noise enhancement problem, the MMSE approach can be employed. The linear MMSE combiner to estimate $\mathbf{s}$, which is derived in Chapter 4, can be used.

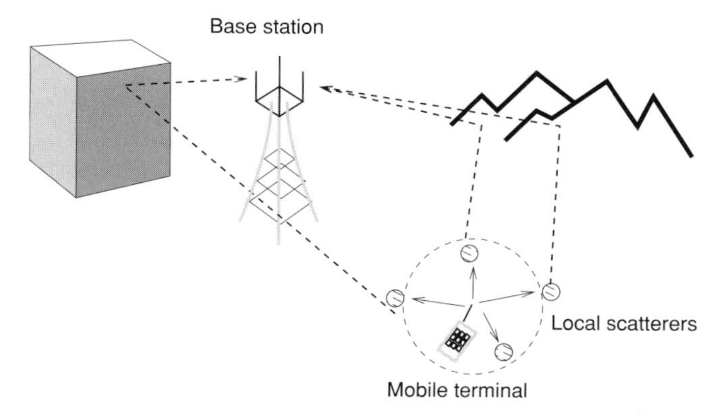

Figure 5.14 Signal propagation in terrestrial wireless communications.

5.5.3 Adaptive array and switched-beam

In order to build beamforming vectors or matrices, we may need to know the ARVs or AoA of the incoming signals. As the locations of the transmitters can vary, the ARVs or AoAs are time-varying. Thus, it is necessary to update beamforming vectors periodically, which means that tracking of ARVs or AoAs is necessary. The resulting system of an antenna array with tracking capability is called adaptive array. For tracking, adaptive beamforming algorithms can be used. In this case, a reference signal should be transmitted periodically to allow updating beamforming vectors. If reference signals are not available, blind adaptive beamforming algorithms can be employed.

In switched-beam systems, multiple fixed beams (or multiple beamforming vectors) are built in advance. In this case, a beam can be chosen for a user, and if the user's location is changed, a different beam can be chosen. Although the implementation of switched-beam systems is easier than that of adaptive array systems, switched-beam systems could suffer from interference in the case that a beam is chosen for multiple users. Thus, there has to be an additional approach to share a common beam to avoid interference. For example, time division multiple access (TDMA) can be used for those users who share a common beam.

5.5.4 Practical issues

There are various practical issues in implementing smart antenna systems. In this subsection, we discuss a few examples.

In terrestrial wireless communications, it is difficult to model a signal source as a single point source. As illustrated in Fig. 5.14, the signal transmitted from a mobile terminal could be scattered by local scatterers. In addition, the scattered signals can experience multipath fading. As a result, the received signal by an antenna array cannot be expressed in terms of a single AoA. Assuming that there are Q scatterers, the received signal is given by

$$\mathbf{r}(t) = \sum_{q=1}^{Q} \mathbf{a}(\theta_{(q)})s(t) + \mathbf{n}(t), \tag{5.83}$$

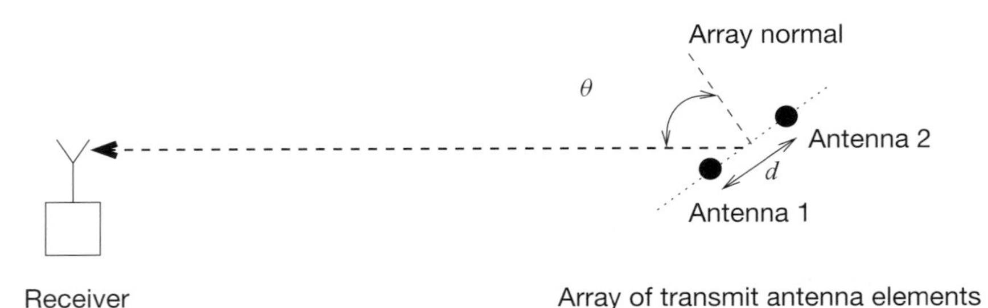

Figure 5.15 An array of two transmit antenna elements and receiver.

where $\theta_{(q)}$ denotes the AoA of the qth scattered signal. In general, the difference between the $\theta_{(q)}$'s is sufficiently small as the local scatterers are in close proximity to the mobile terminal. Moreover, if the signals experience multipath fading, we have

$$\mathbf{r}(t) = \sum_{p=1}^{P} \sum_{q=1}^{Q} \alpha_p \mathbf{a}(\theta_{p,(q)}) s(t - \tau_p) + \mathbf{n}(t), \tag{5.84}$$

where $\theta_{p,(q)}$ denotes the AoA of the qth scattered signal received through the pth multipath and α_p and τ_p represent the fading coefficient and delay of the pth multipath signal, respectively. In the presence of local scatterers and multipath fading, the estimation of AoA becomes difficult and beamforming becomes also complicated. The reader is referred to (Ertel, Cardieri, Sowerby, Rappaport & Reed 1998) and (Liberti & Rappaport 1999) for a detailed account of the characterization of spatial and temporal radio channels in terrestrial wireless communications.

The determination of antenna spacing is crucial in smart antenna systems. For the AoA estimation, the antenna spacing could be a half wavelength with a standard ULA. However, to exploit spatial diversity, the antenna elements on an antenna array should be spaced sufficiently far apart so that the received signals experience independent fading. As shown in (Schwartz *et al.* 1966), the spatial diversity gain is maximized if the received signals experience totally different fading. However, in this case, the AoA estimation becomes difficult because the visible region decreases with antenna spacing as shown in (5.6).

5.5.5 Smart antenna for signal transmission

So far, we have focused on beamforming for signal reception. In smart antenna systems, beamforming can also be used to transmit signals to desired signal directions. In this case, the transmitter needs to be equipped with an antenna array. For example, consider an array of two antenna elements at the transmitter. The angle between the array normal and the direction to the receiver is θ as shown in Fig. 5.15.

Under the assumption of free space propagation, the received signal at the receiver can be given by

$$r(t) = s_1(t) + s_2(t - \tau) + n(t), \tag{5.85}$$

where $s_k(t)$ denotes the transmitted signal from antenna k and $n(t)$ is the background noise. The transmitted signal from antenna 2 arrives at the receiver with an inter-arrival time difference of τ, which is given by

$$\tau = d \sin \theta$$

under the plane wave model, where d is much shorter than the distance between the antenna array and receiver. If $s_k(t)$ are narrowband signals, $s_2(t-\tau)$ is approximated by $\mathrm{e}^{-j2\pi f_c \tau} s_2(t)$. Thus, we have

$$r(t) = \mathbf{a}^{\mathrm{T}}(\theta)\mathbf{s}(t) + n(t), \tag{5.86}$$

where

$$\mathbf{a}(\theta) = [1 \ \mathrm{e}^{-j2\pi f_c d \sin\theta}]^{\mathrm{T}};$$

$$\mathbf{s}(t) = [s_1(t) \ s_2(t)]^{\mathrm{T}}.$$

With more than two antenna elements, the signal model in (5.86) can be straightforwardly extended. For transmit beamforming, the transmitted signals from the antenna array can be given by

$$\mathbf{s}(t) = \mathbf{w}s(t), \tag{5.87}$$

where $\mathbf{w}$ is the transmit beamforming vector and $s(t)$ is the signal to be transmitted. From (5.87), the received signal after sampling becomes

$$r(t) = \mathbf{a}^{\mathrm{T}}(\theta)\mathbf{w}s(t) + n(t).$$

After filtering and sampling the received signal, the SNR at the receiver becomes the function of $\mathbf{w}$ as follows:

$$\mathrm{SNR} = \frac{|\mathbf{a}^{\mathrm{T}}(\theta)\mathbf{w}|^2 \sigma_s^2}{N_0},$$

where $\sigma_s^2 = \mathcal{E}[|s_l|^2]$ and $N_0 = \mathcal{E}[|n_l|^2]$. Here, s_l and n_l denote the lth samples of $s(t)$ and $n(t)$ after filtering, respectively. We can easily show that the SNR is maximized if $\mathbf{w} \propto \mathbf{a}^*(\theta)$. Since the transmission power is constrained by the maximum transmission power, the squared norm of $\mathbf{w}$ is constrained. For normalization, we can consider a unit-norm beamforming vector $\mathbf{w}$. In this case, the MSNR beamforming vector is given by

$$\mathbf{w}_{\mathrm{msnr}} = \frac{\mathbf{a}^*(\theta)}{||\mathbf{a}(\theta)||}.$$

For transmit beamforming, various criteria can be applied. In the presence of multiple receivers, transmit beamforming can be generalized to take into account spatial multiplexing as well as the interference problem. As there are multiple receivers, multiple beams should be built (for spatial multiplexing) and this leads to joint beamforming with multiple beamforming vectors. In general, multiuser transmit beamforming is more complicated than multiuser receive beamforming (e.g. SDMA), because beamforming vectors are to be jointly optimized to avoid interference with transmission power constraints.

Assume that there are K receivers. Let $\mathbf{a}_k^{\mathrm{T}}$ denote the ARV from the transmit antenna array of N transmit antenna elements to receiver k. In addition, let $\mathbf{w}_k$ denote the beamforming vector to the signal to receiver k, which is denoted by $s_k(t)$. Then, the received signal at the receiver k after sampling is given by

$$\begin{aligned} r_{k,l} &= \mathbf{a}_k^{\mathrm{T}} \left(\sum_{q=1}^{K} \mathbf{w}_q s_{q,l} \right) + n_{k,l} \\ &= \mathbf{a}_k^{\mathrm{T}} \mathbf{W} \mathbf{s}_l + n_{k,l}, \end{aligned} \tag{5.88}$$

where $\mathbf{W} = [\mathbf{w}_1\ \mathbf{w}_2\ \ldots\ \mathbf{w}_K]$ and $\mathbf{s}_l = [s_{1,l}\ s_{2,l}\ \ldots\ s_{K,l}]^{\mathrm{T}}$. Here, $s_{k,l}$ is the lth sampled signal of $s_k(t)$. For joint beamforming, we can consider all the received signals together:

$$\begin{aligned} \mathbf{r}_l &= [r_{1,l}\ r_{2,l}\ \ldots\ r_{K,l}]^{\mathrm{T}} \\ &= \mathbf{A}^{\mathrm{T}} \mathbf{W} \mathbf{s}_l + \mathbf{n}_l, \end{aligned} \tag{5.89}$$

where $\mathbf{A} = [\mathbf{a}_1\ \mathbf{a}_2\ \ldots\ \mathbf{a}_K]$ and $\mathbf{n}_l = [n_{1,l}\ n_{2,l}\ \ldots\ n_{K,l}]^{\mathrm{T}}$. If $N \geq K$, then $\mathbf{W}$ can be given by

$$\mathbf{W}_{\mathrm{zf}} = \mathbf{A}^* (\mathbf{A}^{\mathrm{T}} \mathbf{A}^*)^{-1} \tag{5.90}$$

to avoid any interference, which is called ZF transmit beamforming. In this case, the received signal becomes

$$\mathbf{r}_l = \mathbf{s}_l + \mathbf{n}_l.$$

Interestingly, there is no noise enhancement problem. However, the transmit signal power could be arbitrarily high if $\mathbf{A}$ becomes rank deficient. To see this, consider the total transmission power. If the $s_{k,l}$'s are zero-mean and uncorrelated, the total transmit power becomes

$$\begin{aligned} P_{\mathrm{tx}} &= \mathcal{E}[||\mathbf{W}\mathbf{s}_l||^2] \\ &= \sum_{k=1}^{K} ||\mathbf{w}_k||^2 \sigma_k^2, \end{aligned}$$

where $\mathcal{E}[|s_{k,l}|^2] = \sigma_k^2$. If $\sigma_k^2 = \sigma_s^2$ for all k,

$$P_{\mathrm{tx}} = ||\mathbf{W}||_{\mathrm{F}}^2 \sigma_s^2.$$

For the ZF transmit beamforming, we have

$$||\mathbf{W}_{\mathrm{zf}}||_{\mathrm{F}}^2 = \mathrm{tr}\left((\mathbf{A}^{\mathrm{T}} \mathbf{A}^*)^{-1} \right).$$

From this, it can be readily shown that the total transmit power becomes infinity if $\mathbf{A}^{\mathrm{T}}\mathbf{A}^*$ becomes singular. If some receivers are located in the same direction from the transmit antenna array, the column vectors of $\mathbf{A}$ corresponding to those receivers become linearly dependent and $\mathbf{A}^{\mathrm{T}}\mathbf{A}^*$ becomes singular. Therefore, for ZF transmit beamforming, an infinite transmission power is required in this case. To avoid this problem, different approaches should be considered with transmission power constraints.

From (5.88), the SINR becomes

$$\mathrm{SINR}_k = \frac{|\mathbf{a}_k^{\mathrm{T}}\mathbf{w}_k s_{k,l}|^2}{\mathcal{E}\left[\left|\sum_{q \neq k} \mathbf{a}_k^{\mathrm{T}}\mathbf{w}_q s_{q,l} + n_{k,l}\right|^2\right]}.$$

Then, the SINR is given by

$$\mathrm{SINR}_k = \frac{|\mathbf{a}_k^{\mathrm{T}}\mathbf{w}_k|^2 \sigma_k^2}{\sum_{q \neq k} |\mathbf{a}_k^{\mathrm{T}}\mathbf{w}_q|^2 \sigma_q^2 + N_0}. \tag{5.91}$$

With the SINR expression in (5.91), various multiuser transmit beamforming problems can be formulated. For example, the transmit power minimization problem with SINR constraints can be formulated as follows:

$$\begin{aligned} &\min_{\mathbf{W}} ||\mathbf{W}||_{\mathrm{F}}^2 \\ &\mathrm{SINR}_k \geq \Gamma_k, \end{aligned} \tag{5.92}$$

where Γ_k is the target SINR for receiver k. Since the optimization for joint multiuser transmit beamforming is beyond the scope of the book, we do not deal with this any further. For a detailed account, the reader is referred to (Schubert & Boche 2004) and the references therein.

5.6 Summary and notes

We discussed arrays and related signal processing techniques. The application of array processing to smart antennas was also presented.

The textbook (Van Trees 2002) is an encyclopedic source for array processing, while an excellent overview of beamforming can be found in (Veen & Buckley 1988). There is a vast literature on smart antenna systems. However, (Winters 1998) and (Paulraj & Ng 1998) can provide an excellent overview. In addition, Chapter 6 of (Reed 2002) is another excellent source to understand smart antennas in conjunction with software defined radio (SDR).

Problems

Problem 5.1 Derive (5.5) using $\sum_{n=0}^{N-1} x^n = \frac{1-x^N}{1-x}$.

Problem 5.2 The beamwidth can be defined by the main lobe width of the beam pattern. Find the beamwidth of a ULA from (5.5). Note that the main lobe width is the distance between the two first nulls from the origin. (Using the answer to this problem, we can show that the beamwidth decreases with the number of antennas and/or antenna spacing.)

Problem 5.3 Given the AoA θ $(-\pi/2 < \theta < \pi/2)$, define the following function

$$G(\phi, \theta) = |\mathbf{a}^{\mathrm{H}}(\phi)\mathbf{a}(\theta)|, \quad -\pi/2 < \phi < \pi/2.$$

Show that $G(\phi, \theta)$ has the global maximum if $\phi = \theta$ for a standard ULA.

Problem 5.4 Repeat the above problem when $d = \lambda/4$ and $d = \lambda$.

Problem 5.5 (Uniform circular array; UCA) Suppose that there are L antenna elements on a circle of radius r, which are uniformly located. This antenna array is called the uniform circular array (UCA). Find the ARV of this ULA.

Problem 5.6 If the AoA of the desired signal is not precisely estimated, a broader beam is preferable. LCMV beamforming can be used to broaden beamwidth by employing additional derivative constraints (Er & Cantoni 1983). Assume that a standard ULA of $N = 10$ is used. The received signal vector is given by

$$\mathbf{r}_l = \mathbf{a}(\theta)s_l + \mathbf{n}_l,$$

where $\mathbf{n}_l \sim \mathcal{CN}(0, \mathbf{I})$ and $s_l \sim \mathcal{CN}(0, 1)$. Find the LCMV beamforming vector when the AoA is $\theta = 30^\circ$. In addition, to broaden beamwidth, find the LCMV beamforming vector with the additional constraint that

$$\left.\frac{\partial}{\partial \theta}\mathbf{a}(\theta)\right|_{\theta=30^\circ} = 0.$$

Discuss how the beamwidth can be broadened further.

Problem 5.7 Derive Eq. (5.80).

Problem 5.8 Assuming that $\mathbf{w}^H(t)\mathbf{a} = 1$, find a different version of Frost's adaptive algorithm.

6 Optimal combining: multiple-signal

In various signal processing and communication applications, multiple signals can co-exist and a receiver has to detect or estimate multiple signals (or possibly their features) simultaneously. For example, we can consider a multiple-speaker identification system that attempts to identify multiple speakers' voices simultaneously. Another example is a multi-sensor system for multiple-signal classification in radar and sonar applications. The notion of signal combining in Chapter 4 can be extended to the case of multiple signals. Since other signals co-exist, the signal combiner plays a crucial role in not only combining multiple observations, but also mitigating the other signals.

In this chapter, we discuss optimal signal combining to estimate multiple signals simultaneously when multiple observations or received signals are available. Various well-known optimal combiners are introduced. In particular, we mainly focus on the MMSE combiner as it is widely used and has various crucial properties. In signal combining or estimation, no particular constraint on transmitted signals is imposed, while in Chapter 7 we will discuss signal detection (not estimation) for multiple signals under the assumption that each signal is an element of a signal alphabet or constellation, which becomes a crucial constraint in signal detection.

6.1 Systems with multiple signals

Suppose that there are K signal sources transmitting signals simultaneously through a common channel to a receiver as shown in Fig. 6.1.[1] The signals generated from multiple sources could be correlated or not. In general, although the signals are uncorrelated, the receiver can provide a better performance if all the K signals are jointly estimated or detected. In this chapter, we focus on multiple-signal estimation (note that multiple-signal detection problems will be discussed in Chapter 7).

Due to the co-existence of multiple signals, it is often required that the receiver is to be equipped with multiple antennas or sensors (as in Chapter 4, we assume that antenna and sensor are interchangeable). We assume that the receiver is equipped with N antennas. Denote by $s_{k,l}$ the signal transmitted from the kth source during the lth symbol duration. Stacking the signals from N antennas at the receiver, we can form the following $N \times 1$

[1] Although we mainly consider a receiver for digital communications, it is not necessarily true throughout this chapter except for Sections 6.2 and 6.7. In general, a receiver can be understood as an equipment where signals are received and combined for a certain purpose.

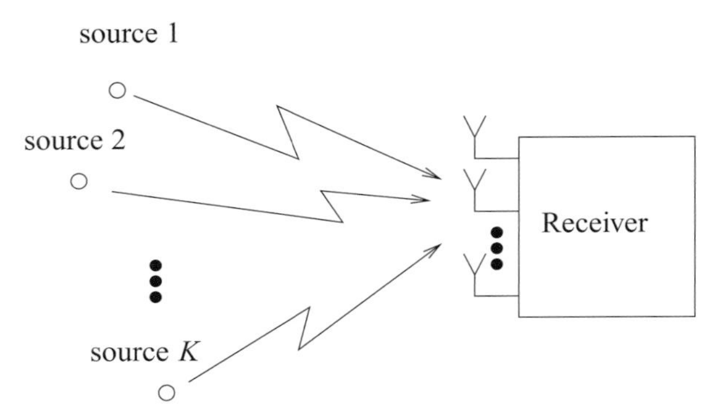

Figure 6.1 Multiple signal sources and a receiver equipped with multiple antennas or sensors.

received signal vector:

$$\begin{aligned}\mathbf{r}_l &= \sum_{k=1}^{K} \mathbf{h}_{k,l} s_{k,l} + \mathbf{n}_l \\ &= \mathbf{H}_l \mathbf{s}_l + \mathbf{n}_l, \quad l = 0, 1, \ldots, L-1, \end{aligned} \tag{6.1}$$

where $\mathbf{s}_l = [s_{1,l} \ s_{2,l} \ \ldots \ s_{K,l}]^{\mathrm{T}}$, L is the length of the signal sequence, and $\mathbf{H}_l = [\mathbf{h}_{1,l} \ \mathbf{h}_{2,l} \ \ldots \ \mathbf{h}_{K,l}]$ is a transfer matrix of size $N \times K$ from the source signals to the receiver's antennas. Here, $\mathbf{h}_{k,l}$ denotes the $N \times 1$ channel vector (or signature vector[2]) from the kth signal source to the receiver during the lth symbol duration. The nth element of $\mathbf{h}_{k,l}$ is the channel attenuation (or gain) from the kth signal source to the nth antenna during the lth symbol duration. In general, the noise vector $\mathbf{n}_l$ is assumed to be a zero-mean independent CSCG random vector with $\mathcal{E}[\mathbf{n}_l \mathbf{n}_l^{\mathrm{H}}] = \mathbf{R}_{\mathbf{n}}$, i.e. $\mathbf{n}_l \sim \mathcal{CN}(\mathbf{0}, \mathbf{R}_{\mathbf{n}})$, and $\mathbf{H}_l$ is called the channel matrix.

For a straightforward processing, it would be desirable that each antenna receives only one signal if possible. In this case, if $N = K$, the channel matrix is diagonal and the channel becomes interference-free (this is the case where there are N parallel independent channels without interference). The received signal from each antenna can be independently processed to extract a desired signal. However, in general, a superposition of multiple signals is received at each antenna. In the presence of multiple signals, the role of a linear combiner is to combine multiple observations as well as mitigate the interference so that the resulting composite channel (channel-plus-combiner) becomes (nearly) interference-free.

6.2 Structures of receiver

The receiver structure depends on the properties of signals and channels. In addition, the receiver's complexity constraint, which is one of the key implementation constraints, is

[2] Since each signal can have a different channel vector and this channel vector can be used to identify the associated signal, it is often called the signature vector.

also important in deriving the receiver structure. In this section, we briefly discuss the receiver structure in the context of digital communications.

For signal sources in digital communications, we can consider two different cases: (i) the signal sources are individually generated and encoded; (ii) the signal sources are jointly generated and encoded. In the first case, $\{s_{k,l}, l = 0, 1, \ldots, L-1\}$ is individually generated by an encoder with the input message sequence $\{m_{k,l}, l = 0, 1, \ldots, L_{\text{mess}} - 1\}$, where L_{mess} is the length of the message sequence. In the second case, $\{\mathbf{s}_l\}$ is the coded signal sequence from an encoder with the input message sequence $\{\mathbf{m}_l\}$. While it is clear that the receiver has to jointly decode the signal for the second case, there could be different receiver structures for the first case. Thus, we focus on the first case in this section.

In the first case, since K signals are transmitted through a common channel, the receiver may need to jointly decode K signals and the ML approach for joint decoding (see Problem 6.1) is given by

$$\{\bar{\mathbf{s}}_l\} = \arg \min_{\{\mathbf{s}_l\} \in \mathcal{S}_{\text{cb}}} \sum_{l=0}^{L-1} (\mathbf{r}_l - \mathbf{H}_l \mathbf{s}_l)^{\text{H}} \mathbf{R}_{\mathbf{n}}^{-1} (\mathbf{r}_l - \mathbf{H}_l \mathbf{s}_l), \tag{6.2}$$

where $\{\bar{\mathbf{s}}_l\}$ represents the decoded sequence and

$$\mathcal{S}_{\text{cb}} = \mathcal{S}_{\text{cb},1} \times \mathcal{S}_{\text{cb},2} \times \cdots \times \mathcal{S}_{\text{cb},K}.$$

Here, $\mathcal{S}_{\text{cb},k}$ denotes the individual codebook for the kth signal, $\mathcal{S}_{\text{cb}}$ denotes the joint codebook, and $\times$ represents the Cartesian product. Figure 6.2 (a) shows the structure of the receiver for joint processing. The complexity of the ML approach depends on the size of $\mathcal{S}_{\text{cb}}$ when an exhaustive search is used. Since the size of $\mathcal{S}_{\text{cb}}$ can grow exponentially with K, the complexity of the ML approach could be prohibitively high for a large K. Thus, low complexity suboptimal approaches are often desirable.

The receiver processing can be divided into two stages for suboptimal processing. The first stage is to estimate $\mathbf{s}_l$ from $\mathbf{r}_l$ without any constraint on $\mathbf{s}_l$ (e.g. $\{\mathbf{s}_l\} \in \mathcal{S}_{\text{cb}}$). In the second stage, the original message sequence can be recovered from the estimated signal sequence, denoted by $\{\hat{\mathbf{s}}_l\}$, by a joint decoder, where $\{\hat{\mathbf{s}}_l\}$ is considered as the signal received through the composite channel of the MIMO channel and linear combiner. Figure 6.2 (b) shows the structure of the receiver in which the estimation of $\mathbf{s}_l$ from $\mathbf{r}_l$ and decoding for $\mathbf{m}_l$ from $\hat{\mathbf{s}}_l$ are separated. For the joint estimation, a linear combiner can be considered within the receiver structure shown in Fig. 6.2 (b). As shown later in this chapter, there is no performance loss due to the separation of the combiner and joint decoder for the case of Gaussian signals and noise. However, the complexity is not reduced if the joint decoding requires an exhaustive search.

In order to reduce the computational complexity significantly, each output of the combiner can be decoded independently as shown in Fig. 6.2 (c). To allow multiple independent decoding, the combiner needs to mitigate the interference so that $\hat{\mathbf{s}}_l$ can be considered as an output of an interference-free channel. For each signal, the decoded

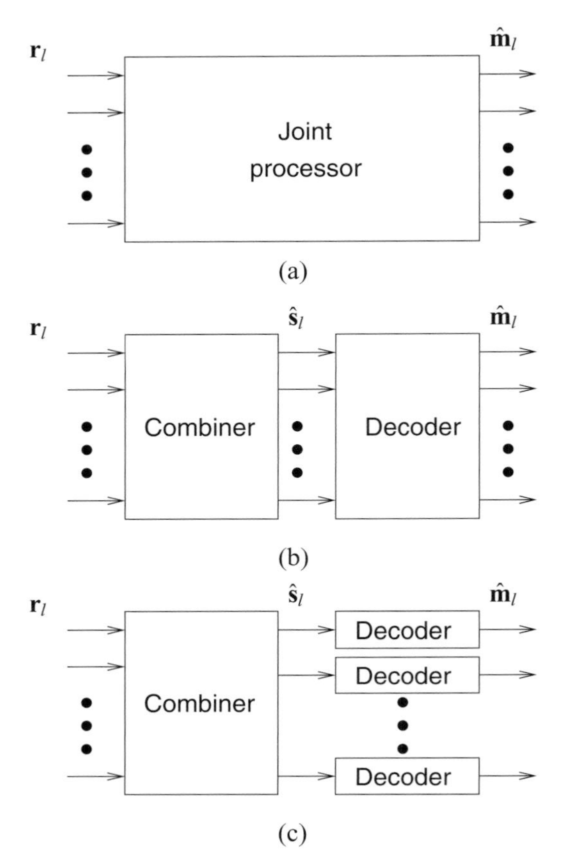

Figure 6.2 Structures of the receiver: (a) Joint processing; (b) separate combining and joint decoding; (c) separate combining and individual decoding.

signal can be obtained as follows:

$$\{\bar{s}_{k,l}\} = \arg \min_{\{s_{k,l}\} \in \mathcal{S}_{\mathrm{cb},k}} \sum_{l=0}^{L-1} |\hat{s}_{k,l} - s_{k,l}|^2, \ k = 1, 2, \ldots, K. \tag{6.3}$$

In this case, there could be a performance loss that is dependent on the channel matrix. To measure the performance loss, we can use the notion of mutual information. In Section 6.7, this performance loss is discussed in detail when the MMSE combiner is employed.

In any case, the role of the signal combiner is crucial when multiple observations are available. In this chapter, we mainly focus on linear combining methods for the combiner within the receiver structure shown in Fig. 6.2 (b) and (c).

6.3 Various combining approaches

A joint combiner or estimator is to produce an estimate of $\mathbf{s}_l$ from $\mathbf{r}_l$, not estimates of individual elements of $\mathbf{s}_l$ separately. As mentioned above, we assume that no constraint on signals is imposed in signal combining. In this section, we discuss well-known combiners. For convenience, we omit the time index l.

6.3.1 Matched filtering approach

From (6.1), we assume that there are $K-1$ interfering signals and one desired signal. Suppose that the first signal, s_1, is the desired signal. Then, the received signal vector is decomposed as

$$\mathbf{r} = \mathbf{h}_1 s_1 + \sum_{k=2}^{K} \mathbf{h}_k s_k + \mathbf{n},$$

where the first term (on the right-hand side) is the desired signal, the second term is the sum of interfering signals, and the third term is the background noise. Suppose that the $\mathbf{h}_k$'s are random. Furthermore, invoking the central limit theorem, suppose that the sum of interfering signals tends to be a white Gaussian random vector as K and N increase. Let

$$\mathbf{u} = \sum_{k=2}^{K} \mathbf{h}_k s_k + \mathbf{n}.$$

Then, for a large K and N, we may assume

$$\mathcal{E}[\mathbf{u}] = \mathbf{0};$$
$$\mathcal{E}[\mathbf{u}\mathbf{u}^{\mathrm{H}}] = I_0 \mathbf{I},$$

where I_0 is the variance of the interference-plus-noise. In this case, since $\mathbf{u}$ is spatially white, the combining vector that maximizes the SNR is given by

$$\mathbf{w}_1 \propto \mathbf{h}_1.$$

The resulting signal combining becomes identical to the matched filtering in the space domain. For convenience, define

$$\mathbf{W} = [\mathbf{w}_1 \ \mathbf{w}_2 \ \ldots \ \mathbf{w}_K].$$

Letting $\mathbf{w}_k = \mathbf{h}_k$, the output vector from a bank of the matched filters is given by

$$\begin{aligned} \hat{\mathbf{s}}_{\mathrm{mf}} &= \begin{bmatrix} \mathbf{w}_1^{\mathrm{H}} \mathbf{r} \\ \mathbf{w}_2^{\mathrm{H}} \mathbf{r} \\ \vdots \\ \mathbf{w}_K^{\mathrm{H}} \mathbf{r} \end{bmatrix} \\ &= \mathbf{W}^{\mathrm{H}} \mathbf{r}, \end{aligned} \tag{6.4}$$

where $\mathbf{W} = \mathbf{W}_{\mathrm{mf}} \triangleq \mathbf{H}$. This approach is called the matched filtering approach. The performance of the matched filtering approach is limited by interference. The kth element of $\hat{\mathbf{s}}_{\text{mf}}$ is given by

$$\hat{s}_{\text{mf},k} = \mathbf{h}_k^{\text{H}} \mathbf{h}_k s_k + \mathbf{h}_k^{\text{H}} \left(\sum_{q \neq k} \mathbf{h}_q s_q + \mathbf{n} \right),$$

where the second term on the right-hand side becomes the interference-plus-noise. From this, we can see that although the SNR approaches infinity (or $\mathbf{n}$ goes to zero), the performance is not significantly improved due to the presence of interference.

6.3.2 Zero-forcing approach

The interference can be suppressed using a linear transformation after the matched filtering in (6.4). Recall the output of the matched filters as follows:

$$\begin{aligned}\hat{\mathbf{s}}_{\text{mf}} &= \mathbf{H}^{\text{H}} \mathbf{r} \\ &= \mathbf{H}^{\text{H}} (\mathbf{H}\mathbf{s} + \mathbf{n}) \\ &= \mathbf{H}^{\text{H}} \mathbf{H}\mathbf{s} + \mathbf{H}^{\text{H}} \mathbf{n}.\end{aligned}$$

In general, $\mathbf{C} = \mathbf{H}^{\text{H}}\mathbf{H}$ is not diagonal unless the $\mathbf{h}_k$'s are orthogonal. If $\mathbf{C}$ is full rank, its inverse exists. Using the inverse of $\mathbf{C}$, we can remove the interference as follows:

$$\begin{aligned}\hat{\mathbf{s}}_{\text{zf}} &= \mathbf{C}^{-1} \hat{\mathbf{s}}_{\text{mf}} \\ &= \mathbf{s} + \mathbf{C}^{-1} \mathbf{H}^{\text{H}} \mathbf{n}.\end{aligned} \tag{6.5}$$

This is called the zero-forcing (ZF) approach and the resulting ZF combining matrix becomes

$$\begin{aligned}\mathbf{W}_{\text{zf}} &= \mathbf{H}\mathbf{C}^{-1} \\ &= \mathbf{H}(\mathbf{H}^{\text{H}}\mathbf{H})^{-1}.\end{aligned} \tag{6.6}$$

Using the projection matrix, we can also derive the ZF combiner. Consider the ZF combiner for the first signal, denoted by $\mathbf{w}_{\text{zf},1}$. The output of the ZF combiner for the first signal is given by

$$\begin{aligned}\hat{s}_{\text{zf},1} &= \mathbf{w}_{\text{zf},1}^{\text{H}} \mathbf{r} \\ &= \mathbf{w}_{\text{zf},1}^{\text{H}} \mathbf{h}_1 s_1 + \mathbf{w}_{\text{zf},1}^{\text{H}} \left(\sum_{k=2}^{K} \mathbf{h}_k s_k + \mathbf{n} \right).\end{aligned} \tag{6.7}$$

To suppress the interference, it is desirable that

$$\mathbf{w}_{\text{zf},1} \perp \mathbf{H}_1 = [\mathbf{h}_2 \ \mathbf{h}_3 \ \dots \ \mathbf{h}_K]. \tag{6.8}$$

Since the orthogonal projection matrix onto $\mathbf{H}_1$ is given by

$$\mathbf{P}_1^{\perp} = \mathbf{I} - \mathbf{H}_1 (\mathbf{H}_1^{\text{H}} \mathbf{H}_1)^{-1} \mathbf{H}_1,$$

$\mathbf{w}_{\mathrm{zf},1}$ can be expressed as

$$\mathbf{w}_{\mathrm{zf},1} = \mathbf{P}_1^{\perp}\mathbf{u},$$

where $\mathbf{u}$ is a certain vector to be determined later. Using the constraint $\mathbf{w}_{\mathrm{zf},1}^{\mathrm{H}}\mathbf{h}_1 = 1$, it can be shown that

$$\mathbf{u}^{\mathrm{H}}\mathbf{P}_1^{\perp}\mathbf{h}_1 = 1. \tag{6.9}$$

There could be multiple solutions of $\mathbf{u}$ that satisfy (6.9). To determine $\mathbf{u}$ with the constraint in (6.9), we can consider the following optimization problem:

$$\begin{aligned} &\min_{\mathbf{u}} ||\mathbf{u}||^2 \\ &\text{subject to } \mathbf{u}^{\mathrm{H}}\mathbf{P}_1^{\perp}\mathbf{h}_1 = 1. \end{aligned}$$

The solution is

$$\mathbf{u}_* = \frac{1}{\mathbf{h}_1^{\mathrm{H}}\mathbf{P}_1^{\perp}\mathbf{h}_1}\mathbf{P}_1^{\perp}\mathbf{h}_1.$$

Thus, we have

$$\mathbf{w}_{\mathrm{zf},1} = \mathbf{P}_1^{\perp}\mathbf{u}_* = \frac{1}{\mathbf{h}_1^{\mathrm{H}}\mathbf{P}_1^{\perp}\mathbf{h}_1}\mathbf{P}_1^{\perp}\mathbf{h}_1. \tag{6.10}$$

We can also verify that $\mathbf{w}_{\mathrm{zf},1}$ in (6.10) is identical to the first column of $\mathbf{W}_{\mathrm{zf}}$ in (6.6) using the matrix inversion lemma directly.

A drawback of the ZF approach is that the noise can be enhanced when $\mathbf{C}$ is close to singular.

Example 6.3.1 Suppose that the channel matrix is given by

$$\mathbf{H} = [\mathbf{h}_1\ \mathbf{h}_2] = \begin{bmatrix} 1 & 1 \\ 2 & 1.99 \end{bmatrix}.$$

This channel matrix is close to singular as its determinant is -0.01. The corresponding ZF combining matrix is

$$\mathbf{W}_{\mathrm{zf}} = \begin{bmatrix} -199 & 200 \\ 100 & -100 \end{bmatrix}.$$

In this case, although the signals can be recovered without interference, we can see that the background noise is significantly enhanced as the values of the elements of $\mathbf{W}_{\mathrm{zf}}$ are large.

As shown in Example 6.3.1, the noise can be significantly enhanced if $\mathbf{H}$ is close to singular or rank-deficient. To see this clearly, the noise term after ZF combining can be considered. To simplify the analysis, we assume that

$$\mathbf{R}_{\mathbf{s}} = \mathrm{Diag}(\sigma_1^2, \sigma_2^2, \ldots, \sigma_K^2);$$

$$\mathbf{R}_{\mathbf{n}} = N_0\mathbf{I},$$

where $\sigma_k^2 = E[|s_k|^2]$. Denote by $\tilde{\mathbf{n}} = \mathbf{C}^{-1}\mathbf{H}^{\mathrm{H}}\mathbf{n}$ the noise vector after ZF combining in (6.5). The ZF combiner's output vector is given by

$$\hat{\mathbf{s}}_{\mathrm{zf}} = \mathbf{s} + \tilde{\mathbf{n}}, \tag{6.11}$$

which shows clearly that the performance depends on $\tilde{\mathbf{n}}$. The noise covariance matrix becomes

$$\begin{aligned}\mathbf{R}_{\tilde{\mathbf{n}}} &= \mathcal{E}[\tilde{\mathbf{n}}\tilde{\mathbf{n}}^{\mathrm{H}}] \\ &= N_0(\mathbf{H}^{\mathrm{H}}\mathbf{H})^{-1}.\end{aligned} \tag{6.12}$$

The individual SNR after ZF combining is given by

$$\begin{aligned}\mathrm{SNR}_{\mathrm{zf},k} &= \frac{\sigma_k^2}{N_0[(\mathbf{H}^{\mathrm{H}}\mathbf{H})^{-1}]_{k,k}} \\ &= \frac{\sigma_k^2}{N_0}\mathbf{h}_k^{\mathrm{H}}\mathbf{P}_k^{\perp}\mathbf{h}_k,\end{aligned} \tag{6.13}$$

where

$$\mathbf{P}_k^{\perp} = \mathbf{I} - \mathbf{H}_k(\mathbf{H}_k^{\mathrm{H}}\mathbf{H}_k)^{-1}\mathbf{H}_k^{\mathrm{H}}.$$

Here, $\mathbf{H}_k$ is the submatrix obtained from $\mathbf{H}$ after deleting the kth column vector. We can also show that

$$\begin{aligned}\mathrm{tr}(\mathbf{R}_{\tilde{\mathbf{n}}}) &= N_0 \sum_{k=1}^{K} \frac{1}{\lambda_k(\mathbf{H}^{\mathrm{H}}\mathbf{H})}; \\ \det(\mathbf{R}_{\tilde{\mathbf{n}}}) &= N_0 \prod_{k=1}^{K} \frac{1}{\lambda_k(\mathbf{H}^{\mathrm{H}}\mathbf{H})},\end{aligned} \tag{6.14}$$

where $\lambda_k(\mathbf{H}^{\mathrm{H}}\mathbf{H})$ denotes the kth largest eigenvalue of $\mathbf{H}^{\mathrm{H}}\mathbf{H}$. It can be readily shown that if $\mathbf{H}^{\mathrm{H}}\mathbf{H}$ is rank-deficient or any eigenvalues become zeros, both $\mathrm{tr}(\mathbf{R}_{\tilde{\mathbf{n}}})$ and $\det(\mathbf{R}_{\tilde{\mathbf{n}}})$ approach infinity. This indicates that the SNR after ZF combining can approach zero if $\mathbf{H}^{\mathrm{H}}\mathbf{H}$ is rank-deficient due to the noise enhancement.

In deriving the ZF combiner, it is not necessary to know the statistical properties of the background noise, $\mathbf{n}$, as the main aim of ZF combining is to suppress the interference. However, if the covariance matrix of $\mathbf{n}$ is available, it is often desirable to make the background noise spatially white before combining. Now, we consider the ZF combiner after spatial noise whitening. We assume that $\mathbf{R}_{\mathbf{n}}$ is full rank. Consider the factorization of $\mathbf{R}_{\mathbf{n}}$ as follows:

$$\mathbf{R}_{\mathbf{n}} = \mathbf{R}_{\mathbf{n}}^{1/2}\mathbf{R}_{\mathbf{n}}^{\mathrm{H}/2},$$

where $\mathbf{R}_{\mathbf{n}}^{1/2}$ is a square-root of $\mathbf{R}_{\mathbf{n}}$, which is positive semi-definite. For spatial noise whitening, pre-multiplying $\mathbf{R}_{\mathbf{n}}^{-1/2}$ to $\mathbf{r}$ results in

$$\begin{aligned}\mathbf{x} &= \mathbf{R}_{\mathbf{n}}^{-1/2}\mathbf{r} \\ &= \mathbf{R}_{\mathbf{n}}^{-1/2}\mathbf{H}\mathbf{s} + \mathbf{R}_{\mathbf{n}}^{-1/2}\mathbf{n}.\end{aligned} \tag{6.15}$$

Let $\mathbf{G} = \mathbf{R}_\mathbf{n}^{-1/2}\mathbf{H}$. Clearly, the noise is whitened since the covariance matrix of $\mathbf{R}_\mathbf{n}^{-1/2}\mathbf{n}$ is $\mathbf{I}$. The ZF combining matrix for $\mathbf{x}$ becomes

$$\begin{aligned}\tilde{\mathbf{W}}_{\text{zf}} &= \mathbf{G}(\mathbf{G}^{\text{H}}\mathbf{G})^{-1} \\ &= \mathbf{R}_\mathbf{n}^{-1/2}\mathbf{H}(\mathbf{H}^{\text{H}}\mathbf{R}_\mathbf{n}^{-1}\mathbf{H})^{-1}.\end{aligned}$$

The ZF combiner's output is given by

$$\begin{aligned}\hat{\mathbf{s}}_{\text{zf}} &= \tilde{\mathbf{W}}_{\text{zf}}^{\text{H}}\mathbf{x} \\ &= (\mathbf{H}^{\text{H}}\mathbf{R}_\mathbf{n}^{-1}\mathbf{H})^{-1}\mathbf{H}^{\text{H}}\mathbf{R}_\mathbf{n}^{-\text{H}/2}\mathbf{R}_\mathbf{n}^{-1/2}\mathbf{r} \\ &= (\mathbf{H}^{\text{H}}\mathbf{R}_\mathbf{n}^{-1}\mathbf{H})^{-1}\mathbf{H}^{\text{H}}\mathbf{R}_\mathbf{n}^{-1}\mathbf{r} \\ &= \mathbf{W}_{\text{w-zf}}^{\text{H}}\mathbf{r},\end{aligned} \tag{6.16}$$

where $\mathbf{W}_{\text{w-zf}}$ is the ZF combining matrix after noise whitening and given by

$$\mathbf{W}_{\text{w-zf}} = \mathbf{R}_\mathbf{n}^{-1}\mathbf{H}(\mathbf{H}^{\text{H}}\mathbf{R}_\mathbf{n}^{-1}\mathbf{H})^{-1}. \tag{6.17}$$

If $\mathbf{R}_\mathbf{n} = N_0\mathbf{I}$, $\mathbf{W}_{\text{w-zf}}$ reduces to $\mathbf{W}_{\text{zf}}$. ZF combining with noise whitening is related to ML combining, which will be shown later.

6.3.3 MMSE approach

To avoid the noise enhancement in ZF combining, we can take into account the background noise and interference together in deriving a combining matrix, which is the main idea of the MMSE combiner.

We assume that $\mathcal{E}[\mathbf{s}] = \mathbf{0}$ for convenience. For a given linear estimate of s_k which is given by

$$\hat{s}_k = \mathbf{w}_k^{\text{H}}\mathbf{r},\ k = 1, 2, \dots, K,$$

the MSE of $\hat{s}_k$ becomes

$$\text{MSE}_k = \mathcal{E}[|s_k - \hat{s}_k|^2]$$

and the total MSE is given by

$$\begin{aligned}\text{MSE} &= \sum_{k=1}^{K}\text{MSE}_k \\ &= \mathcal{E}[||\mathbf{s} - \hat{\mathbf{s}}||^2],\end{aligned} \tag{6.18}$$

where

$$\begin{aligned}\hat{\mathbf{s}} &= [\hat{s}_1\ \hat{s}_2\ \dots\ \hat{s}_K]^{\text{T}} \\ &= \mathbf{W}^{\text{H}}\mathbf{r}.\end{aligned} \tag{6.19}$$

The solution of the $\mathbf{w}_k$'s which minimizes the total MSE can be easily found by noting that the total MSE in (6.18) is a quadratic function of the $\mathbf{w}_k$'s. We also note that each

MSE term is a function of its own combining vector. That is, MSE_k is a quadratic function of $\mathbf{w}_k$. Based on this observation, we can see that the minimization of the total MSE can be obtained by minimizing the individual MSEs as follows.

Theorem 6.3.1 *Let*

$$\mathbf{w}_{\mathrm{mmse},k} = \arg\min_{\mathbf{w}_k} \mathcal{E}[|s_k - \hat{s}_k|^2],\ k = 1, 2, \ldots, K.$$

Then, we have

$$\begin{aligned}\mathbf{W}_{\mathrm{mmse}} &= \arg\min_{\mathbf{W}} \mathcal{E}[||\mathbf{s} - \hat{\mathbf{s}}||^2] \\ &= [\mathbf{w}_{\mathrm{mmse},1}\ \mathbf{w}_{\mathrm{mmse},2}\ \cdots\ \mathbf{w}_{\mathrm{mmse},K}].\end{aligned} \tag{6.20}$$

Let us find the MMSE solution of $\mathbf{W}$ using the orthogonality principle. Applying the orthogonality principle in Section 3.6, we have

$$\mathbf{W}_{\mathrm{mmse}} = \mathbf{R}_{\mathbf{r}}^{-1}\mathbf{R}_{\mathbf{rs}}, \tag{6.21}$$

where

$$\begin{aligned}\mathbf{R}_{\mathbf{r}} &= \mathcal{E}[\mathbf{r}\mathbf{r}^{\mathrm{H}}]; \\ \mathbf{R}_{\mathbf{rs}} &= \mathcal{E}[\mathbf{r}\mathbf{s}^{\mathrm{H}}].\end{aligned}$$

Finally, the MMSE estimate of $\mathbf{s}$ is given by

$$\hat{\mathbf{s}}_{\mathrm{mmse}} = \mathbf{W}_{\mathrm{mmse}}^{\mathrm{H}}\mathbf{r}.$$

If $\mathbf{s}$ and $\mathbf{n}$ are uncorrelated, we have

$$\begin{aligned}\mathbf{W}_{\mathrm{mmse}} &= \left(\mathbf{H}\mathbf{R}_{\mathbf{s}}\mathbf{H}^{\mathrm{H}} + \mathbf{R}_{\mathbf{n}}\right)^{-1}\mathbf{H}\mathbf{R}_{\mathbf{s}} \\ &= \mathbf{R}_{\mathbf{n}}^{-1}\mathbf{H}(\mathbf{H}^{\mathrm{H}}\mathbf{R}_{\mathbf{n}}^{-1}\mathbf{H} + \mathbf{R}_{\mathbf{s}}^{-1})^{-1},\end{aligned} \tag{6.22}$$

because

$$\begin{aligned}\mathbf{R}_{\mathbf{r}} &= \mathbf{H}\mathbf{R}_{\mathbf{s}}\mathbf{H}^{\mathrm{H}} + \mathbf{R}_{\mathbf{n}}; \\ \mathbf{R}_{\mathbf{rs}} &= \mathbf{H}\mathbf{R}_{\mathbf{s}},\end{aligned}$$

where $\mathbf{R}_{\mathbf{s}} = \mathcal{E}[\mathbf{s}\mathbf{s}^{\mathrm{H}}]$.

Example 6.3.2 Let

$$\mathbf{H} = [\mathbf{h}_1\ \mathbf{h}_2] = \begin{bmatrix} 1 & 1 \\ 2 & 1.99 \end{bmatrix},\ \mathbf{R}_{\mathbf{s}} = \begin{bmatrix} 2 & 1 \\ 1 & 2 \end{bmatrix},\ \text{and}\ \mathbf{R}_{\mathbf{n}} = \begin{bmatrix} 1 & 0 \\ 0 & 1 \end{bmatrix}.$$

The MMSE combining matrix becomes

$$\mathbf{W}_{\mathrm{mmse}} = \begin{bmatrix} 0.0952 & 0.0991 \\ 0.1949 & 0.1927 \end{bmatrix}.$$

Compared with the result in Example 6.3.1, we can see that the noise cannot be enhanced.

Suppose that $\mathbf{R_s} = \sigma_s^2\mathbf{I}$ and $\mathbf{R_n} = N_0\mathbf{I}$. If the SNR, σ_s^2/N_0, increases, we can show that the MMSE combining matrix approaches the ZF combining matrix, $\mathbf{W}_{\text{zf}} = \mathbf{H}(\mathbf{H}^{\text{H}}\mathbf{H})^{-1}$. To show this, from (6.22), when $\mathbf{R_s} = \sigma_s^2\mathbf{I}$ and $\mathbf{R_n} = N_0\mathbf{I}$, we can re-derive the MMSE combining matrix as follows:

$$\begin{aligned}\mathbf{W}_{\text{mmse}} &= N_0^{-1}\mathbf{H}\left(\mathbf{H}^{\text{H}}\mathbf{H}N_0^{-1} + \frac{1}{\sigma_s^2}\mathbf{I}\right)^{-1}\\ &= \mathbf{H}\left(\mathbf{H}^{\text{H}}\mathbf{H} + \frac{N_0}{\sigma_s^2}\mathbf{I}\right)^{-1}.\end{aligned}$$

As $\sigma_s^2/N_0 \to \infty$, $\mathbf{W}_{\text{mmse}}$ approaches $\mathbf{W}_{\text{zf}}$. This behavior results from the minimization of the MSE. As σ_s^2/N_0 increases, the dominant term of the MSE becomes the interference rather than the background noise. Therefore, the MMSE combiner behaves as the ZF combiner to suppress the interference when the background noise is negligible. This fact is often useful in performance analysis. In general, the performance of the MMSE combiner is more difficult than that of the ZF combiner. Thus, we often use the performance analysis of the ZF combiner to derive performance bounds and asymptotic results for the MMSE combiner.

To illustrate the performance difference between the MMSE and ZF combiners, consider an example. Suppose that the channel matrix is given by

$$\mathbf{H} = \begin{bmatrix} 1 & \rho \\ \rho & 1 \end{bmatrix},$$

where ρ represents the cross-talk coefficient. If $\rho = 0$, there is no interference. If $\rho = 1$ or -1, the rank of $\mathbf{H}$ becomes 1, which means that the two signals are not able to be decoupled from $\mathbf{r}$ even if there is no background noise. Assume that $\mathbf{s} \sim \mathcal{CN}(\mathbf{0}, \sigma_s^2\mathbf{I})$ and $\mathbf{n} \sim \mathcal{CN}(\mathbf{0}, N_0\mathbf{I})$. Furthermore, we normalize the signal power, i.e. $\sigma_s^2 = 1$. Figure 6.3 shows the sum of MSEs (SMSEs), which is given by

$$\text{SMSE} = \mathcal{E}[(\mathbf{s} - \mathbf{W}^{\text{H}}\mathbf{r})^{\text{H}}(\mathbf{s} - \mathbf{W}^{\text{H}}\mathbf{r})].$$

For the ZF combiner, the SMSE is $\text{tr}(\mathbf{R}_{\tilde{\mathbf{n}}}) = N_0\text{tr}((\mathbf{H}^{\text{H}}\mathbf{H})^{-1})$ from (6.12). The MSE of the MMSE combiner will be derived in Section 6.4. From the results in Fig. 6.3, we can see that the performance difference between the MMSE and ZF combiners decreases as N_0 decreases or the SNR ($= \sigma_s^2/N_0$) increases. The cross-talk coefficient also affects the performance. As ρ increases, the level of interference increases and the MSE also increases.

6.3.4 ML and MAP combining

In deriving linear combiners, we notice that different types of side information of channels, signals, and noise have been used. For example, the ZF combiner only needs to know the channel matrix, while the MMSE combiner requires various second-order statistics of signals and noise. If we have a complete description of signals and noise

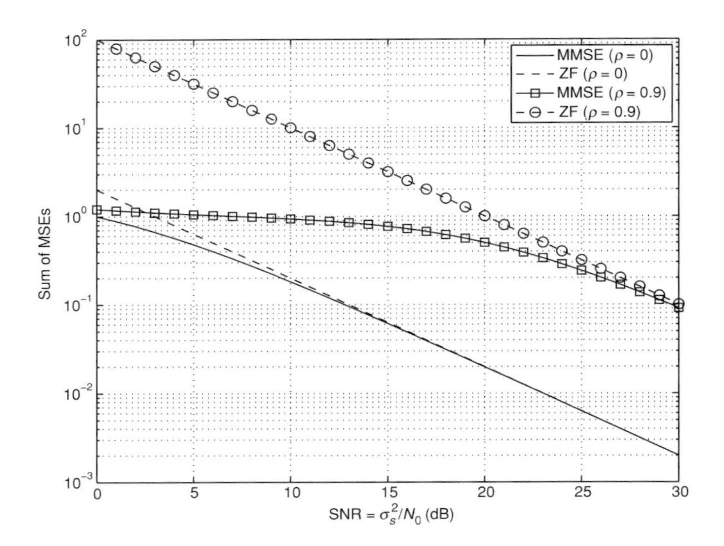

Figure 6.3 The SMSEs of the MMSE and ZF combiners.

(i.e. the pdfs of $\mathbf{s}$ and $\mathbf{n}$), other statistical signal combining approaches, including the ML and MAP combining, can also be found.

To derive the ML and MAP combiners, we assume that $\mathbf{n} \sim \mathcal{CN}(0, \mathbf{R_n})$. If we take $\mathbf{s}$ as a parameter vector to be estimated, the ML and MAP signal combiners are equivalent to the ML and MAP estimators, respectively. For the ML estimation, we can show that

$$\begin{aligned} \hat{\mathbf{s}}_{\text{ml}} &= \arg\max_{\mathbf{s}} f(\mathbf{r}|\mathbf{s}) \\ &= \arg\max_{\mathbf{s}} \exp\left(-(\mathbf{r}-\mathbf{Hs})^{\text{H}}\mathbf{R}_{\mathbf{n}}^{-1}(\mathbf{r}-\mathbf{Hs})\right) \\ &= \arg\min_{\mathbf{s}} (\mathbf{r}-\mathbf{Hs})^{\text{H}}\mathbf{R}_{\mathbf{n}}^{-1}(\mathbf{r}-\mathbf{Hs}) \\ &= \left(\mathbf{H}^{\text{H}}\mathbf{R}_{\mathbf{n}}^{-1}\mathbf{H}\right)^{-1}\mathbf{H}^{\text{H}}\mathbf{R}_{\mathbf{n}}^{-1}\mathbf{r}. \end{aligned} \tag{6.23}$$

From this, we can observe that the ML estimate can be obtained by linear combining for Gaussian noise. The resulting ML combining matrix is given by

$$\begin{aligned} \mathbf{W}_{\text{ml}} &= \left(\left(\mathbf{H}^{\text{H}}\mathbf{R}_{\mathbf{n}}^{-1}\mathbf{H}\right)^{-1}\mathbf{H}^{\text{H}}\mathbf{R}_{\mathbf{n}}^{-1}\right)^{\text{H}} \\ &= \mathbf{R}_{\mathbf{n}}^{-1}\mathbf{H}\left(\mathbf{H}^{\text{H}}\mathbf{R}_{\mathbf{n}}^{-1}\mathbf{H}\right)^{-1}. \end{aligned} \tag{6.24}$$

As shown in (6.17), we can see that the ML combining matrix is identical to the ZF combining matrix after noise whitening, $\mathbf{W}_{\text{w-zf}}$.

The MAP estimate of $\mathbf{s}$ can be found as follows:

$$\begin{aligned} \hat{\mathbf{s}}_{\text{map}} &= \arg\max_{\mathbf{s}} f(\mathbf{s}|\mathbf{r}) \\ &= \arg\max_{\mathbf{s}} f(\mathbf{r}|\mathbf{s})f(\mathbf{s}), \end{aligned} \tag{6.25}$$

where $f(\mathbf{s})$ denotes the a priori probability of $\mathbf{s}$. If $\mathbf{s}$ is a Gaussian random vector, i.e. $\mathbf{s} \sim \mathcal{CN}(\bar{\mathbf{s}}, \mathbf{R}_\mathbf{s})$, we have

$$f(\mathbf{s}|\mathbf{r}) \propto \mathrm{e}^{-(\mathbf{r}-\mathbf{Hs})^{\mathrm{H}}\mathbf{R}_\mathbf{n}^{-1}(\mathbf{r}-\mathbf{Hs})}\mathrm{e}^{-(\mathbf{s}-\bar{\mathbf{s}})^{\mathrm{H}}\mathbf{R}_\mathbf{s}^{-1}(\mathbf{s}-\bar{\mathbf{s}})}$$
$$\propto \mathrm{e}^{-\mathbf{s}^{\mathrm{H}}(\mathbf{H}^{\mathrm{H}}\mathbf{R}_\mathbf{n}^{-1}\mathbf{H}+\mathbf{R}_\mathbf{s}^{-1})\mathbf{s}+2\Re\{(\mathbf{r}^{\mathrm{H}}\mathbf{R}_\mathbf{n}^{-1}\mathbf{H}+\bar{\mathbf{s}}^{\mathrm{H}}\mathbf{R}_\mathbf{s}^{-1})\mathbf{s}\}}.$$

Finally, the MAP estimate of $\mathbf{s}$ is given by

$$\begin{aligned}\hat{\mathbf{s}}_{\mathrm{map}} &= \arg\max_{\mathbf{s}} \left\{\mathbf{s}^{\mathrm{H}}(\mathbf{H}^{\mathrm{H}}\mathbf{R}_\mathbf{n}^{-1}\mathbf{H} + \mathbf{R}_\mathbf{s}^{-1})\mathbf{s} - 2\Re\{(\mathbf{r}^{\mathrm{H}}\mathbf{R}_\mathbf{n}^{-1}\mathbf{H} + \bar{\mathbf{s}}^{\mathrm{H}}\mathbf{R}_\mathbf{s}^{-1})\mathbf{s}\}\right\} \\ &= (\mathbf{H}^{\mathrm{H}}\mathbf{R}_\mathbf{n}^{-1}\mathbf{H} + \mathbf{R}_\mathbf{s}^{-1})^{-1}(\mathbf{H}^{\mathrm{H}}\mathbf{R}_\mathbf{n}^{-1}\mathbf{r} + \mathbf{R}_\mathbf{s}^{-1}\bar{\mathbf{s}}).\end{aligned} \tag{6.26}$$

The MAP estimate can be expressed as an output of a linear combiner with an offset as follows:

$$\hat{\mathbf{s}}_{\mathrm{map}} = \mathbf{W}_{\mathrm{map}}^{\mathrm{H}}\mathbf{r} + \tilde{\mathbf{s}}_{\mathrm{offset}},$$

where

$$\mathbf{W}_{\mathrm{map}} = \mathbf{R}_\mathbf{n}^{-1}\mathbf{H}(\mathbf{H}^{\mathrm{H}}\mathbf{R}_\mathbf{n}^{-1}\mathbf{H} + \mathbf{R}_\mathbf{s}^{-1})^{-1}$$

and the offset vector, $\tilde{\mathbf{s}}_{\mathrm{offset}}$, is given by

$$\tilde{\mathbf{s}}_{\mathrm{offset}} = (\mathbf{H}^{\mathrm{H}}\mathbf{R}_\mathbf{n}^{-1}\mathbf{H} + \mathbf{R}_\mathbf{s}^{-1})^{-1}\mathbf{R}_\mathbf{s}^{-1}\bar{\mathbf{s}}.$$

Then, we can see that $\mathbf{W}_{\mathrm{map}}$ is identical to $\mathbf{W}_{\mathrm{mmse}}$ in (6.22). This implies that the MMSE combiner's output is the MAP estimate if $\mathbf{s} \sim \mathcal{CN}(0, \mathbf{R}_\mathbf{s})$.

6.4 Properties of MMSE combining

In this section, we focus on the properties and performance of MMSE combining. If signal sources are uncorrelated, the statistical analysis results are generally identical to those of MMSE combining for single-signal in Chapter 4. This implies that we can deal with each signal individually for performance analysis. However, if signal sources are correlated, a multivariate approach is required for performance analysis.

6.4.1 Individual MMSE combining

We assume that

$$\mathcal{E}[\mathbf{s}] = \mathbf{0};$$
$$\mathbf{R}_\mathbf{s} = \mathrm{Diag}(\sigma_1^2, \sigma_2^2, \ldots, \sigma_K^2).$$

That is, all the signal sources have zero-mean and are uncorrelated. In addition, $\mathcal{E}[\mathbf{n}] = \mathbf{0}$ and $\mathcal{E}[\mathbf{nn}^{\mathrm{H}}] = N_0\mathbf{I}$. If s_1 is the desired signal, the interference-plus-noise vector is

given by

$$\mathbf{u} = \sum_{k=2}^{K} \mathbf{h}_k s_k + \mathbf{n}.$$

If the s_k's and $\mathbf{n}$ are mutually uncorrelated, the covariance matrix of $\mathbf{u}$ becomes

$$\begin{aligned}\mathbf{R_u} &= \mathcal{E}[\mathbf{u}\mathbf{u}^{\mathrm{H}}] \\ &= \mathbf{H}_1 \mathbf{\Sigma}_1 \mathbf{H}_1^{\mathrm{H}} + N_0 \mathbf{I},\end{aligned} \tag{6.27}$$

where

$$\mathbf{\Sigma}_1 = \mathrm{Diag}(\sigma_2^2, \sigma_3^2, \ldots, \sigma_K^2).$$

The MMSE estimate of s_1 becomes

$$\begin{aligned}\hat{s}_1 &= \mathbf{w}_{\mathrm{mmse},1}^{\mathrm{H}} \mathbf{r} \\ &= \mathbf{w}_{\mathrm{mmse},1}^{\mathrm{H}} \mathbf{h}_1 s_1 + \mathbf{w}_{\mathrm{mmse},1}^{\mathrm{H}} \left(\sum_{k=2}^{K} \mathbf{h}_k s_k + \mathbf{n} \right),\end{aligned} \tag{6.28}$$

where

$$\begin{aligned}\mathbf{w}_{\mathrm{mmse},1} &= \mathbf{R_r}^{-1} \mathbf{h}_1 \sigma_1^2 \\ &= (\mathbf{h}_1 \mathbf{h}_1^{\mathrm{H}} \sigma_1^2 + \mathbf{R_u})^{-1} \mathbf{h}_1 \sigma_1^2.\end{aligned} \tag{6.29}$$

The individual MMSE is given by

$$\begin{aligned}\epsilon_1^2 &= \mathcal{E}[|s_1 - \hat{s}_1|^2] \\ &= \sigma_1^2 (1 - \sigma_1^2 \mathbf{h}^{\mathrm{H}} \mathbf{R_r}^{-1} \mathbf{h}) \\ &= \frac{\sigma_1^2}{1 + \sigma_1^2 \mathbf{h}_1^{\mathrm{H}} \mathbf{R_u}^{-1} \mathbf{h}_1}.\end{aligned} \tag{6.30}$$

It is noteworthy that the MMSE vector is also given by

$$\mathbf{w}_{\mathrm{mmse},1} = \epsilon_1^2 \mathbf{R_u}^{-1} \mathbf{h}_1. \tag{6.31}$$

The results are identical to those of the MMSE combining for single-signal in Chapter 4.

6.4.2 Biased and unbiased SINR

There could be different definitions of the SINR after MMSE combining or filtering. From (6.28), we have

$$\hat{s}_1 = a_1 s_1 + \mathbf{w}_{\mathrm{mmse},1}^{\mathrm{H}} \mathbf{u},$$

where $a_1 = \mathbf{w}_{\text{mmse},1}^{\text{H}} \mathbf{h}_1$. We assume that the signal sources are uncorrelated. With scaling by $1/a_1$, an unbiased estimate of s_1 can be obtained as

$$\begin{aligned}\hat{s}_{\text{u},1} &= \frac{\hat{s}_1}{a_1} \\ &= s_1 + \frac{\mathbf{w}_{\text{mmse},1}^{\text{H}} \mathbf{u}}{a_1}.\end{aligned}$$

The resulting unbiased SINR is given by

$$\begin{aligned}\text{SINR}_{\text{mmse},1} &= \sigma_1^2 \mathbf{h}_1^{\text{H}} \mathbf{R}_{\mathbf{u}}^{-1} \mathbf{h}_1 \\ &= \frac{\sigma_1^2}{\epsilon_1^2} - 1.\end{aligned} \tag{6.32}$$

As shown in Chapter 4, MMSE combining also maximizes the SINR. For convenience, the SINR after MMSE combining will be denoted by SINR_k instead of $\text{SINR}_{\text{mmse},k}$.

Without scaling by $1/a_1$, we can define the biased SINR using the following expression:

$$\hat{s}_1 = s_1 + (a_1 - 1)s_1 + \mathbf{w}_{\text{mmse},1}^{\text{H}} \mathbf{u},$$

where $(a_1 - 1)s_1 + \mathbf{w}_1^{\text{H}} \mathbf{u}$ is the error between s_1 and its estimate, $\hat{s}_1$. The biased SINR becomes

$$\begin{aligned}\text{SINR}_{\text{b},1} &= \frac{\sigma_1^2}{(a_1 - 1)^2 \sigma_1^2 + \mathbf{w}_{\text{mmse},1}^{\text{H}} \mathbf{R}_{\mathbf{u}} \mathbf{w}_{\text{mmse},1}} \\ &= \frac{\sigma_1^2}{\epsilon_1^2} \\ &= \text{SINR}_1 + 1.\end{aligned} \tag{6.33}$$

Although the biased SINR is higher than the unbiased SINR by 1 as shown in (6.33), this improvement is an artifact due to different definitions of the SINR. While the SINR could have different definitions, the mutual information cannot. In Section 6.7, we can see that the mutual information remains the same after MMSE combining under a certain condition.

6.4.3 Joint MMSE combining

It is straightforward to characterize the performance of MMSE combining for individual signals using the notion of the SINR although there could be multiple definitions of the SINR. However, when the joint MMSE estimation is considered for correlated signal sources, it is not straightforward to define the SINR as the signal is in a vector form (the SINR for vector signals will also be discussed in Section 6.7). Thus, we mainly focus on the error covariance matrix of the joint MMSE estimation in this subsection.

The MMSE error covariance matrix is given by

$$\begin{aligned}\mathbf{Q_s} &= \mathcal{E}[(\mathbf{s} - \hat{\mathbf{s}})(\mathbf{s} - \hat{\mathbf{s}})^{\mathrm{H}}] \\ &= \mathbf{R_s} - \mathbf{R_{sr}} \mathbf{R_r}^{-1} \mathbf{R_{rs}},\end{aligned} \tag{6.34}$$

where $\mathbf{R_{xy}}$ denotes $\mathcal{E}[\mathbf{xy}^{\mathrm{H}}]$. Noting that

$$\begin{aligned}\mathbf{R_{sr}} &= \mathbf{R_s H}^{\mathrm{H}}; \\ \mathbf{R_r} &= \mathbf{H R_s H}^{\mathrm{H}} + N_0 \mathbf{I},\end{aligned}$$

we have

$$\mathbf{Q_s} = \mathbf{R_s} - \mathbf{R_s H}^{\mathrm{H}} \left(\mathbf{H R_s H}^{\mathrm{H}} + N_0 \mathbf{I} \right)^{-1} \mathbf{H R_s}. \tag{6.35}$$

The total MMSE is given by

$$\mathrm{MMSE} = \mathrm{tr}(\mathbf{Q_s}).$$

Certainly, as MMSE combining is to minimize the MSE, we have

$$\begin{aligned}\mathrm{tr}(\mathbf{Q_s}) &= \mathcal{E}[(\mathbf{s} - \mathbf{W}_{\mathrm{mmse}}^{\mathrm{H}} \mathbf{r})^{\mathrm{H}} (\mathbf{s} - \mathbf{W}_{\mathrm{mmse}}^{\mathrm{H}} \mathbf{r})] \\ &\leq \mathcal{E}[(\mathbf{s} - \mathbf{W}^{\mathrm{H}} \mathbf{r})^{\mathrm{H}} (\mathbf{s} - \mathbf{W}^{\mathrm{H}} \mathbf{r})]\end{aligned}$$

for any combining matrix $\mathbf{W}$. If we define a total SINR as follows:

$$\mathrm{SINR} = \frac{\text{the sum of signal powers}}{\text{the sum of MSEs}} = \frac{\mathrm{tr}(\mathbf{R_s})}{\mathrm{tr}(\mathbf{Q_s})}, \tag{6.36}$$

we can easily show that the MMSE combining also maximizes the SINR.

The normalized MMSE error covariance matrix is defined as

$$\begin{aligned}\bar{\mathbf{Q}}_{\mathbf{s}} &= \mathbf{R_s}^{-1/2} \mathbf{Q_s} \mathbf{R_s}^{-\mathrm{H}/2} \\ &= \mathbf{I} - \mathbf{R_s}^{\mathrm{H}/2} \mathbf{H}^{\mathrm{H}} \left(\mathbf{H R_s H}^{\mathrm{H}} + N_0 \mathbf{I} \right)^{-1} \mathbf{H R_s}^{1/2}.\end{aligned} \tag{6.37}$$

Here, the signal covariance matrix is factorized as follows:

$$\mathbf{R_s} = \mathbf{R_s}^{1/2} \mathbf{R_s}^{\mathrm{H}/2}.$$

From (6.37), the sum of normalized MMSEs is given by

$$\mathrm{tr}(\bar{\mathbf{Q}}_{\mathbf{s}}) = K - \mathrm{tr}\left(\left(\mathbf{H R_s H}^{\mathrm{H}} + N_0 \mathbf{I} \right)^{-1} \mathbf{H R_s H}^{\mathrm{H}} \right). \tag{6.38}$$

If $\mathbf{s} \sim \mathcal{CN}(\mathbf{0}, \mathbf{R_s})$ and $\mathbf{n} \sim \mathcal{CN}(\mathbf{0}, \mathbf{R_n})$, the MMSE estimate of $\mathbf{s}$ can be characterized as

$$\hat{\mathbf{s}}_{\mathrm{mmse}} \sim \mathcal{CN}\left(\mathbf{0}, \mathbf{R_s H}^{\mathrm{H}} (\mathbf{H R_s H}^{\mathrm{H}} + \mathbf{R_n})^{-1} \mathbf{H R_s} \right).$$

Let $\mathbf{R}_{\hat{\mathbf{s}}} = \mathbf{R_s H}^{\mathrm{H}} (\mathbf{H R_s H}^{\mathrm{H}} + \mathbf{R_n})^{-1} \mathbf{H R_s}$ denote the covariance matrix of $\hat{\mathbf{s}}_{\mathrm{mmse}}$. It can be shown that

$$\mathbf{R_s} \geq \mathbf{R}_{\hat{\mathbf{s}}}. \tag{6.39}$$

Using the matrix inversion lemma, $\bar{\mathbf{Q}}_s$ in (6.37) can also be expressed as

$$\bar{\mathbf{Q}}_s = \left(\mathbf{I} + \frac{1}{N_0}\mathbf{R}_s^{H/2}\mathbf{H}^H\mathbf{H}\mathbf{R}_s^{1/2}\right)^{-1}. \tag{6.40}$$

If we denote by $\lambda_k(\mathbf{A})$ the kth eigenvalue of $\mathbf{A}$, we can show that the total normalized MMSE becomes

$$\text{tr}(\bar{\mathbf{Q}}_s) = \sum_{k=1}^{K} \frac{1}{1 + \frac{1}{N_0}\lambda_k\left(\mathbf{R}_s^{H/2}\mathbf{H}^H\mathbf{H}\mathbf{R}_s^{1/2}\right)}. \tag{6.41}$$

If $\mathbf{R}_s = \sigma_s^2\mathbf{I}$, the total normalized MMSE is reduced to

$$\text{tr}(\bar{\mathbf{Q}}_s) = \sum_{k=1}^{K} \frac{1}{1 + \frac{\sigma_s^2}{N_0}\lambda_k(\mathbf{H}^H\mathbf{H})}. \tag{6.42}$$

6.5 Dimension and suppression capability

An important feature of the MMSE combiner is that it can mitigate interfering signals. In order to see how the MMSE combiner can work to suppress interfering signals, we consider the case where the Kth signal, s_K, which is an interfering signal, becomes stronger. That is, $\sigma_K^2 \to \infty$. In this section, we assume that the signals are uncorrelated and independently generated.

The covariance matrix of $\mathbf{u}$ is given by

$$\begin{aligned}\mathbf{R}_\mathbf{u} &= \sum_{k=2}^{K} \mathbf{h}_k\mathbf{h}_k^H\sigma_k^2 + N_0\mathbf{I} \\ &= \underbrace{\sum_{k=2}^{K-1} \mathbf{h}_k\mathbf{h}_k^H\sigma_k^2 + N_0\mathbf{I}}_{=\mathbf{R}_K} + \sigma_K^2\mathbf{h}_K\mathbf{h}_K^H.\end{aligned}$$

Then, using the matrix inversion lemma, we have

$$\mathbf{R}_\mathbf{u}^{-1} = \mathbf{R}_K^{-1} - \frac{\sigma_K^2}{1 + \sigma_K^2\mathbf{h}_K^H\mathbf{R}_K^{-1}\mathbf{h}_K}\mathbf{R}_K^{-1}\mathbf{h}_K\mathbf{h}_K^H\mathbf{R}_K^{-1}.$$

The MMSE combining vector for s_1 is given by

$$\begin{aligned}\mathbf{w}_1 &= \epsilon_1^2\mathbf{R}_\mathbf{u}^{-1}\mathbf{h}_1 \\ &= \epsilon_1^2\left(\mathbf{R}_K^{-1} - \frac{\sigma_K^2}{1 + \sigma_K^2\mathbf{h}_K^H\mathbf{R}_K^{-1}\mathbf{h}_K}\mathbf{R}_K^{-1}\mathbf{h}_K\mathbf{h}_K^H\mathbf{R}_K^{-1}\right)\mathbf{h}_1.\end{aligned} \tag{6.43}$$

If σ_K^2 approaches infinity and $\mathbf{h}_K^H \mathbf{R}_K^{-1} \mathbf{h}_K > 0$, we have

$$\begin{aligned}
\mathbf{w}_{\infty,1} &= \lim_{\sigma_K^2 \to \infty} \mathbf{w}_1 \\
&= \lim_{\sigma_K^2 \to \infty} \epsilon_1^2 \left(\mathbf{R}_K^{-1} - \frac{\sigma_K^2}{1 + \sigma_K^2 \mathbf{h}_K^H \mathbf{R}_K^{-1} \mathbf{h}_K} \mathbf{R}_K^{-1} \mathbf{h}_K \mathbf{h}_K^H \mathbf{R}_K^{-1} \right) \mathbf{h}_1 \\
&= \epsilon_1^2 \left(\mathbf{R}_K^{-1} - \frac{1}{\mathbf{h}_K^H \mathbf{R}_K^{-1} \mathbf{h}_K} \mathbf{R}_K^{-1} \mathbf{h}_K \mathbf{h}_K^H \mathbf{R}_K^{-1} \right) \mathbf{h}_1.
\end{aligned}$$

Then, it can be shown that

$$\begin{aligned}
\mathbf{w}_{\infty,1}^H \mathbf{h}_K &= \epsilon_1^2 \mathbf{h}_1^H \left(\mathbf{R}_K^{-1} - \frac{1}{\mathbf{h}_K^H \mathbf{R}_K^{-1} \mathbf{h}_K} \mathbf{R}_K^{-1} \mathbf{h}_K \mathbf{h}_K^H \mathbf{R}_K^{-1} \right) \mathbf{h}_K \\
&= 0.
\end{aligned}$$

This implies that

$$\mathbf{w}_{\infty,1} \perp \mathbf{h}_K.$$

This is a natural consequence of MMSE combining. Since the contribution of the strong Kth signal to the MSE is significant if $\mathbf{w}_{\infty,1}$ is not orthogonal to $\mathbf{h}_K$ (in this case, the MSE can approach infinity as $\sigma_K^2 \to \infty$), $\mathbf{w}_{\infty,1}$ should be orthogonal to $\mathbf{h}_K$ so that the Kth signal is completely suppressed. The observation above can be generalized as follows.

Theorem 6.5.1 *Suppose that the signature vectors, $\mathbf{h}_k$, $k = 1, 2, \ldots, K$, are linearly independent and $N \geq K$. Then, as $\sigma_k^2 \to \infty$, $k = 2, 3, \ldots, K$, we have*

$$\mathbf{w}_1 \perp \mathbf{h}_k, \ k = 2, 3, \ldots, K. \tag{6.44}$$

Example 6.5.1 Let

$$\mathbf{h}_1 = \begin{bmatrix} 1 \\ 1 \end{bmatrix}, \mathbf{h}_2 = \begin{bmatrix} 1 \\ 0 \end{bmatrix}, \ \mathbf{R}_\mathbf{s} = \begin{bmatrix} 1 & 0 \\ 0 & \alpha \end{bmatrix}, \text{ and } \mathbf{R}_\mathbf{n} = \begin{bmatrix} 1 & 0 \\ 0 & 1 \end{bmatrix}.$$

If $\alpha = 1$,

$$\mathbf{W}_{\text{mmse}} = \begin{bmatrix} 0.2 & 0.4 \\ 0.4 & -0.2 \end{bmatrix}.$$

On the other hand, as $\alpha \to \infty$, we have

$$\mathbf{W}_{\text{mmse}} = \begin{bmatrix} 0.0 & 1.0 \\ 0.5 & -0.5 \end{bmatrix}.$$

In this case, we can see that $\mathbf{w}_{\text{mmse},1}^H \mathbf{h}_2 = 0$ to suppress the strong interference from s_2.

6.6 Asymptotic analysis for random signature vectors

For a given $\mathbf{H}$, various closed-form expressions for some performance metrics are available to analyze the MMSE combiner. As they depend on a realization of $\mathbf{H}$, however, it is often difficult to see general performance behaviors of the MMSE combiner for an arbitrary $\mathbf{H}$. To overcome this difficulty, in (Tse & Hanly 1999), asymptotic approaches to analyze the performance of the MMSE receiver over random channel matrices are investigated using the notion of random matrices. These approaches are now widely used as they can provide insights into various MMSE-based algorithms in a number of applications. In this section, we derive statistical properties of the MMSE combiner for a class of random matrices based on (Tse & Hanly 1999).

The SINR of the MMSE combiner's output is a random variable for random signature vectors. Obviously, statistical properties of the SINR depend on those of random signature vectors. In this section, in order to derive statistical properties of SINR, we consider an example where the elements of signature vector are iid. In addition, we assume that the s_k's are uncorrelated and their means are all zeros. That is, $\mathcal{E}[\mathbf{s}] = \mathbf{0}$ and $\mathcal{E}[\mathbf{s}\mathbf{s}^{\mathrm{H}}] = \mathrm{Diag}(\sigma_1^2, \sigma_2^2, \ldots, \sigma_K^2)$.

Let

$$\mathbf{h}_k = \frac{1}{\sqrt{N}}[\tilde{h}_{1,k}\ \tilde{h}_{2,k}\ \ldots\ \tilde{h}_{N,k}]^{\mathrm{T}},$$

where the $\tilde{h}_{n,k}$'s are iid random variables with $\mathcal{E}[\tilde{h}_{n,k}] = 0$ and $\mathcal{E}[|\tilde{h}_{n,k}|^2] = 1$. Then, we can show that

$$\begin{aligned}\mathcal{E}[\mathbf{h}_k^{\mathrm{H}}\mathbf{h}_k] &= 1;\\ \mathcal{E}[\mathbf{h}_k\mathbf{h}_k^{\mathrm{H}}] &= \frac{1}{N}\mathbf{I}.\end{aligned} \tag{6.45}$$

Note that since the $\mathbf{h}_k$'s are random, $\mathbf{R}_{\mathbf{u}}$ in (6.27) is also a random matrix. Consider the following eigen-decomposition:

$$\mathbf{H}_1\boldsymbol{\Sigma}_1\mathbf{H}_1^{\mathrm{H}} = \mathbf{U}\boldsymbol{\Lambda}\mathbf{U}^{\mathrm{H}}, \tag{6.46}$$

where

$$\begin{aligned}\mathbf{U} &= [\mathbf{u}_1\ \mathbf{u}_2\ \ldots\ \mathbf{u}_N];\\ \boldsymbol{\Lambda} &= \mathrm{Diag}(\lambda_1, \lambda_2, \ldots, \lambda_N).\end{aligned}$$

Here, $\mathbf{u}_n$ denotes the nth eigenvector corresponding to the nth largest eigenvalue, λ_n, of $\mathbf{H}_1\boldsymbol{\Sigma}_1\mathbf{H}_1^{\mathrm{H}}$. As $\mathbf{H}_1$ is random, the eigenvectors and eigenvalues of $\mathbf{H}_1\boldsymbol{\Sigma}_1\mathbf{H}_1^{\mathrm{H}}$ are also random. Using (6.27) and (6.46), $\mathbf{R}_{\mathbf{u}}$ is re-written as

$$\begin{aligned}\mathbf{R}_{\mathbf{u}} &= \mathbf{H}_1\boldsymbol{\Sigma}_1\mathbf{H}_1^{\mathrm{H}} + N_0\mathbf{I}\\ &= \mathbf{U}(\boldsymbol{\Lambda} + N_0\mathbf{I})\mathbf{U}^{\mathrm{H}},\end{aligned}$$

because $\mathbf{U}^{\mathrm{H}}\mathbf{U} = \mathbf{U}\mathbf{U}^{\mathrm{H}} = \mathbf{I}$ ($\mathbf{U}$ is unitary).

The SINR is given by

$$\begin{aligned}\mathrm{SINR}_1 &= \sigma_1^2 \mathbf{h}_1^{\mathrm{H}} \mathbf{R}_{\mathbf{u}}^{-1} \mathbf{h}_1 \\ &= \sigma_1^2 (\mathbf{U}\mathbf{h}_1)^{\mathrm{H}} (\mathbf{\Lambda} + N_0 \mathbf{I})^{-1} (\mathbf{U}\mathbf{h}_1).\end{aligned}$$

It is noteworthy that the statistical properties of $\mathbf{h}_1$ and $\mathbf{U}\mathbf{h}_1$ are the same as $\mathbf{U}$ is unitary. This implies from (6.45) that

$$\begin{aligned}\mathcal{E}[\mathbf{c}_1^{\mathrm{H}} \mathbf{c}_1] &= 1; \\ \mathcal{E}[\mathbf{c}_1 \mathbf{c}_1^{\mathrm{H}}] &= \frac{1}{N} \mathbf{I},\end{aligned} \tag{6.47}$$

where $\mathbf{c}_1 = \mathbf{U}\mathbf{h}_1$. The SINR is now given by

$$\begin{aligned}\mathrm{SINR}_1 &= \sigma_1^2 \mathbf{c}_1^{\mathrm{H}} (\mathbf{\Lambda} + N_0 \mathbf{I})^{-1} \mathbf{c}_1 \\ &= \sigma_1^2 \sum_{n=1}^{N} \frac{|c_{n,1}|^2}{\lambda_n + N_0},\end{aligned} \tag{6.48}$$

where $c_{n,1}$ denotes the nth component of $\mathbf{c}_1$. Since $|c_{n,1}|^2$ and λ_n are random, SINR_1 in (6.48) is also random. To characterize the SINR, we will derive its mean and variance.

Let us consider the mean of the SINR. Noting that $c_{n,1}$ and λ_n are statistically independent, the mean can be given by

$$\begin{aligned}\mathcal{E}[\mathrm{SINR}_1] &= \sigma_1^2 \mathcal{E}[\mathbf{c}_1^{\mathrm{H}} (\mathbf{\Lambda} + N_0 \mathbf{I})^{-1} \mathbf{c}_1] \\ &= \sigma_1^2 \sum_{n=1}^{N} \mathcal{E}\left[\frac{|c_{n,1}|^2}{\lambda_n + N_0}\right] \\ &= \sigma_1^2 \sum_{n=1}^{N} \mathcal{E}[|c_{n,1}|^2] \mathcal{E}\left[\frac{1}{\lambda_n + N_0}\right].\end{aligned} \tag{6.49}$$

From (6.47), we have

$$\mathcal{E}[|c_{n,1}|^2] = \left[\mathcal{E}[\mathbf{c}_1 \mathbf{c}_1^{\mathrm{H}}]\right]_{n,n} = \frac{1}{N}.$$

It follows that

$$\mathcal{E}[\mathrm{SINR}_1] = \sigma_1^2 \frac{1}{N} \sum_{n=1}^{N} \mathcal{E}\left[\frac{1}{\lambda_n + N_0}\right]. \tag{6.50}$$

To find the mean SINR, we need to find $\frac{1}{N}\sum_{n=1}^{N} \mathcal{E}\left[\frac{1}{\lambda_n + N_0}\right]$. A straightforward computation requires the N pdfs of the λ_n's, whose expressions are usually complicated. To avoid this, an asymptotic approach can be used as follows.

Let $G_N(\lambda)$ denote the empirical distribution of the eigenvalues. Formally,

$$G_N(\lambda) = \frac{1}{N}\left|\{\lambda_n(\mathbf{H}_1 \mathbf{\Sigma}_1 \mathbf{H}_1^{\mathrm{H}}) \mid \lambda_n(\mathbf{H}_1 \mathbf{\Sigma}_1 \mathbf{H}_1^{\mathrm{H}}) \le \lambda\}\right|,$$

where $\lambda_n(\mathbf{A})$ denotes the nth largest eigenvalue of $\mathbf{A}$ and $|\{\cdot\}|$ represents the cardinality or size of the set $\{\cdot\}$. Since the empirical distribution depends on a realization of $\mathbf{H}_1$, it is random. Using $G_N(\lambda)$, it can be shown that

$$\frac{1}{N}\sum_{n=1}^{N}\frac{1}{\lambda_n + N_0} = \int \frac{1}{\lambda + N_0}\mathrm{d}G_N(\lambda).$$

In Fig. 6.4, two empirical distributions are presented when $N = 16$ and $N = 640$ with a fixed ratio of $(K-1)/N = 0.5$. It is assumed that the elements of $\mathbf{H}_1$ are independent CSCG random variables with zero mean and variance $1/N$. In addition, $\mathbf{R}_s = \mathbf{I}$ and $N_0 = 0.1$. Note that since $(K-1)/N = 0.5$, a half of the eigenvalues are zero. Thus, $G_N(0) = 1/2$.

In (Silverstein & Bai 1995), it is shown that the empirical distribution tends to a certain deterministic (i.e. non-random) distribution, denoted by $G(\lambda)$, as N and $K-1$ grow with a fixed ratio (i.e. $(K-1)/N$ is fixed) (as shown in Fig. 6.4, the empirical distribution converges to a certain deterministic distribution as $N \to \infty$). From this,

$$\frac{1}{N}\sum_{n=1}^{N}\frac{1}{\lambda_n + N_0} \to \int \frac{1}{\lambda + N_0}\mathrm{d}G(\lambda)$$

and

$$\mathcal{E}[\mathrm{SINR}_1] \to \bar{\beta}_1 = \sigma_1^2 \int \frac{1}{\lambda + N_0}\mathrm{d}G(\lambda) \tag{6.51}$$

as $N, K-1 \to \infty$ with $(K-1)/N$ fixed. Here, $\bar{\beta}_1$ is the mean SINR for the asymptotic case where $N, K-1 \to \infty$ with $(K-1)/N$ fixed. It is also useful to notice that for any n

$$\mathcal{E}\left[\frac{1}{\lambda_n + N_0}\right] = \int \frac{1}{\lambda + N_0}\mathrm{d}G(\lambda) \tag{6.52}$$

in the asymptotic case. As shown in (Tulino & Verdu 2004), a closed-form expression for $G(\lambda)$ is available for a number of cases. Thus, the integration in (6.52) could be carried out to find a closed-form expression for the mean. Furthermore, as shown in (Tse & Hanly 1999), the Stieltjes transform can be used to perform the expectation in (6.52) without integration.

The variance of the SINR is given by

$$\begin{aligned}\mathrm{Var}(\mathrm{SINR}_1) &= \mathcal{E}\left[\left(\sigma_1^2 \sum_{n=1}^{N}\frac{|c_{n,1}|^2}{\lambda_n + N_0} - \mathcal{E}[\mathrm{SINR}_1]\right)^2\right] \\ &= \sigma_1^4 \sum_{k=1}^{N}\sum_{n=1}^{N}\mathcal{E}\left[\frac{|c_{k,1}|^2|c_{n,1}|^2}{(\lambda_k + N_0)(\lambda_n + N_0)}\right] - \mathcal{E}^2[\mathrm{SINR}_1]. \end{aligned} \tag{6.53}$$

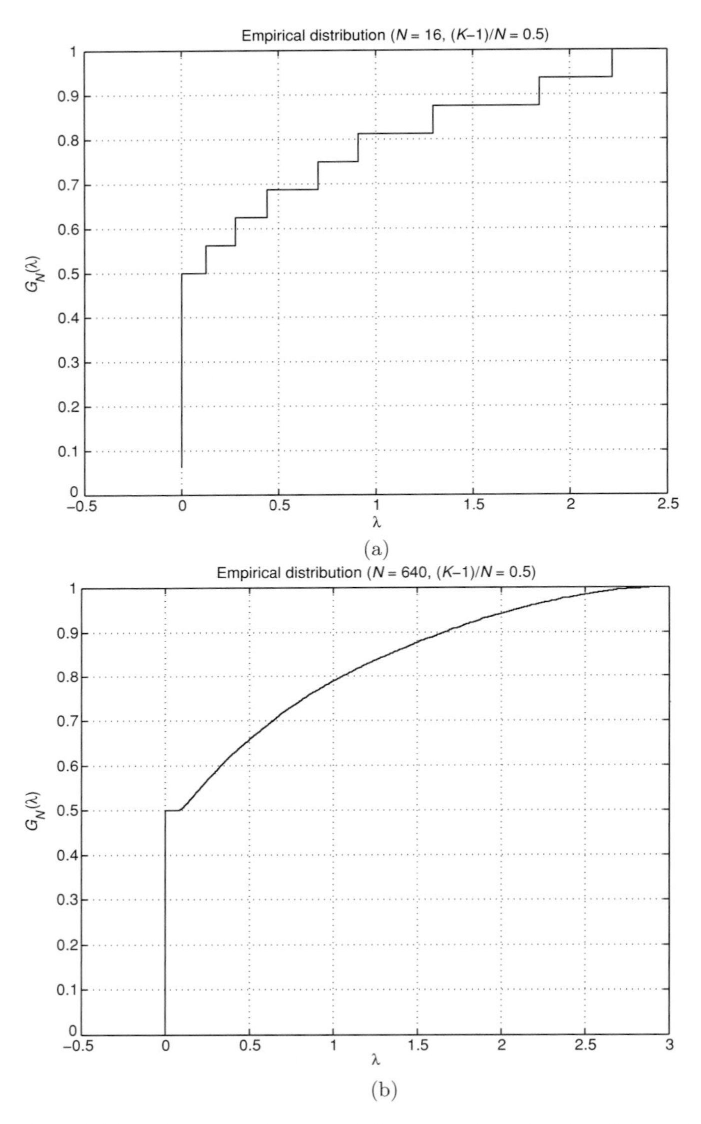

Figure 6.4 Empirical distributions of the eigenvalues, λ, of random matrices (a) $N = 16$ and $K = 9$; (b) $N = 640$ and $K = 321$.

Since $c_{k,1}$ and $c_{n,1}$ are uncorrelated if $k \neq n$, we can have

$$\sum_{k=1}^{N}\sum_{n=1}^{N} \mathcal{E}\left[\frac{|c_{k,1}|^2|c_{l,1}|^2}{(\lambda_k + N_0)(\lambda_n + N_0)}\right] = \sum_{k=1}^{N}\sum_{n=1}^{N} \mathcal{E}\left[\frac{|c_{k,1}|^2}{\lambda_k + N_0}\right] \mathcal{E}\left[\frac{|c_{n,1}|^2}{\lambda_n + N_0}\right] + \sum_{k=1}^{N} \mathcal{E}\left[\frac{|c_{k,1}|^4}{(\lambda_k + N_0)^2}\right] - \left(\mathcal{E}\left[\frac{|c_{k,1}|^2}{\lambda_k + N_0}\right]\right)^2. \tag{6.54}$$

Noting that $\mathcal{E}[|c_{k,1}|^2] = 1/N$, the first term on the right-hand side in (6.54) becomes

$$\begin{aligned}
\sum_{k=1}^{N}\sum_{n=1}^{N} \mathcal{E}\left[\frac{|c_{k,1}|^2}{\lambda_k + N_0}\right] \mathcal{E}\left[\frac{|c_{n,1}|^2}{\lambda_n + N_0}\right] &= \left(\frac{1}{N}\sum_{n=1}^{N} \mathcal{E}\left[\frac{1}{\lambda_n + N_0}\right]\right)^2 \\
&= \left(\mathcal{E}\left[\frac{1}{N}\sum_{k=1}^{N} \frac{1}{\lambda_k + N_0}\right]\right)^2 \\
&= \left(\mathcal{E}\left[\int \frac{1}{\lambda + N_0} \mathrm{d}G_N(\lambda)\right]\right)^2 .
\end{aligned}$$

Let $m_4 = \mathcal{E}[|c_{k,1}|^4]$. Then, the second term becomes

$$\begin{aligned}
\sum_{k=1}^{N} \mathcal{E}\left[\frac{|c_{k,1}|^4}{(\lambda_k + N_0)^2}\right] &= m_4 \sum_{k=1}^{N} \mathcal{E}\left[\frac{1}{(\lambda_k + N_0)^2}\right] \\
&= m_4 N \mathcal{E}\left[\frac{1}{N}\sum_{k=1}^{N} \frac{1}{(\lambda_k + N_0)^2}\right] \\
&= m_4 N \mathcal{E}\left[\int \frac{1}{(\lambda + N_0)^2} \mathrm{d}G_N(\lambda)\right] .
\end{aligned}$$

Prior to dealing with the third term, we need to consider the asymptotic case, where $G_N(\lambda)$ approaches $G(\lambda)$. We have

$$\begin{aligned}
\left(\int \frac{1}{\lambda + N_0} \mathrm{d}G_N(\lambda)\right)^2 &\to \frac{\bar{\beta}_1^2}{\sigma_1^4}; \\
\int \frac{1}{(\lambda + N_0)^2} \mathrm{d}G_N(\lambda) &\to \frac{v_1}{\sigma_1^4},
\end{aligned}$$

where

$$v_1 = \sigma_1^4 \int \frac{1}{(\lambda + N_0)^2} \mathrm{d}G(\lambda).$$

Note that both $\bar{\beta}_1^2$ and v_1 are deterministic as $G(\lambda)$ is.

Using (6.52), the third term for the asymptotic case becomes

$$\begin{aligned}
\sum_{k=1}^{N} \left(\mathcal{E}\left[\frac{|c_{k,1}|^2}{\lambda_k + N_0}\right]\right)^2 &= \frac{1}{N}\left(\int \frac{1}{\lambda + \sigma^2} \mathrm{d}G(\lambda)\right)^2 \\
&= \frac{1}{N}\frac{\bar{\beta}_1^2}{\sigma_1^4} .
\end{aligned}$$

Now, it can be shown that

$$\sum_{k=1}^{N}\sum_{n=1}^{N} \mathcal{E}\left[\frac{|c_{k,1}|^2 |c_{n,1}|^2}{(\lambda_k + N_0)(\lambda_n + N_0)}\right] \to \frac{1}{\sigma_1^4}\left(\left(1 - \frac{1}{N}\right)\bar{\beta}_1^2 + N m_4 v_1\right) . \tag{6.55}$$

Substituting (6.55) into (6.53), the asymptotic variance of the SINR is given by

$$\begin{aligned}\mathrm{Var}(\mathrm{SINR}_1) \to \bar{\varphi}_1 &= \left(\left(1 - \frac{1}{N}\right)\bar{\beta}_1^2 + N m_4 v_1\right) - \bar{\beta}_1^2 \\ &= N m_4 v_1 - \frac{\bar{\beta}_1^2}{N}.\end{aligned} \tag{6.56}$$

Example 6.6.1 Suppose that $\tilde{h}_{n,k} \in \{-1, 1\}$ and $\Pr(\tilde{h}_{n,k} = \pm 1) = 1/2$. Then, we can show that

$$m_4 = \mathcal{E}[|c_{k,1}|^4] = \frac{1}{N^2}.$$

In this case, we have

$$\begin{aligned}\mathrm{Var}(\mathrm{SINR}_1) &= \frac{v_1 - \bar{\beta}_1^2}{N} \\ &= \frac{\sigma_1^4}{N}\mathrm{Var}\left(\frac{1}{\lambda + N_0}\right),\end{aligned} \tag{6.57}$$

where λ represents the (random) eigenvalue of $\mathbf{H}_1\boldsymbol{\Sigma}_1\mathbf{H}_1^{\mathrm{H}}$. This shows that, for binary signature vectors, the variance of the SINR depends only on the distribution of the eigenvalue of the interference covariance matrix, $\mathbf{H}_1\boldsymbol{\Sigma}_1\mathbf{H}_1^{\mathrm{H}}$.

From Example 6.6.1, we can see that the variance of the SINR can approach zero as $N \to \infty$. In this case,

$$\lim_{N\to\infty} \mathrm{SINR}_1 = \bar{\beta}_1$$

in probability.

The asymptotic normalized MMSE can also be characterized by the asymptotic distribution. In (6.42), suppose that there exists the asymptotic distribution of $\lambda_k(\mathbf{H}^{\mathrm{H}}\mathbf{H})$, which is denoted by $F(\lambda)$, as follows:

$$F(\lambda) = \lim_{N,K\to\infty,\ \frac{K}{N}=\alpha} \frac{1}{K}\left|\{\lambda_k(\mathbf{H}^{\mathrm{H}}\mathbf{H}) \mid \lambda_k(\mathbf{H}^{\mathrm{H}}\mathbf{H}) \le \lambda\}\right|,$$

where $K/N = \alpha$ is a fixed ratio. Then, the asymptotic normalized MMSE is given by

$$\begin{aligned}\lim_{N,K\to\infty,\ \frac{K}{N}=\alpha} \frac{1}{K}\mathrm{tr}(\bar{\mathbf{Q}}_s) &= \lim_{N,K\to\infty,\ \frac{K}{N}=\alpha} \frac{1}{K}\sum_{k=1}^{K} \frac{1}{1 + \frac{\sigma_s^2}{N_0}\lambda_k(\mathbf{H}^{\mathrm{H}}\mathbf{H})} \\ &= \int \frac{1}{1 + \frac{\sigma_s^2}{N_0}\lambda}\mathrm{d}F(\lambda),\end{aligned} \tag{6.58}$$

where $\sigma_s^2 = \sigma_k^2 = E[|s_k|^2]$ for all k.

As shown above, the statistical properties of the MMSE combiner depend on the asymptotic distribution for large N and K. In (Tulino & Verdu 2004), closed-form expressions for the asymptotic distributions of the eigenvalues of random matrices are presented. Using them, the statistical properties of the MMSE combiner can be obtained. The distributions of the SINR and MMSE are more involved. The reader is referred to (Li, Paul, Narasimhan & Cioffi 2006) for a detailed account of approximated distributions of the SINR.

The performance analysis of the ZF combiner for random signature vectors is relatively straightforward. Related examples can be found in Problems 6.8 and 6.9.

6.7 Mutual information with MMSE combiner

For coded signals, the mutual information becomes an important measure to understand the information loss due to combining. The mutual information between $\mathbf{r}$ and $\mathbf{s}$ is determined by the channel conditions (e.g., channel matrix and noise distribution) and distribution of $\mathbf{s}$. After combining, an estimate of $\mathbf{s}$, $\hat{\mathbf{s}}$, is available. If there is any information loss due to combining, the mutual information between $\hat{\mathbf{s}}$ and $\mathbf{s}$ becomes smaller than that between $\mathbf{r}$ and $\mathbf{s}$. Furthermore, in order to see the information loss due to individual decoding, the sum of individual mutual information between $\hat{s}_k$ and s_k can be considered. In this section, we focus on the mutual information after the MMSE combining.

6.7.1 Mutual information of joint MMSE

Throughout this section, we assume that $\mathbf{s} \sim \mathcal{CN}(\mathbf{0}, \mathbf{R}_\mathbf{s})$ and $\mathbf{n} \sim \mathcal{CN}(\mathbf{0}, N_0\mathbf{I})$. Furthermore, $\mathbf{s}$ and $\mathbf{n}$ are independent.

The mutual information between $\mathbf{r}$ and $\mathbf{s}$ is given by

$$\begin{aligned} \mathsf{I}(\mathbf{r}; \mathbf{s}) &= \mathsf{h}(\mathbf{s}) - \mathsf{h}(\mathbf{s}|\mathbf{r}) \\ &= \mathsf{h}(\mathbf{r}) - \mathsf{h}(\mathbf{r}|\mathbf{s}). \end{aligned} \tag{6.59}$$

Since $\mathbf{s}$ is assumed to be Gaussian, we have

$$\mathsf{h}(\mathbf{r}) = \log\det(\pi e \mathbf{R}_\mathbf{r}),$$

where

$$\mathbf{R}_\mathbf{r} = \mathbf{H}\mathbf{R}_\mathbf{s}\mathbf{H}^\mathrm{H} + N_0\mathbf{I}.$$

Throughout this section, logarithms are taken to the base 2. Noting that

$$\mathsf{h}(\mathbf{r}|\mathbf{s}) = \mathsf{h}(\mathbf{n}) = \log\det(\pi e N_0 \mathbf{I}),$$

we can show that the mutual information is given by

$$\mathsf{I}(\mathbf{r};\mathbf{s}) = \log\det\left(\mathbf{I} + \frac{1}{N_0}\mathbf{H}\mathbf{R}_\mathbf{s}\mathbf{H}^\mathrm{H}\right). \tag{6.60}$$

The MMSE estimate of $\mathbf{s}$ is given by

$$\hat{\mathbf{s}} = \mathbf{s} + \tilde{\mathbf{s}},$$

where $\tilde{\mathbf{s}}$ denotes the MMSE error. The MMSE estimate can be seen as the output of an additive noise channel where $\mathbf{s}$ is the input vector and $\tilde{\mathbf{s}}$ is the additive noise vector. The mutual information between $\mathbf{s}$ and $\hat{\mathbf{s}}$ is given by

$$\mathsf{I}(\hat{\mathbf{s}};\mathbf{s}) = \mathsf{h}(\mathbf{s}) - \mathsf{h}(\mathbf{s}|\hat{\mathbf{s}}). \tag{6.61}$$

Since $\mathbf{s}$ is Gaussian,

$$\mathsf{h}(\mathbf{s}) = \log\det(\pi e\mathbf{R}_\mathbf{s}). \tag{6.62}$$

Furthermore, we have

$$\mathsf{h}(\hat{\mathbf{s}}|\mathbf{s}) = \mathsf{h}(\tilde{\mathbf{s}}), \tag{6.63}$$

because $\tilde{\mathbf{s}}$ and $\hat{\mathbf{s}}$ are orthogonal, i.e. $\mathcal{E}[\hat{\mathbf{s}}\tilde{\mathbf{s}}^\mathrm{H}] = \mathbf{0}$ (due to the orthogonality principle). Then, from (6.34), it follows that

$$\mathsf{h}(\mathbf{s}|\hat{\mathbf{s}}) = \mathsf{h}(\tilde{\mathbf{s}}) = \log\det(\pi e\mathbf{Q}_\mathbf{s}).$$

The mutual information is given by

$$\begin{aligned}\mathsf{I}(\hat{\mathbf{s}};\mathbf{s}) &= \log\det(\pi e\mathbf{R}_\mathbf{s}) - \log\det(\pi e\mathbf{Q}_\mathbf{s}) \\ &= \log\frac{\det(\mathbf{R}_\mathbf{s})}{\det(\mathbf{Q}_\mathbf{s})}.\end{aligned} \tag{6.64}$$

If we consider the relationship between the SINR and mutual information over an AWGN channel as follows:

$$\mathsf{I} = \log(1 + \mathrm{SINR}),$$

according to (6.64), the SINR of the joint MMSE estimation could be defined as

$$\mathrm{SINR} = \frac{\det(\mathbf{R}_\mathbf{s})}{\det(\mathbf{Q}_\mathbf{s})} - 1. \tag{6.65}$$

As in Subsection 6.4.2, $\mathrm{SINR}_\mathrm{b} = \det(\mathbf{R}_\mathbf{s})/\det(\mathbf{Q}_\mathbf{s})$ can be considered as the biased SINR. It is noteworthy that this biased SINR is the SINR for multi-channels and different from that in (6.36). The biased multi-channel SINR, $\det(\mathbf{R}_\mathbf{s})/\det(\mathbf{Q}_\mathbf{s})$, in (6.65) is the equivalent SINR that can provide the same mutual information over a single AWGN channel, while the multi-channel SINR in (6.36) is the ratio of the total signal power to the total noise (or error) power. Thus, although the SINR in (6.36) is widely used in signal processing, it may not be a proper definition in the applications of digital communications.

Example 6.7.1 Suppose that $N = K = 2$ and the channel matrix is given by

$$\mathbf{H} = \begin{bmatrix} 1 & 0.2 \\ -0.5 & 1 \end{bmatrix}.$$

In addition, assume that the signal covariance matrix is $\mathbf{R}_\mathrm{s} = \mathrm{Diag}(\sigma_1^2, \sigma_2^2)$ and the noise covariance matrix is $\mathbf{R}_\mathbf{n} = \mathbf{I}$. We consider two different signal covariance matrices:

$$\text{Case (a)}: \ \sigma_1^2 = 1, \ \sigma_1^2 = 4;$$

$$\text{Case (b)}: \ \sigma_1^2 = 2.5, \ \sigma_1^2 = 2.5.$$

In both the cases, the total signal power is the same. We can find that

$$\begin{aligned} \text{Case (a)}: \ & \frac{\det(\mathbf{R}_\mathrm{s})}{\det(\mathbf{Q}_\mathrm{s})} = 11.25, \ \frac{\mathrm{tr}(\mathbf{R}_\mathrm{s})}{\mathrm{tr}(\mathbf{Q}_\mathrm{s})} = 3.97; \\ \text{Case (b)}: \ & \frac{\det(\mathbf{R}_\mathrm{s})}{\det(\mathbf{Q}_\mathrm{s})} = 14.29, \ \frac{\mathrm{tr}(\mathbf{R}_\mathrm{s})}{\mathrm{tr}(\mathbf{Q}_\mathrm{s})} = 3.70. \end{aligned} \tag{6.66}$$

The preferable signal power allocation can be different depending on the SINR definition. For the SINR that provides higher mutual information, the power allocation according to Case (a) is better, while the power allocation according to Case (b) is better if the SINR is defined as in (6.36).

We will show that the mutual information between $\hat{\mathbf{s}}$ and $\mathbf{s}$ is identical to that between $\mathbf{r}$ and $\mathbf{s}$. Let

$$\mathbf{R}_\mathrm{s} = \mathbf{R}_\mathrm{s}^{1/2}\mathbf{R}_\mathrm{s}^{\mathrm{H}/2}.$$

The error covariance matrix can be rewritten as

$$\begin{aligned} \mathbf{Q}_\mathrm{s} &= \mathbf{R}_\mathrm{s} - \mathbf{R}_\mathrm{sr}\mathbf{R}_\mathrm{r}^{-1}\mathbf{R}_\mathrm{rs} \\ &= \mathbf{R}_\mathrm{s}^{1/2}\left(\mathbf{I} - \mathbf{R}_\mathrm{s}^{-1/2}\mathbf{R}_\mathrm{sr}\mathbf{R}_\mathrm{r}^{-1}\mathbf{R}_\mathrm{rs}\mathbf{R}_\mathrm{s}^{-\mathrm{H}/2}\right)\mathbf{R}_\mathrm{s}^{\mathrm{H}/2} \end{aligned} \tag{6.67}$$

Substituting (6.67) into (6.64), we have

$$\mathsf{I}(\hat{\mathbf{s}}; \mathbf{s}) = \log\det\left(\mathbf{I} - \mathbf{R}_\mathrm{s}^{-1/2}\mathbf{R}_\mathrm{sr}\mathbf{R}_\mathrm{r}^{-1}\mathbf{R}_\mathrm{rs}\mathbf{R}_\mathrm{s}^{-\mathrm{H}/2}\right)^{-1}. \tag{6.68}$$

Using the matrix inversion lemma, the mutual information is given by

$$\begin{aligned} \mathsf{I}(\hat{\mathbf{s}}; \mathbf{s}) &= \log\det\left(\mathbf{I} + \frac{1}{N_0}\mathbf{R}_\mathrm{s}^{\mathrm{H}/2}\mathbf{H}^\mathrm{H}\mathbf{H}\mathbf{R}_\mathrm{s}^{1/2}\right) \\ &= \log\det\left(\mathbf{I} + \frac{1}{N_0}\mathbf{H}\mathbf{R}_\mathrm{s}\mathbf{H}^\mathrm{H}\right). \end{aligned} \tag{6.69}$$

This shows that joint MMSE combining does not lose information as the mutual information is the same as that in (6.60). There are also other linear combiners that do not lose information.

Consider a linear combiner, $\mathbf{W}$, of size $N \times K$, and its output that is given by

$$\mathbf{x} = \mathbf{W}^{\mathrm{H}}\mathbf{r}. \tag{6.70}$$

In addition, assume that $K \leq N$. Then, we can show that

$$\begin{aligned}
\mathsf{I}(\mathbf{x}; \mathbf{s}) &= \log\det\left(\mathbf{I} + \frac{1}{N_0}(\mathbf{W}^{\mathrm{H}}\mathbf{W})^{-1}\mathbf{W}^{\mathrm{H}}\mathbf{H}\mathbf{R}_{\mathbf{s}}\mathbf{H}^{\mathrm{H}}\mathbf{W}\right) \\
&= \log\det\left(\mathbf{I} + \frac{1}{N_0}\mathbf{W}(\mathbf{W}^{\mathrm{H}}\mathbf{W})^{-1}\mathbf{W}^{\mathrm{H}}\mathbf{H}\mathbf{R}_{\mathbf{s}}\mathbf{H}^{\mathrm{H}}\right).
\end{aligned} \tag{6.71}$$

We note that $\mathbf{W}(\mathbf{W}^{\mathrm{H}}\mathbf{W})^{-1}\mathbf{W}^{\mathrm{H}}$ is a projection. Thus, if $\mathrm{Range}(\mathbf{W}) = \mathrm{Range}(\mathbf{H})$, we have

$$\mathbf{W}(\mathbf{W}^{\mathrm{H}}\mathbf{W})^{-1}\mathbf{W}^{\mathrm{H}}\mathbf{H} = \mathbf{H}.$$

This implies that the mutual information between $\mathbf{W}^{\mathrm{H}}\mathbf{r}$ and $\mathbf{s}$ is identical to that between $\mathbf{r}$ and $\mathbf{s}$ if $\mathrm{Range}(\mathbf{W}) = \mathrm{Range}(\mathbf{H})$. From this, we can also see that there is no information loss for the case of $\mathbf{W} = \mathbf{H}$ (i.e. the matched filtering).

From (6.64), the normalized MMSE covariance matrix in (6.37) can also be used to express the mutual information as follows:

$$\mathsf{I}(\hat{\mathbf{s}}; \mathbf{s}) = -\log\det(\bar{\mathbf{Q}}_{\mathbf{s}}). \tag{6.72}$$

6.7.2 Mutual information of individual MMSE

As long as joint decoding is carried out, linear combining is usually information lossless as shown in Subsection 6.7.1. We now consider a different case where individual decoding is carried out for each estimated symbol.

The estimate of s_1 from the output of the MMSE combiner is given by

$$\hat{s}_1 = s_1 + \tilde{s}_1, \tag{6.73}$$

where $\tilde{s}_1 = (a_1 - 1)s_1 + \mathbf{w}_1^{\mathrm{H}}\mathbf{u}$ denotes the error. From (6.73), we can model the output of the MMSE combiner as the output of the additive noise channel as shown in Fig. 6.5, where $\hat{s}_1$ is the channel output, s_1 is the channel input, and $\tilde{s}_1$ is the additive noise. Assuming that s_1 and $\tilde{s}_1$ are Gaussian, the mutual information becomes

$$\begin{aligned}
\mathsf{I}(\hat{s}_1; s_1) &= \mathsf{h}(s_1) - \mathsf{h}(s_1|\hat{s}_1) \\
&= \log(\pi e \sigma_1^2) - \log(\pi e \epsilon_1^2) \\
&= \log\left(\frac{\sigma_1^2}{\epsilon_1^2}\right) \\
&= \log(\mathrm{SINR}_{\mathrm{b},1}) \\
&= \log(1 + \mathrm{SINR}_1).
\end{aligned} \tag{6.74}$$

As discussed in Chapter 4, $\tilde{s}_1$ is uncorrelated with $\hat{s}_1$ due to the orthogonality principle. This results in $\mathsf{h}(s_1|\hat{s}_1) = \mathsf{h}(\hat{s}_1 - \tilde{s}_1|\hat{s}_1) = \mathsf{h}(\tilde{s}_1)$.

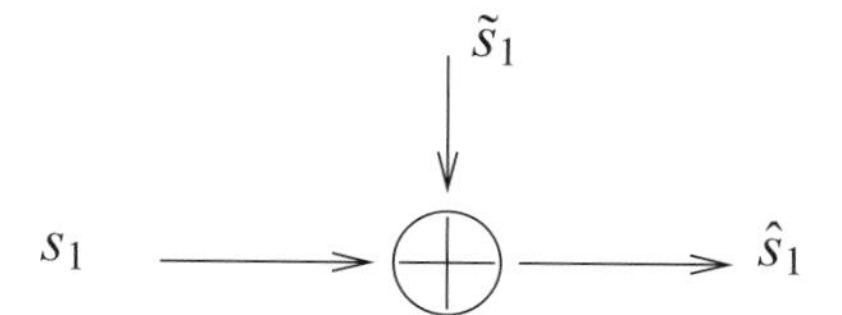

Figure 6.5 An equivalent additive noise channel model for the output of the MMSE combiner. Note that this channel model is slightly different from the conventional AWGN channel model where the channel output $r = s + n$ is correlated with the noise n, while in the above channel model, the channel output, $\hat{s}_1$, is not correlated with the noise, $\tilde{s}_1$. However, if we consider the reverse case where s_1 and $\hat{s}_1$ become the channel output and input, respectively, it becomes the conventional AWGN channel model.

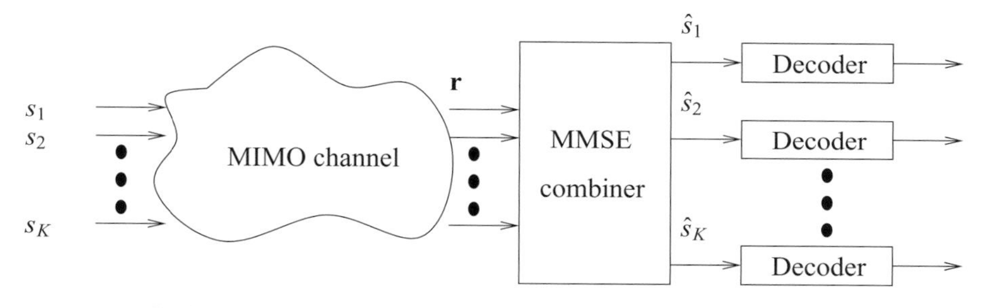

Figure 6.6 Individual decoding after MMSE combining.

Since $\mathsf{I}(\hat{s}_1; s_1)$ is the mutual information between s_1 and $\hat{s}_1$, this becomes the achievable rate if s_1 is decoded from $\hat{s}_1$ only. From (6.74), the sum mutual information when each signal is decoded individually after MMSE combining as shown in Fig. 6.6 is given by

$$\begin{aligned}\mathsf{I}_{\text{i-sum}} &= \sum_{k=1}^{K} \mathsf{I}(\hat{s}_k; s_k) \\ &= \sum_{k=1}^{K} \log\left(\frac{\sigma_k^2}{\epsilon_k^2}\right) \\ &= \log\left(\frac{\prod_{k=1}^{K} \sigma_k^2}{\prod_{k=1}^{K} \epsilon_k^2}\right) \\ &= \sum_{k=1}^{K} \log(1 + \text{SINR}_k).\end{aligned} \tag{6.75}$$

If $\mathbf{R}_s = \text{Diag}(\sigma_1^2, \sigma_2^2, \ldots, \sigma_K^2)$, we can show that

$$\mathsf{I}_{\text{i-sum}} = \log\left(\frac{\det(\mathbf{R}_s)}{\prod_{k=1}^{K} \epsilon_k^2}\right). \tag{6.76}$$

We will show that $\mathsf{I}_{\text{i-sum}}$ is less than or equal to the mutual information between $\mathbf{r}$ and $\mathbf{s}$, which can be achieved when joint decoding is used. This indicates that there is information loss if signals are individually decoded after MMSE combining.

Theorem 6.7.1 *If* $\mathbf{s}$ *and* $\mathbf{n}$ *are independent zero-mean Gaussian random vectors,*

$$\mathsf{I}(\hat{\mathbf{s}}; \mathbf{s}) \geq \mathsf{I}_{\text{i-sum}}. \tag{6.77}$$

Proof: For the proof, we need the following determinant inequality.

Lemma 6.7.1 *If* $\mathbf{A}$ *(of size* $N \times N$*) is Hermitian positive semi-definite,*

$$\det(\mathbf{A}) \leq \prod_{q=1}^{N} a_{qq} \tag{6.78}$$

with equality if and only if $\mathbf{A}$ *is diagonal.*

Using (6.78), we can show that

$$\det(\mathbf{Q}_\mathbf{s}) \leq \prod_{k=1}^{K} \epsilon_k^2, \tag{6.79}$$

since ϵ_k^2 becomes the kth diagonal element of $\mathbf{Q}_\mathbf{s}$. Substituting (6.79) into (6.64) and comparing with (6.76), we can show (6.77). □

The result in (6.77) shows that the mutual information decreases if individual decoding is employed after MMSE combining. According to Theorem 6.7.1, although the s_k's are uncorrelated (or independently encoded), it may be necessary to decode the signals jointly to avoid any information loss.

In Fig. 6.7, we show the mutual information for two different channels when $\mathbf{R}_\mathbf{s} = \sigma_s^2 \mathbf{I}$ (this is the case where the signals are uncorrelated or independently encoded) and $\mathbf{R}_\mathbf{n} = N_0 \mathbf{I}$ with $N = K = 2$. In Fig. 6.7 (a), $\mathbf{H}$ is given by

$$\mathbf{H} = \begin{bmatrix} 1 & 0.3 \\ 0.3 & 1 \end{bmatrix},$$

while, in Fig. 6.7 (b),

$$\mathbf{H} = \begin{bmatrix} 1 & 0.9 \\ 0.9 & 1 \end{bmatrix}.$$

The channel matrix in Fig. 6.7 (b) has more interference than that in Fig. 6.7 (a). It is shown that the individual decoding results in information loss although the signal sources are uncorrelated and the information loss depends on the channel and increases with the level of the interference induced by the channel.

6.8 Summary and notes

Signal combiners can be extended to the case where multiple signals co-exist. We discussed linear combiners for multiple signals in this chapter with a great deal of

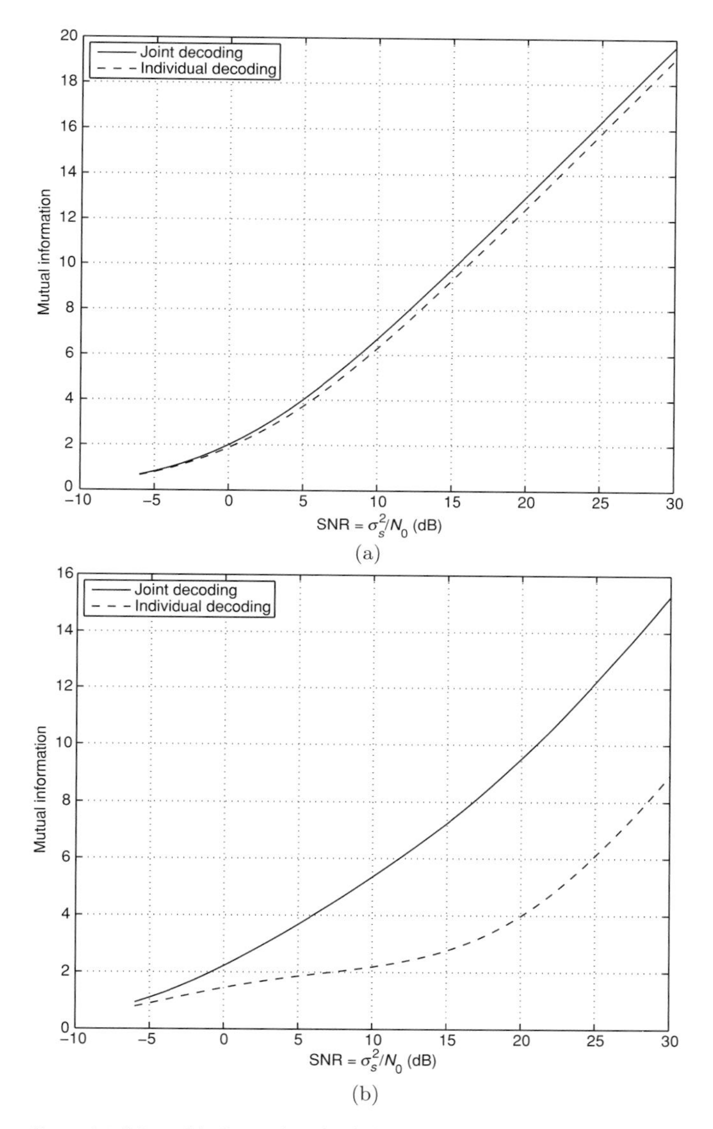

Figure 6.7 Mutual information for joint and individual decoding after MMSE combining: (a) channel of less interference; (b) channel of more interference.

emphasis on the MMSE combiner. Taking **s** as a parameter vector to be estimated, we noticed that signal combiners become estimators. In this view, a number of the books on estimation theory (e.g. (Scharf 1991), (Porat 1994), (Kay 1993)) could be helpful for in-depth understanding of signal combiners.

The MMSE combiner has been employed as a preprocessor for various receivers in multiuser communication systems. In particular, in code division multiple access (CDMA) systems, the MMSE receiver is widely studied (see (Hong, Choi, Jung & Kim 2004), (Madhow & Honig 1994), (Rapajic & Vucetic 1994), (Tse & Hanly 1999)).

Problems

Problem 6.1 ML decoding (MLD) is to find the codeword that maximizes the likelihood function. Let $\mathcal{S}_{\text{code}} = \{\mathbf{s}_1, \mathbf{s}_2, \ldots, \mathbf{s}_M\}$ denote the code, where $\mathbf{s}_m$ stands for the mth codeword. In digital communications, each codeword corresponds to a message (bit) sequence. If the mth codeword is transmitted, the received signal over an AWGN channel is given by

$$\mathbf{r} = \mathbf{s}_m + \mathbf{n},$$

where $\mathbf{n} \sim \mathcal{CN}(\mathbf{0}, \mathbf{R}_\mathbf{n})$, the MLD is as follows:

$$\begin{aligned}\hat{\mathbf{s}}_{\text{ml}} &= \arg \max_{\mathbf{s} \in \mathcal{S}_{\text{code}}} f(\mathbf{r}|\mathbf{s}) \\ &= \arg \min_{\mathbf{s} \in \mathcal{S}_{\text{code}}} (\mathbf{r} - \mathbf{s})^{\mathrm{H}} \mathbf{R}_\mathbf{n}^{-1} (\mathbf{r} - \mathbf{H}\mathbf{s}),\end{aligned}$$

where $f(\mathbf{r}|\mathbf{s})$ is the likelihood function of $\mathbf{s}$ for given $\mathbf{r}$. Using the notion of MLD above, show that MLD for multiple signals can be performed as in (6.2).

Problem 6.2 Suppose that the channel matrix is given by

$$\mathbf{H} = \begin{bmatrix} 1 & \rho \\ \rho & 1 \end{bmatrix},$$

where $\rho = |\rho| e^{j\theta}$. Provided that $|\rho| > 0$ is fixed, find θ that minimizes the error variance of the ZF combiner, i.e., $\text{tr}(\mathbf{R}_{\tilde{\mathbf{n}}})$, where $\mathbf{R}_{\tilde{\mathbf{n}}}$ is given in (6.12). Show that the column vectors of the resulting channel matrix with the θ that minimizes the error variance of the ZF combiner are orthogonal.

Problem 6.3 Derive (6.22) using the matrix inversion lemma.

Problem 6.4 Derive (6.31) using the matrix inversion lemma.

Problem 6.5 Derive (6.39).

Problem 6.6 Show that (6.38) and (6.41) are the same.

Problem 6.7 Suppose that Z denotes the eigenvalue of a random symmetric positive semi-definite matrix and its distribution is denoted by $G(\lambda)$, $\lambda \geq 0$. Define

$$m(x) = \mathcal{E}\left[\frac{1}{Z - x}\right] = \int_0^\infty \frac{1}{\lambda - x} \mathrm{d}G(\lambda),$$

which is called the Stieltjes transform of $G(\lambda)$. Then, show that

$$\frac{1}{n!} \frac{\mathrm{d}^n m(x)}{\mathrm{d}x^n} = \int_0^\infty \frac{1}{(\lambda - x)^{(n+1)}} \mathrm{d}G(\lambda).$$

(Once the Stieltjes transform is available, we can easily find the asymptotic mean and variance of the SINR in (6.51) and (6.56), respectively, for the MMSE combiner.)

Problem 6.8 Suppose that $\mathbf{R}_\mathbf{s} = \sigma_s^2 \mathbf{I}$. Furthermore, the $h_{k,l}$'s are iid and $h_{k,l} \sim \mathcal{CN}(0, 1/N)$. Show that the SNR after ZF combining can be expressed as

$$\mathrm{SNR}_{\mathrm{zf},k} = \frac{\sigma_s^2}{N_0} \sum_{l=1}^{N-K+1} |X_l|^2, \tag{6.80}$$

where the X_l's are independent and $X_l \sim \mathcal{CN}(0, \frac{1}{N})$.

Problem 6.9 Show that the $\mathrm{SNR}_{\mathrm{zf},k}$'s in (6.80) are independent of each other.

Problem 6.10 Suppose that $\mathbf{R}_\mathbf{s} = \sigma_s^2 \mathbf{I}$ and $N \geq K$. The eigendecomposition of $\mathbf{H}_{\text{l}} \mathbf{H}_{\text{l}}^{\mathrm{H}}$ is given by

$$\mathbf{H}_{\text{l}} \mathbf{H}_{\text{l}}^{\mathrm{H}} = \mathbf{U} \mathbf{\Lambda} \mathbf{U}^{\mathrm{H}},$$

where $\mathbf{U} = [\mathbf{u}_{(1)}\ \mathbf{u}_{(2)}\ \cdots\ \mathbf{u}_{(N)}]$ and $\mathbf{\Lambda} = \mathrm{Diag}(\lambda_{(1)}, \lambda_{(2)}, \ldots, \lambda_{(N)})$. Here, $\lambda_{(l)}$ represents the lth largest eigenvalue and its associated eigenvector is denoted by $\mathbf{u}_{(l)}$. Show that

$$\mathrm{SINR}_1 = \mathrm{SNR}_{\mathrm{mf},1} + \frac{\sigma_s^2}{N_0} \sum_{l=1}^{K-1} \frac{|c_{1,(l)}|^2}{1 + \frac{\lambda_{(l)}}{N_0}}, \tag{6.81}$$

where $c_{1,(l)} = \mathbf{u}_{(l)}^{\mathrm{H}} \mathbf{h}_1$.

Problem 6.11 Show that

$$\det(\mathbf{I} + \mathbf{A}\mathbf{A}^{\mathrm{H}}) = \det(\mathbf{I} + \mathbf{A}^{\mathrm{H}}\mathbf{A})$$

using the SVD of $\mathbf{A}$. From this equality, the second equality in (6.69) can be obtained.

7 Multiple signal detection in vector space: MIMO detection

Multiple signal detection is a generalization of single signal detection that is discussed in Chapter 2. In theory, this generalization is readily accomplished by extending the signal alphabet to include all the possible transmitted signal combinations from multiple signals. However, since the size of this extended signal alphabet grows exponentially with the number of signals, most optimal detectors have prohibitively high computational complexity, which becomes the major problem in terms of implementation.

In this chapter, we discuss various optimal and suboptimal detectors in the presence of multiple signals. In general, suboptimal detectors have low complexity at the expense of degraded performance. In addition, we present various approaches to analyze the performance of various detectors over random channels.

7.1 SIMO detection

In this section, we consider signal detection when multiple observations are available in the presence of a single signal, which is called single-input multiple-output (SIMO) detection. It is assumed that the receiver is equipped with N antennas for multiple observations. Alternatively, we can assume that the received signal is oversampled by a factor of N. Suppose that the qth received signal is given by

$$r_q = h_q s + n_q, \ q = 1, 2 \ldots, N,$$

where h_q and n_q represent the channel gain and noise of the qth received signal, respectively, and s denotes the desired signal, which is a data symbol transmitted by a transmitter for digital communications. Let $\mathbf{h} = [h_1 \ h_2 \ \ldots \ h_N]^{\mathrm{T}}$ and $\mathbf{n} = [n_1 \ n_2 \ \ldots \ n_N]^{\mathrm{T}}$. Then, the received signal vector becomes

$$\begin{aligned} \mathbf{r} &= [r_1 \ r_2 \ \ldots \ r_N]^{\mathrm{T}} \\ &= \mathbf{h}s + \mathbf{n}. \end{aligned} \tag{7.1}$$

Suppose that $s \in \mathcal{S}$, where $\mathcal{S}$ denotes the signal alphabet of M symbols for M-ary communications. It is assumed that s is randomly chosen from $\mathcal{S}$ and, therefore, a random variable. For convenience, the mean of s is assumed to be zero and the variance is given by

$$E_s = \mathcal{E}[|s|^2] \\ = \sum_{s \in \mathcal{S}} |s|^2 \Pr(s),$$

where $\Pr(s)$ is the (a priori) probability of s. Note that we use the notation E_s rather than σ_s^2 to emphasize that E_s is the variance of a data symbol which is a discrete random variable. Since E_s is also the symbol energy, the relation between it and the bit energy, denoted by E_b, is given by

$$E_b = \frac{E_s}{\log_2 M},$$

because a symbol consists of $\log_2 M$ bits. Thus, for binary signaling (i.e. $M = 2$), E_s becomes E_b.

Throughout this chapter, we assume that $\mathbf{h}$, s, and $\mathbf{n}$ are all complex-valued. Furthermore, it is assumed that $\mathbf{n} \sim \mathcal{CN}(\mathbf{0}, \mathbf{R_n})$.

7.1.1 ML and MAP detection

The ML detection is to find the symbol in $\mathcal{S}$ that maximizes the likelihood function. That is,

$$\hat{s}_{\text{ml}} = \arg\max_{s \in \mathcal{S}} f(\mathbf{r}|s),$$

where $f(\mathbf{r}|s)$ denotes the likelihood function of s for given $\mathbf{r}$. Since $\mathbf{n}$ is assumed to be a CSCG random vector, we have

$$f(\mathbf{r}|s) = \frac{1}{\det(\pi \mathbf{R_n})} \exp\left(-(\mathbf{r} - \mathbf{h}s)^{\text{H}} \mathbf{R_n}^{-1} (\mathbf{r} - \mathbf{h}s)\right).$$

From this, the ML detection becomes

$$\hat{s}_{\text{ml}} = \arg\min_{s \in \mathcal{S}} (\mathbf{r} - \mathbf{h}s)^{\text{H}} \mathbf{R_n}^{-1} (\mathbf{r} - \mathbf{h}s).$$

If the a priori probability of s is available at the receiver, the MAP detection can be formulated as follows:

$$\hat{s}_{\text{map}} = \arg\max_{s \in \mathcal{S}} \Pr(s|\mathbf{r}) \\ = \arg\max_{s \in \mathcal{S}} f(\mathbf{r}|s) \Pr(s). \quad (7.2)$$

7.1.2 Soft-decision using LLR

In general, the ML or MAP detector produces a hard-decision. In some cases, however, a soft-decision is more desirable. For instance, if signals are encoded by a channel encoder, soft-decisions are more appropriate for the input to a channel decoder as information

Table 7.1 An example of mapping for $M = 4$.

Symbol, s	Bit sequence
$s^{(1)}$	00
$s^{(2)}$	01
$s^{(3)}$	10
$s^{(4)}$	11

loss can be minimized. In this subsection, we consider signal detection that produces a soft-decision.

Suppose that $\mathcal{S} = \{-1, +1\}$. Then, for the ML detection of s, the likelihood ratio (LR) can be used. The LR is given by

$$\text{LR} = \frac{f(\mathbf{r}|s = +1)}{f(\mathbf{r}|s = -1)} = \frac{\exp\left(-(\mathbf{r} - \mathbf{h})^{\mathrm{H}}\mathbf{R}_{\mathbf{n}}^{-1}(\mathbf{r} - \mathbf{h})\right)}{\exp\left(-(\mathbf{r} + \mathbf{h})^{\mathrm{H}}\mathbf{R}_{\mathbf{n}}^{-1}(\mathbf{r} + \mathbf{h})\right)}. \tag{7.3}$$

For a soft-decision, the log-likelihood ratio (LLR) can be used:

$$\text{LLR} = \log(\text{LR}) = 4\Re\{\mathbf{h}^{\mathrm{H}}\mathbf{R}_{\mathbf{n}}^{-1}\mathbf{r}\}, \tag{7.4}$$

where log stands for the natural logarithm. The sign of the LLR is identical to the hard-decision from the ML detection, while the absolute value of the LLR can be used to measure the reliability of decision. Since the LLR is a function of $\mathbf{r}$, it is a random variable. In addition, the LLR is a sufficient statistic for $s \in \{-1, +1\}$. For coded signals, the LLR is also a sufficient statistic for channel decoding (Hagenauer, Offer & Papke 1996), (Richardson & Urbanke 2008).

In above, we consider a special case that $\mathcal{S} = \{-1, +1\}$. For a general case, we can consider the following signal alphabet for M-ary signaling:

$$\mathcal{S} = \{s^{(1)}, s^{(2)}, \ldots, s^{(M)}\}, \tag{7.5}$$

where $s^{(m)}$ denotes the mth element of $\mathcal{S}$. Each symbol in $\mathcal{S}$ can represent a sequence of $\log_2 M$ bits (if M is a power of 2) and there is a one-to-one mapping between symbols and bit sequences. Let $\mathcal{S}_l^0$ and $\mathcal{S}_l^1$ denote the sets of the symbols whose lth bit is 0 and 1, respectively. For example, if $M = 4$ and the mapping is given in Table 7.1, we have $\mathcal{S}_1^0 = \{s^{(1)},\ s^{(2)}\}$ and $\mathcal{S}_2^0 = \{s^{(1)},\ s^{(3)}\}$.

Using $\mathcal{S}_l^0$ and $\mathcal{S}_l^1$, the LLR for the lth bit is given by

$$\begin{aligned}\text{LLR}_l &= \log \frac{f(\mathbf{r}|\mathcal{S}_l^0)}{f(\mathbf{r}|\mathcal{S}_l^1)} \\ &= \log \frac{\Pr(\mathcal{S}_l^0|\mathbf{r})}{\Pr(\mathcal{S}_l^1|\mathbf{r})} - \log \frac{\Pr(\mathcal{S}_l^0)}{\Pr(\mathcal{S}_l^1)},\end{aligned} \tag{7.6}$$

where $\Pr(\mathcal{S}_l^i | \mathbf{r})$ and $\Pr(\mathcal{S}_l^i)$ denote the a posteriori and a priori probabilities of $\mathcal{S}_l^i$, $i = 0, 1$, respectively. Using Bayes' rule, it can be shown that

$$\Pr(\mathcal{S}_l^i | \mathbf{r}) = \sum_{s \in \mathcal{S}_l^i} \Pr(s | \mathbf{r}) \propto \sum_{s \in \mathcal{S}_l^i} f(\mathbf{r}|s) \Pr(s).$$

Then, it follows that

$$\text{LLR}_l = \log \frac{\sum_{s \in \mathcal{S}_l^0} f(\mathbf{r}|s) \Pr(s)}{\sum_{s \in \mathcal{S}_l^1} f(\mathbf{r}|s) \Pr(s)} - \log \frac{\Pr(\mathcal{S}_l^0)}{\Pr(\mathcal{S}_l^1)}. \tag{7.7}$$

If s is equally probable, the LLR reduces to

$$\text{LLR}_l = \log \frac{\sum_{s \in \mathcal{S}_l^0} f(\mathbf{r}|s)}{\sum_{s \in \mathcal{S}_l^1} f(\mathbf{r}|s)}. \tag{7.8}$$

If we take the sign of the LLR in (7.8), the result is the same as that of the ML decision.

Using the following approximation:

$$\mathrm{e}^{-x_1} + \mathrm{e}^{-x_2} + \cdots + \mathrm{e}^{-x_N} \simeq \max_q \mathrm{e}^{-x_q}, \quad x_n > 0, \; n = 1, 2, \ldots, q.$$

we can show that

$$\begin{aligned}
\log \sum_{s \in \mathcal{S}_l^i} f(\mathbf{r}|s) \Pr(s) &\simeq \log \max_{s \in \mathcal{S}_l^i} f(\mathbf{r}|s) \Pr(s) \\
&= \max_{s \in \mathcal{S}_l^i} \{\log f(\mathbf{r}|s) + \log \Pr(s)\} \\
&= -\min_{s \in \mathcal{S}_l^i} \{(\mathbf{r} - \mathbf{h}s)^{\mathrm{H}} \mathbf{R}_{\mathbf{n}}^{-1} (\mathbf{r} - \mathbf{h}s) - \log \Pr(s)\} + C,
\end{aligned} \tag{7.9}$$

where C is a constant. The resulting approximate LLR is given by

$$\begin{aligned}
\text{LLR}_l \simeq\; & \min_{s \in \mathcal{S}_l^1} \{(\mathbf{r} - \mathbf{h}s)^{\mathrm{H}} \mathbf{R}_{\mathbf{n}}^{-1} (\mathbf{r} - \mathbf{h}s) - \log \Pr(s)\} \\
& - \min_{s \in \mathcal{S}_l^0} \{(\mathbf{r} - \mathbf{h}s)^{\mathrm{H}} \mathbf{R}_{\mathbf{n}}^{-1} (\mathbf{r} - \mathbf{h}s) - \log \Pr(s)\} - \log \frac{\Pr(\mathcal{S}_l^0)}{\Pr(\mathcal{S}_l^1)}.
\end{aligned} \tag{7.10}$$

Example 7.1.1 Let $M = 4$ and the mapping is given in Table 7.1. In addition, let $s^{(m)} = \mathrm{e}^{j(\frac{\pi}{2}(m-1)+\frac{\pi}{4})}$, $m = 1, 2, 3, 4$, as shown in Fig. 7.1. Suppose that

$$\mathbf{h} = \begin{bmatrix} j \\ (1+j)/2 \end{bmatrix} \quad \text{and} \quad \mathbf{n} \sim \mathcal{CN}(\mathbf{0}, \mathbf{I}).$$

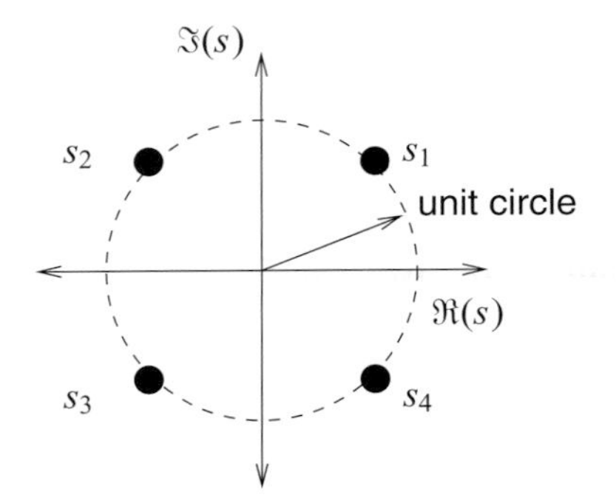

Figure 7.1 Signal constellation for Example 7.1.1.

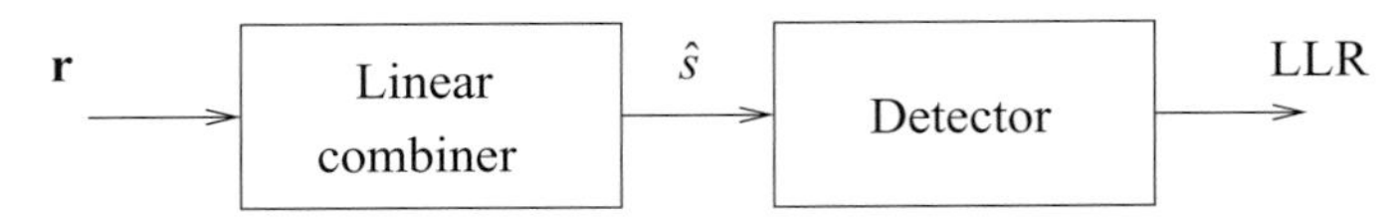

Figure 7.2 Combiner and detector for vector signal detection.

If $\mathbf{r} = [1 \ -j]^{\mathrm{T}}$ and s is equally likely (thus, S_l^i is also equally likely), we have

$$\begin{aligned}\mathrm{LLR}_1 &= \log \frac{\mathrm{e}^{-||\mathbf{r}-\mathbf{h}s^{(1)}||^2} + \mathrm{e}^{-||\mathbf{r}-\mathbf{h}s^{(2)}||^2}}{\mathrm{e}^{-||\mathbf{r}-\mathbf{h}s^{(3)}||^2} + \mathrm{e}^{-||\mathbf{r}-\mathbf{h}s^{(4)}||^2}} \\ &= -4.2426; \\ \mathrm{LLR}_2 &= \log \frac{\mathrm{e}^{-||\mathbf{r}-\mathbf{h}s^{(1)}||^2} + \mathrm{e}^{-||\mathbf{r}-\mathbf{h}s^{(3)}||^2}}{\mathrm{e}^{-||\mathbf{r}-\mathbf{h}s^{(2)}||^2} + \mathrm{e}^{-||\mathbf{r}-\mathbf{h}s^{(4)}||^2}} \\ &= 1.3604.\end{aligned}$$

The approximate LLR in (7.10) provides good approximates as follows:

$$\begin{aligned}\mathrm{LLR}_1 &\simeq -4.2426; \\ \mathrm{LLR}_2 &\simeq 1.4142.\end{aligned}$$

7.1.3 Optimality of MMSE combiner

The SIMO detection can be performed by two steps: (i) linear combining to produce an estimate of s and (ii) detection with the estimate of s to produce the LLR as shown in Fig. 7.2.

Theorem 7.1.1 *Let $\mathbf{w}_{\rm mmse}$ denote the linear MMSE combining vector for signal s, which is given by*

$$\mathbf{w}_{\rm mmse} = \arg\min_{\mathbf{w}} \mathcal{E}[|s - \mathbf{w}^{\rm H}\mathbf{r}|^2] = \left(\mathbf{h}\mathbf{h}^{\rm H} + E_{\rm s}^{-1}\mathbf{R}_{\mathbf{n}}\right)^{-1}\mathbf{h}$$

from Section 4.2. Consider the output of the MMSE combiner:

$$\hat{s} = \mathbf{w}_{\rm mmse}^{\rm H}\mathbf{r} = \mu s + u, \tag{7.11}$$

where $\mu = \mathbf{w}_{\rm mmse}^{\rm H}\mathbf{h}$ and $u = \mathbf{w}_{\rm mmse}^{\rm H}\mathbf{n}$. If $\mathbf{n}$ is a zero-mean CSCG random vector, the LLR from $\hat{s}$ in (7.11) is identical to that in (7.7).

Proof: If $\mathbf{n}$ is a zero-mean CSCG random vector, we can show that the ratio of the two likelihood functions is not a function of s, i.e. constant:

$$f(\mathbf{r}|s)/f(\hat{s}|s) = constant. \tag{7.12}$$

This completes the proof. □

From the result of Theorem 7.1.1, we can see that ML or MAP detection with the output from the linear MMSE combiner is identical to ML or MAP detection with the received signal vector, $\mathbf{r}$, respectively. Therefore, in implementation, an ML (or MAP) detector that has been built for the case of $N = 1$ shown in Fig. 7.2 can be used after linear combining for SIMO detection.

Example 7.1.2 We repeat Example 7.1.1 with the output of the MMSE combiner. The MMSE weighting vector is given by

$$\mathbf{w}_{\rm mmse} = \epsilon_{\rm mmse}^2\mathbf{R}_{\mathbf{n}}^{-1}\mathbf{h} = \begin{bmatrix} \frac{2j}{5} & \frac{(1+j)}{5} \end{bmatrix}^{\rm T},$$

where $\epsilon_{\rm mmse}^2$ is the MMSE that is given by

$$\epsilon_{\rm mmse}^2 = \frac{E_{\rm s}}{1 + E_{\rm s}\mathbf{h}^{\rm H}\mathbf{R}_{\mathbf{n}}^{-1}\mathbf{h}}.$$

Then,

$$\mu = \frac{3}{5};$$

$$u \sim \mathcal{CN}\left(0, \frac{6}{25}\right).$$

The LLRs are obtained as follows:

$$\begin{aligned}\mathrm{LLR}_1 &= \log \frac{\mathrm{e}^{-|\hat{s}-\mu s^{(1)}|^2/\sigma_u^2} + \mathrm{e}^{-|\hat{s}-\mu s^{(2)}|^2/\sigma_u^2}}{\mathrm{e}^{-|\hat{s}-\mu s^{(3)}|^2/\sigma_u^2} + \mathrm{e}^{-|\hat{s}-\mu s^{(4)}|^2/\sigma_u^2}} \\ &= -4.2426; \\ \mathrm{LLR}_2 &= \log \frac{\mathrm{e}^{-|\hat{s}-\mu s^{(1)}|^2/\sigma_u^2} + \mathrm{e}^{-|\hat{s}-\mu s^{(3)}|^2/\sigma_u^2}}{\mathrm{e}^{-|\hat{s}-\mu s^{(2)}|^2/\sigma_u^2} + \mathrm{e}^{-|\hat{s}-\mu s^{(4)}|^2/\sigma_u^2}} \\ &= 1.3604,\end{aligned}$$

where $\sigma_u^2 = \mathcal{E}[|u|^2] = 6/25$. The results are identical to those in Example 7.1.1.

Suppose that there are two received signal vectors as follows:

$$\begin{aligned}\mathbf{r}_1 &= \mathbf{h}_1 s + \mathbf{n}_1; \\ \mathbf{r}_2 &= \mathbf{h}_2 s + \mathbf{n}_2.\end{aligned}$$

The LLR can be obtained from each received vector separately. Denote by $\mathrm{LLR}(\mathbf{r}_l)$ the LLR from $\mathbf{r}_l$. Furthermore, let $\mathrm{LLR}(\mathbf{r}_1, \mathbf{r}_2)$ denote the LLR obtained jointly from both $\mathbf{r}_1$ and $\mathbf{r}_2$ or

$$\begin{aligned}\mathbf{r} &= [\mathbf{r}_1^{\mathrm{T}}\ \mathbf{r}_2^{\mathrm{T}}]^{\mathrm{T}} \\ &= \begin{bmatrix}\mathbf{h}_1 \\ \mathbf{h}_2\end{bmatrix} s + \begin{bmatrix}\mathbf{n}_1 \\ \mathbf{n}_2\end{bmatrix}.\end{aligned}$$

If the background noise vectors are independent, we can readily show that

$$\mathrm{LLR}(\mathbf{r}_1, \mathbf{r}_2) = \mathrm{LLR}(\mathbf{r}_1) + \mathrm{LLR}(\mathbf{r}_2),$$

because $f(\mathbf{r}|s) = f(\mathbf{r}_1|s) f(\mathbf{r}_2|s)$. A generalization is given below.

Theorem 7.1.2 *Suppose that there are L observations as follows:*

$$\mathbf{r}_l = \mathbf{h}_l s + \mathbf{n}_l,\ l = 0, 1, \ldots, L-1.$$

If the $\mathbf{n}_l$'s are independent, we have

$$\mathrm{LLR}(\mathbf{r}_0, \mathbf{r}_1, \ldots, \mathbf{r}_{L-1}) = \sum_{l=0}^{L-1} \mathrm{LLR}(\mathbf{r}_l). \tag{7.13}$$

This shows that the joint LLR is simply a sum of individual LLRs.

7.2 Performance of optimal SIMO detection

In this section, we analyze the performance of the ML detector for SIMO detection. For performance analysis, it is assumed that $\mathbf{n} \sim \mathcal{CN}(0, N_0\mathbf{I})$ throughout this section.

Using PEPs and union bound, we can find an upper bound on the symbol error probability. To find a PEP, we assume that there are only two symbols, say $s^{(1)}$ and $s^{(2)}$, where $s^{(1)}, s^{(2)} \in \mathcal{S}$. Provided that $s^{(1)}$ is transmitted, the probability that the symbol $s^{(2)} \neq s^{(1)}$ is erroneously detected is as follows:

$$\begin{aligned} P(s^{(1)} \rightarrow s^{(2)}|\mathbf{h}) &= \Pr(||\mathbf{r} - \mathbf{h}s^{(2)}||^2 \leq ||\mathbf{r} - \mathbf{h}s^{(1)}||^2) \\ &= \Pr(||\mathbf{h}(s^{(1)} - s^{(2)}) + \mathbf{n}||^2 \leq ||\mathbf{n}||^2) \\ &= \Pr(||\mathbf{h}d||^2 \leq -2\Re(\mathbf{n}^{\mathrm{H}}\mathbf{h}d)), \end{aligned} \tag{7.14}$$

where $d = s^{(1)} - s^{(2)}$. For given $\mathbf{h}$ and d, $\mathbf{n}^{\mathrm{H}}\mathbf{h}d$ is a CSCG random variable. We can show that

$$\begin{aligned} \mathcal{E}[\mathbf{n}^{\mathrm{H}}\mathbf{h}d] &= 0; \\ \mathcal{E}[|\mathbf{n}^{\mathrm{H}}\mathbf{h}d|^2] &= \mathcal{E}[\mathbf{h}^{\mathrm{H}}\mathbf{n}\mathbf{n}^{\mathrm{H}}\mathbf{h}|d|^2] \\ &= \mathbf{h}^{\mathrm{H}}\mathcal{E}[\mathbf{n}\mathbf{n}^{\mathrm{H}}]\mathbf{h}|d|^2 \\ &= N_0||\mathbf{h}||^2|d|^2; \\ \mathcal{E}[(\mathbf{n}^{\mathrm{H}}\mathbf{h}d)^2] &= \mathcal{E}[\mathbf{h}^{\mathrm{T}}\mathbf{n}^*\mathbf{n}^{\mathrm{H}}\mathbf{h}d^2] \\ &= \mathbf{h}^{\mathrm{T}}\mathcal{E}[\mathbf{n}^*\mathbf{n}^{\mathrm{H}}]\mathbf{h}d^2 \\ &= 0. \end{aligned}$$

Define a random variable as (its distribution can be characterized by the above results)

$$Z = -2\Re(\mathbf{n}^{\mathrm{H}}\mathbf{h}d) \sim \mathcal{N}(0, 2N_0||\mathbf{h}||^2|d|^2).$$

Using Z, from (7.14), the PEP becomes

$$\begin{aligned} P(s^{(1)} \rightarrow s^{(2)}|\mathbf{h}) &= \Pr(Z \geq ||\mathbf{h}||^2|d|^2) \\ &= \mathcal{Q}\left(\sqrt{\frac{||\mathbf{h}||^2|d|^2}{2N_0}}\right). \end{aligned} \tag{7.15}$$

So far, the PEP, $P(s^{(1)} \rightarrow s^{(2)}|\mathbf{h})$, is a conditional probability given $\mathbf{h}$. Thus, the PEP in (7.15) depends on $\mathbf{h}$. Since the channel vector, $\mathbf{h}$, can be arbitrary, we may assume that $\mathbf{h}$ is a random vector. From this, we can derive the average PEP by taking the expectation with respect to $\mathbf{h}$. The distribution of $\mathbf{h}$ depends on channel environments. In this section, we assume

$$\mathbf{h} \sim \mathcal{CN}(0, \sigma_h^2\mathbf{I}). \tag{7.16}$$

In the context of wireless communications, the corresponding channel model is called the Rayleigh fading channel, because the absolute value of an element of $\mathbf{h}$ is Rayleigh distributed (Schwartz *et al.* 1966).

There are various techniques to derive the average PEP in a closed-form. In this section, we consider three different approaches.

7.2.1 An exact closed-form for PEP

To find the average PEP, we can use a closed-form expression derived in (Proakis 1995).

As h_q is assumed to be a CSCG random variable, $|h_q|^2$ is a chi-square random variable with two degrees of freedom, which is also an exponential random variable. The pdf of $Y = |h_q|^2$ is given by

$$f(y = |h_q|^2) = \frac{1}{\sigma_h^2} e^{-\frac{y}{\sigma_h^2}}, \ y \geq 0.$$

The moment generating function (mgf) of $|h_q|^2$ is given by

$$\mathcal{E}[e^{s|h_q|^2}] = \frac{1}{1 - s\sigma_h^2}. \tag{7.17}$$

Noting

$$\psi = ||\mathbf{h}||^2 = \sum_{q=1}^{N} |h_q|^2$$

and using (7.17), the mgf of $||\mathbf{h}||^2$ is given by

$$\begin{aligned}\mathcal{E}[e^{s\psi}] &= \prod_{q=1}^{N} \mathcal{E}[e^{s|h_q|^2}] \\ &= \prod_{q=1}^{N} \frac{1}{1 - s\sigma_h^2} \\ &= \frac{1}{(1 - s\sigma_h^2)^N},\end{aligned} \tag{7.18}$$

where the first equality results from the assumption that the elements of $\mathbf{h}$ are independent as in (7.16). From this, we have

$$\mathcal{E}[e^{-s||\mathbf{h}||^2}] = \frac{1}{(1 + s\sigma_h^2)^N}.$$

The pdf of $\psi = ||\mathbf{h}||^2$ can be obtained by taking the inverse Laplace transform from (7.18):

$$f(\psi) = \frac{1}{(N-1)!\sigma_h^2} \psi^{N-1} e^{-\frac{\psi}{\sigma_h^2}}.$$

From (Proakis 1995), a closed-form expression for the average PEP is available as

$$\begin{aligned}\mathcal{E}[P(s^{(1)} \to s^{(2)}|\mathbf{h})] &= \int_0^\infty \mathcal{Q}\left(\sqrt{\frac{\psi|d|^2}{2N_0}}\right) f(\psi) d\psi \\ &= \left(\frac{1-\mu}{2}\right)^N \sum_{q=0}^{N-1} \binom{N-1+q}{q} \left(\frac{1+\mu}{2}\right)^q,\end{aligned} \tag{7.19}$$

where

$$\mu = \sqrt{\frac{\gamma_d}{1+\gamma_d}}$$

and

$$\gamma_d = \frac{\sigma_h^2 |d|^2}{4N_0}.$$

Once the average PEPs are obtained as in (7.19), an upper bound on the average symbol error probability can be found using the union bound. Let $d_{p,q} = s^{(p)} - s^{(q)}$, where $s^{(p)}, s^{(q)} \in \mathcal{S}$. Then, the average symbol error probability is upper-bounded as

$$P_s \le \sum_p \sum_{q \ne p} \mathcal{E}\left[P(s^{(1)} \to s^{(2)}|\mathbf{h})\right]. \tag{7.20}$$

7.2.2 Chernoff bound

An upper bound on the PEP can be obtained using the Chernoff bound:

$$\mathcal{Q}(\sqrt{x}) \le \exp\left(-\frac{x}{2}\right).$$

Then, from (7.15), an upper bound on the PEP is given by

$$P(s^{(1)} \to s^{(2)}|\mathbf{h}) \le \exp\left(-\frac{||\mathbf{h}||^2 |d|^2}{4N_0}\right). \tag{7.21}$$

Using (7.18), the average PEP in (7.21) is upper-bounded as

$$\begin{aligned} \mathcal{E}[P(s^{(1)} \to s^{(2)}|\mathbf{h})] &\le \mathcal{E}\left[\mathrm{e}^{-\frac{||\mathbf{h}||^2|d|^2}{4N_0}}\right] \\ &= \left(1 + \frac{|d|^2\sigma_h^2}{4N_0}\right)^{-N}. \end{aligned} \tag{7.22}$$

Using the union and Chernoff bounds, we can have an upper bound on the average symbol error probability as follows:

$$P_s \le \sum_p \sum_{q \ne p} \left(1 + \frac{|d_{p,q}|^2\sigma_h^2}{4N_0}\right)^{-N}. \tag{7.23}$$

This shows that an N-fold diversity gain is achieved. Since N is the number of receive antennas, it is said that a full receive antenna diversity gain is achieved.

As a special case, we can consider a binary signaling with $\mathcal{S} = \{\pm\sqrt{E_b}\}$. Since $|d|^2 = 4E_b$, the PEP becomes the bit error probability and its upper bound is given by

$$\begin{aligned} P_b &= \mathcal{E}\left[\mathcal{Q}\left(\sqrt{\frac{2||\mathbf{h}||^2 E_b}{N_0}}\right)\right] \\ &\le (1+\bar{\gamma})^{-N}. \end{aligned} \tag{7.24}$$

where $\bar{\gamma} = E_b\sigma_h^2/N_0$ is the average SNR.

7.2.3 Approximate PEP based on Gauss–Chebyshev quadrature

In (Ventura-Traveset, Caire, Biglieri & Taricco 1997), an approach based on Gauss–Chebyshev quadrature is introduced to derive a closed-form expression for the average PEP. In this technique, the PEP is first expressed by an inverse Laplace transform. Then, the Gauss–Chebyshev quadrature rule is applied to the inverse Laplace transform to approximate the PEP.

In order to illustrate this approach, consider (7.14). From (7.14), the average PEP can be given by

$$\begin{aligned} P(s^{(1)} \to s^{(2)}) &= \mathcal{E}\left[\mathcal{Q}\left(\sqrt{\frac{||\mathbf{h}||^2|d|^2}{2N_0}}\right)\right] \\ &= \Pr(Z \geq ||\mathbf{h}||^2|d|^2) \\ &= \Pr(X < 0), \end{aligned} \tag{7.25}$$

where

$$X = 2\Re(\mathbf{n}^{\mathrm{H}}\mathbf{h}d) + ||\mathbf{h}||^2|d|^2.$$

Let

$$\Delta = \frac{X}{\sqrt{N_0|d|^2}}. \tag{7.26}$$

Since $|d|\sqrt{N_0} \geq 0$, $\Pr(X < 0) = \Pr(\Delta < 0)$. Let $\beta = |d|^2/N_0$. Using the mgf of Δ, it can be shown that

$$\begin{aligned} \Phi(s) &= \mathcal{E}\left[\mathrm{e}^{-s\Delta}\right] \\ &= \mathcal{E}_{\mathbf{h}}\left[\exp\left(||\mathbf{h}||^2(s^2 - \sqrt{\beta}s)\right)\right] \\ &= \left(\frac{1}{1-\sigma_h^2(s^2-\sqrt{\beta}s)}\right)^N, \end{aligned} \tag{7.27}$$

since $\Delta \sim \mathcal{N}\left(||\mathbf{h}||^2\sqrt{\beta}, 2||\mathbf{h}||^2\right)$ (see Problem 7.6). Using the inverse (bilateral) Laplace transform (especially, the property of integration), we can have

$$\Pr(\Delta < 0) = \frac{1}{2\pi j}\int_{c-j\infty}^{c+j\infty} \frac{1}{s}\Phi(s)\mathrm{d}s. \tag{7.28}$$

We now need to find the inverse Laplace transform of the term on the right-hand side in (7.28).

Let $s = \sqrt{\beta}/2 + j\omega$. We can show that

$$\mathrm{d}s = j\mathrm{d}\omega;$$
$$s^2 - \sqrt{\beta}s = -\left(\frac{\beta}{4} + \omega^2\right);$$
$$\frac{1}{s} = \frac{\frac{\sqrt{\beta}}{2} - j\omega}{\frac{\beta}{4} + \omega^2}.$$

From this, it follows that

$$\frac{1}{2\pi j}\int_{c-j\infty}^{c+j\infty} \frac{1}{s}\Phi(s)\mathrm{d}s = \frac{1}{2\pi}\int_{-\infty}^{\infty} \frac{\frac{\sqrt{\beta}}{2} - j\omega}{\frac{\beta}{4} + \omega^2}\Psi(\omega)\mathrm{d}\omega, \tag{7.29}$$

where

$$\Psi(\omega) = \left(\frac{1}{1 + \sigma_h^2\left(\frac{\beta}{4} + \omega^2\right)}\right)^N. \tag{7.30}$$

Since $\Psi(\omega)$ is an even function, (7.29) is reduced to

$$\begin{aligned}\Pr(\Delta < 0) &= \frac{1}{2\pi j}\int_{c-j\infty}^{c+j\infty} \frac{1}{s}\Phi(s)\mathrm{d}s \\ &= \frac{1}{2\pi}\int_0^{\infty} \frac{\sqrt{\beta}}{\frac{\beta}{4} + \omega^2}\Psi(\omega)\mathrm{d}\omega.\end{aligned} \tag{7.31}$$

To use the Gauss–Chebyshev quadrature rule, let

$$x = \frac{1}{\sqrt{1 + \frac{4\omega^2}{\beta}}}.$$

This transform maps $\omega \in [0, \infty)$ into $x \in [0, 1)$. Then, after some manipulations, we can show that

$$\int_0^{\infty} \frac{\sqrt{\beta}}{\frac{\beta}{4} + \omega^2}\Psi(\omega)\mathrm{d}\omega = 2\int_0^1 \frac{F(x)}{\sqrt{1 - x^2}}\mathrm{d}x, \tag{7.32}$$

where

$$F(x) = \left(\frac{x^2}{x^2 + \frac{\beta\sigma_h^2}{4}}\right)^N.$$

Noting that $F(x)$ is an even function, we can show that

$$2\int_0^1 \frac{F(x)}{\sqrt{1 - x^2}}\mathrm{d}x = \int_{-1}^1 \frac{F(x)}{\sqrt{1 - x^2}}\mathrm{d}x.$$

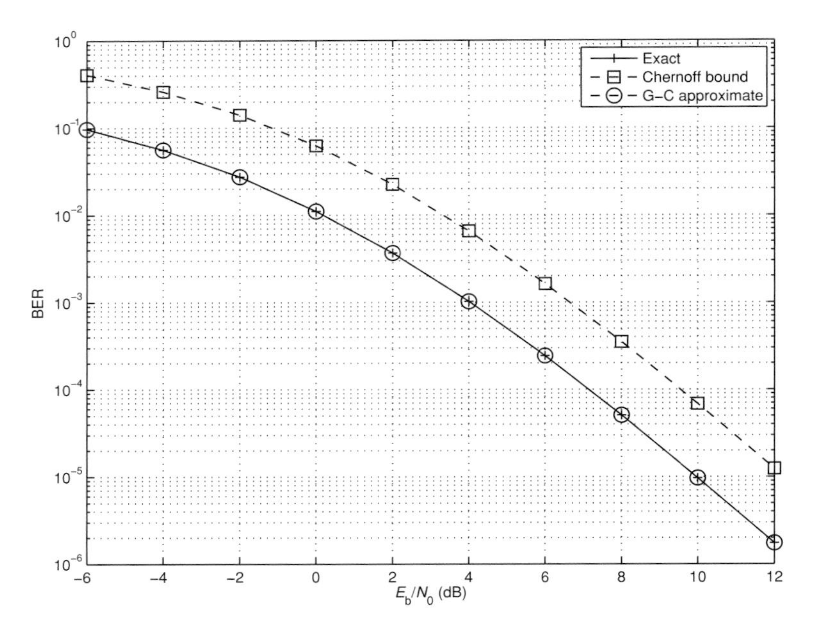

Figure 7.3 BER of optimal SIMO combining when $N = 4$.

The Gauss–Chebyshev quadrature rule is given by

$$\int_{-1}^{1} \frac{F(x)}{\sqrt{1-x^2}} \mathrm{d}x = \lim_{M \to \infty} \frac{\pi}{M} \sum_{m=1}^{M} F\left(\cos\left(\frac{(2m-1)\pi}{2M}\right)\right). \tag{7.33}$$

Finally, for a large M, from (7.31), (7.32), and (7.33), we have

$$P(s^{(1)} \to s^{(2)}) = \Pr(\Delta < 0) \simeq \frac{1}{2M} \sum_{m=1}^{M} F\left(\cos\left(\frac{(2m-1)\pi}{2M}\right)\right). \tag{7.34}$$

Figure 7.3 shows the BER when $\mathcal{S} = \{\pm\sqrt{E_b}\}$ and $N = 4$. It is shown that the Gauss–Chebyshev quadrature rule can provide a good approximate, while the Chernoff bound has about 2–3 dB gap from the exact BER for a range BER of 10^{-2}–10^{-5}.

The Gaussian–Chebyshev quadrature rule is a powerful tool to derive a closed-form expression for the average PEP. This technique can be extended to more complicated random channel models than that in (7.16). The reader is referred to (Ventura-Traveset *et al.* 1997) for a detailed account.

7.3 Optimal and suboptimal MIMO detection

When multiple signals are transmitted simultaneously, they can be jointly detected. In general, joint multiple signal detection outperforms individual single signal detection where the other signals are regarded as interferers.

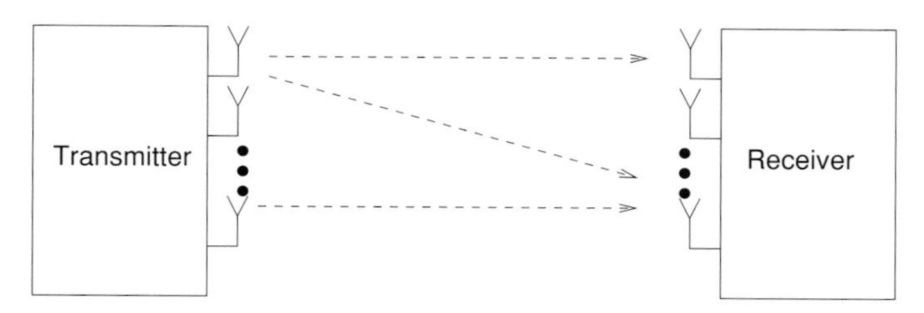

Figure 7.4 Transmitter and receiver equipped with multiple antennas.

Suppose that there are K signals and N receive antennas. The received signal at the qth receive antenna is a superposition of K signals and given by

$$r_q = \sum_{k=1}^{K} h_{q,k} s_k + n_q, \; q = 1, 2, \ldots, N,$$

where $h_{q,k}$ represents the channel gain from the kth signal, denoted by s_k, to the qth receive antenna. We assume that an M-ary signaling is used, i.e. $|\mathcal{S}| = M$. Stacking the N received signals, we have

$$\begin{aligned} \mathbf{r} &= [r_1 \, r_2 \, \ldots \, r_N]^{\mathrm{T}} \\ &= \mathbf{Hs} + \mathbf{n}, \end{aligned} \tag{7.35}$$

where $[\mathbf{H}]_{q,k} = h_{q,k}$, $\mathbf{s} = [s_1 \, s_2 \, \ldots \, s_K]^{\mathrm{T}}$, and $\mathbf{n} = [n_1 \, n_2 \, \ldots \, n_N]^{\mathrm{T}}$. Using the ML or MAP criterion, the symbol vector or K signals can be jointly detected. As we need to find K signals jointly, the range of $\mathbf{s}$ becomes $\mathcal{S}^K = \mathcal{S} \times \mathcal{S} \times \cdots \times \mathcal{S}$, where $\times$ represents the Cartesian product. The multiple signal detection in a vector space generated by multiple receive antennas is called multiple-input multiple-output (MIMO) detection.

7.3.1 MIMO systems

Multiple signals can be generated independently or jointly. In multiuser communication systems, multiple users can transmit signals simultaneously and independently. Then, a receiver can jointly detect multiple signals. The resulting detector is called a multiuser detector (Verdu 1998). It is shown that a multiuser detector outperforms a single-user detector where only one desired signal is detected.

It is also possible that a transmitter transmits multiple signals simultaneously through multiple transmit antennas. In this case, the transmitted signals can be jointly encoded. At a receiver, multiple signals can be jointly detected and decoded as shown in Fig. 7.4. It is shown in (Telatar 1999) and (Foschini & Gans 1998) that if a transmitter is equipped with multiple transmit antennas and a receiver is also equipped with multiple receive antennas, the channel capacity can grow linearly with the number of antennas. This linear capacity increase is a significant observation. Although one can readily expect that the total capacity of multiple parallel channels where each channel is isolated from the others can grow linearly with the number of parallel channels, the same or similar

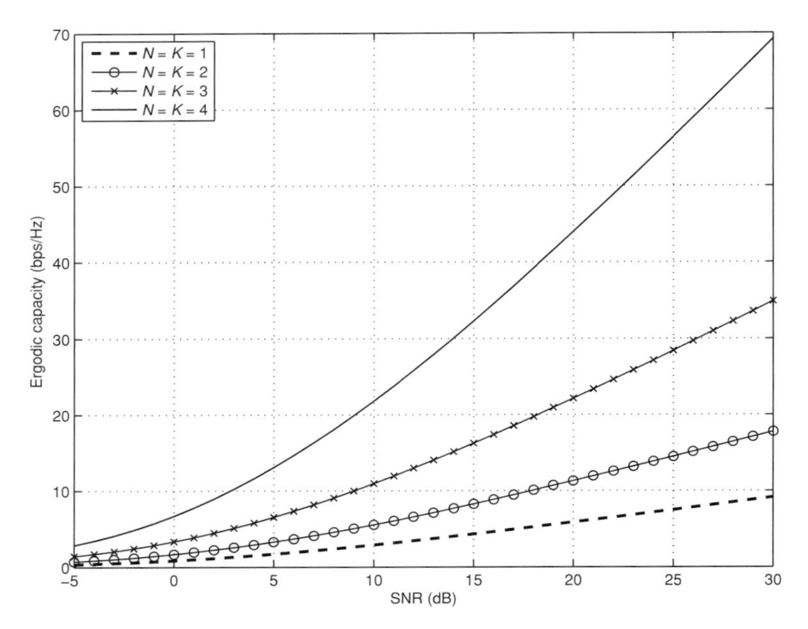

Figure 7.5 MIMO ergodic capacity versus SNR.

behavior of the MIMO channel capacity may not be easily predicted due to the existence of interference from other signals.

The channel capacity of the MIMO channel in (7.35) is given by

$$C = \mathcal{E}\left[\log_2 \det\left(\mathbf{I} + \frac{1}{N_0}\mathbf{H}\mathbf{R}_\mathbf{s}\mathbf{H}^\mathrm{H}\right)\right], \tag{7.36}$$

where $\mathbf{R}_\mathbf{s} = \mathcal{E}[\mathbf{s}\mathbf{s}^\mathrm{H}]$ and $\mathbf{R}_\mathbf{n} = N_0\mathbf{I}$ and the expectation is carried out with respect to random channel matrix $\mathbf{H}$. The channel capacity in (7.36), which is the average capacity over random channel realizations, is called ergodic capacity because it can be seen as a temporal average over random time-varying channels. Figure 7.5 shows the ergodic capacity for various numbers of antennas when $N = K$. We assume that $\mathbf{R}_\mathbf{s} = \frac{P_\mathrm{total}}{K}\mathbf{I}$, where P_total represents the total power as $P_\mathrm{total} = \mathrm{tr}(\mathbf{R}_\mathbf{s})$, and define the SNR as

$$\gamma = \frac{P_\mathrm{total}}{K N_0}.$$

In (Shin & Lee 2003), a closed-form expression for the ergodic capacity in (7.36) is derived when each element of $\mathbf{H}$ is an independent zero-mean CSCG random variable with unit variance, i.e. $[\mathbf{H}]_{n,k} \sim \mathcal{CN}(0, 1)$. The results in Fig. 7.5 are shown using the closed-form expression in (Shin & Lee 2003). We can observe that the capacity can be significantly improved as the number of antennas increases.

Figure 7.6 shows the ergodic capacity for various numbers of antennas. It is shown that the capacity grows linearly with the number of antennas. This behavior of the ergodic capacity can be explained using the law of large numbers. As $N = K$ increases, we can

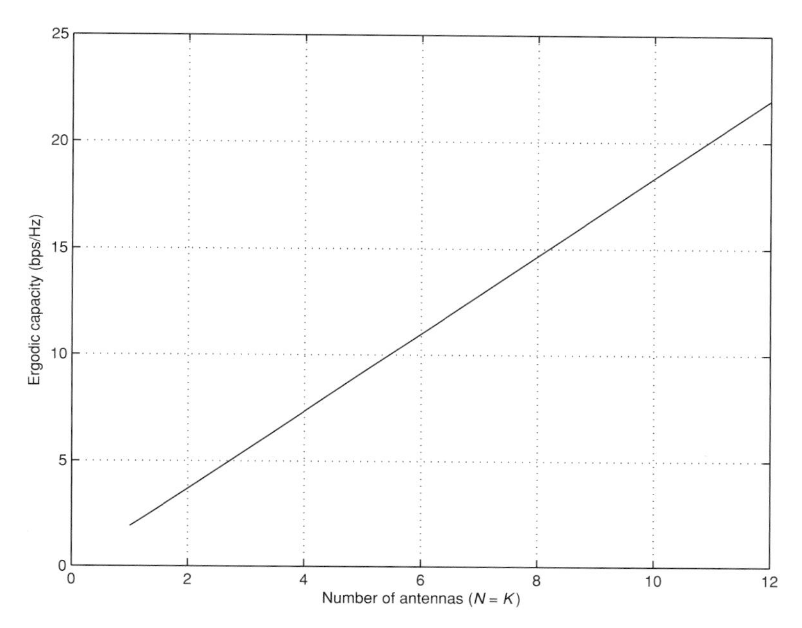

Figure 7.6 MIMO ergodic capacity versus $N = K$ (the number of antennas).

show that each element of $\frac{1}{K}\mathbf{H}\mathbf{H}^{\mathrm{H}}$ converges to its mean value as follows:

$$\left[\frac{1}{K}\mathbf{H}\mathbf{H}^{\mathrm{H}}\right]_{k,n} \rightarrow \begin{cases} 1, & \text{if } k = n; \\ 0, & \text{if } k \neq n. \end{cases}$$

From this, it can be shown that

$$C = \mathcal{E}\left[\log_2 \det\left(1 + \frac{P_{\text{total}}}{K N_0}\mathbf{H}\mathbf{H}^{\mathrm{H}}\right)\right] \rightarrow K \log_2\left(1 + \frac{P_{\text{total}}}{N_0}\right) \tag{7.37}$$

for a sufficiently large $N = K$. This shows that the ergodic capacity is a linear function of K with the slope $\log_2(1 + P_{\text{total}}/N_0)$, which is the single-input single-output (SISO) channel capacity. In other words, the MIMO channel for large $N = K$ can be seen as K parallel SISO channels with no interference at all.

In general, the MIMO ergodic capacity can be achieved when the length of a codeword is sufficiently long so that the variation of channel during the transmission of a codeword can be sufficiently rich to have the following averaging effect by invoking the ergodicity theorem:

$$\mathcal{E}\left[\log_2 \det\left(\mathbf{I} + \frac{1}{N_0}\mathbf{H}\mathbf{R}_{\mathrm{s}}\mathbf{H}^{\mathrm{H}}\right)\right] \simeq \frac{1}{L}\sum_{l=1}^{L}\log_2 \det\left(\mathbf{I} + \frac{1}{N_0}\mathbf{H}_l\mathbf{R}_{\mathrm{s}}\mathbf{H}_l^{\mathrm{H}}\right),$$

where $\mathbf{H}_l$ represents the lth realization of random time-varying channel $\mathbf{H}$.

If the channel variation is not fast enough or the codeword length is short, the ergodic capacity cannot be achieved. As an extreme case, the channel may not vary during the transmission of a codeword. In this case, instead of the ergodic capacity, the information outage probability can be considered as a performance index, which is defined as

$$P_{\text{out}}(R) = \Pr\left(\log_2 \det\left(\mathbf{I} + \frac{1}{N_0}\mathbf{H}\mathbf{R}_{\mathbf{s}}\mathbf{H}^{\text{H}}\right) \le R\right). \tag{7.38}$$

Using the outage probability, ϵ-capacity or outage capacity is defined as

$$C_\epsilon = \arg\max_R \{P_{\text{out}}(R) \le \epsilon\}, \tag{7.39}$$

which is the maximum transmission rate for a given outage probability ϵ.

From the observation that the ergodic channel capacity grows linearly with the number of antennas, we can conclude that the transmission rate should also grow linearly with the number of antennas to fully exploit the MIMO capacity. Thus, it would be natural for MIMO systems to employ a higher-order modulation. Furthermore, in order to achieve a performance close to the ergodic capacity or outage capacity, capacity-achieving channel codes have to be used. From these facts, the receiver design for MIMO systems is very challenging, because the complexity of an optimal receiver where joint ML detection and decoding is performed is prohibitively high.

To reduce the complexity, detection and decoding can be separately carried out. Even in this case, the complexity of optimal MIMO detectors could be exceedingly high due to a higher-order modulation. Thus, we may need to devise suboptimal but low complexity MIMO detectors. This is the key issue in this chapter as well as the next two chapters.

Note that the performance degradation due to this separation can be mitigated by using iterative approaches (Choi 2006).

More details of MIMO channel analysis are beyond the scope of the book. Thus, we do not discuss this further. Excellent sources for MIMO channel analyses and applications can be found in (Tse & Viswanath 2005), (Paulraj, Gore & Nabar 2003), (Paulraj, Gore, Nabar & Bolcskei 2004), and (Gesbert, Shafi, Shiu, Smith & Naguib 2003).

7.3.2 ML and MAP detection

ML detection is to find the symbol vector that maximizes the likelihood function as follows:

$$\hat{\mathbf{s}}_{\text{ml}} = \arg\max_{\mathbf{s}\in\mathcal{S}^K} f(\mathbf{r}|\mathbf{s}). \tag{7.40}$$

Since

$$f(\mathbf{r}|\mathbf{s}) = \frac{1}{\det(\pi\mathbf{R}_{\mathbf{n}})} \exp\left(-(\mathbf{r} - \mathbf{H}\mathbf{s})^{\text{H}}\mathbf{R}_{\mathbf{n}}^{-1}(\mathbf{r} - \mathbf{H}\mathbf{s})\right),$$

we can show that

$$\hat{\mathbf{s}}_{\text{ml}} = \arg\min_{\mathbf{s}\in\mathcal{S}^K} (\mathbf{r} - \mathbf{H}\mathbf{s})^{\text{H}}\mathbf{R}_{\mathbf{n}}^{-1}(\mathbf{r} - \mathbf{H}\mathbf{s}). \tag{7.41}$$

A straightforward implementation of this ML detection using an exhaustive search requires the complexity that grows exponentially with K as $|\mathcal{S}^K| = M^K$. For example, if $M = 64$ and $K = 4$, $M^K = 16, 777, 216$. Therefore, an exhaustive search for ML detection is not practical.

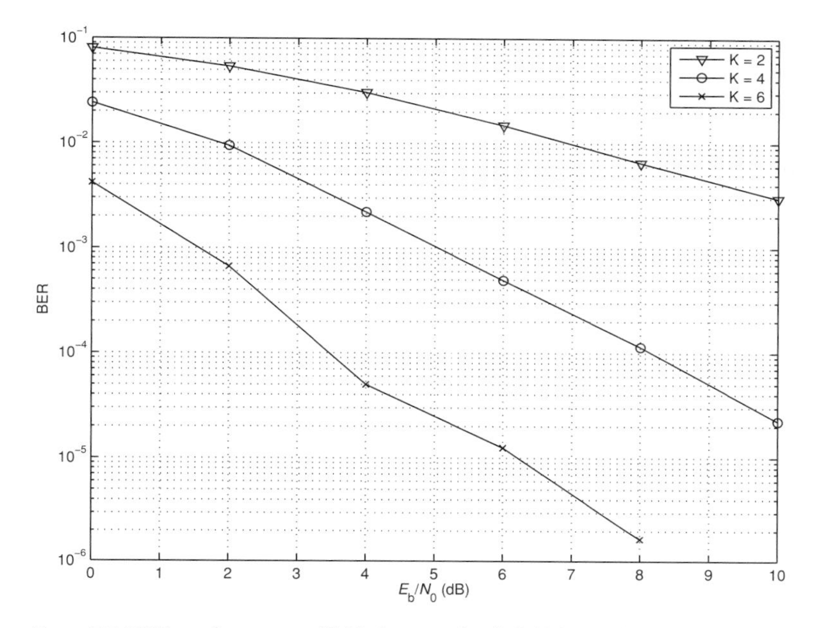

Figure 7.7 BER performance of ML detector for 4-QAM over $K \times K$ MIMO channels.

Example 7.3.1 Let $s \in \mathcal{S} = \{-3, -1, 1, 3\}$ and $N = K = 2$. The channel matrix is given by

$$\mathbf{H} = \begin{bmatrix} 2 & 0.5 \\ 1 & 2 \end{bmatrix}.$$

Suppose that the received signal vector is $\mathbf{r} = [1\ 0.9]^{\mathrm{T}}$. Using an exhaustive search with $4^2 = 16$ candidates of $\mathbf{s}$, the ML solution can be found as

$$\hat{\mathbf{s}}_{\mathrm{ml}} = [1\ -1]^{\mathrm{T}}.$$

Figure 7.7 shows a BER performance of the ML detector over $K \times K$ MIMO channels when 4-QAM is used for modulation. It is assumed that each element of $\mathbf{H}$ is an independent zero-mean CSCG random variable with unit variance (i.e., $h_{n,k} \sim \mathcal{CN}(0, 1)$) and $\mathbf{R_n} = N_0\mathbf{I}$. In Fig. 7.7, E_{b} represents the bit energy to transmit one bit per transmit antenna. A BER performance with 16-QAM is also shown in Fig. 7.8. It is shown that the performance is degraded if the size of the symbol alphabet, M, increases. On the other hand, the performance is improved as more antennas are used. This performance improvement results from the increase of receive diversity gain.

If $\mathbf{R_n} = N_0\mathbf{I}$ and the column vectors of the channel matrix, $\mathbf{H}$, are orthogonal, the MIMO ML detection problem can be decomposed into K SIMO ML detection problems. To see this, assume that the column vectors of $\mathbf{H}$ are orthogonal. Then, $\mathbf{H}$

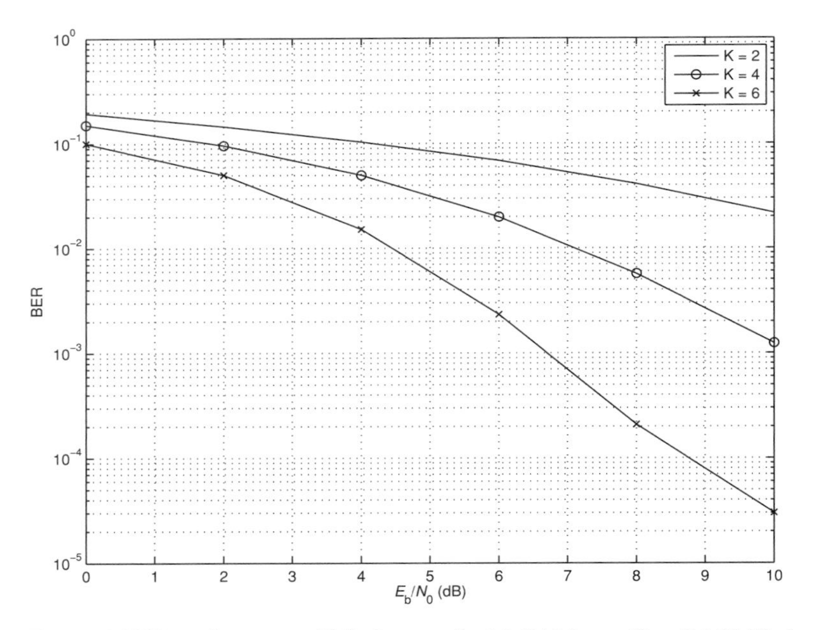

Figure 7.8 BER performance of ML detector for 16-QAM over $K \times K$ MIMO channels.

can be expressed as

$$\mathbf{H} = \mathbf{Q}\boldsymbol{\Sigma},$$

where $\mathbf{Q}$ is unitary and $\boldsymbol{\Sigma}$ is diagonal. That is,

$$\mathbf{H} = [\mathbf{q}_1 \; \mathbf{q}_2 \; \cdots \; \mathbf{q}_K] \begin{bmatrix} \sigma_1 & 0 & \cdots & 0 \\ 0 & \sigma_2 & \cdots & 0 \\ \vdots & \vdots & \ddots & \vdots \\ 0 & 0 & \cdots & \sigma_K \end{bmatrix},$$

where $\mathbf{q}_k$ is the orthonormal vector and σ_k is the kth diagonal element of $\boldsymbol{\Sigma}$. It can be shown that

$$\begin{aligned}(\mathbf{r} - \mathbf{Hs})^{\mathrm{H}} \mathbf{R}_{\mathbf{n}}^{-1} (\mathbf{r} - \mathbf{Hs}) &= \frac{1}{N_0} ||\mathbf{r} - \mathbf{Hs}||^2 \\ &= \frac{1}{N_0} ||\mathbf{r} - \mathbf{Q}\boldsymbol{\Sigma}\mathbf{s}||^2 \\ &= \frac{1}{N_0} ||\mathbf{Q}^{\mathrm{H}}\mathbf{r} - \boldsymbol{\Sigma}\mathbf{s}||^2. \end{aligned}$$

Let $\mathbf{x} = \mathbf{Q}^{\mathrm{H}}\mathbf{r}$. Then,

$$||\mathbf{Q}^{\mathrm{H}}\mathbf{r} - \boldsymbol{\Sigma}\mathbf{s}||^2 = \sum_{k=1}^{K} |x_k - \sigma_k s_k|^2.$$

Due to the decomposition, it follows that

$$\hat{s}_{\mathrm{ml},k} = \arg\min_{s\in\mathcal{S}} |x_k - \sigma_k s_k|^2, \ k = 1, 2, \ldots, K.$$

This shows that ML detection can be carried out individually in each dimension. In this case, the complexity of ML detection grows linearly with K. A total complexity becomes $\mathcal{O}(K|\mathcal{S}|)$ for orthogonal channels, while an exhaustive search for ML detection over an arbitrary channel matrix requires a complexity of $\mathcal{O}(|\mathcal{S}|^K)$. Consequently, we can see that the complexity of MIMO detection can be significantly reduced if channel matrices are orthogonal.

In general, channel matrices are not orthogonal. However, using a linear transformation prior to transmitting $\mathbf{s}$, the channel can be orthogonalized. Consider the singular value decomposition (SVD) of $\mathbf{H}$ as follows:

$$\mathbf{H} = \mathbf{U}\mathbf{\Sigma}\mathbf{V}^{\mathrm{H}},$$

where $\mathbf{U}$ and $\mathbf{V}$ are unitary matrices and $\mathbf{\Sigma}$ is diagonal. We can transmit $\mathbf{c}$ instead of $\mathbf{s}$:

$$\mathbf{c} = \mathbf{V}\mathbf{s}. \tag{7.42}$$

In this case, the received signal vector is rewritten as

$$\begin{aligned}\mathbf{r} &= \mathbf{H}\mathbf{c} + \mathbf{n}\\ &= \mathbf{H}\mathbf{V}\mathbf{s} + \mathbf{n}\\ &= \mathbf{U}\mathbf{\Sigma}\underbrace{(\mathbf{V}^{\mathrm{H}}\mathbf{V})}_{=\mathbf{I}}\mathbf{s} + \mathbf{n}\\ &= \mathbf{U}\mathbf{\Sigma}\mathbf{s} + \mathbf{n}.\end{aligned} \tag{7.43}$$

Thus, the new channel matrix, $\mathbf{U}\mathbf{\Sigma}$, is orthogonal. Therefore, the complexity of MIMO detection becomes $\mathcal{O}(K|\mathcal{S}|)$.

The transformation in (7.42) is an example of linear precoding. For orthogonalization in (7.42), the transmitter needs to know $\mathbf{V}$ of the channel matrix. This could be impractical in some cases. In Chapter 9, we discuss a different approach that makes the column vectors of an MIMO channel matrix nearly orthogonal without precoding.

We now consider MAP detection for MIMO channels. Suppose that the a priori probability of $\mathbf{s}$, denoted by $\Pr(\mathbf{s})$, is available. MAP detection is to find the symbol vector that maximizes the a posteriori probability as follows:

$$\begin{aligned}\hat{\mathbf{s}}_{\mathrm{map}} &= \arg\max_{\mathbf{s}\in\mathcal{S}^K} \Pr(\mathbf{s}|\mathbf{r})\\ &= \arg\max_{\mathbf{s}\in\mathcal{S}^K} f(\mathbf{r}|\mathbf{s})\Pr(\mathbf{s}).\end{aligned} \tag{7.44}$$

If the symbols are equally likely, $\Pr(\mathbf{s})$ becomes a constant for all $\mathbf{s} \in \mathcal{S}^K$. In this case, MAP detection reduces to ML detection.

Both the ML and MAP detectors provide a hard-decision of $\mathbf{s}$. For a soft-decision, as in Section 7.1, the LLR can be used. Let $\mathcal{S}_{k,l}^0$ and $\mathcal{S}_{k,l}^1$ denote the sets of the symbols whose lth bit of s_k is 0 and 1, respectively. Then, the LLR of the lth bit of s_k becomes

$$\begin{aligned}\mathrm{LLR}_{k,l} &= \log \frac{f(\mathbf{r}|\mathcal{S}^0_{k,l})}{f(\mathbf{r}|\mathcal{S}^1_{k,l})} \\ &= \log \frac{\Pr(\mathcal{S}^0_{k,l}|\mathbf{r})}{\Pr(\mathcal{S}^1_{k,l}|\mathbf{r})} - \log \frac{\Pr(\mathcal{S}^0_{k,l})}{\Pr(\mathcal{S}^1_{k,l})},\end{aligned} \tag{7.45}$$

where $\Pr(\mathcal{S}^i_{k,l}|\mathbf{r})$ and $\Pr(\mathcal{S}^i_{k,l})$ denote the a posteriori and a priori probabilities of $\mathcal{S}^i_{k,l}$, $i = 0, 1$, respectively. Alternatively, the LLR is given by

$$\mathrm{LLR}_{k,l} = \log \frac{\sum_{\mathbf{s}\in\mathcal{S}^0_{k,l}} f(\mathbf{r}|\mathbf{s}) \Pr(\mathbf{s})}{\sum_{\mathbf{s}\in\mathcal{S}^1_{k,l}} f(\mathbf{r}|\mathbf{s}) \Pr(\mathbf{s})} - \log \frac{\Pr(\mathcal{S}^0_{k,l})}{\Pr(\mathcal{S}^1_{k,l})}. \tag{7.46}$$

If the data symbols are equally likely, the LLR reduces to

$$\mathrm{LLR}_{k,l} = \log \frac{\sum_{\mathbf{s}\in\mathcal{S}^0_{k,l}} f(\mathbf{r}|\mathbf{s})}{\sum_{\mathbf{s}\in\mathcal{S}^1_{k,l}} f(\mathbf{r}|\mathbf{s})}, \quad k = 1, 2, \ldots, K; \; l = 1, 2, \ldots, \log_2 M. \tag{7.47}$$

7.3.3 Gaussian approximation

Due to a prohibitively high complexity, ML or MAP detection could be impractical. However, using reasonable approximations, an affordable complexity can be achieved.

The received signal vector in (7.35) can be rewritten as

$$\begin{aligned}\mathbf{r} &= \mathbf{h}_k s_k + \sum_{q\neq k} \mathbf{h}_q s_q + \mathbf{n} \\ &= \mathbf{h}_k s_k + \mathbf{u}_k,\end{aligned} \tag{7.48}$$

where $\mathbf{u}_k$ is the interference-plus-noise vector that is given by

$$\mathbf{u}_k = \sum_{q\neq k} \mathbf{h}_q s_q + \mathbf{n}.$$

Suppose that the s_k's are independent and the a priori probability, $\Pr(s_k)$, is available. Furthermore, let us assume that the interfering vector, $\mathbf{u}_k$, is Gaussian, which is called Gaussian approximation. This would be a good approximation when the background noise is dominant or K is sufficiently large so that the central limit theorem can be applied. If the distribution of s_k is close to a Gaussian distribution due to signal shaping (Fischer 2002), this approximation would also be reasonably good. Under the Gaussian approximation, $\mathbf{u}_k$ can be characterized by its mean vector and covariance matrix. The mean vector of $\mathbf{u}_k$ can be found as

$$\begin{aligned}\bar{\mathbf{u}}_k &= \mathcal{E}[\mathbf{u}_k] \\ &= \mathcal{E}\left[\sum_{q\neq k} \mathbf{h}_q s_q + \mathbf{n}\right] \\ &= \sum_{q\neq k} \mathbf{h}_q \bar{s}_q,\end{aligned} \tag{7.49}$$

where $\bar{s}_q = \mathcal{E}[s_q]$. The expectation of s_q can be readily obtained from the a priori probability of s_q, $\Pr(s_q)$. The covariance matrix of $\mathbf{u}_k$ is given by

$$\begin{aligned}\mathbf{R}_{\mathbf{u}_k} &= \mathcal{E}\left[(\mathbf{u}_k - \bar{\mathbf{u}}_k)(\mathbf{u}_k - \bar{\mathbf{u}}_k)^{\mathrm{H}}\right] \\ &= \mathcal{E}\left[\left(\sum_{q\neq k} \mathbf{h}_q(s_q - \bar{s}_q) + \mathbf{n}\right)\left(\sum_{q\neq k} \mathbf{h}_q(s_q - \bar{s}_q) + \mathbf{n}\right)^{\mathrm{H}}\right] \\ &= \sum_{q\neq k} \mathbf{h}_q \mathbf{h}_q^{\mathrm{H}} \mathcal{E}\left[|s_q - \bar{s}_q|^2\right] + \mathbf{R}_{\mathbf{n}}.\end{aligned} \tag{7.50}$$

With the Gaussian approximation of $\mathbf{u}_k$, the detection of s_k from (7.48) becomes a SIMO detection. The Gaussian approximation based ML detection is now given by

$$\hat{s}_{\mathrm{ga-ml},k} = \arg\max_{s_k\in\mathcal{S}} f_{\mathrm{ga},k}(\mathbf{r}_k|s_k), \ k = 1, 2, \ldots, K, \tag{7.51}$$

where $f_{\mathrm{ga},k}(\mathbf{r}_k|s_k)$ is the likelihood function of s_k for given $\mathbf{r}_k = \mathbf{r} - \bar{\mathbf{u}}_k$ based on the Gaussian approximation:

$$f_{\mathrm{ga},k}(\mathbf{r}_k|s_k) = \frac{1}{\det(\pi\mathbf{R}_{\mathbf{u}_k})} \exp\left(-(\mathbf{r}_k - \mathbf{h}_k s_k)^{\mathrm{H}} \mathbf{R}_{\mathbf{u}_k}^{-1}(\mathbf{r}_k - \mathbf{h}_k s_k)\right).$$

Finally, we have

$$\hat{s}_{\mathrm{ga-ml},k} = \arg\min_{s_k\in\mathcal{S}} (\mathbf{r}_k - \mathbf{h}_k s_k)^{\mathrm{H}} \mathbf{R}_{\mathbf{u}_k}^{-1}(\mathbf{r}_k - \mathbf{h}_k s_k), \ k = 1, 2, \ldots, K. \tag{7.52}$$

Assuming that $\mathcal{E}[s_k] = 0$, according to Theorem 7.1.1, the detection can also be carried out with the output of the MMSE combiner that is given by

$$\begin{aligned}\hat{s}_k &= \mathbf{w}_{\mathrm{mmse},k}^{\mathrm{H}} \mathbf{r}_k \\ &= \mathbf{w}_{\mathrm{mmse},k}^{\mathrm{H}} (\mathbf{r} - \bar{\mathbf{u}}_k),\end{aligned}$$

where

$$\mathbf{w}_{\mathrm{mmse},k} = (\mathbf{h}_k \mathbf{h}_k^{\mathrm{H}} \sigma_k^2 + \mathbf{R}_{\mathbf{u}_k})^{-1} \mathbf{h}_k \sigma_k^2,$$

where $\sigma_k^2 = \mathcal{E}[|s_k|^2]$.

The Gaussian approximation based MAP detection can also be formulated as follows:

$$\hat{s}_{\mathrm{ga-map},k} = \arg\max_{s_k\in\mathcal{S}} f_{\mathrm{ga},k}(\mathbf{r}_k|s_k)\Pr(s_k), \ k = 1, 2, \ldots, K. \tag{7.53}$$

Alternatively, using Theorem 7.1.1, we can also show that

$$\hat{s}_{\text{ga-map},k} = \arg\max_{s_k \in \mathcal{S}} f_k(\hat{s}_k | s_k) \Pr(s_k), \ k = 1, 2, \ldots, K, \tag{7.54}$$

where $f_k(\hat{s}_k | s_k) = \mathcal{CN}(\mathbf{w}_{\text{mmse},k}^{\text{H}} \mathbf{h}_k s_k, \mathbf{w}_{\text{mmse},k}^{\text{H}} \mathbf{R}_{\mathbf{u}_k} \mathbf{w}_{\text{mmse},k})$.

7.3.4 Linear detection

In general, the optimal detectors (e.g. the ML and MAP detectors) require prohibitively high computational complexity. Thus, low complexity suboptimal detectors are often preferable when the detector's computational complexity is limited.

For low complexity suboptimal detectors, we consider a class of linear detectors under the assumption that $N \geq K$. The operation of a linear detector consists of two steps as follows. A linear combiner as a linear preprocessor is first applied to the received signal vector. The preprocessor output is given by

$$\hat{\mathbf{s}} = \mathbf{W}^{\text{H}} \mathbf{r}, \tag{7.55}$$

where $\mathbf{W}$ denotes the linear combiner or preprocessor. Then, each element of $\hat{\mathbf{s}}$ is considered as the received signal in the absence of other signals and from which the associated signal is independently detected. Therefore, the linear combiner should suppress the other signals as much as possible. As explained above, we can see that a linear detector consists of a linear preprocessor and K (single-signal) detectors. Due to the two-step approach in a linear detector, we can see that its complexity grows linearly with K if the complexity for preprocessing is ignored.

The ZF detector has the linear preprocessor that can suppress the other signals completely. The preprocessor output is given by

$$\begin{aligned} \hat{\mathbf{s}}_{\text{zf}} &= \mathbf{W}_{\text{zf}}^{\text{H}} \mathbf{r} \\ &= \mathbf{H}^{\dagger} \mathbf{r} \\ &= \mathbf{s} + \mathbf{H}^{\dagger} \mathbf{n}, \end{aligned} \tag{7.56}$$

where $\mathbf{W}_{\text{zf}}^{\text{H}} = \mathbf{H}^{\dagger} = (\mathbf{H}^{\text{H}} \mathbf{H})^{-1} \mathbf{H}^{\text{H}}$ is the pseudo inverse of $\mathbf{H}$.

Example 7.3.2 Consider Example 7.3.1 again. The ZF detector's output is given by

$$\hat{\mathbf{s}}_{\text{zf}} = \mathbf{H}^{\dagger} \mathbf{r} = [0.5 \ 0.2]^{\text{T}}.$$

Thus, the hard decision for $\mathbf{s}$ becomes $[1 \ 1]^{\text{T}}$, which is different from the ML solution. This difference results from different decision regions that are illustrated in Fig. 7.9. While $\mathbf{r} = [1 \ 0.9]^{\text{T}}$ belongs to the decision region for $[1 \ -1]^{\text{T}}$ according to the ML detection, it belongs to the decision region for $[1 \ 1]^{\text{T}}$ according to the ZF detection.

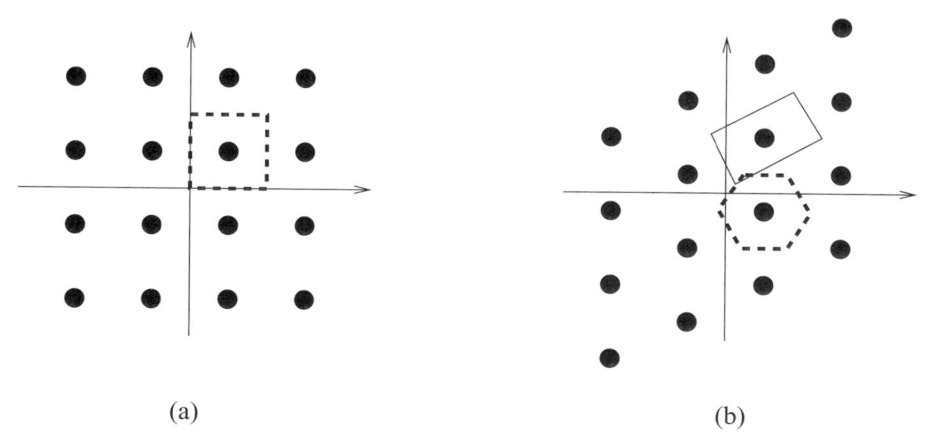

Figure 7.9 The signal constellations: (a) before transmission, $\mathbf{s}$; (b) after transmission, $\mathbf{Hs}$. The decision regions for the ZF and ML detectors are shown by the dashed and solid lines in (b), respectively.

Assume that $\mathcal{E}[\mathbf{s}] = 0$, $\mathcal{E}[\mathbf{ss}^{\mathrm{H}}] = E_s\mathbf{I}$, and $\mathbf{R_n} = N_0\mathbf{I}$. The MMSE detector is another linear detector, whose preprocessor output is given by

$$\hat{\mathbf{s}}_{\mathrm{mmse}} = \mathbf{W}_{\mathrm{mmse}}^{\mathrm{H}}\mathbf{r}, \tag{7.57}$$

where

$$\begin{aligned}\mathbf{W}_{\mathrm{mmse}} &= \arg\min_{\mathbf{W}} \mathcal{E}[||\mathbf{s} - \mathbf{W}^{\mathrm{H}}\mathbf{r}||^2] \\ &= \left(E_s\mathbf{HH}^{\mathrm{H}} + N_0\mathbf{I}\right)^{-1}\mathbf{H}E_s \\ &= \left(\mathbf{HH}^{\mathrm{H}} + \frac{N_0}{E_s}\mathbf{I}\right)^{-1}\mathbf{H}.\end{aligned} \tag{7.58}$$

Using the matrix inversion lemma, it can be further shown that

$$\mathbf{W}_{\mathrm{mmse}} = \mathbf{H}\left(\frac{N_0}{E_s}\mathbf{I} + \mathbf{H}^{\mathrm{H}}\mathbf{H}\right)^{-1}. \tag{7.59}$$

Furthermore, the ZF and MMSE detectors can also be derived by the following unified approach. To derive the unified approach that can encompass the ZF and MMSE detectors, we can consider the extended channel matrix as follows:

$$\mathbf{H}_{\mathrm{ex}} = \begin{bmatrix}\mathbf{H} \\ c\mathbf{I}\end{bmatrix}, \tag{7.60}$$

where $c \geq 0$ is a constant. The pseudo-inverse of $\mathbf{H}_{\mathrm{ex}}$ is given by

$$\mathbf{H}_{\mathrm{ex}}^{\dagger} = (c^2\mathbf{I} + \mathbf{H}^{\mathrm{H}}\mathbf{H})^{-1}[\mathbf{H}^{\mathrm{H}}\ c\mathbf{I}]. \tag{7.61}$$

Let

$$[\mathbf{H}_{\mathrm{ex}}^{\dagger}] = (\mathbf{H}^{\mathrm{H}}\mathbf{H} + c^2\mathbf{I})^{-1}\mathbf{H}^{\mathrm{H}}.$$

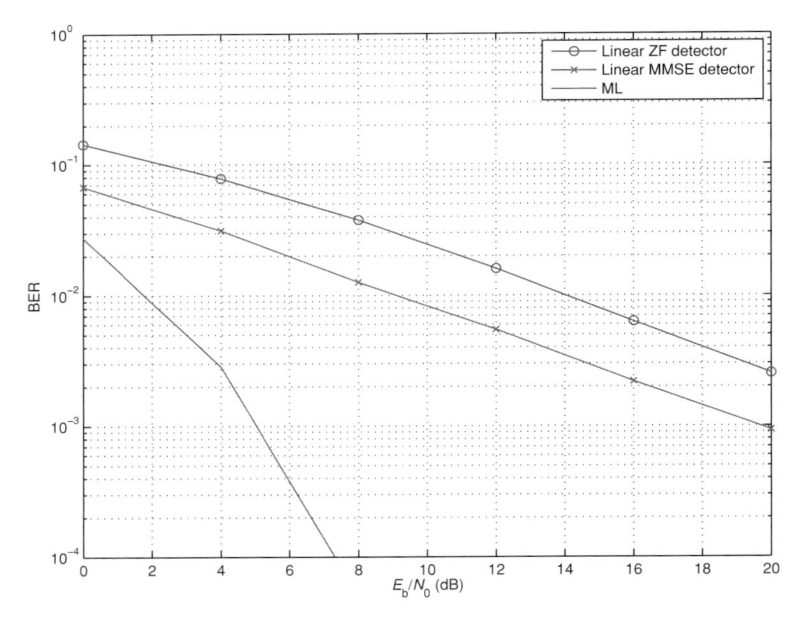

Figure 7.10 BER performance of linear detectors for 4-QAM over 4×4 MIMO channels.

Then, we can re-derive the preprocessors for the ZF and MMSE detectors as follows:

$$\begin{aligned} \mathbf{W}^{\mathrm{H}} &= [\mathbf{H}_{\mathrm{ex}}^{\dagger}] \\ &= \begin{cases} \mathbf{W}_{\mathrm{zf}}^{\mathrm{H}} = \mathbf{H}^{\dagger}, & \text{if } c = 0; \\ \mathbf{W}_{\mathrm{mmse}}^{\mathrm{H}}, & \text{if } c = \sqrt{N_0/E_s}. \end{cases} \end{aligned} \tag{7.62}$$

Figure 7.10 and 7.11 show BER performance of linear detectors with 4-QAM and 16-QAM for modulation, respectively, over 4×4 MIMO channels. For comparison purposes, the ML detector's BER performance is also shown. Although the performance difference is small when the SNR (or E_b/N_0) is low, the difference increases significantly as the SNR grows. This results from different diversity gain. The ML detector exploits a full receive antenna diversity gain, while linear detectors cannot.

7.4 Space-time codes

In this section, we briefly present the main idea of space-time codes that were originally proposed in (Tarokh, Seshadri & Calderbank 1998) and (Guey, Fitz, Bell & Kuo 1999) with key properties and design criteria. Space-time codes are now widely studied and employed by various wireless standards.

A space-time code is a set of codeword matrices:

$$\mathcal{S}_{\mathrm{st}} = \{\mathbf{S}^{(1)}, \mathbf{S}^{(2)}, \ldots, \mathbf{S}^{(M)}\},$$

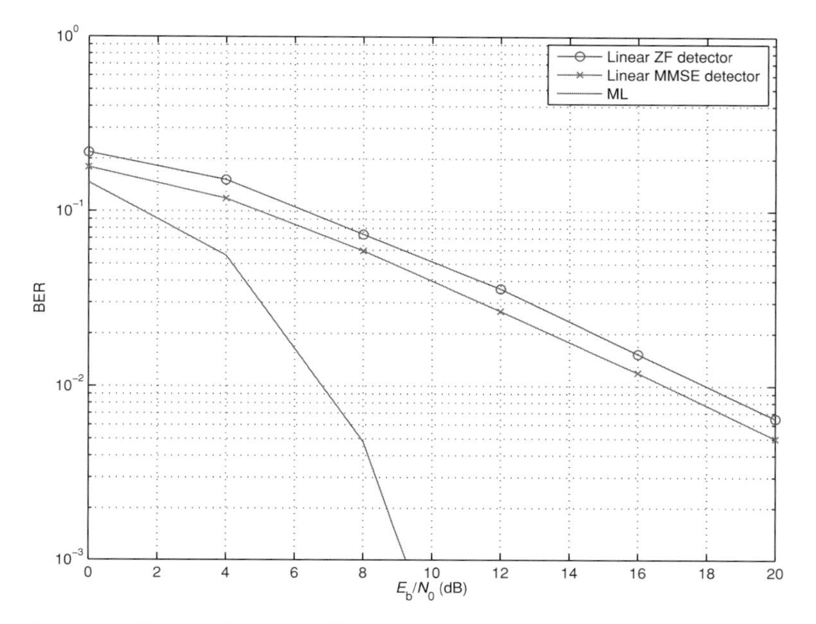

Figure 7.11 BER performance of linear detectors for 16-QAM over 4×4 MIMO channels.

where $\mathcal{S}_{\text{st}}$ denotes the space-time code of which codeword matrices are denoted by $\mathbf{S}^{(m)}$. While a codeword of a conventional code is transmitted over time, a codeword of a space-time code is transmitted in both the space and time domains, which allows to exploit the diversity gain in the space and time domains simultaneously.

Suppose that the size of $\mathbf{S}^{(m)}$ is $K \times L$. This implies that an L-symbol period is required to transmit one space-time codeword matrix through K transmit antennas. Each codeword matrix can be generated by a systematical approach. In (Alamouti 1998), for two transmit antennas ($K = 2$), a space-time code is proposed, which is called the Alamouti code. Its codeword matrix is given by

$$\mathbf{S}(s_1, s_2) = \begin{bmatrix} s_1 & -s_2^* \\ s_2 & s_1^* \end{bmatrix}, \tag{7.63}$$

where $s_1, s_2 \in \mathcal{S}$ and $\mathcal{S}$ is a signal alphabet. If $|\mathcal{S}| = 4$, then the size of the space-time code with codewords in (7.63) is 16. This codeword matrix has the following orthogonality property:

$$\left(\mathbf{S}(s_1, s_2) - \mathbf{S}(s_1', s_2')\right)\left(\mathbf{S}(s_1, s_2) - \mathbf{S}(s_1', s_2')\right)^{\mathrm{H}} = (|s_1 - s_1'|^2 + |s_2 - s_2'|^2)\mathbf{I},$$

where $s_i, s_i' \in \mathcal{S}$, $i = 1, 2$. If a space-time code has this orthogonality property, it is called orthogonal space-time code.

There are various aspects of space-time codes and performance issues over space-time fading channels. Since those topics are beyond the scope of the book, we do not discuss further. For a detailed account, the reader is referred to various books on space-time codes, e.g. (Jafarkhani 2005).

7.5 Performance of detectors

In general, the derivation of an exact error probability of joint detection is quite involved. Thus, we often consider bounds or approximations of error probability that can be relatively easily obtained. Some approximations or bounds can provide useful insight into performance.

7.5.1 Performance of linear detectors

We assume that $N \geq K$ to analyze the performance of linear detectors. If $K > N$, the performance of linear detectors is generally poor even if the SNR is high due to the limited interference suppression capability.

In general, the preprocessor output of a linear detector is given by

$$\hat{\mathbf{s}} = \mathbf{W}^{\mathrm{H}}\mathbf{r}. \tag{7.64}$$

As $\hat{\mathbf{s}}$ is not a hard-decision yet, each element of $\hat{\mathbf{s}}$ can be used for individual detection. The kth element of $\hat{\mathbf{s}}$ is given by

$$\begin{aligned} \hat{s}_k &= \mathbf{w}_k^{\mathrm{H}}\mathbf{r} \\ &= \mathbf{w}_k^{\mathrm{H}}\mathbf{h}_k s_k + \mathbf{w}_k^{\mathrm{H}}\left(\sum_{q=1,q\neq k}^{K} \mathbf{h}_q s_q + \mathbf{n}\right). \end{aligned} \tag{7.65}$$

For convenience, let $k = 1$. For the ZF detector, the interference is completely suppressed, which implies

$$\mathbf{w}_1^{\mathrm{H}} \sum_{q=2}^{K} \mathbf{h}_q s_q = 0.$$

Now, for performance analysis, we need to develop the relationship between $\mathbf{w}_1$ and the signature vectors of the interfering signals. To this end, let

$$\mathbf{H}_1 = [\mathbf{h}_2 \; \mathbf{h}_3 \ldots \mathbf{h}_K].$$

To suppress the $K - 1$ interfering signals, $\mathbf{w}_1$ has to be orthogonal to the subspace of $\mathbf{H}_1$:

$$\mathbf{w}_1 \perp \mathrm{Range}(\mathbf{H}_1).$$

To see the properties of the ZF detector further, we can derive an explicit expression for $\mathbf{w}_1$. Since $\mathbf{W}_{\mathrm{zf}} = \mathbf{H}(\mathbf{H}^{\mathrm{H}}\mathbf{H})^{-1}$, we can show that

$$\mathbf{w}_1 = \mathbf{H}\mathbf{g}_1,$$

where $\mathbf{g}_1$ is the first column vector of $(\mathbf{H}^{\mathrm{H}}\mathbf{H})^{-1}$. Using the matrix inversion lemma, it can be shown that

$$\mathbf{g}_1 \propto \begin{bmatrix} 1 \\ -(\mathbf{H}_1^{\mathrm{H}}\mathbf{H}_1)^{-1}\mathbf{H}_1^{\mathrm{H}}\mathbf{h}_1 \end{bmatrix}.$$

From this, it follows

$$\mathbf{w}_1 \propto \left(\mathbf{I} - \mathbf{H}_1(\mathbf{H}_1^{\mathrm{H}}\mathbf{H}_1)^{-1}\mathbf{H}_1^{\mathrm{H}}\right)\mathbf{h}_1. \tag{7.66}$$

Let $\mathbf{P}_1$ denote the orthogonal projection onto Span($\mathbf{H}_1$), i.e.

$$\mathbf{P}_1 = \mathbf{H}_1(\mathbf{H}_1^{\mathrm{H}}\mathbf{H}_1)^{-1}\mathbf{H}_1^{\mathrm{H}}.$$

Then, it follows that

$$\mathbf{w}_1 \propto (\mathbf{I} - \mathbf{P}_1)\mathbf{h}_1.$$

From this, we can also confirm that $\mathbf{w}_1^{\mathrm{H}}\mathbf{H}_1 = \mathbf{0}^{\mathrm{T}}$. This result will be used for performance analysis.

For convenience, let $s_1 \in \{\sqrt{\frac{E_s}{2}}(\pm 1 \pm j)\}$ for 4-QAM. For equally likely symbols, we can easily show that

$$E_{\mathrm{s}} = \mathcal{E}[|s_1|^2].$$

Let $||\mathbf{w}_1||^2 = 1$. Then, from (7.66), we have

$$\mathbf{w}_1 = \frac{1}{||(\mathbf{I} - \mathbf{P}_1)\mathbf{h}_1||}(\mathbf{I} - \mathbf{P}_1)\mathbf{h}_1.$$

The preprocessor output of the ZF detector is given by

$$\hat{s}_1 = \mathbf{w}_1^{\mathrm{H}}\mathbf{h}_1 s_1 + \mathbf{w}_1^{\mathrm{H}}\mathbf{n}. \tag{7.67}$$

We can show that $\mathbf{w}^{\mathrm{H}}\mathbf{n} \sim \mathcal{CN}(0, N_0)$. Thus, the bit error probability conditioned on $\mathbf{h}_1$ and $\mathbf{P}_1$ becomes

$$\begin{aligned} P_{\mathrm{b}}(\mathbf{h}_1, \mathbf{P}_1) &= \mathcal{Q}\left(\sqrt{\frac{2E_{\mathrm{b}}}{N_0}}\mathbf{w}^{\mathrm{H}}\mathbf{h}_1\right) \\ &= \mathcal{Q}\left(\sqrt{\frac{2E_{\mathrm{b}}\mathbf{h}_1^{\mathrm{H}}(\mathbf{I} - \mathbf{P}_1)\mathbf{h}_1}{N_0}}\right) \\ &\le \exp\left(-\frac{E_{\mathrm{b}}\mathbf{h}_1^{\mathrm{H}}(\mathbf{I} - \mathbf{P}_1)\mathbf{h}_1}{N_0}\right), \end{aligned} \tag{7.68}$$

where $E_{\mathrm{b}} = E_{\mathrm{s}}/2$ becomes the bit energy as one symbol can represent two bits in 4-QAM. Let

$$Z = \mathbf{h}_1^{\mathrm{H}}(\mathbf{I} - \mathbf{P}_1)\mathbf{h}_1.$$

To continue the analysis further, we assume as follows:

(**A1**) The $\mathbf{h}_k$'s are independent random vectors and

$$\mathbf{h}_k \sim \mathcal{CN}(0, \sigma_h^2\mathbf{I}). \tag{7.69}$$

Since $\mathbf{P}_1$ and $\mathbf{h}_1$ are independent, $(\mathbf{I} - \mathbf{P}_1)\mathbf{h}_1$ is a projection of $\mathbf{h}_1$ onto an arbitrary subspace of $(N - K + 1)$-dimension. Based on Assumption (**A1**), this implies that

$Z = ||(\mathbf{I} - \mathbf{P}_1)\mathbf{h}_1||^2$ is a chi-square random variable with $2(N - K + 1)$ degrees of freedom, while $||\mathbf{h}_1||^2$ is a chi-square random variable with $2N$ degrees of freedom. This results from the suppression of $K - 1$ interfering signals. That is, a subspace of $(K - 1)$-dimension is not exploited for signal combining as $K - 1$ interferers should be suppressed.

The average bit error probability is given by

$$\begin{aligned} P_b &= \mathcal{E}[P_b(\mathbf{h}_1, \mathbf{P}_1)] \\ &= \mathcal{E}\left[\mathcal{Q}\left(\sqrt{\frac{2E_b Z}{N_0}}\right)\right] \\ &\leq \mathcal{E}\left[\exp\left(-\frac{E_b Z}{N_0}\right)\right] \\ &= \left(\frac{1}{1+\bar{\gamma}}\right)^{N-K+1}, \end{aligned} \tag{7.70}$$

where

$$\bar{\gamma} = \frac{E_b \sigma_h^2}{N_0}.$$

From this, we can observe that the diversity order becomes $N - K + 1$.

Example 7.5.1 In the above, we only consider 4-QAM. For a higher-order modulation, the analysis can be extended with PEPs and the union bound. For instance, consider M-ary signaling with the signal alphabet $\mathcal{S} = \{s^{(1)}, s^{(2)}, \ldots, s^{(M)}\}$. Taking the preprocessor output of the ZF detector in (7.67) as the received signal, the PEP, which is the probability that the symbol $s^{(2)} \neq s^{(1)}$ is erroneously detected provided that $s^{(1)}$ is transmitted, is as follows:

$$\begin{aligned} P(s^{(1)} \to s^{(2)}|\mathbf{h}_1, \mathbf{P}_1) &= \Pr(|\hat{s}_1 - \mathbf{w}_1^{\mathrm{H}}\mathbf{h}_1 s^{(2)}|^2 \leq |\hat{s}_1 - \mathbf{w}_1^{\mathrm{H}}\mathbf{h}_1 s^{(1)}|^2) \\ &= \Pr(|\mathbf{w}_1^{\mathrm{H}}\mathbf{h}_1(s^{(1)} - s^{(2)}) + \mathbf{w}_1^{\mathrm{H}}\mathbf{n}|^2 \leq |\mathbf{w}_1^{\mathrm{H}}\mathbf{n}|^2) \\ &= \mathcal{Q}\left(\sqrt{\frac{|d|^2 \mathbf{h}_1^{\mathrm{H}}(\mathbf{I} - \mathbf{P}_1)\mathbf{h}_1}{2N_0}}\right) \\ &= \mathcal{Q}\left(\sqrt{\frac{|d|^2 Z}{2N_0}}\right), \end{aligned} \tag{7.71}$$

where $d = s^{(1)} - s^{(2)}$. Then, after taking the expectation with respect to Z, we can apply the union bound to find an upper bound on the average error probability.

Now, we focus on the performance analysis of the MMSE detector with Assumption (**A1**). For MMSE detection, the combining vector becomes

$$\mathbf{w}_1 \propto \left(\mathbf{H}_1\mathbf{H}_1^{\mathrm{H}} + \frac{N_0}{E_{\mathrm{b}}}\mathbf{I}\right)^{-1} \mathbf{h}_1.$$

With 4-QAM, the bit error probability conditioned on $\mathbf{H}_1$ and $\mathbf{h}_1$ is given by

$$\begin{aligned} P_{\mathrm{b}}(\mathbf{h}_1, \mathbf{P}_1) &= \mathcal{Q}\left(\sqrt{2\mathbf{h}_1^{\mathrm{H}}\left(\mathbf{H}_1\mathbf{H}_1^{\mathrm{H}} + \tfrac{N_0}{E_{\mathrm{b}}}\mathbf{I}\right)^{-1}\mathbf{h}_1}\right) \\ &\leq \exp\left(-\mathbf{h}_1^{\mathrm{H}}\left(\mathbf{H}_1\mathbf{H}_1^{\mathrm{H}} + \tfrac{N_0}{E_{\mathrm{b}}}\mathbf{I}\right)^{-1}\mathbf{h}_1\right). \end{aligned} \tag{7.72}$$

Consider the eigen-decomposition of $\mathbf{H}_1\mathbf{H}_1^{\mathrm{H}}$ as

$$\mathbf{H}_1\mathbf{H}_1^{\mathrm{H}} = \mathbf{E}\boldsymbol{\Lambda}\mathbf{E}^{\mathrm{H}},$$

where $\mathbf{E} = [\mathbf{e}_1\ \mathbf{e}_2\ \dots\ \mathbf{e}_N]$ and $\boldsymbol{\Lambda} = \mathrm{Diag}(\lambda_1, \lambda_2, \dots, \lambda_N)$. Here, λ_q is the qth smallest eigenvalue and $\mathbf{e}_q$ is its corresponding eigenvector. We also note that

$$\mathbf{H}_1\mathbf{H}_1^{\mathrm{H}} + \frac{N_0}{E_{\mathrm{b}}}\mathbf{I} = \mathbf{E}\left(\boldsymbol{\Lambda} + \frac{N_0}{E_{\mathrm{b}}}\mathbf{I}\right)\mathbf{E}^{\mathrm{H}}.$$

Let

$$\mathbf{a} = \mathbf{E}^{\mathrm{H}}\mathbf{h}_1.$$

Then, we have

$$\mathbf{h}_1^{\mathrm{H}}\left(\mathbf{H}_1\mathbf{H}_1^{\mathrm{H}} + \frac{N_0}{E_{\mathrm{b}}}\mathbf{I}\right)^{-1}\mathbf{h}_1 = \sum_{q=1}^{N}\left(\lambda_q + \frac{N_0}{E_{\mathrm{b}}}\right)^{-1}|a_q|^2,$$

where a_q denotes the qth element of $\mathbf{a}$. Since the rank of $\mathbf{H}_1$ is $K - 1 (\leq N)$, the smallest $N - K + 1$ eigenvalues of $\mathbf{H}_1\mathbf{H}_1^{\mathrm{H}}$ are zero. From this, it follows

$$\mathbf{h}_1^{\mathrm{H}}\left(\mathbf{H}_1\mathbf{H}_1^{\mathrm{H}} + \frac{N_0}{E_{\mathrm{b}}}\mathbf{I}\right)^{-1}\mathbf{h}_1 = \frac{E_{\mathrm{b}}}{N_0}\sum_{q=1}^{N-K+1}|a_q|^2 + \sum_{q=N-K+2}^{N}\left(\lambda_q + \frac{N_0}{E_{\mathrm{b}}}\right)^{-1}|a_q|^2. \tag{7.73}$$

Substituting (7.73) into (7.72), we have

$$\begin{aligned} P_{\mathrm{b}}(\mathbf{h}_1, \mathbf{P}_1) \leq &\exp\left(-\frac{E_{\mathrm{b}}}{N_0}\sum_{q=1}^{N-K+1}|a_q|^2\right) \\ &\times \exp\left(-\sum_{q=N-K+2}^{N}\left(\lambda_q + \frac{N_0}{E_{\mathrm{b}}}\right)^{-1}|a_q|^2\right). \end{aligned} \tag{7.74}$$

Since $\mathbf{E}$ is unitary, $\mathbf{E}\mathbf{h}_1 \sim \mathcal{CN}(0, \sigma_h^2 \mathbf{I})$ from Assumption (**A1**). Therefore, a_q is independent of each other, which implies that the two terms on the right-hand side in (7.74) are statistically independent. Thus, in order to obtain the average bit error probability, we can find the mean of each term. The mean of the first term is given by

$$\mathcal{E}\left[\exp\left(-\frac{E_\mathrm{b}}{N_0}\sum_{q=1}^{N-K+1}|a_q|^2\right)\right] = \left(\frac{1}{1+\bar{\gamma}}\right)^{N-K+1}. \tag{7.75}$$

Given λ_q's, the mean of the second term becomes

$$\mathcal{E}\left[\exp\left(-\sum_{q=N-K+2}^{N}\left(\lambda_q + \frac{N_0}{E_\mathrm{b}}\right)^{-1}|a_q|^2\right)\right] = \prod_{q=N-K+2}^{N}\frac{\lambda_q E_\mathrm{b} + N_0}{\lambda_q E_\mathrm{b} + \sigma_h^2 E_\mathrm{b} + N_0}. \tag{7.76}$$

To derive the average error probability, we need to take the expectation with respect to the λ_q's, which are the eigenvalues of the random matrix. Since it requires a $(K-1)$-fold integration, it is usually difficult to find a closed-form expression by completing the integration. We now consider an approximation.

Using the inequality of arithmetic and geometric means, it can be shown that

$$\prod_{q=N-K+2}^{N}\frac{\lambda_q + a}{\lambda_q + b} \le \left(\frac{1}{K-1}\sum_{q=N-K+2}^{N}\frac{\lambda_q + a}{\lambda_q + b}\right)^{K-1},$$

where $a = N_0/E_\mathrm{b}$ and $b = \sigma_h^2 + N_0/E_\mathrm{b}$. The arithmetic mean is now approximated by the statistical mean as follows:

$$\frac{1}{K-1}\sum_{q=N-K+2}^{N}\frac{\lambda_q + a}{\lambda_q + b} \simeq \mathcal{E}\left[\frac{\lambda_q + a}{\lambda_q + b}\right]. \tag{7.77}$$

Since $\frac{\lambda_q + a}{\lambda_q + b}$ is concave (in terms of λ_q) if $b > a > 0$, using Jensen's inequality, we have

$$\mathcal{E}\left[\frac{\lambda_q + a}{\lambda_q + b}\right] \le \frac{\mathcal{E}[\lambda_q] + a}{\mathcal{E}[\lambda_q] + b}.$$

It can also be shown that

$$\begin{aligned}\mathcal{E}[\lambda_q] &= \frac{1}{K-1}\mathcal{E}\left[\sum_{q=N-K+2}^{N}\lambda_q\right] \\ &= \frac{1}{K-1}\mathcal{E}\left[\mathrm{tr}(\mathbf{H}_1^\mathrm{H}\mathbf{H}_1)\right] \\ &= N\sigma_h^2.\end{aligned} \tag{7.78}$$

From this, we have

$$\frac{\mathcal{E}[\lambda_q]+a}{\mathcal{E}[\lambda_q]+b}=\frac{N+\frac{1}{\bar{\gamma}}}{N+1+\frac{1}{\bar{\gamma}}}. \tag{7.79}$$

Substituting (7.79) into (7.76), we have

$$\prod_{q=N-K+2}^{N}\frac{\lambda_q E_\mathrm{b}+N_0}{\lambda_q E_\mathrm{b}+\sigma_h^2 E_\mathrm{b}+N_0}\simeq\left(\frac{N+\frac{1}{\bar{\gamma}}}{N+1+\frac{1}{\bar{\gamma}}}\right)^{K-1}. \tag{7.80}$$

Finally, the average error probability is given by

$$P_\mathrm{b}\simeq\left(\frac{1}{1+\bar{\gamma}}\right)^{N-K+1}\left(\frac{N+\frac{1}{\bar{\gamma}}}{N+1+\frac{1}{\bar{\gamma}}}\right)^{K-1}. \tag{7.81}$$

This shows that the average error probability is lower than that of the ZF detector in (7.70), but both the MMSE and ZF detectors have the same diversity order, $N-K+1$. Since the diversity order is obtained when the SNR approaches infinity, it is useful to consider the case of a high SNR. The performance of a linear detector is generally limited by the interference when the SNR is sufficiently high. Thus, in the presence of $K-1$ interfering signals, only the desired signal components in a $(N-K+1)$-dimensional space become interference-free and can contribute to the diversity order as the SNR approaches infinity. This implies that the diversity order of linear detectors becomes the difference between the total degree of freedom (or the number of receive antennas) and the number of interferers, which is $N-(K-1)$.

Note that the average error probability in (7.81) is an approximation because of the step in (7.77). If the statistical mean is close to the arithmetic mean, the average error probability in (7.81) could be a good upper bound.

If $\bar{\gamma}$ is sufficiently large, while $\beta=(K-1)/N$ is fixed, we have

$$\begin{aligned}\lim_{K\to\infty}\left(\frac{N+\frac{1}{\bar{\gamma}}}{N+1+\frac{1}{\bar{\gamma}}}\right)^{K-1}&\simeq\lim_{K\to\infty}\left(\frac{1}{1+\frac{1}{N}}\right)^{K-1}\\&=\mathrm{e}^{-\beta}.\end{aligned}$$

Thus, for large $(K-1)$ and N, we can also have the following approximate average error probability of the MMSE detector:

$$P_\mathrm{b}\simeq\left(\frac{1}{1+\bar{\gamma}}\right)^{N-K+1}\mathrm{e}^{-(K-1)/N}. \tag{7.82}$$

Figure 7.12 shows simulation results and upper bounds on the bit error probabilities of the ZF and MMSE detectors for 4-QAM over MIMO channels where $N=6$ and $K=4$. As the bounds are based on the Chernoff bound (c.f. (7.70) and (7.72)), the bounds are not tight, but reveal the diversity order.

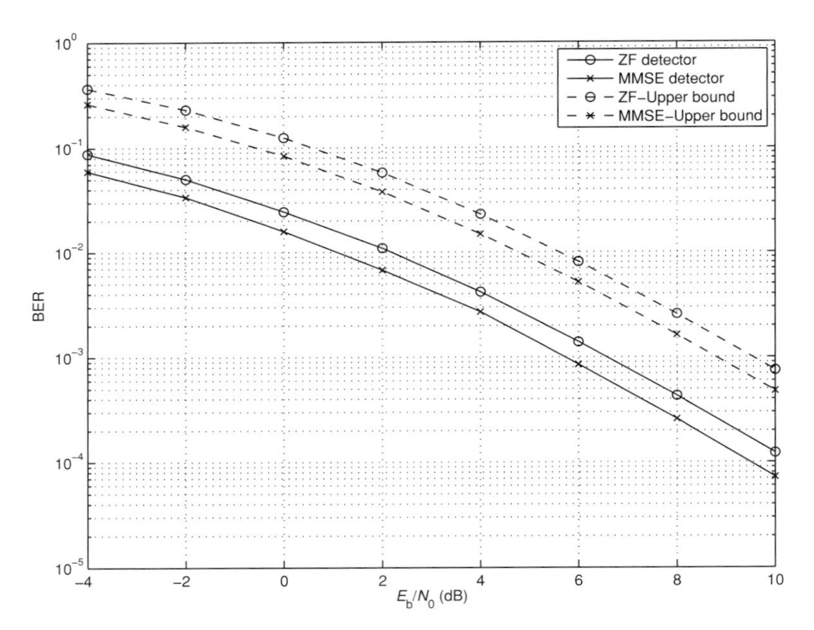

Figure 7.12 Bit error probability of linear detectors with their bounds for 4-QAM over 6×4 MIMO channels (i.e., $N = 6$ and $K = 4$).

For comparison, we also consider the case of $K = N = 4$ and the results are shown in Fig. 7.13. Compared with the results in Fig. 7.12, it can be shown that the diversity order is reduced as there are less receive antennas.

7.5.2 Performance of ML detector

In linear detection, since each data symbol in a transmitted symbol vector, $\mathbf{s}$, is independently detected, we were interested in finding the average bit or symbol error probability in Subsection 7.5.1. However, in ML detection, since a symbol vector or even a space-time codeword (for space-time coded signals) is directly detected, we are interested in finding the average symbol vector (or space-time matrix) error probability in this subsection.

An upper bound on the symbol vector error probability can be obtained by using the union bound of PEPs whose closed-form expression can be easily found in many cases. The PEP is the error probability that can be obtained by assuming that there are only two possible signal vectors, while the rest of the signal vectors in $\mathcal{S}^K$ are ignored.

To include space-time coded signals, we consider the following received signal:

$$\begin{aligned}\mathbf{R} &= [\mathbf{r}_1 \ \mathbf{r}_2 \ \dots \ \mathbf{r}_L] \\ &= \mathbf{HS} + \mathbf{N},\end{aligned} \tag{7.83}$$

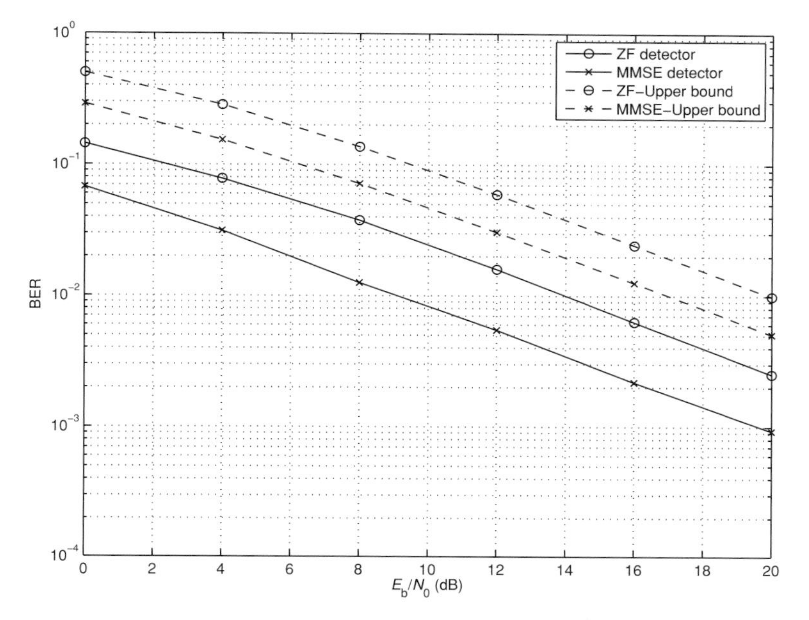

Figure 7.13 Bit error probability of linear detectors with their bounds for 4-QAM over 4 × 4 MIMO channels (i.e., $N = 4$ and $K = 4$).

where $\mathbf{r}_l$ denotes the lth received signal and $\mathbf{N} = [\mathbf{n}_1\ \mathbf{n}_2\ \ldots\ \mathbf{n}_L]$, where $\mathbf{n}_l \sim \mathcal{CN}(\mathbf{0}, N_0\mathbf{I})$ is independent background noise. Here, $\mathbf{S} = [\mathbf{s}_{(1)}\ \mathbf{s}_{(2)}\ \ldots\ \mathbf{s}_{(L)}]$ is a space-time codeword that is transmitted over an L-symbol duration, where $\mathbf{s}_{(l)}$ denotes the lth column vector of $\mathbf{S}$. This is the case where the channel matrix is invariant during the transmission of a space-time codeword. Thus, we only have spatial diversity gain. To exploit both spatial and temporal diversity gains, the channel needs to be different for each symbol duration.

First, we consider the case of $L = 1$ to derive the PEP. That is, the following received signal vector is considered:

$$\mathbf{r} = \mathbf{H}\mathbf{s} + \mathbf{n}.$$

To find the PEP, we assume that there are only two symbol vectors, say $\mathbf{s}_1$ and $\mathbf{s}_2$. Provided that $\mathbf{s}_1$ is transmitted, the probability that the symbol vector $\mathbf{s}_2$ ($\neq \mathbf{s}_1$) is erroneously detected is as follows:

$$\begin{aligned} P(\mathbf{s}_1 \to \mathbf{s}_2) &= \Pr(||\mathbf{r} - \mathbf{H}\mathbf{s}_2||^2 \le ||\mathbf{r} - \mathbf{H}\mathbf{s}_1||^2) \\ &= \Pr(||\mathbf{H}(\mathbf{s}_1 - \mathbf{s}_2) + \mathbf{n}||^2 \le ||\mathbf{n}||^2) \\ &= \Pr(||\mathbf{H}\boldsymbol{\Delta}||^2 \le -2\Re(\mathbf{n}^{\mathrm{H}}\mathbf{H}\boldsymbol{\Delta})), \end{aligned} \tag{7.84}$$

where $\boldsymbol{\Delta} = \mathbf{s}_1 - \mathbf{s}_2$. Let

$$Z = \Re(\mathbf{n}^{\mathrm{H}}\mathbf{H}\boldsymbol{\Delta}).$$

Since $\mathbf{n}$ is a CSCG random vector, $\mathbf{n}^{\mathrm{H}}\mathbf{H}\boldsymbol{\Delta}$ is also a CSCG random vector with the following properties:

$$\begin{aligned}
\mathcal{E}[\mathbf{n}^{\mathrm{H}}\mathbf{H}\boldsymbol{\Delta}] &= \mathbf{0};\\
\mathcal{E}[\mathbf{n}^{\mathrm{H}}\mathbf{H}\boldsymbol{\Delta}\boldsymbol{\Delta}^{\mathrm{H}}\mathbf{H}^{\mathrm{H}}\mathbf{n}] &= \mathcal{E}[\mathrm{tr}(\mathbf{H}^{\mathrm{H}}\mathbf{n}\mathbf{n}^{\mathrm{H}}\mathbf{H}\boldsymbol{\Delta}\boldsymbol{\Delta}^{\mathrm{H}})]\\
&= N_0\mathrm{tr}(\mathbf{H}^{\mathrm{H}}\mathbf{H}\boldsymbol{\Delta}\boldsymbol{\Delta}^{\mathrm{H}});\\
\mathcal{E}[\mathbf{n}^{\mathrm{H}}\mathbf{H}\boldsymbol{\Delta}\boldsymbol{\Delta}^{\mathrm{T}}\mathbf{H}^{\mathrm{T}}\mathbf{n}^*] &= \mathcal{E}[\mathrm{tr}(\mathbf{H}^{\mathrm{T}}\mathbf{n}^*\mathbf{n}^{\mathrm{H}}\mathbf{H}\boldsymbol{\Delta}\boldsymbol{\Delta}^{\mathrm{T}})]\\
&= \mathbf{0}.
\end{aligned} \tag{7.85}$$

From (7.85), it can be shown that

$$Z \sim \mathcal{N}\left(0, \frac{N_0}{2}\mathrm{tr}(\mathbf{H}^{\mathrm{H}}\mathbf{H}\boldsymbol{\Delta}\boldsymbol{\Delta}^{\mathrm{H}})\right).$$

Note that

$$\mathrm{tr}(\mathbf{A}\mathbf{A}^{\mathrm{H}}) = \mathrm{tr}(\mathbf{A}^{\mathrm{H}}\mathbf{A}) = ||\mathbf{A}||_{\mathrm{F}}^2.$$

Thus, the variance of Z is given by

$$\begin{aligned}
\frac{N_0}{2}\mathrm{tr}(\mathbf{H}^{\mathrm{H}}\mathbf{H}\boldsymbol{\Delta}\boldsymbol{\Delta}^{\mathrm{H}}) &= \frac{N_0}{2}||\mathbf{H}\boldsymbol{\Delta}||_{\mathrm{F}}^2\\
&= \frac{N_0}{2}||\mathbf{H}\boldsymbol{\Delta}||^2.
\end{aligned}$$

Using the Q-function, the PEP is now given by

$$\begin{aligned}
P(\mathbf{s}_1 \to \mathbf{s}_2) &= \Pr(||\mathbf{H}\boldsymbol{\Delta}||^2 \le -2Z)\\
&= \mathcal{Q}\left(\frac{||\mathbf{H}\boldsymbol{\Delta}||}{\sqrt{2N_0}}.\right)\\
&\le \exp\left(-\frac{||\mathbf{H}\boldsymbol{\Delta}||^2}{4N_0}\right),
\end{aligned} \tag{7.86}$$

where the last inequality is due to the Chernoff bound.

When $L > 1$, the PEP is given by

$$P(\mathbf{S}_1 \to \mathbf{S}_2) = \Pr(||\mathbf{R} - \mathbf{H}\mathbf{S}_2||_{\mathrm{F}}^2 \le ||\mathbf{R} - \mathbf{H}\mathbf{S}_1||_{\mathrm{F}}^2). \tag{7.87}$$

Let $\boldsymbol{\Delta} = \mathbf{S}_1 - \mathbf{S}_2$. Denote by $\boldsymbol{\Delta}_l$ and $\mathbf{n}_l$ the lth column vectors of $\boldsymbol{\Delta}$ and $\mathbf{N}$, respectively. Letting

$$Z_l = \Re(\mathbf{n}_l^{\mathrm{H}}\mathbf{H}\boldsymbol{\Delta}_l),$$

the PEP is rewritten as

$$P(\mathbf{S}_1 \to \mathbf{S}_2) = \Pr\left(||\mathbf{H}\boldsymbol{\Delta}||_{\mathrm{F}}^2 \le 2\sum_{l=0}^{L-1} Z_l\right). \tag{7.88}$$

Since the $\mathbf{n}_l$'s are the independent,

$$Z = \sum_{l=0}^{L-1} Z_l \sim \mathcal{N}\left(0, \frac{N_0}{2} \sum_{l=0}^{L-1} ||\mathbf{H}\boldsymbol{\Delta}_l||^2\right).$$

Noting that

$$\sum_{l=0}^{L-1} ||\mathbf{H}\boldsymbol{\Delta}_l||^2 = ||\mathbf{H}\boldsymbol{\Delta}||_{\mathrm{F}}^2,$$

we can show that

$$\begin{aligned} P(\mathbf{S}_1 \to \mathbf{S}_2) &= \mathcal{Q}\left(\frac{||\mathbf{H}\boldsymbol{\Delta}||_{\mathrm{F}}}{\sqrt{2N_0}}.\right) \\ &\le \exp\left(-\frac{||\mathbf{H}\boldsymbol{\Delta}||_{\mathrm{F}}^2}{4N_0}\right). \end{aligned} \tag{7.89}$$

The upper bound in (7.89) is a general expression that can also include the upper bound in (7.86) for $L = 1$.

A special case can be considered when $\mathbf{H}^{\mathrm{H}}\mathbf{H} = \mathbf{I}$, where we can assume that there are K interference-free parallel channels. In this case, the PEP is given by

$$P(\mathbf{S}_1 \to \mathbf{S}_2) = \mathcal{Q}\left(\frac{||\boldsymbol{\Delta}||_{\mathrm{F}}}{\sqrt{2N_0}}\right). \tag{7.90}$$

In general, the average PEP can be obtained for a channel matrix by taking the expectation with respect to $\mathbf{H}$. The statistical properties of $\mathbf{H}$ depend on a given channel environment. Suppose that the elements of $\mathbf{H}$ are independent zero-mean CSCG random variables with variance unity. That is, $h_{n,q} \sim \mathcal{CN}(0, 1)$. This is the case where the channel has a rich scattering environment. To find an upper bound on the average PEP, we need the following result.

Lemma 7.5.1 *Let* $\mathbf{h} \sim \mathcal{CN}(\mathbf{0}, \mathbf{R}_{\mathbf{h}})$. *Then, for a symmetric and positive semi-definite matrix,* $\mathbf{A}$, *we have*

$$\mathcal{E}[\exp(-\mathbf{h}^{\mathrm{H}}\mathbf{A}\mathbf{h})] = \det(\mathbf{I} + \mathbf{A}\mathbf{R}_{\mathbf{h}})^{-1}. \tag{7.91}$$

Proof: We can show that

$$\begin{aligned} \mathcal{E}[\exp(-\mathbf{h}^{\mathrm{H}}\mathbf{A}\mathbf{h})] &= \int \exp(-\mathbf{h}^{\mathrm{H}}\mathbf{A}\mathbf{h}) \frac{1}{\det(\pi\mathbf{R}_{\mathbf{h}})} \exp(-\mathbf{h}^{\mathrm{H}}\mathbf{R}_{\mathbf{h}}^{-1}\mathbf{h})d\mathbf{h} \\ &= \frac{1}{\det(\pi\mathbf{R}_{\mathbf{h}})} \int \exp(-\mathbf{h}^{\mathrm{H}}(\mathbf{A} + \mathbf{R}_{\mathbf{h}}^{-1})\mathbf{h})d\mathbf{h} \\ &= \frac{c}{\det(\pi\mathbf{R}_{\mathbf{h}})} \int \frac{1}{c} \exp(-\mathbf{h}^{\mathrm{H}}(\mathbf{A} + \mathbf{R}_{\mathbf{h}}^{-1})\mathbf{h})d\mathbf{h}, \end{aligned} \tag{7.92}$$

where $c > 0$ is any constant. Let

$$c = \int \exp(-\mathbf{h}^{\mathrm{H}}(\mathbf{A} + \mathbf{R}_{\mathbf{h}}^{-1})\mathbf{h})d\mathbf{h} = \det(\pi(\mathbf{A} + \mathbf{R}_{\mathbf{h}}^{-1})^{-1}).$$

It follows that

$$\begin{aligned}\mathcal{E}[\exp(-\mathbf{h}^{\mathrm{H}}\mathbf{A}\mathbf{h})] &= \frac{\det(\pi(\mathbf{A}+\mathbf{R}_{\mathbf{h}}^{-1})^{-1})}{\det(\pi\mathbf{R}_{\mathbf{h}})} \\ &= \frac{1}{\det(\mathbf{I}+\mathbf{A}\mathbf{R}_{\mathbf{h}})} \\ &= \det(\mathbf{I}+\mathbf{A}\mathbf{R}_{\mathbf{h}})^{-1}. \end{aligned} \tag{7.93}$$

This completes the proof. □

Using the following vectorization operation:

$$\mathrm{Vec}(\mathbf{H}\boldsymbol{\Delta}) = (\boldsymbol{\Delta}^{\mathrm{T}} \otimes \mathbf{I}_N)\mathrm{Vec}(\mathbf{H}),$$

the upper bound on the PEP in (7.89) is given by

$$\begin{aligned}P(\mathbf{S}_1 \to \mathbf{S}_2) &\le \exp\left(-\frac{||\mathbf{H}\boldsymbol{\Delta}||_{\mathrm{F}}^2}{4N_0}\right) \\ &= \exp\left(-\frac{(\mathrm{Vec}(\mathbf{H}))^{\mathrm{H}}(\boldsymbol{\Delta}^{\mathrm{T}}\otimes\mathbf{I}_N)^{\mathrm{H}}(\boldsymbol{\Delta}^{\mathrm{T}}\otimes\mathbf{I}_N)\mathrm{Vec}(\mathbf{H})}{4N_0}\right).\end{aligned} \tag{7.94}$$

Taking the expectation with respect to Vec(**H**) using the result in (7.91), we have

$$\begin{aligned}\bar{P}(\mathbf{S}_1 \to \mathbf{S}_2) &= \mathcal{E}[P(\mathbf{S}_1 \to \mathbf{S}_2)] \\ &\le \det\left(\mathbf{I}+\frac{1}{4N_0}(\boldsymbol{\Delta}^{\mathrm{T}}\otimes\mathbf{I}_N)^{\mathrm{H}}(\boldsymbol{\Delta}^{\mathrm{T}}\otimes\mathbf{I}_N)\right)^{-1}.\end{aligned} \tag{7.95}$$

Since

$$(\boldsymbol{\Delta}^{\mathrm{T}}\otimes\mathbf{I}_N)^{\mathrm{H}}(\boldsymbol{\Delta}^{\mathrm{T}}\otimes\mathbf{I}_N) = (\boldsymbol{\Delta}^*\boldsymbol{\Delta}^{\mathrm{T}}\otimes\mathbf{I}_N),$$

and

$$\mathbf{I}+\frac{1}{4N_0}(\boldsymbol{\Delta}^*\boldsymbol{\Delta}^{\mathrm{T}}\otimes\mathbf{I}_N) = (\mathbf{I}+\frac{1}{4N_0}\boldsymbol{\Delta}^*\boldsymbol{\Delta}^{\mathrm{T}})\otimes\mathbf{I}_N,$$

we can show that

$$\det\left(\left(\mathbf{I}+\frac{1}{4N_0}\boldsymbol{\Delta}^*\boldsymbol{\Delta}^{\mathrm{T}}\right)\otimes\mathbf{I}_N\right)^{-1} = \det\left(\mathbf{I}+\frac{1}{4N_0}\boldsymbol{\Delta}^*\boldsymbol{\Delta}^{\mathrm{T}}\right)^{-N}. \tag{7.96}$$

Consequently, the upper bound on the average PEP is given by

$$\begin{aligned}\bar{P}(\mathbf{S}_1 \to \mathbf{S}_2) &\le \det\left(\mathbf{I}+\frac{1}{4N_0}\boldsymbol{\Delta}^*\boldsymbol{\Delta}^{\mathrm{T}}\right)^{-N} \\ &= \det\left(\mathbf{I}+\frac{1}{4N_0}\boldsymbol{\Delta}\boldsymbol{\Delta}^{\mathrm{H}}\right)^{-N}.\end{aligned} \tag{7.97}$$

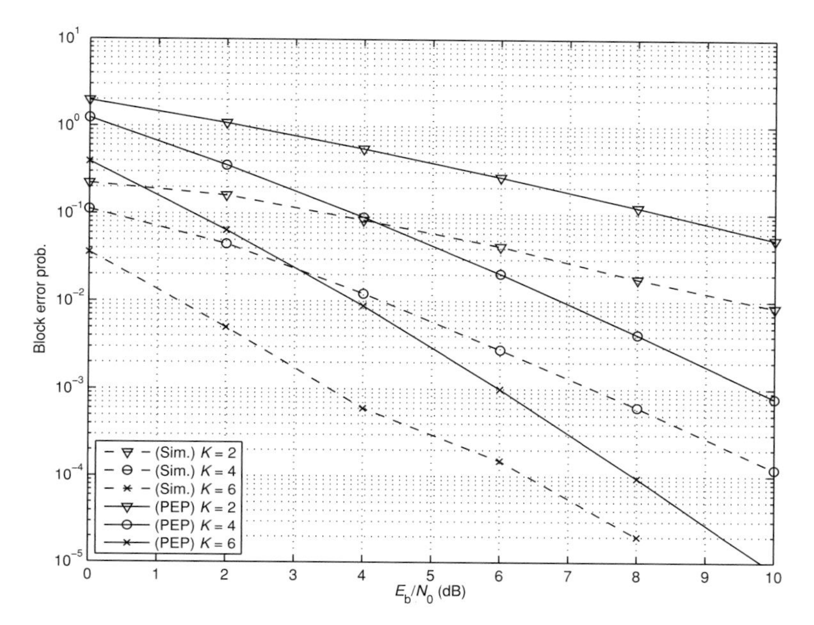

Figure 7.14 Symbol block error probability of ML detector for 4-QAM over $K \times K$ MIMO channels.

Finally, an upper bound on the symbol block error probability is given by

$$P_{\text{block}} \leq \frac{1}{|\mathcal{S}_{\text{stc}}|} \sum_{\mathbf{S}_1} \sum_{\mathbf{S}_2, \mathbf{S}_2 \neq \mathbf{S}_1} \det \left(\mathbf{I} + \frac{1}{4N_0} (\mathbf{S}_1 - \mathbf{S}_2)(\mathbf{S}_1 - \mathbf{S}_2)^{\text{H}} \right)^{-N}, \tag{7.98}$$

where $\mathcal{S}_{\text{stc}}$ is the symbol block alphabet (or code) of $\mathbf{S}$.

In (Tarokh *et al.* 1998), the PEP in (7.97) is derived and used to design space-time codes. As shown in (7.97), a full diversity is achieved if $\mathbf{\Delta}$ is full rank. This rank condition becomes one of the design criteria for space-time codes in (Tarokh *et al.* 1998). (For a detailed account of design criteria, see also (Jafarkhani 2005).)

If $L = 1$, the upper bound on the average PEP in (7.97) is reduced to

$$\begin{aligned} \bar{P}(\mathbf{s}_1 \to \mathbf{s}_2) &\leq \det \left(\mathbf{I} + \frac{1}{4N_0} \mathbf{\Delta} \mathbf{\Delta}^{\text{H}} \right)^{-N} \\ &= \left(1 + \frac{||\mathbf{\Delta}||^2}{4N_0} \right)^{-N}, \end{aligned} \tag{7.99}$$

where $\mathbf{\Delta} = \mathbf{s}_1 - \mathbf{s}_2$. This clearly shows that the diversity order is N, which means a full receive antenna diversity is achieved.

7.6 Summary and notes

We presented well-known MIMO detection methods in this chapter. It was shown that the complexity of optimal detectors grows exponentially with the number of multiple signals, while that of linear detectors grows linearly with the number of multiple signals. The performance of MIMO detectors was also discussed over random MIMO channels.

Early work on the performance analysis of linear MIMO detectors can be found in (Salz 1985). The diversity gain of linear MIMO detectors over wireless fading channels is studied in (Winters, Salz & Gitlin 1994). In (Chiani, Win, Zanella, Mallik & Winters 2003), a detailed analysis for the performance of the MMSE detector with interfering signals is presented. A general approach for the performance analysis of MIMO detectors is discussed with space-time coded signals in (Biglieri, Taricco & Tulino 2002).

Problems

Problem 7.1 Let $\ell = \text{LLR}$ in (7.4). Show that

$$f(\mathbf{r}|s = +1) = \mathrm{e}^{\ell} f(-\mathbf{r}|s = +1).$$

Note that since ℓ is a sufficient statistic, it can also be shown that

$$f(\ell|s = +1) = \mathrm{e}^{\ell} f(-\ell|s = +1),$$

where $f(\ell|s)$ denotes the conditional pdf of ℓ given s.

Problem 7.2 Consider the following received signal vector:

$$\mathbf{r} = \mathbf{h}s + \mathbf{n},$$

where $\mathbf{n}$ is noise vector, $\mathbf{h}$ is the channel vector, and $s \in \{-1, 1\}$ is the data symbol.

(i) Show that

$$\text{LLR} = \log \frac{\Pr(s = +1|\mathbf{r})}{\Pr(s = -1|\mathbf{r})}$$

if s is equally likely.

(ii) Show that

$$\begin{aligned} \mathcal{E}[s|\mathbf{r}] &= 2\Pr(s = +1|\mathbf{r}) - 1 \\ &= \frac{\text{LR} - 1}{\text{LR} + 1} \\ &= \tanh(\text{LLR}), \end{aligned}$$

where $\text{LR} = \exp(\text{LLR}) = \frac{\Pr(s=+1|\mathbf{r})}{\Pr(s=-1|\mathbf{r})}$.

Problem 7.3 Suppose that $f(\mathbf{r}_l|x_l)$ is the likelihood function of $x_l \in \mathbb{F}_2 = \{0, 1\}$, where $\mathbb{F}_2$ denotes the binary field, for given received signal vector $\mathbf{r}_l$. We assume that x_l is a binary symbol over the field $\mathbb{F}_2$ where the exclusive OR (XOR) operation $\oplus$ is defined as $x \oplus 0 = x$ and $x \oplus 1 = \bar{x} = 1 - x$ for $x \in \mathbb{F}_2$. It is also assumed that x_l is equally likely and $f(\mathbf{r}_1, \mathbf{r}_2, \ldots, \mathbf{r}_L | x_1, x_2, \ldots, x_L) = \prod_{l=1}^{L} f(\mathbf{r}_l | x_l)$.

(i) Show that if

$$p = 2\Pr(x_1 = 0|\mathbf{r}_1) - 1;$$

$$q = 2\Pr(x_2 = 0|\mathbf{r}_2) - 1,$$

then

$$2\Pr(x_1 \oplus x_2 = 0|\mathbf{r}_1, \mathbf{r}_2) - 1 = pq.$$

(ii) Show that

$$\begin{aligned} \text{LLR} &= \log \frac{\Pr(x_1 \oplus x_2 \oplus \cdots \oplus x_L = 0|\mathbf{r}_1, \mathbf{r}_2, \ldots \mathbf{r}_L)}{\Pr(x_1 \oplus x_2 \oplus \cdots \oplus x_L = 1|\mathbf{r}_1, \mathbf{r}_2, \ldots \mathbf{r}_L)} \\ &= \log \frac{1 + \prod_{l=1}^{L} \tanh(\text{LLR}_l)}{1 - \prod_{l=1}^{L} \tanh(\text{LLR}_l)}, \end{aligned}$$

where

$$\text{LLR}_l = \log \frac{\Pr(x_l = 0|\mathbf{r}_l)}{\Pr(x_l = 1|\mathbf{r}_l)}.$$

Problem 7.4 Let $N = 2$ and $K = 3$. Suppose that $s_1, s_2, s_3 \in \{-1, 1\}$ and the channel matrix is given by

$$\mathbf{H} = \begin{bmatrix} 1 & 1 & 1 \\ 1 & 0.5 & -1 \end{bmatrix}.$$

In addition, assume that $\mathbf{n} \sim \mathcal{N}(0, \mathbf{I}_{2\times 2})$ and the received signal vector is given by $\mathbf{r} = [r_1\ r_2]^{\mathrm{T}} = [1\ -0.5]^{\mathrm{T}}$. Note that all the quantities are real-valued.

(i) Find the ML decision of $\{s_1, s_2, s_3\}$.
(ii) Derive the MMSE combiner matrix and find the MMSE detector's output.
(iii) Repeat (a) and (b) when $\mathbf{n} \sim \mathcal{N}(0, \epsilon\mathbf{I}_{2\times 2})$, $\epsilon \to 0$. Discuss the limitation of the MMSE detector.

Problem 7.5 Consider the 2×3 MIMO system in Problem 7.4. Suppose that the a priori probability of s_k is given by

$$\Pr(s_1 = +1) = 0.8;\ \Pr(s_2 = +1) = 0.2;\ \Pr(s_3 = +1) = 0.55.$$

(i) Perform the individual ML detection of s_k using Gaussian approximation.
(ii) Perform the MMSE detection (note that the mean of s_k is not zero).

Problem 7.6 Derive (7.27) using the mgf of a Gaussian random variable, $X \sim \mathcal{N}(\mu, \sigma^2)$, which is given by

$$\mathcal{E}[\mathrm{e}^{sX}] = \exp\left(\mu s + \frac{\sigma^2 s^2}{2}\right)$$

and $\mathcal{E}[\mathrm{e}^{-s\Delta}] = \mathcal{E}_{\mathbf{h}}\left[\mathcal{E}_{\mathbf{n}}[\mathrm{e}^{-s\Delta}|\mathbf{h}]\right]$.

Problem 7.7 Using the Gauss–Chebyshev quadrature rule, derive the (approximate) average error probability of the ZF and MMSE detector for MIMO channels (assuming 4-QAM is employed for modulation).

Problem 7.8 Show that (7.59) using the matrix inversion lemma in (3.1) in Appendix 1.

Problem 7.9 Let $\mathbf{A}$ denote a complex-valued $N \times L$ matrix. In addition, let $\mathbf{N}$ be an $N \times L$ random matrix whose elements are independent and zero-mean CSCG random variables with variance σ^2. Using

$$\mathrm{tr}(\mathbf{A}^{\mathrm{H}}\mathbf{N}) = \sum_{l=1}^{L} \mathbf{a}_l^{\mathrm{H}}\mathbf{n}_l, \tag{7.100}$$

where $\mathbf{a}_l$ and $\mathbf{n}_l$ denote the lth column vectors of $\mathbf{A}$ and $\mathbf{N}$, respectively, show that

$$\mathrm{Var}\left(\mathrm{tr}(\mathbf{A}^{\mathrm{H}}\mathbf{N})\right) = \sigma^2 ||\mathbf{A}||_{\mathrm{F}}^2. \tag{7.101}$$

Problem 7.10 Show that the diversity order of the ML detector (or decoder) becomes $2N$ if the Alamouti code in (7.63) is used for space-time coding.

Problem 7.11 Suppose that a space-time codeword is transmitted for a period of $L = 2$ symbol durations. In addition, the channel is different for each symbol duration. Thus, the received signal becomes

$$\mathbf{r}_l = \mathbf{H}_l \mathbf{s}_{(l)} + \mathbf{n}_l, \ l = 1, 2, \tag{7.102}$$

where $\mathbf{H}_l$ and $\mathbf{n}_l$ are the MIMO channel matrix and independent background noise during the lth symbol duration, respectively. Here, $\mathbf{S} = [\mathbf{s}_{(1)} \ \mathbf{s}_{(2)}]$ is a space-time codeword. Derive the PEP if the elements of $\mathbf{H}_l$ are independent CSCG random variables with mean zero and unit variance and $\mathbf{n}_l \sim \mathcal{CN}(0, N_0\mathbf{I})$.

8 MIMO detection with successive interference cancellation

In the presence of multiple signals, an optimal detection approach is usually based on joint detection. Unfortunately, joint detection requires a prohibitively high computational complexity, because the size of the signal alphabet can grow exponentially with the number of multiple signals. Thus, in the case that the computational complexity of receivers is limited, it is necessary to find low complexity suboptimal detectors. To this end, successive interference cancellation (SIC) can be employed for a suboptimal detector to detect multiple signals sequentially with low complexity. In this chapter, we discuss various SIC detectors and their performance.

8.1 Successive interference cancellation

If a receiver receives multiple signals simultaneously, the detection performance is generally poor when the signals are individually detected. For example, suppose that the received signal is given by

$$r = h_1 s_1 + h_2 s_2 + n,$$

where s_k and h_k represent the kth signal and its associated channel gain, respectively, and n is the background noise. When s_1 is to be detected, the signal-to-interference-plus-noise ratio (SINR) becomes

$$\mathrm{SINR}_1 = \frac{|h_1| E_1}{|h_2|^2 E_2 + N_0}, \tag{8.1}$$

where $\mathcal{E}[|s_k|^2] = E_k$ and $\mathcal{E}[|n|^2] = N_0$, because s_2 becomes the interference. If $E_1 = E_2$ and $|h_1|^2 = |h_2|^2$, the SINR is less than 1 (or 0 dB) and the resulting detection performance becomes poor.

To avoid poor detection performance, a joint detection can be employed, i.e. the two signals, s_1 and s_2, can be jointly detected as in Chapter 7. In this case, however, the detection complexity increases.

An alternative approach to detect multiple signals is to use successive interference cancellation (SIC). Suppose that it is possible to detect s_1 as SINR_1 in (8.1) is sufficiently high (this could happen if $E_1 > E_2$). Let $\hat{s}_1$ denote the detected signal of s_1. Assuming that $\hat{s}_1$ is correct, it is possible to cancel the interference from s_1 in detecting s_2. That is, if $\hat{s}_1 = s_1$, s_2 can be detected from

$$\begin{aligned} u_2 &= r - h_1 \hat{s}_1 \\ &= h_2 s_2 + n, \end{aligned}$$

which is interference-free. This detection approach is called SIC detection and can be generalized for any number of multiple signals. When we have K transmitted signals, the received signal is given by

$$r = \sum_{k=1}^{K} h_k s_k + n. \tag{8.2}$$

If the SIC detection is carried out in the ascending order, the SINR for the kth signal becomes

$$\text{SINR}_k = \frac{|h_k| E_k}{\sum_{q=k+1}^{K} |h_q|^2 E_q + N_0}, \quad k = 1, 2, \ldots, K, \tag{8.3}$$

where $E_k = \mathcal{E}[|s_k|^2]$. This SINR is achievable under the assumption that the previous detection is successful and SIC is perfect. As seen in (8.3), the SINR could increase with k since the interference decreases with k by SIC.

The success of SIC depends on cancellation performance. If any signals in the previous detection are not correctly detected, the cancellation cannot be perfect and there could be cancellation error, which will result in a worse performance than the expected performance with perfect SIC. For coded signals, SIC can be carried out with more reliably decoded signals to minimize erroneous interference cancellation (in the case of coded signals, s_k is considered as a coded *sequence*, rather than a symbol).

To perform SIC detection/decoding without cancellation errors or with negligible cancellation errors, the SINRs in (8.3) can provide a guide to decide the transmission rate for each signal. If the SINR is low, the corresponding signal should have a low transmission rate, and vice versa. For uncoded signals, this implies that the signal constellation size could be different for each signal and it could be small for the signal detected earlier. For example, when $K = 2$, we can use BPSK for s_1 and 16-QAM for s_2. For coded signals, if each signal is independently encoded, the code rate can be low for the signals to be detected earlier.

There are a number of communication systems where the received signal can be modeled as in (8.2). As shown in Chapter 7, MIMO systems are a typical example. When a signal is transmitted over an inter-symbol interference (ISI) channel, the received signal can be represented as in (8.2), where s_k is a delayed version of the transmitted signal. In the context of ISI channels, SIC detection becomes decision feedback equalizer (DFE).

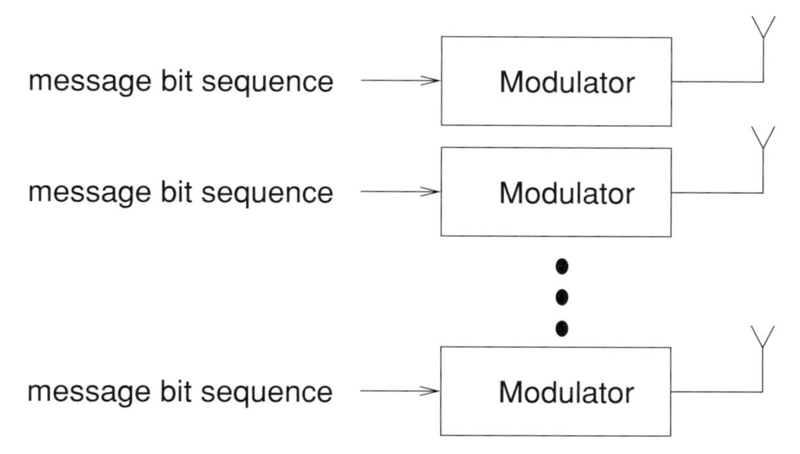

Figure 8.1 A layered transmission scheme of K layers.

8.2 SIC detectors

In this section, we discuss SIC detectors in the context of MIMO channels. Since the receiver is equipped with multiple antennas, spatial processing or filtering is available to mitigate interference in conjunction with SIC.

Recall the received signal vector over MIMO channels as follows:

$$\mathbf{r} = \mathbf{H}\mathbf{s} + \mathbf{n}, \tag{8.4}$$

where $\mathbf{H}$, $\mathbf{s}$, and $\mathbf{n}$ are the channel matrix, transmitted signal vector, and noise vector, respectively. We assume that the transmitter and receiver are equipped with K and N antennas, respectively. Thus, the size of $\mathbf{H}$ is $N \times K$. In addition, throughout this chapter, we assume that $\mathbf{n} \sim \mathcal{CN}(\mathbf{0}, N_0\mathbf{I})$ and the signal alphabet for the s_k's are the same and denoted by $\mathcal{S}$, i.e. $s_k \in \mathcal{S}$ for all k. Thus, the symbol energy of s_k, $\mathcal{E}[|s_k|^2] = E_\mathrm{s}$, is the same for all k unless stated otherwise.

For SIC detection, we adopt the notion of layered transmission. It is assumed that the signal transmitted by each transmit antenna is independently encoded and modulated. Thus, the transmitter consists of multiple independent sub-transmitters that have one transmit antenna. Each sub-transmitter builds a layer as shown in Fig. 8.1. As a result, there are K layers. For convenience, s_k is called the (transmitted) signal in layer k. This layered transmission scheme is introduced in (Foschini, Chizhik, Gans, Papadias & Valenzuela 2003). The layered transmission scheme can transmit K data sequences simultaneously, which implies that the spatial multiplexing gain is K. Since each sub-transmitter has individual design parameters (e.g. transmission power, transmission rate, etc.), the layered transmission scheme could be flexible and adjustable to different MIMO channel conditions by optimizing design parameters. This certainly becomes an advantage of layered transmission.

8.2.1 ZF-SIC detector

To build a SIC detector for MIMO channels, we can combine zero-forcing (ZF) interference suppression and SIC (the resulting detector is no longer linear when SIC is carried out with hard-decision), which results in ZF-SIC detection. In SIC, the first signal detection is the most vulnerable to interference. Therefore, spatial filtering should be used to suppress interfering signals as much as possible. Since there are $K - 1$ interfering signals in the first signal detection, a linear combining vector that is orthogonal to the subspace constructed by $K - 1$ interfering signals can be used to completely suppress the interference. This linear suppression requires $N \geq K$. At the receiver (equipped with N receive antennas), a $(K - 1)$-dimension is used to suppress the interference and the other $(N - (K - 1))$-dimension is used to combine the desired signal. This implies that $N - (K - 1) \geq 1$ or $N \geq K$. If k signals are already detected, they can be used for SIC in the $(k + 1)$th signal detection. In this case, there are $K - (k + 1)$ interfering signals, and more dimension becomes available for signal combining. Consequently, a better detection performance is expected as k increases.

ZF-SIC detection can be explained with the QR factorization of $\mathbf{H}$, because the QR factorization is an effective means to perform ZF interference suppression with SIC. Assume that $K = N$ for convenience. The QR factorization of the channel matrix becomes

$$\begin{aligned}\mathbf{H} &= \mathbf{QR} \\ &= \mathbf{Q}\begin{bmatrix} r_{1,1} & r_{1,2} & \cdots & r_{1,K} \\ 0 & r_{2,2} & \cdots & r_{2,K} \\ \vdots & \vdots & \ddots & \vdots \\ 0 & 0 & \cdots & r_{K,K} \end{bmatrix},\end{aligned}$$

where $\mathbf{Q}$ is unitary and $\mathbf{R}$ is upper triangular. Pre-multiplying $\mathbf{Q}^{\mathrm{H}}$ to $\mathbf{r}$ results in

$$\begin{aligned}\mathbf{Q}^{\mathrm{H}}\mathbf{r} &= \mathbf{Q}^{\mathrm{H}}\mathbf{H}\mathbf{s} + \mathbf{Q}^{\mathrm{H}}\mathbf{n} \\ &= \mathbf{R}\mathbf{s} + \mathbf{Q}^{\mathrm{H}}\mathbf{n}.\end{aligned} \tag{8.5}$$

Let $\mathbf{x} = \mathbf{Q}^{\mathrm{H}}\mathbf{r}$. We now have two important observations from (8.5). First, since $\mathbf{R}$ is upper triangular, we can see that more and more interfering signals are suppressed in x_k as k increases. For example, x_K becomes interference-free as it has only the s_K term (ignoring the background noise). Thus, s_K can be detected first. Second, we note that $\mathbf{Q}^{\mathrm{H}}\mathbf{n}$ has the same statistical properties as $\mathbf{n}$: both $\mathbf{Q}^{\mathrm{H}}\mathbf{n}$ and $\mathbf{n}$ are zero-mean CSCG random vectors with the same covariance matrix as

$$\begin{aligned}\mathcal{E}[\mathbf{Q}^{\mathrm{H}}\mathbf{n}\mathbf{n}^{\mathrm{H}}\mathbf{Q}] &= \mathbf{Q}^{\mathrm{H}}\underbrace{\mathcal{E}[\mathbf{n}\mathbf{n}^{\mathrm{H}}]}_{=N_0\mathbf{I}}\mathbf{Q} \\ &= N_0\mathbf{I}.\end{aligned}$$

As $\mathbf{Q}^{\mathrm{H}}\mathbf{n}$ has the same statistical properties as $\mathbf{n}$, we simply use $\mathbf{n}$ to denote $\mathbf{Q}^{\mathrm{H}}\mathbf{n}$. The resulting signal is given by

$$\mathbf{x} = \mathbf{R}\mathbf{s} + \mathbf{n}. \tag{8.6}$$

From (8.6), we can show that

$$\begin{aligned} x_K &= r_{K,K}s_K + n_K \\ x_{K-1} &= r_{K-1,K}s_K + r_{K-1,K-1}s_{K-1} + n_{K-1} \\ &\vdots \end{aligned} \tag{8.7}$$

This suggests a sequential detection procedure in SIC.

The signal in the lowest layer or layer K, s_K, can be detected from x_K, as x_K is interference-free (thus, no joint detection is required). A hard-decision of s_K can be used to cancel the interference caused by s_K in detecting the signal in the second lowest layer, s_{K-1}, from x_{K-1}. Let $\hat{s}_K$ denote a hard-decision from x_K. For example, if symbol-wise ML detection is used, we have

$$\hat{s}_K = \arg\max_{s\in\mathcal{S}} \exp\left(-\frac{1}{N_0}|x_K - r_{K,K}s|^2\right).$$

Assuming that $\hat{s}_K = s_K$, we can find the following interference-free signal for the detection of s_{K-1}:

$$\begin{aligned} u_{K-1} &= x_{K-1} - r_{K-1,K}\hat{s}_K \\ &= r_{K-1,K-1}s_{K-1} + n_{K-1}. \end{aligned}$$

A hard-decision of s_{K-1} can be obtained from u_{K-1} and used for the next detection.

This detection and cancellation operation can be successively carried out to upper layers (i.e. toward layer 1). For the detection of the kth data symbol, s_k, after canceling $K-k$ data symbols (i.e., $s_{k+1}, s_{k+2}, \ldots, s_K$), we can use the following signal:

$$u_k = x_k - \sum_{q=k+1}^{K} r_{k,q}\hat{s}_q, \ k \in \{1, 2, \ldots, K-1\},$$

where $\hat{s}_q$ denotes a hard-decision of s_q from u_q. If the ML approach is employed for the sub-detection, the hard-decision can be found as

$$\hat{s}_k = \arg\max_{s\in\mathcal{S}} \exp\left(-\frac{1}{N_0}|u_k - r_{k,k}s|^2\right)$$

under the assumption that the signals in the lower layers, $s_{k+1}, s_{k+2}, \ldots, s_K$, are perfectly canceled. This is ZF-SIC detection.

Unfortunately, this approach has a few drawbacks. The performance is degraded by error propagation. If incorrect decision(s) are made in the early stages, the subsequent cancellation becomes erroneous and more incorrect decisions can follow. This implies that the overall performance is strongly dependent on the first detection for s_K. A detailed performance analysis can be found in Section 8.5 (see also (Choi 2005*a*)).

Example 8.2.1 Let

$$\mathbf{H} = \frac{1}{\sqrt{2}} \begin{bmatrix} 1 & 1 \\ 1 & 0.5 \end{bmatrix}.$$

Then, the QR factorization of $\mathbf{H}$ is given by

$$\mathbf{H} = \mathbf{QR} = \begin{bmatrix} \frac{1}{\sqrt{2}} & \frac{1}{\sqrt{2}} \\ \frac{1}{\sqrt{2}} & -\frac{1}{\sqrt{2}} \end{bmatrix} \begin{bmatrix} 1 & \frac{3}{4} \\ 0 & \frac{1}{4} \end{bmatrix}$$

and $\mathbf{x} = \mathbf{Q}^{\mathrm{H}}\mathbf{r}$ becomes

$$x_1 = s_1 + \frac{3}{4}s_2 + n_1;$$

$$x_2 = \frac{1}{4}s_2 + n_2.$$

Here, x_2 is interference-free and x_1 can also be interference-free if s_2 can be canceled.

In the above, we consider the case of $N = K$. ZF-SIC detection is also available when $N \neq K$. If $N > K$, the size of the $\mathbf{R}$ matrix becomes $N \times K$ and the last $N - K$ rows are all zero. Therefore, the last $N - K$ elements of $\mathbf{x} = \mathbf{Q}^{\mathrm{H}}\mathbf{r}$ become useless because they do not have any signal component. Taking the first K elements of $\mathbf{x}$ (this operation is equivalent to the dimension reduction where the signal subspace is preserved, while the noise subspace is discarded), SIC detection can be carried out as the case of $N = K$.

The case of $N < K$ is a bit tricky. The last element of $\mathbf{x}$, x_N, which is given by

$$x_N = r_{N,N}s_N + r_{N,N+1}s_{N+1} + \cdots + r_{N,K}s_K + n_N,$$

is not interference-free. Thus, a joint detection for $\{s_N, s_{N+1}, \ldots, s_K\}$ is required for the first detection in SIC. Then, the following detection is the same as that in the case of $N = K$.

8.2.2 MMSE-SIC detector

In the ZF-SIC detector, the transformation, $\mathbf{Q}^{\mathrm{H}}$, performs nulling to suppress the signals in higher layers. Since the ZF-SIC is just a different form of the ZF-DFE over ISI channels, we can also derive the MMSE-SIC under the same principle for the MMSE-DFE over ISI channels by accounting for the background noise in the interference suppression.

Suppose that the signal in the lowest layer is detected first as in the ZF-SIC. The MMSE combining vector to suppress interference-plus-noise in detecting s_K from $\mathbf{r}$ is

given by

$$\mathbf{w}_{\text{mmse-sic},K} = \arg\min_{\mathbf{w}} \mathcal{E}[|s_K - \mathbf{w}^{\mathrm{H}}\mathbf{r}|^2]$$
$$= (E_{\mathrm{s}}\mathbf{H}\mathbf{H}^{\mathrm{H}} + N_0\mathbf{I})^{-1}\mathbf{h}_K E_{\mathrm{s}}$$
$$= \left(\mathbf{H}\mathbf{H}^{\mathrm{H}} + \frac{N_0}{E_{\mathrm{s}}}\mathbf{I}\right)^{-1} \mathbf{h}_K, \quad (8.8)$$

where $\mathbf{h}_k$ stands for the kth column of $\mathbf{H}$. Once s_K is estimated through the MMSE combiner, a hard-decision is made and used to cancel its contribution in detecting the other symbols. With a hard-decision of s_K, $\hat{s}_K$, which is given by

$$\hat{s}_K = \mathrm{HD}(\hat{c}_K),$$

where $\mathrm{HD}(\cdot)$ represents a hard-decision operation whose output is an element of $\mathcal{S}$ and $\hat{c}_k$ is the MMSE-SIC estimate, which is the MMSE estimate after SIC and given by

$$\hat{c}_K = \mathbf{w}^{\mathrm{H}}_{\text{mmse-sic},K}\mathbf{r}.$$

If $k = K$, since no SIC is available, the MMSE-SIC estimate is identical to the (conventional linear) MMSE estimate. To avoid confusion, we denote a hard-decision of s_k by $\hat{s}_k$ and the MMSE-SIC estimate (which is a soft-decision) by $\hat{c}_k$ in this section.

The signal vector after cancellation becomes

$$\mathbf{r}_{K-1} = \mathbf{r} - \mathbf{h}_K\hat{s}_K. \quad (8.9)$$

Assuming that $\hat{s}_K = s_K$, we have

$$\mathbf{r}_{K-1} = \sum_{k=1}^{K-1} \mathbf{h}_k s_k + \mathbf{n}. \quad (8.10)$$

With $\mathbf{r}_{K-1}$, MMSE combining to detect s_{K-1} can be carried out. Repeating cancellation and MMSE combining, the detection of the s_k's can be performed. Under the assumption that the SIC is perfect (i.e. $\hat{s}_k = s_k$ in the previous detection), the MMSE-SIC combining vector is given by

$$\mathbf{w}_{\text{mmse-sic},k} = \arg\min_{\mathbf{w}} \mathcal{E}[|s_k - \mathbf{w}^{\mathrm{H}}\mathbf{r}_k|^2]$$
$$= \left(E_{\mathrm{s}}\sum_{q=1}^{k}\mathbf{h}_q\mathbf{h}_q^{\mathrm{H}} + N_0\mathbf{I}\right)^{-1}\mathbf{h}_k E_{\mathrm{s}}$$
$$= \left(\sum_{q=1}^{k}\mathbf{h}_q\mathbf{h}_q^{\mathrm{H}} + \frac{N_0}{E_{\mathrm{s}}}\mathbf{I}\right)^{-1}\mathbf{h}_k, \quad (8.11)$$

where

$$\mathbf{r}_k = \mathbf{r} - \sum_{q=k+1}^{K} \mathbf{h}_q s_q + \mathbf{n}$$
$$= \sum_{q=1}^{k} \mathbf{h}_q s_q + \mathbf{n}.$$

A hard-decision of s_k is given by

$$\hat{s}_k = \mathrm{HD}(\hat{c}_k),$$

where $\hat{c}_k$ is the MMSE-SIC estimate of s_k and given by

$$\hat{c}_k = \mathbf{w}_{\text{mmse-sic},k}^{\mathrm{H}} \mathbf{r}_k.$$

The resulting MMSE-SIC detector can be summarized as follows.

(S0) Let $k = K$ and $\mathbf{r}_K = \mathbf{r}$.
(S1) Find the MMSE combining vector, $\mathbf{w}_{\text{mmse-sic},k}$ as in (8.11).
(S2) Detect the kth symbol as follows:

$$\hat{s}_k = \mathrm{HD}(\mathbf{w}_{\text{mmse-sic},k}^{\mathrm{H}} \mathbf{r}_k).$$

(S3) Cancel the kth symbol's contribution as follows:

$$\mathbf{r}_{k-1} = \mathbf{r}_k - \mathbf{h}_k \hat{s}_k.$$

(S4) If $k = 1$, stop. Otherwise, set $k \leftarrow k - 1$ and go to (S1).

Figure 8.2 shows BER performance of the ZF-SIC and MMSE-SIC detectors for 16-QAM over 4×4 channels. For comparison, we also show BER performance of the ML and two linear detectors (ZF and MMSE detectors). The SIC detectors can perform better than the linear detectors and the performance gap increases with the SNR. It is noteworthy that the performance of SIC detectors is not significantly better than that of linear detectors when the SNR is low.

In SIC detection, it is assumed that the cancellation is perfect. However, in practice, an erroneous decision can be made and the subsequent cancellation becomes imperfect. This imperfect cancellation results in erroneous detection in higher layers. This phenomenon is called error propagation. Consequently, SIC detectors become vulnerable to error propagation. In Section 8.5, we analyze the impact of error propagation on performance.

8.3 Square-root algorithm for MMSE-SIC detector

The MMSE-SIC detector can also be derived by other approaches, including the square-root algorithm. The square-root algorithm is derived in (Hassibi 2000) to perform MMSE-SIC detection with low computational complexity. The conventional approach for MMSE-SIC detection in Subsection 8.2.2 requires a matrix inversion for each layer

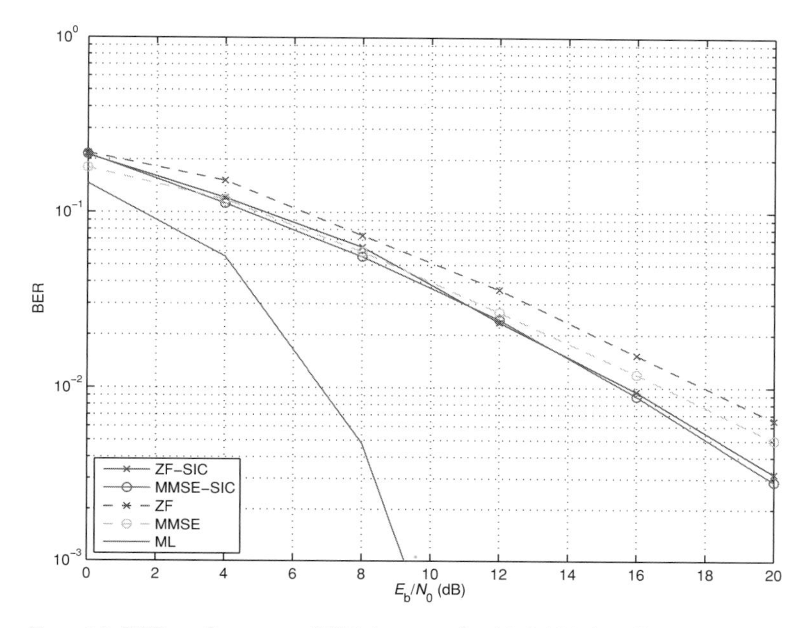

Figure 8.2 BER performance of SIC detectors for 16-QAM signaling over 4×4 MIMO channels.

to find the MMSE combining vector. In the square-root algorithm, however, it is not necessary to perform a matrix inversion for each layer and, therefore, the computational complexity becomes low.

Recall the MMSE estimate of $\mathbf{s}$, which is now denoted by $\hat{\mathbf{c}}_K$:

$$\begin{aligned}\hat{\mathbf{c}}_K &= \left(\mathbf{H}\mathbf{H}^{\mathrm{H}} + \frac{N_0}{E_{\mathrm{s}}}\mathbf{I}\right)^{-1} \mathbf{H}^{\mathrm{H}}\mathbf{r} \\ &= \left(\mathbf{H}\mathbf{H}^{\mathrm{H}} + \frac{N_0}{E_{\mathrm{s}}}\mathbf{I}\right)^{-1} \left[\mathbf{H}^{\mathrm{H}} \ \sqrt{\frac{N_0}{E_{\mathrm{s}}}}\mathbf{I}\right] \begin{bmatrix}\mathbf{r} \\ \mathbf{0}\end{bmatrix},\end{aligned} \tag{8.12}$$

where the second expression is an extended form. Define the following extended channel matrix as

$$\mathbf{H}_{\mathrm{ex}} = \begin{bmatrix}\mathbf{H} \\ c\mathbf{I}\end{bmatrix},$$

where $c = \sqrt{N_0/E_{\mathrm{s}}}$. Accordingly, the received signal vector can be extended by appending K zeros as

$$\begin{aligned}\mathbf{r}_{\mathrm{ex}} &= \begin{bmatrix}\mathbf{r} \\ \mathbf{0}\end{bmatrix} \\ &= \mathbf{H}_{\mathrm{ex}}\mathbf{s} + \mathbf{n}_{\mathrm{ex}},\end{aligned} \tag{8.13}$$

where

$$\mathbf{n}_{\text{ex}} = \begin{bmatrix} \mathbf{n} \\ -c\mathbf{s} \end{bmatrix}.$$

From (8.12), we can show that

$$\hat{\mathbf{s}}_K = \mathbf{H}_{\text{ex}}^{\dagger} \mathbf{r}_{\text{ex}}, \tag{8.14}$$

where † denotes the pseudo-inverse, since

$$\begin{aligned}
\mathbf{H}_{\text{ex}}^{\dagger} &= (\mathbf{H}_{\text{ex}}^{\text{H}} \mathbf{H}_{\text{ex}})^{-1} \mathbf{H}_{\text{ex}}^{\text{H}} \\
&= \left(\mathbf{H}\mathbf{H}^{\text{H}} + \frac{N_0}{E_{\text{s}}} \mathbf{I} \right)^{-1} \left[\mathbf{H}^{\text{H}} \ \sqrt{\frac{N_0}{E_{\text{s}}}} \mathbf{I} \right].
\end{aligned}$$

Note that the Kth row of $\bar{\mathbf{H}}^{\dagger}$ is identical to $\mathbf{w}_{\text{mmse-sic},K}^{\text{H}}$ in (8.11).

It can be shown that ZF-SIC detection with the extended channel matrix can be used to perform MMSE-SIC detection. Consider the QR factorization of $\mathbf{H}_{\text{ex}}$:

$$\mathbf{H}_{\text{ex}} = \mathbf{Q}_{\text{ex}} \mathbf{R}.$$

Premultiplying $\mathbf{Q}_{\text{ex}}^{\text{H}}$ to $\mathbf{r}_{\text{ex}}$ results in

$$\mathbf{Q}_{\text{ex}}^{\text{H}} \mathbf{r}_{\text{ex}} = \mathbf{R}\mathbf{s} + \mathbf{Q}_{\text{ex}}^{\text{H}} \mathbf{n}_{\text{ex}}.$$

Since the pseudo-inverse of $\mathbf{H}_{\text{ex}}$ is given by

$$\mathbf{H}_{\text{ex}}^{\dagger} = \mathbf{R}^{-1} \mathbf{Q}_{\text{ex}}^{\text{H}} \tag{8.15}$$

and $[\mathbf{R}^{-1}]_{k,k} = ([\mathbf{R}]_{k,k})^{-1}$ (because $\mathbf{R}$ is upper triangular), $k = 1, 2, \ldots, K$, the MMSE-SIC estimate of s_K can be found as follows:

$$\begin{aligned}
\hat{c}_K &= [\hat{\mathbf{c}}_K]_K \\
&= \mathbf{w}_{\text{mmse-sic},K}^{\text{H}} \mathbf{r} \\
&= ([\mathbf{R}]_{K,K})^{-1} [\mathbf{Q}_{\text{ex}}^{\text{H}} \mathbf{r}_{\text{ex}}]_K.
\end{aligned} \tag{8.16}$$

We can repeat the same approach to obtain $\hat{c}_k = \mathbf{w}_{\text{mmse-sic},k}^{\text{H}} \mathbf{r}_k$, $k = 1, 2, \ldots, K-1$, using the QR factorization. Define the following $N \times k$ MMSE combining matrix after the SIC with the last $K - k$ symbols, $\{s_{k+1}, s_{k+2}, \ldots, s_K\}$:

$$\begin{aligned}
\mathbf{W}_{\text{mmse-sic},k} &= \arg\min_{\mathbf{W}} \mathcal{E}[||\mathbf{s}_k - \mathbf{W}^{\text{H}} \mathbf{r}_k||^2] \\
&= \left(\mathbf{H}_k \mathbf{H}_k^{\text{H}} + \frac{N_0}{E_{\text{s}}} \mathbf{I} \right)^{-1} \mathbf{H}_k \\
&= \mathbf{H}_k \left(\frac{N_0}{E_{\text{s}}} \mathbf{I} + \mathbf{H}_k^{\text{H}} \mathbf{H}_k \right)^{-1},
\end{aligned}$$

where $\mathbf{s}_k = [s_1 \ s_2 \ \ldots \ s_k]^{\text{T}}$ and $\mathbf{H}_k = [\mathbf{h}_1 \ \mathbf{h}_2 \ \ldots \ \mathbf{h}_k]$. Then, we can readily show that the kth column vector of $\mathbf{W}_{\text{mmse-sic},k}$ is the same as $\mathbf{w}_{\text{mmse-sic},k}$ in (8.11) and the MMSE-SIC estimate of s_k is given by

$$\hat{c}_k = [\hat{\mathbf{c}}_k]_k,$$

where

$$\hat{\mathbf{c}}_k = \mathbf{W}^{\mathrm{H}}_{\text{mmse-sic},k}\mathbf{r}_k.$$

To find $\hat{\mathbf{c}}_k$ using the QR factorization, consider the following extended matrix:

$$\mathbf{H}_{\mathrm{ex},k} = \begin{bmatrix} \mathbf{H}_k \\ c\mathbf{I} \end{bmatrix}.$$

The QR factorization of the extended sub-channel matrix is given by

$$\begin{bmatrix} \mathbf{H}_k \\ c\mathbf{I} \end{bmatrix} = \mathbf{Q}_{\mathrm{ex},k}\mathbf{R}_k.$$

Then, it follows

$$\begin{aligned} \hat{c}_k &= [\hat{\mathbf{c}}_k]_k \\ &= \mathbf{w}^{\mathrm{H}}_{\text{mmse-sic},k}\mathbf{r}_k \\ &= [\mathbf{H}^{\dagger}_{\mathrm{ex},k}\mathbf{r}_{\mathrm{ex},k}]_k \\ &= ([\mathbf{R}_k]_{k,k})^{-1}[\mathbf{Q}^{\mathrm{H}}_{\mathrm{ex},k}\mathbf{r}_{\mathrm{ex},k}]_k. \end{aligned} \tag{8.17}$$

where

$$\mathbf{r}_{\mathrm{ex},k} = \begin{bmatrix} \mathbf{r}_k \\ \mathbf{0} \end{bmatrix}.$$

Although (8.17) shows that the MMSE-SIC estimate of s_k can be obtained using the QR factorization of the corresponding extended subchannel matrix, we do not see yet that this approach can be computationally efficient, because one QR factorization is required for each layer. However, this approach is useful in deriving a computationally efficient algorithm that requires only one QR factorization.

The key for the computationally efficient algorithm is the following result:

$$\begin{aligned} \begin{bmatrix} \mathbf{H}_k \\ c\mathbf{I} \end{bmatrix} &= [\mathbf{Q}_{\mathrm{ex}}]_{1:N+k,1:k}[\mathbf{R}]_{1:k,1:k} \\ &= \mathbf{Q}_{\mathrm{ex},k}\mathbf{R}_k, \ k = 1, 2, \ldots, K. \end{aligned} \tag{8.18}$$

This implies that the QR factorization of $\mathbf{H}_{\mathrm{ex}}$ is sufficient to find $\mathbf{Q}_{\mathrm{ex},k}$ and $\mathbf{R}_k$ for all k, because they are the submatrices of $\mathbf{Q}_{\mathrm{ex}}$ and $\mathbf{R}$. The rest of this section will be devoted to prove that (8.18) is true.

Let $\mathbf{P}_k = (c^2\mathbf{I} + \mathbf{H}_k^{\mathrm{H}}\mathbf{H}_k)^{-1}$. Then, we can show that

$$\begin{aligned} \mathbf{P}_k &= \left[(\mathbf{Q}_{\mathrm{ex},k}\mathbf{R}_k)^{\mathrm{H}}(\mathbf{Q}_{\mathrm{ex},k}\mathbf{R}_k)\right]^{-1} \\ &= (\mathbf{R}_k^{\mathrm{H}}\mathbf{R}_k)^{-1} \\ &= \mathbf{R}_k^{-1}\mathbf{R}_k^{-\mathrm{H}}. \end{aligned} \tag{8.19}$$

From this, we can see that $\mathbf{R}_k^{-1}$ is a square-root of $\mathbf{P}_k$. That is,

$$\mathbf{P}_k^{1/2} = \mathbf{R}_k^{-1}.$$

Thus,

$$\begin{aligned} \begin{bmatrix} \mathbf{H}_k \\ c\mathbf{I} \end{bmatrix} &= \mathbf{Q}_{\mathrm{ex},k}\mathbf{R}_k \\ &= \mathbf{Q}_{\mathrm{ex},k}\mathbf{P}_k^{-1/2}. \end{aligned} \tag{8.20}$$

Theorem 8.3.1 $\mathbf{P}_{k-1}^{1/2}$ *is a square-root factor of* $\mathbf{P}_k^{1/2}$. *That is,*

$$\mathbf{P}_k^{1/2} = \begin{bmatrix} \mathbf{P}_{k-1}^{1/2} & \mathbf{a} \\ \mathbf{0} & x \end{bmatrix}, \tag{8.21}$$

where $\mathbf{a}$ *and* x *are certain vector and constant, respectively.*

Proof: We can show that

$$\begin{aligned} \mathbf{P}_k^{-1} &= c\mathbf{I} + \mathbf{H}_k^{\mathrm{H}}\mathbf{H}_k \\ &= c\mathbf{I} + \begin{bmatrix} \mathbf{H}_{k-1}^{\mathrm{H}} \\ \mathbf{h}_k^{\mathrm{H}} \end{bmatrix} [\mathbf{H}_{k-1}\ \mathbf{h}_k] \\ &= \begin{bmatrix} c\mathbf{I} + \mathbf{H}_{k-1}^{\mathrm{H}}\mathbf{H}_{k-1} & \mathbf{H}_{k-1}^{\mathrm{H}}\mathbf{h}_k \\ \mathbf{h}_k^{\mathrm{H}}\mathbf{H}_{k-1} & c + \mathbf{h}_k^{\mathrm{H}}\mathbf{h}_k \end{bmatrix}. \end{aligned} \tag{8.22}$$

Note that

$$c\mathbf{I} + \mathbf{H}_{k-1}^{\mathrm{H}}\mathbf{H}_{k-1} = (\mathbf{P}_{k-1}^{1/2}\mathbf{P}_{k-1}^{\mathrm{H}/2})^{-1}. \tag{8.23}$$

Let

$$\mathbf{P}_k^{1/2} = \begin{bmatrix} \mathbf{A} & \mathbf{a} \\ \mathbf{0} & x \end{bmatrix},$$

where $\mathbf{A}$ is upper triangular. Then, we have

$$\begin{aligned} \mathbf{P}_k^{-1} &= (\mathbf{P}_k^{1/2}\mathbf{P}_k^{\mathrm{H}/2})^{-1} \\ &= \mathbf{P}_k^{-\mathrm{H}/2}\mathbf{P}_k^{-1/2} \\ &= \begin{bmatrix} \mathbf{A}^{\mathrm{H}} & \mathbf{0} \\ \mathbf{a}^{\mathrm{H}} & x^* \end{bmatrix}^{-1} \begin{bmatrix} \mathbf{A} & \mathbf{a} \\ \mathbf{0} & x \end{bmatrix}^{-1} \\ &= \begin{bmatrix} \mathbf{A}^{-\mathrm{H}} & \mathbf{0} \\ \times & \times \end{bmatrix} \begin{bmatrix} \mathbf{A}^{-1} & \times \\ \mathbf{0} & \times \end{bmatrix} \\ &= \begin{bmatrix} (\mathbf{A}\mathbf{A}^{\mathrm{H}})^{-1} & \times \\ \times & \times \end{bmatrix}, \end{aligned} \tag{8.24}$$

where $\times$ denotes the terms that are not relevant. From (8.22), (8.23), and (8.24), it follows that

$$(\mathbf{A}\mathbf{A}^{\mathrm{H}})^{-1} = c\mathbf{I} + \mathbf{H}_{k-1}^{\mathrm{H}}\mathbf{H}_{k-1} = (\mathbf{P}_{k-1}^{1/2}\mathbf{P}_{k-1}^{\mathrm{H}/2})^{-1}.$$

Thus, $\mathbf{A} = \mathbf{P}_{k-1}^{1/2}$ is a square-root of $c\mathbf{I} + \mathbf{H}_{k-1}^{\mathrm{H}}\mathbf{H}_{k-1}$ and (8.21) holds. □

Let $\mathbf{P}^{1/2} = \mathbf{R}^{-1}$. From (8.21), we can show that

$$\begin{aligned} [\mathbf{P}^{1/2}]_{1:k,1:k} &= \mathbf{P}_k^{1/2} \\ [\mathbf{P}^{-1/2}]_{1:k,1:k} &= [\mathbf{R}]_{1:k,1:k} = \mathbf{P}_k^{-1/2}. \end{aligned} \tag{8.25}$$

Theorem 8.3.2 *(8.18) is true.*

Proof: Note that

$$\begin{aligned} \begin{bmatrix} \mathbf{H} \\ c\mathbf{I} \end{bmatrix} = \begin{bmatrix} \mathbf{H}_k & \times \\ c\mathbf{I} & \times \\ \mathbf{0} & \times \end{bmatrix} &= \mathbf{Q}_{\mathrm{ex}}\mathbf{R} \\ &= \mathbf{Q}_{\mathrm{ex}}\mathbf{P}^{-1/2} \\ &= \mathbf{Q}_{\mathrm{ex}} \begin{bmatrix} \mathbf{P}_k^{1/2} & \times \\ \mathbf{0} & \times \end{bmatrix}^{-1} \\ &= \mathbf{Q}_{\mathrm{ex}} \begin{bmatrix} \mathbf{P}_k^{-1/2} & \times \\ \mathbf{0} & \times \end{bmatrix}. \end{aligned} \tag{8.26}$$

Finally, using (8.25), we can show that

$$\begin{aligned} \begin{bmatrix} \mathbf{H}_k \\ c\mathbf{I} \end{bmatrix} &= [\mathbf{Q}_{\mathrm{ex}}]_{1:N+k,:} \begin{bmatrix} \mathbf{P}_k^{-1/2} \\ \mathbf{0} \end{bmatrix} \\ &= [\mathbf{Q}_{\mathrm{ex}}]_{1:N+k,1:k}\mathbf{P}_k^{-1/2} \\ &= [\mathbf{Q}_{\mathrm{ex}}]_{1:N+k,1:k}[\mathbf{R}]_{1:k,1:k}. \end{aligned} \tag{8.27}$$

This completes the proof. □

From the results above, we can see that MMSE-SIC detection can be performed by ZF-SIC detection with $\mathbf{H}_{\mathrm{ex}} = \mathbf{Q}_{\mathrm{ex}}\mathbf{R}$ and $\mathbf{r}_{\mathrm{ex}}$. Since only one QR factorization is required, this approach becomes computationally efficient compared with the straightforward implementation in Subsection 8.2.2. Furthermore, this approach becomes a generalized SIC since it becomes the ZF-SIC detector when $c = 0$ and MMSE-SIC detector when $c = \sqrt{N_0/E_{\mathrm{s}}}$.

8.4 Principle of partial MAP detection

Although SIC detection is computationally efficient, it is suboptimal and its performance could be much worse than the ML performance due to error propagation. However, under a certain condition, SIC detection can be used to solve the ML detection problem. In this section, we derive this condition.

If the channel matrix is upper triangular, the ML detection problem can be converted into a partial MAP detection problem, which could be solved by the SIC detection approach. To illustrate the idea, consider the received signal vector over a 2×2 upper triangular channel matrix from (8.6) as follows:

$$\begin{bmatrix} x_1 \\ x_2 \end{bmatrix} = \begin{bmatrix} r_{1,1} & r_{1,2} \\ 0 & r_{2,2} \end{bmatrix} \begin{bmatrix} s_1 \\ s_2 \end{bmatrix} + \begin{bmatrix} n_1 \\ n_2 \end{bmatrix}. \tag{8.28}$$

Since the channel matrix is upper triangular, the ML detection problem for (8.28) is given by

$$\begin{aligned} \{\hat{s}_{\mathrm{ml},1}, \hat{s}_{\mathrm{ml},2}\} &= \arg \max_{\{s_1, s_2\}} f(x_1, x_2 | s_1, s_2) \\ &= \arg \max_{\{s_1, s_2\}} f(x_1 | s_1, s_2) f(x_2 | s_2) \end{aligned} \tag{8.29}$$

if n_1 and n_2 are independent. Suppose that s_1 and s_2 are equally likely. With x_2, we can find the a posteriori probability of s_2 as follows:

$$\Pr(s_2) = \frac{f(x_2 | s_2)}{\sum_{s_2 \in \mathcal{S}} f(x_2 | s_2)}. \tag{8.30}$$

Now, taking $\Pr(s_2)$ as a priori probability of s_2, we can formulate a detection problem to detect s_1 and s_2 with the observation x_1 from (8.29) as follows:

$$\begin{aligned} \{\hat{s}_{\mathrm{ml},1}, \hat{s}_{\mathrm{ml},2}\} &= \arg \max_{\{s_1, s_2\}} f(x_1 | s_1, s_2) \Pr(s_2) \\ &= \arg \max_{\{s_1, s_2\}} f(x_1 | s_1, s_2) \Pr(s_2) \Pr(s_1) \\ &= \arg \max_{\{s_1, s_2\}} \Pr(s_1, s_2 | x_1). \end{aligned}$$

From this, we can see that the ML detection problem to detect $\{s_1, s_2\}$ from $\{x_1, x_2\}$ becomes a partial MAP detection problem to detect $\{s_1, s_2\}$ from $\{x_1\}$, where there is no prior information of s_1, while the a priori probability of s_2 is available.

The approach in the above can be generalized to any block upper triangular channels. Consider the received signal vector as follows:

$$\begin{bmatrix} \mathbf{x}_1 \\ \mathbf{x}_2 \end{bmatrix} = \begin{bmatrix} \mathbf{R}_{1,1} & \mathbf{R}_{1,2} \\ \mathbf{0} & \mathbf{R}_{2,2} \end{bmatrix} \begin{bmatrix} \mathbf{s}_1 \\ \mathbf{s}_2 \end{bmatrix} + \begin{bmatrix} \mathbf{n}_1 \\ \mathbf{n}_2 \end{bmatrix},$$

where $\mathbf{n}_1$ and $\mathbf{n}_2$ are independent. Let

$$\Pr(\mathbf{s}_2) = \frac{f(\mathbf{x}_2 | \mathbf{s}_2)}{\sum_{\mathbf{s}_2} f(\mathbf{x}_2 | \mathbf{s}_2)}.$$

Then, to solve the ML detection problem, we can define the following partial MAP detection problem:

$$\{\hat{\mathbf{s}}_{\mathrm{ml},1}, \hat{\mathbf{s}}_{\mathrm{ml},2}\} = \arg \max_{\{\mathbf{s}_1, \mathbf{s}_2\}} f(\mathbf{x}_1 | \mathbf{s}_1, \mathbf{s}_2) \Pr(\mathbf{s}_2), \tag{8.31}$$

where $\mathbf{s}_1$ and $\mathbf{s}_2$ are the symbol vectors to be detected.

The partial MAP detection problem in (8.31), can be solved using the notion of SIC. To show this, consider the following received signal:

$$\mathbf{x}_1 = \mathbf{A}_1\mathbf{s}_1 + \mathbf{A}_2\mathbf{s}_2 + \mathbf{n},$$

where $\mathbf{A}_1$ and $\mathbf{A}_2$ are the channel matrices for the input vectors $\mathbf{s}_1$ and $\mathbf{s}_2$, respectively, and $\mathbf{n} \sim \mathcal{CN}(\mathbf{0}, N_0\mathbf{I})$. The likelihood function of $\mathbf{s}_1$ and $\mathbf{s}_2$ is given by

$$f(\mathbf{x}_1|\mathbf{s}_1, \mathbf{s}_2) \propto \exp\left(-\frac{1}{N_0}||\mathbf{x}_1 - (\mathbf{A}_1\mathbf{s}_1 + \mathbf{A}_2\mathbf{s}_2)||^2\right).$$

The partial MAP detection problem is formulated as follows:

$$\begin{aligned}\{\hat{\mathbf{s}}_{\text{pmap},1}, \hat{\mathbf{s}}_{\text{pmap},2}\} &= \arg\max_{\{\mathbf{s}_1,\mathbf{s}_2\}} f(\mathbf{x}_1|\mathbf{s}_1, \mathbf{s}_2)\Pr(\mathbf{s}_2) \\ &= \arg\min_{\{\mathbf{s}_1,\mathbf{s}_2\}} \left\{\frac{1}{N_0}||\mathbf{x}_1 - (\mathbf{A}_1\mathbf{s}_1 + \mathbf{A}_2\mathbf{s}_2)||^2 - \log\Pr(\mathbf{s}_2)\right\}. \quad (8.32)\end{aligned}$$

In this section, logarithms are taken to the base e.

Theorem 8.4.1 *Suppose that $\mathbf{s}_1$ and $\mathbf{s}_2$ are independent and $\mathbf{s}_1$ is equally likely. Let $\hat{\mathbf{s}}_2 = \arg\max_{\mathbf{s}_2}\Pr(\mathbf{s}_2)$ and suppose that*

$$\min_{\mathbf{s}_1,\mathbf{s}_2}\frac{1}{N_0}||\mathbf{x}_1 - (\mathbf{A}_1\mathbf{s}_1 + \mathbf{A}_2\mathbf{s}_2)||^2 \geq C, \quad (8.33)$$

where C is a constant. In addition, let $\mathbf{x}_{12} = \mathbf{x}_1 - \mathbf{A}_2\hat{\mathbf{s}}_2$. If

$$\min_{\mathbf{s}_1}\frac{1}{N_0}||\mathbf{x}_{12} - \mathbf{A}_1\mathbf{s}_1||^2 \leq C + \min_{\mathbf{s}_2\neq\hat{\mathbf{s}}_2}\log\frac{\Pr(\hat{\mathbf{s}}_2)}{\Pr(\mathbf{s}_2)}, \quad (8.34)$$

then the partial MAP detection problem in (8.32) *has the following solution:*

$$\begin{aligned}\hat{\mathbf{s}}_{\text{pmap},1} &= \arg\min_{\mathbf{s}_1}\frac{1}{N_0}||\mathbf{x}_{12} - \mathbf{A}_1\mathbf{s}_1||^2; \\ \hat{\mathbf{s}}_{\text{pmap},2} &= \hat{\mathbf{s}}_2. \quad (8.35)\end{aligned}$$

Proof: To prove it, we need to show the following inequality holds:

$$\begin{aligned}&\min_{\mathbf{s}_1}\tfrac{1}{N_0}||\mathbf{x}_1 - (\mathbf{A}_1\mathbf{s}_1 + \mathbf{A}_2\hat{\mathbf{s}}_2)||^2 + \log\tfrac{1}{\Pr(\hat{\mathbf{s}}_2)} \\ &\leq \min_{\mathbf{s}_1,\mathbf{s}_2\neq\hat{\mathbf{s}}_2}\tfrac{1}{N_0}||\mathbf{x}_1 - (\mathbf{A}_1\mathbf{s}_1 + \mathbf{A}_2\mathbf{s}_2)||^2 + \log\tfrac{1}{\Pr(\mathbf{s}_2)}. \quad (8.36)\end{aligned}$$

If the above inequality is true, this implies that $\hat{\mathbf{s}}_2$ is the optimal solution. Since

$$\min_{\mathbf{s}_1,\mathbf{s}_2\neq\hat{\mathbf{s}}_2}\frac{1}{N_0}||\mathbf{x}_1 - (\mathbf{A}_1\mathbf{s}_1 + \mathbf{A}_2\mathbf{s}_2)||^2 + \log\frac{1}{\Pr(\mathbf{s}_2)} \geq C + \min_{\mathbf{s}_2\neq\hat{\mathbf{s}}_2}\log\frac{1}{\Pr(\mathbf{s}_2)},$$

a sufficient condition for the inequality in (8.36) is

$$\min_{\mathbf{s}_1}\frac{1}{N_0}||\mathbf{x}_1 - (\mathbf{A}_1\mathbf{s}_1 + \mathbf{A}_2\hat{\mathbf{s}}_2)||^2 + \log\frac{1}{\Pr(\hat{\mathbf{s}}_2)} \leq C + \min_{\mathbf{s}_2\neq\hat{\mathbf{s}}_2}\log\frac{1}{\Pr(\mathbf{s}_2)}$$

or

$$\min_{\mathbf{s}_1} \frac{1}{N_0} ||\mathbf{x}_1 - (\mathbf{A}_1\mathbf{s}_1 + \mathbf{A}_2\hat{\mathbf{s}}_2)||^2 \leq C + \min_{\mathbf{s}_2 \neq \hat{\mathbf{s}}_2} \log \frac{\Pr(\hat{\mathbf{s}}_2)}{\Pr(\mathbf{s}_2)}.$$

This completes the proof. □

Note that C in (8.33) can be 0 as the term on the left-hand side in (8.33) is always greater than 0. Then, the sufficient condition in (8.34) becomes

$$\min_{\mathbf{s}_1} \frac{1}{N_0} ||\mathbf{x}_1 - \mathbf{A}_1\mathbf{s}_1 - \mathbf{A}_2\hat{\mathbf{s}}_2||^2 \leq \min_{\mathbf{s}_2 \neq \hat{\mathbf{s}}_2} \log \frac{\Pr(\hat{\mathbf{s}}_2)}{\Pr(\mathbf{s}_2)}.$$

According to Theorem 8.4.1, if the condition in (8.34) is satisfied, the partial MAP detection problem can be solved as follows:

$$\begin{aligned} \hat{\mathbf{s}}_{\text{pmap},2} &= \arg\max_{\mathbf{s}_2} \Pr(\mathbf{s}_2); \\ \hat{\mathbf{s}}_{\text{pmap},1} &= \arg\max_{\mathbf{s}_1} f(\mathbf{x}_1|\mathbf{s}_1, \hat{\mathbf{s}}_2). \end{aligned} \tag{8.37}$$

They are also the ML solution when $\Pr(\mathbf{s}_2)$ is the a posteriori probability of $\mathbf{s}_2$ for given $\mathbf{x}_2$. This implies that SIC detection can solve the ML detection problem if (8.34) holds. Thus, the condition (8.34) can be regarded as the SIC optimality condition. For a detailed account, the reader is referred to (Choi 2005*b*).

Example 8.4.1 In this example, assume that all the quantities are real-valued for convenience. Let $s_1, s_2 = \mathcal{S} = \{-3, -1, 1, 3\}$. The received signal is given by

$$\begin{bmatrix} x_1 \\ x_2 \end{bmatrix} = \begin{bmatrix} 1 & \frac{3}{4} \\ 0 & \frac{1}{4} \end{bmatrix} \begin{bmatrix} s_1 \\ s_2 \end{bmatrix} + \begin{bmatrix} n_1 \\ n_2 \end{bmatrix}, \tag{8.38}$$

where $n_q \sim \mathcal{N}(0, \sigma_{\text{n}}^2)$ and $\sigma_{\text{n}}^2 = 0.5$. If $\mathbf{x} = [x_1\ x_2]^{\text{T}} = [-1.5\ 1.2]^{\text{T}}$, after an exhaustive search, the ML solution is obtained as follows:

$$\begin{aligned} \mathbf{s}_{\text{ml}} &= \arg\min_{s_1, s_2} ||\mathbf{x} - \mathbf{R}\mathbf{s}||^2 \\ &= [-3\ 3]^{\text{T}}. \end{aligned}$$

To apply the partial MAP, we need to verify the condition in (8.34) with $C = 0$. Since

$$\Pr(s_2) \propto \exp\left(-\frac{1}{2\sigma_{\text{n}}^2}|x_2 - r_{2,2}s_2|^2\right),$$

we have

$$\min_{s_2 \neq \hat{s}_2} \log \frac{\Pr(\hat{s}_2)}{\Pr(s_2)} = 0.7,$$

where $\hat{s}_2 = 3$. Furthermore,

$$\min_{s_1 \in \mathcal{S}} \frac{1}{2\sigma_{\text{n}}^2} |x_1 - r_{1,1}s_1 - r_{1,2}\hat{s}_2|^2 = 0.5625.$$

This minimum is achieved when $\hat{s}_1 = -3$. As shown above, we can check that the condition in (8.34) is satisfied. This implies that the ML solution is $\mathbf{s}_{\text{ml}} = [-3 \ 3]^{\text{T}}$, which agrees with the result obtained by an exhaustive search.

8.5 Performance analysis of SIC detection

Although SIC detectors have low computational complexity, they suffer from error propagation in general, which leads to a poor performance. In this section, we analyze the performance of SIC detectors to understand the impact of error propagation for uncoded signals. In particular, we focus on the ZF-SIC detector over Rayleigh fading channels as its analysis is tractable.

We assume that each element of $\mathbf{H}$ is an independent CSCG random variable with mean zero and unit variance, i.e. $[\mathbf{H}]_{n,k} \sim \mathcal{CN}(0, 1)$. It is also assumed that $\mathbf{n} \sim \mathcal{CN}(\mathbf{0}, N_0\mathbf{I})$. The following result from (Edelman 1989) is necessary to find average error probability.

Lemma 8.5.1 *Suppose that each element of* $\mathbf{H}$ *is an independent CSCG random variable with mean zero and unit variance. Consider the QR factorization of* $\mathbf{H}$ *of size* $N \times K$*, where* $N \geq K$*, as*

$$\mathbf{H} = \mathbf{QR}.$$

Then, we have the following results.

(i) The squared absolute values of the diagonal elements of $\mathbf{R}$*,* $|r_{m,m}|^2$ *are independent chi-square distributed random variables with* $2(N - m + 1)$ *degrees of freedom and the pdf is given by*

$$f(z = |r_{m,m}|^2) = \frac{1}{(N-m)!} z^{N-m} \mathrm{e}^{-z}, \ z \geq 0. \tag{8.39}$$

(ii) The upper off-diagonal elements $r_{m,l}$*,* $l < m$*, are independent CSCG random variables with mean zero and unit variance.*

Consider a 2×2 MIMO system for illustrating an approach to find average error probability. Recall the received signal in (8.28) with $s_k \in \{\pm\sqrt{E_k}\}$ (i.e., BPSK for modulation). We assume that $r_{1,1}$, $r_{2,2}$, and $r_{1,2}$ are mutually independent (this is true for Rayleigh fading channels as shown in Lemma 8.5.1). The bit error probability in detecting s_2 with x_2 is given by

$$P_2(r_{2,2}) = \mathcal{Q}\left(\sqrt{\frac{2E_2|r_{2,2}|^2}{N_0}}\right). \tag{8.40}$$

For convenience, define the average bit error probability of s_2 as

$$P_2 = \mathcal{E}[P_2(r_{2,2})],$$

where the expectation is carried out with respect to $r_{2,2}$. The detection performance of s_1 with x_1 depends on the bit error probability of s_2 due to error propagation. Provided that s_1 is correctly detected, the bit error probability of s_1 is given by

$$P_{1,0}(r_{1,1}) = \mathcal{Q}\left(\sqrt{\frac{2E_1|r_{1,1}|^2}{N_0}}\right). \tag{8.41}$$

However, if s_2 is erroneously detected, the signal after cancellation is given by

$$x_1 - r_{1,2}\hat{s}_2 = r_{1,1}s_1 \pm 2r_{1,2}\sqrt{E_2} + n_1.$$

Let

$$\bar{n}_1 = \pm 2r_{1,2}\sqrt{E_2} + n_1.$$

According to Lemma 8.5.1, $r_{1,2} \sim \mathcal{CN}(0, 1)$. Therefore, we have

$$\bar{n}_1 \sim \mathcal{CN}(0, 4E_2 + N_0).$$

From this, if s_2 is erroneously detected, the error probability of s_1 can be obtained as follows:

$$P_{1,1}(r_{1,1}) = \mathcal{Q}\left(\sqrt{\frac{2E_1|r_{1,1}|^2}{4E_2 + N_0}}\right). \tag{8.42}$$

Finally, the average error probability of s_1 is

$$\begin{aligned} P_1 &= \mathcal{E}[P_{1,0}(r_{1,1})(1 - P_2) + P_{1,1}(r_{1,1})P_2] \\ &= \mathcal{E}[P_{1,0}(r_{1,1})](1 - P_2) + \mathcal{E}[P_{1,1}(r_{1,1})]P_2, \end{aligned}$$

where the expectation is carried out with respect to $r_{1,1}$. Certainly, this error probability is greater than that in (8.41) (which is the error probability without error propagation) and shows that error propagation increases the error probability.

To clearly see the impact of error propagation, the diversity gain can be considered using the Chernoff bound. From (7.24), we can have an upper bound on P_2 as

$$\begin{aligned} P_2 &= \mathcal{E}\left[\mathcal{Q}\left(\sqrt{\frac{2E_2|r_{2,2}|^2}{N_0}}\right)\right] \\ &\le \mathcal{E}\left[\exp\left(-\frac{E_2|r_{2,2}|^2}{N_0}\right)\right] \\ &= (1 + \bar{\gamma}_2)^{-1}, \end{aligned}$$

where $\bar{\gamma}_k = E_k/N_0$. Note that according to Lemma 8.5.1, $|r_{2,2}|^2$ is exponentially distributed. In addition, we can show that

$$\begin{aligned} P_{1,0} &= \mathcal{E}[P_{1,0}(r_{1,1})] \\ &\le (1 + \bar{\gamma}_{1,0})^{-2}; \\ P_{1,1} &= \mathcal{E}[P_{1,1}(r_{1,1})] \\ &\le (1 + \bar{\gamma}_{1,1})^{-2}, \end{aligned}$$

where

$$\bar{\gamma}_{1,0} = \frac{E_1}{N_0} = \bar{\gamma}_1;$$

$$\bar{\gamma}_{1,1} = \frac{E_1}{4E_2 + N_0} = \frac{\bar{\gamma}_1}{4\bar{\gamma}_2 + 1}.$$

It follows that

$$\begin{aligned} P_1 &\le (1 - P_2)(1 + \bar{\gamma}_{1,0})^{-2} + P_2(1 + \bar{\gamma}_{1,1})^{-2} \\ &= \left(\frac{\bar{\gamma}_2}{1+\bar{\gamma}_2}\right)\left(\frac{1}{1+\bar{\gamma}_{1,0}}\right)^2 + \left(\frac{1}{1+\bar{\gamma}_2}\right)\left(\frac{1}{1+\bar{\gamma}_{1,1}}\right)^2. \end{aligned} \tag{8.43}$$

If $E_1 = E_2 = E_b$ and $\bar{\gamma} = E_b/N_0$ grows, we can show that the second term on the right-hand side in (8.43) becomes dominant and

$$\left(\frac{\bar{\gamma}_2}{1+\bar{\gamma}_2}\right)\left(\frac{1}{1+\bar{\gamma}_{1,0}}\right)^2 + \left(\frac{1}{1+\bar{\gamma}_2}\right)\left(\frac{1}{1+\bar{\gamma}_{1,1}}\right)^2 \simeq \left(\frac{4}{5}\right)^2 \bar{\gamma}^{-1}, \quad \bar{\gamma} \gg 1.$$

From this result, we can observe that the diversity order of layer 1 is the same as that of layer 2 due to error propagation, although the error probability of layer 1 signals is slightly lower than that of layer 2 signals. As a result, the overall diversity order is the same as that of layer 2. This observation is also applicable to a general case as shown in (Choi 2005*a*).

Figure 8.3 shows the average BER for each layer when $N = K = 4$. The results show that each layer has the same diversity gain that is decided by the lowest layer (in which signals are detected first).

8.6 Approaches to mitigate error propagation

Various approaches are proposed for SIC detectors to overcome error propagation. In this section, we introduce some approaches.

8.6.1 Ordering

Since the overall performance of SIC detection is limited by the first detection as shown in Section 8.5, the symbol that can be the most reliably detected could be detected first. This implies that the performance could be improved by optimizing detection order.

In MMSE-SIC detection, the first symbol to be detected is the symbol that has the smallest MSE or (equivalently) highest SINR:

$$\{\alpha(1), \mathbf{w}_{\alpha(1)}\} = \arg\min_k \min_{\mathbf{w}} \mathcal{E}[|s_k - \mathbf{w}^{\mathrm{H}}\mathbf{r}|^2],$$

where $\alpha(1)$ denotes the index of the data symbol that has the smallest MSE and $\mathbf{w}_{\alpha(1)}$ denotes the corresponding MMSE combining vector. Then, the cancellation is carried

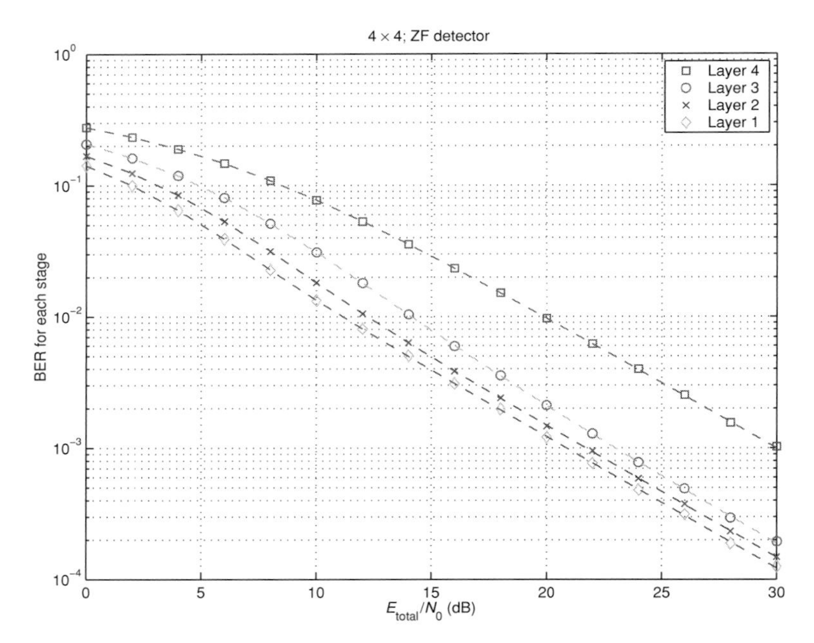

Figure 8.3 BER performance of the ZF-SIC detector for each layer (4 transmit and 4 receive antennas; marks represent the simulation results, while dashed curves represent the theoretical results).

out as follows:

$$\mathbf{r}_1 = \mathbf{r}_0 - \mathbf{h}_{\alpha(1)}\hat{s}_{\alpha(1)}, \tag{8.44}$$

where $\mathbf{r}_0 = \mathbf{r}$ and $\hat{s}_{\alpha(1)}$ denotes a hard-decision of $s_{\alpha(1)}$ from $\mathbf{w}^H_{\alpha(1)}\mathbf{r}_0$. With $\mathbf{r}_1$, the next symbol to be detected is found as

$$\{\alpha(2), \mathbf{w}_{\alpha(2)}\} = \arg \min_{k \in I_1} \min_{\mathbf{w}} \mathcal{E}[|s_k - \mathbf{w}^{\mathrm{H}}\mathbf{r}_1|^2],$$

where $I_1 = I_0 \setminus \alpha(1)$ and $I_0 = \{1, 2, \ldots, K\}$. The cancellation and MMSE combining can be repeated until all symbols are detected. The resulting SIC detector is called the ordered MMSE-SIC detector. It can be summarized as follows.

(S0) Let $\mathbf{r}_0 = \mathbf{r}$ and $I_0 = \{1, 2, \ldots, K\}$. Set $k = 1$.

(S1) Solve the following optimization problem:

$$\{\alpha(k), \mathbf{w}_{\alpha(k)}\} = \arg \min_{q \in I_{k-1}} \min_{\mathbf{w}} \mathcal{E}[|s_q - \mathbf{w}^{\mathrm{H}}\mathbf{r}_{k-1}|^2], \tag{8.45}$$

(S2) Cancel the detected signal:

$$\mathbf{r}_k = \mathbf{r}_{k-1} - \mathbf{h}_{\alpha(k)}\hat{s}_{\alpha(k)} \tag{8.46}$$

and update $I_k = I_{k-1} \setminus \alpha(k)$.

(S3) If $k = K$, stop. Otherwise, $k = k + 1$ and go to (S1).

This ordered MMSE-SIC detector is proposed in (Wolniansky, Foschini, Golden & Valenzuela 1998). A detailed discussion on this SIC detector and its application can be found in (Foschini *et al.* 2003).

Example 8.6.1 Let $\mathcal{S} = \{-1, +1\}$. Consider a 3×3 MIMO channel as follows:

$$\mathbf{H} = \begin{bmatrix} 1 & \frac{1}{2} & 2 \\ 2 & -1 & 1 \\ 0 & 1 & 3 \end{bmatrix}.$$

With $N_0 = 1$, the MMSE combining vectors can be found as

$$\begin{aligned} \mathbf{W}_{\text{mmse}} &= \left[\mathbf{w}_{\text{mmse},1}\ \mathbf{w}_{\text{mmse},2}\ \mathbf{w}_{\text{mmse},3}\right] \\ &= (\mathbf{H}\mathbf{H}^{\text{H}} + \mathbf{I})^{-1}\mathbf{H} \\ &= \begin{bmatrix} 0.1884 & 0.2013 & 0.0428 \\ 0.2355 & -0.2484 & 0.0535 \\ -0.1542 & 0.0171 & 0.2377 \end{bmatrix}. \end{aligned}$$

The MSE's are found as $\{0.3405, 0.6338, 0.1478\}$. Since the MSE of s_3 is the smallest, which is 0.1478, we have $\alpha(1) = 3$, i.e. the third signal is chosen for the first detection and cancellation. Then, after canceling the third signal, the MMSE combining vectors for the first and second signals are found as

$$\begin{aligned} \mathbf{W}_{\text{mmse}} &= \left[\mathbf{w}_{\text{mmse},1}\ \mathbf{w}_{\text{mmse},2}\right] \\ &= ([\mathbf{h}_1\ \mathbf{h}_2][\mathbf{h}_1\ \mathbf{h}_2]^{\text{H}} + \mathbf{I})^{-1}[\mathbf{h}_1\ \mathbf{h}_2] \\ &= \begin{bmatrix} 0.2319 & 0.2609 \\ 0.2899 & -0.1739 \\ 0.0870 & 0.3478 \end{bmatrix}. \end{aligned}$$

The MSE's are given by $\{0.1884, 0.3478\}$. Then, the first signal is chosen for the next (i.e., second) detection and cancellation, i.e., $\alpha(2) = 1$. As $\alpha(3) = 2$ (the second signal is the last signal to be detected), the MMSE combining vector for the second signal is now given by

$$\mathbf{w}_{\text{mmse},2} = (\mathbf{h}_2\mathbf{h}_2^{\text{H}} + \mathbf{I})^{-1}\mathbf{h}_2 = [0.1538\ \ -0.3077\ 0.3077]^{\text{T}}.$$

The MSE becomes 0.3077.

We note that, without cancellation, the MSE's are $\{0.3405, 0.6338, 0.1478\}$ which are given in the first stage. If the cancellation was perfect, the MSE's of the MMSE-SIC detector become $\{0.3077, 0.1884, 0.1478\}$.

Although the optimal ordering can improve the performance, it does not improve the diversity order as the ordering cannot completely avoid error propagation.

8.6.2 SIC-list detection

To mitigate error propagation, multiple candidate symbols (rather than a single detected symbol) can be used for SIC. To illustrate this approach, consider a 2×2 MIMO system whose received signal is given in (8.28). Let $s_k \in \mathcal{S} = \{s^{(1)}, s^{(2)}, \ldots, s^{(M)}\}$. Using the ML criterion, let

$$|x_2 - r_{2,2}\hat{s}_{2,(1)}|^2 \leq |x_2 - r_{2,2}\hat{s}_{2,(2)}|^2 \leq \ldots \leq |x_2 - r_{2,2}\hat{s}_{2,(M)}|^2, \ \hat{s}_{2,(m)} \in \mathcal{S},$$

where $\hat{s}_{2,(m)}$ denotes the mth likely signal for s_2. Clearly, $\hat{s}_{2,(1)}$ is the ML solution and the cancellation is carried out with $s_2 = \hat{s}_{2,(1)}$ to detect s_1 in the conventional ZF-SIC detection as follows:

$$\hat{s}_1 = \arg\min_{s \in \mathcal{S}} |x_1 - r_{1,1}s - r_{1,2}\hat{s}_{2,(1)}|^2.$$

The probability of error propagation is the same as the error probability of s_2. To minimize the probability of error propagation, more candidates can be taken into account for cancellation. For example, a list of $\bar{M}$ ($\leq M$) candidate symbols can be considered for cancellation. Then, the detection for s_1 becomes

$$\hat{s}_{1,(m)} = \arg\min_{s \in \mathcal{S}} |x_1 - r_{1,1}s - r_{1,2}\hat{s}_{2,(m)}|^2, \ m = 1, 2, \ldots, \bar{M}.$$

As a result, we have a list (or set) of the following $\bar{M}$ candidate pairs:

$$\{(\hat{s}_{1,(1)}, \hat{s}_{2,(1)}), (\hat{s}_{1,(2)}, \hat{s}_{2,(2)}), \ldots, (\hat{s}_{1,(\bar{M})}, \hat{s}_{2,(\bar{M})})\}.$$

Among the $\bar{M}$ candidates, the best pair that minimizes the squared distance, $||\mathbf{x} - \mathbf{R}\mathbf{s}||^2$, can be used as a hard-decision. As $\bar{M}$ increases, error propagation is more mitigated and it leads to a better performance. It can be easily shown that this best pair becomes the ML solution if $\bar{M} = M$.

SIC detection with a list of multiple symbols as above can be generalized to any size of MIMO channels. This approach is an example of the QR decomposition/M-algorithm (QRD-M) to MIMO detection (Choi 2006) (Kim, Yue, Iltis & Gibson 2005).

8.6.3 Channel coding

Another approach to mitigate error propagation in SIC is to use channel codes. To effectively perform SIC with a sufficiently low probability of error propagation, each layer should transmit an independently encoded signal as shown in Fig. 8.4. At the receiver, once all the coded signals are received, SIC starts. The signal sequence in the lowest layer is decoded first. After decoding, the decoded signal in the lowest layer (i.e. layer K) is stripped off from the received signals and then the signal in the second lowest layer is decoded. This operation is repeated to the highest layer (i.e. layer 1). Since the error probability can be significantly reduced by channel coding, compared to the error probability of uncoded signals, this results in a low probability of error propagation.

In (Foschini 1996), (Foschini, Golden, Valenzuela & Wolniansky 1999), and (Foschini *et al.* 2003), various space-time layered approaches, called Bell Labs layered space-time (BLAST) architectures, are proposed to exploit spatial multiplexing and/or transmit

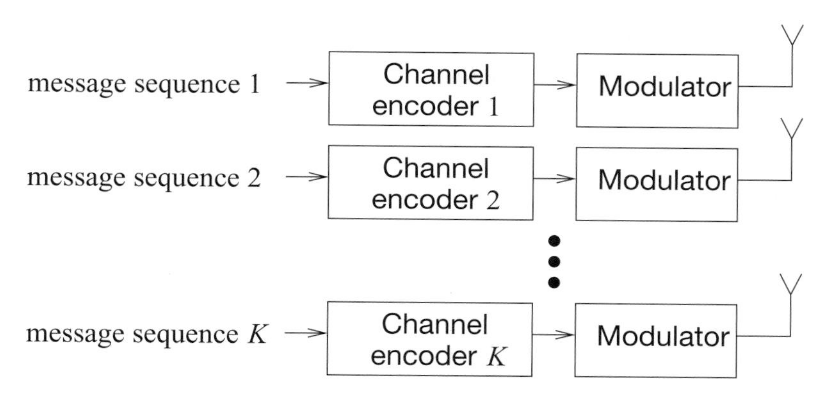

Figure 8.4 Block diagram for layered transmission with coded signals.

diversity of MIMO channels. Among them, the horizontal BLAST (H-BLAST) has the transmitter structure which is identical to that in Fig. 8.4. A slightly different version of H-BLAST is diagonal BLAST (D-BLAST), in which an encoded signal sequence is transmitted diagonally across transmit antennas. Each coded signal in D-BLAST can fully exploit spatial diversity gain. However, D-BLAST suffers from rate loss.

8.7 SIC receiver for coded signals

In this section, we discuss the performance of SIC from an information-theoretic point of view where the performance is considered with mutual information. It is assumed that the layered transmission in Fig. 8.4 is employed with capacity-achieving channel codes.

8.7.1 Layered transmission for SIC receiver

Throughout this section, we assume that each signal can have a different power, which is denoted by $E_k = \mathcal{E}[|s_k|^2]$, $k = 1, 2, \ldots, K$. Together with rate, power allocation across layers, which results in joint power and rate allocation, can be considered to improve the performance at a system level.

Suppose that the received signal is given by

$$r_l = \sum_{k=1}^{K} h_k s_{l,k} + n_l, \; l = 0, 1, \ldots, L-1, \tag{8.47}$$

where $s_{l,k}$ and h_k are the lth signal symbol from the kth coded sequence of length L and its corresponding channel gain, respectively, and n_l is a white Gaussian noise. We assume that $\{s_{l,k}, \; l = 0, 1, \ldots\}$ is generated from a random Gaussian codebook and each coded signal is independently encoded and statistically independent. If the sum of the K signals in (8.47) is regarded as the desired signal, the SNR becomes

$$\mathrm{SNR}_{\mathrm{sum}} = \frac{\sum_{k=1}^{K} |h_k|^2 E_k}{N_0},$$

where $s_{l,k} \sim \mathcal{CN}(0, E_k)$ and $n_l \sim \mathcal{CN}(0, N_0)$, and the (sum) channel capacity is given by

$$\mathsf{I}_{\mathrm{sum}} = \log_2 \left(1 + \frac{\sum_{k=1}^{K} |h_k|^2 E_k}{N_0} \right). \tag{8.48}$$

This capacity is achievable if the receiver can jointly decode all the K signals although they are encoded independently. Fortunately, this sum capacity is also achievable using SIC, whose complexity is much lower than that of joint decoding.

Suppose that the Kth signal is decoded first. Taking the remaining $K - 1$ signals as interference, the capacity of layer K can be found as

$$\mathsf{I}_K = \log_2 \left(1 + \frac{|h_K|^2 E_K}{\sum_{k=1}^{K-1} |h_k|^2 E_k + N_0} \right). \tag{8.49}$$

Once the Kth signal is decoded, this signal can be stripped off from r_l. The resulting signal is given by

$$r_l - s_{l,K} = \sum_{k=1}^{K-1} h_k s_{l,k} + n_l, \; l = 0, 1, \ldots, L - 1.$$

With this signal, we can decode the next signal, $s_{l,K-1}$. The resulting capacity becomes

$$\mathsf{I}_{K-1} = \log_2 \left(1 + \frac{|h_{K-1}|^2 E_{K-1}}{\sum_{k=1}^{K-2} |h_k|^2 E_k + N_0} \right). \tag{8.50}$$

We can repeat SIC and decoding to the last signal in layer 1 whose capacity is given by

$$\mathsf{I}_1 = \log_2 \left(1 + \frac{|h_1|^2 E_1}{N_0} \right). \tag{8.51}$$

Interestingly, it can be shown that

$$\mathsf{I}_{\mathrm{sum}} = \sum_{k=1}^{K} \mathsf{I}_k. \tag{8.52}$$

This implies that the sum capacity or mutual information can be achieved using SIC and individual decoding (not joint decoding). In general, since the implementation of SIC and individual decoding for each coded signal can be carried out with relatively low computational complexity, this observation is significant in terms of computational complexity. It is noteworthy that individual channel capacity depends on decoding order, although the sum capacity is invariant with respect to decoding order.

In order to achieve the sum capacity with SIC and individual decoding, however, there are some additional requirements as follows:

- Each coded signal should be encoded at a rate less than each layer's capacity given as follows:

$$\mathsf{I}_k = \log_2\left(1 + \frac{|h_k|^2 E_k}{\sum_{q=1}^{k-1} |h_q|^2 E_q + N_0}\right), \quad k = 1, 2, \ldots, K.$$

That is, $R_k < \mathsf{I}_k$, where R_k is the rate for the coded signal of layer k. Since the capacity for each layer can vary depending on decoding order in SIC, the decoding order has to be pre-determined and known to the transmitter so that the transmitter can decide the rates. Alternatively, the receiver can provide the transmitter with the required rate less than I_k according to decoding order.
- For ideal SIC, capacity-achieving codes should be employed, which may be impractical. Thus, with practical channel codes, there would be error propagation.

Example 8.7.1 Suppose that there are $K = 2$ coded signals with $E_1 = 5$ and $E_2 = 10$. In addition, $N_0 = 1$. The sum capacity becomes

$$\mathsf{I}_{\text{sum}} = \log_2(1 + (5 + 10)) = 4.$$

Consider the decoding order from signal 2 to signal 1. Then, we have

$$\begin{aligned} \mathsf{I}_2 &= \log_2\left(1 + \frac{10}{5+1}\right) \\ &= 1.4150; \\ \mathsf{I}_1 &= \log_2(1 + 5) \\ &= 2.5850. \end{aligned}$$

For the reverse decoding order, we have

$$\begin{aligned} \mathsf{I}_1 &= \log_2\left(1 + \frac{5}{10+1}\right) \\ &= 0.5406; \\ \mathsf{I}_2 &= \log_2(1 + 10) \\ &= 3.4594. \end{aligned}$$

We can easily see that the sum capacity is invariant with respect to decoding order, but individual capacity for each layer depends on decoding order.

Using the notion of SIC, we can also characterize the capacity region of multiple access channel (MAC). MAC is a common channel where multiple users can transmit signals simultaneously. Thus, the received signal can be represented as in (8.47). In code division multiple access (CDMA) as an example of MAC, each user has his/her dedicated signature waveform. Signals are spread by signature waveforms and transmitted to a common channel. At a receiver, each user's signal can be distinguished from the others

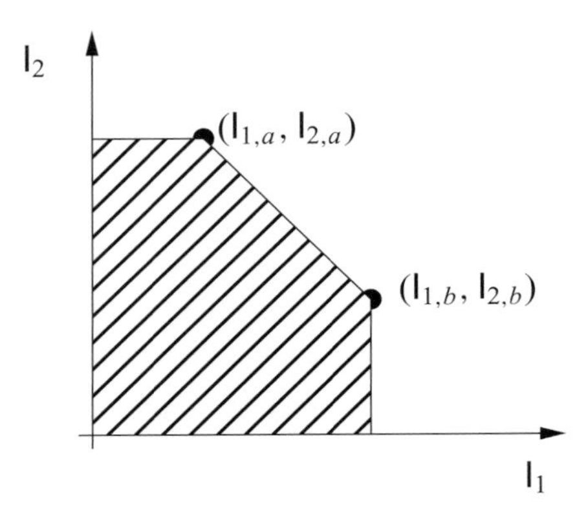

Figure 8.5 Capacity region for 2-user MAC. The rates corresponding to any point in the shaded region are achievable.

using signature waveforms. To see the capacity region for CDMA or MAC, consider the case of $K = 2$. Assuming that the receiver uses SIC, we can have two pairs of the channel capacity depending on decoding order as follows:

$$\begin{aligned} I_{1,a} &= \log_2 \left(1 + \frac{|h_1|^2 E_1}{|h_2|^2 E_2 + N_0}\right); \\ I_{2,a} &= \log_2 \left(1 + \frac{|h_2|^2 E_2}{N_0}\right) \end{aligned}$$

and

$$\begin{aligned} I_{1,b} &= \log_2 \left(1 + \frac{|h_1|^2 E_1}{N_0}\right); \\ I_{2,b} &= \log_2 \left(1 + \frac{|h_2|^2 E_2}{|h_1|^2 E_1 + N_0}\right), \end{aligned}$$

where $I_{k,a}$ is the channel capacity that is obtained with the ascending order for SIC and $I_{k,b}$ with the descending order. Figure 8.5 shows the capacity region for 2-user MAC. Any point on the straight line between points a and b can be achieved by time-sharing of a and b. For a detailed account, the reader is referred to (Cover & Thomas 1991).

8.7.2 Information-theoretical optimality of MMSE-SIC receiver

For MIMO channels, SIC receivers can also achieve channel capacity. Due to multiple receive antennas, a SIC receiver needs a combiner. We can show that the MMSE combiner can be used together with SIC to achieve channel capacity. Recall the received signal vector:

$$\begin{aligned} \mathbf{r} &= \mathbf{H}\mathbf{s} + \mathbf{n} \\ &= \sum_{k=1}^{K} \mathbf{h}_k s_k + \mathbf{n}, \end{aligned} \tag{8.53}$$

where $\mathbf{h}_k$ denotes the kth column vector of $\mathbf{H}$. The mutual information between $\mathbf{r}$ and $\mathbf{s}$ is

$$\begin{aligned}\mathsf{I}(\mathbf{r};\mathbf{s}) &= \mathsf{I}(\mathbf{r}; s_1, s_2, \ldots, s_K) \\ &= \mathsf{I}(\mathbf{r}; s_K) + \mathsf{I}(\mathbf{r}; s_{K-1}|s_K) + \cdots + \mathsf{I}(\mathbf{r}; s_1|s_2, s_3, \ldots, s_K). \end{aligned} \tag{8.54}$$

The second equality is due to the chain rule of mutual information.

Let

$$\mathbf{r}_k = \mathbf{r} - \sum_{q=k+1}^{K} \mathbf{h}_q s_q = \mathbf{h}_k s_k + \sum_{q=1}^{k-1} \mathbf{h}_q s_q + \mathbf{n}. \tag{8.55}$$

Theorem 8.7.1 *If* $\sum_{q=1}^{k-1} \mathbf{h}_q s_q + \mathbf{n}$ *is Gaussian,*

$$\begin{aligned}\mathsf{I}(\mathbf{r}; s_k|s_{k+1}, s_{k+2}, \ldots, s_K) &= \mathsf{I}(\mathbf{r}_k; s_k) \\ &= \mathsf{I}(\hat{c}_k; s_k), \end{aligned} \tag{8.56}$$

where $\hat{c}_k = \mathbf{w}_k^{\mathrm{H}} \mathbf{r}_k$ *and* $\mathbf{w}_k$ *denotes the MMSE vector that is given by*

$$\mathbf{w}_k = \alpha_k \mathbf{R}_k^{-1} \mathbf{h}_k.$$

Here,

$$\alpha_k = \frac{E_k}{1 + E_k \mathbf{h}_k^H \mathbf{R}_k^{-1} \mathbf{h}_k};$$

$$\mathbf{R}_k = \mathcal{E}\left[\left(\sum_{q=1}^{k-1} \mathbf{h}_q s_q + \mathbf{n}\right)\left(\sum_{q=1}^{k-1} \mathbf{h}_q s_q + \mathbf{n}\right)^{\mathrm{H}}\right].$$

Note that $\hat{c}_k$ *is the MMSE-SIC estimate of* s_k *after the cancellation of* $s_{k+1}, s_{k+2}, \ldots, s_K$.

This implies that the combination of SIC (which results in $\mathbf{r}_k$) and MMSE estimation (which results in $\hat{c}_k$) can achieve the mutual information, $\mathsf{I}(\mathbf{r};\mathbf{s})$. Note that the cancellation in (8.55) should be perfect, which means the signals are encoded using capacity-achieving codes and the rate of the signal, s_k, is given by

$$R_k \le \mathsf{I}(\mathbf{r}; s_k|s_{k+1}, s_{k+2}, \ldots, s_K),\ k = 1, 2, \ldots, K. \tag{8.57}$$

While the MMSE-SIC receiver can achieve $\mathsf{I}(\mathbf{r};\mathbf{s})$, the ZF-SIC receiver cannot. If the ZF-SIC receiver in Subsection 8.2.1 is used, the sum rate can be given by

$$\begin{aligned}\mathsf{I}_{\mathrm{zf-sic,sum}} &= \sum_{k=1}^{K} \mathsf{I}(x_k; s_k|s_{k+1}, s_{k+2}, \ldots, s_K) \\ &= \sum_{k=1}^{K} \mathsf{I}(u_k; s_k) \end{aligned} \tag{8.58}$$

and it is smaller than or equal to $\mathsf{I}(\mathbf{r};\mathbf{s})$.

8.8 Summary and notes

SIC detection was studied in this chapter. SIC is applicable to both uncoded and coded signals. For uncoded signals, since hard-decisions of detected signals are used for cancellation, the error propagation became the major drawback of SIC detectors. For Rayleigh MIMO fading channels, it was shown that the diversity order is decided by the first signal detection that cannot fully exploit spatial diversity gain due to interference suppression. This shows why the performance of SIC detectors is not much better than that of linear detectors in MIMO systems. For coded signals, however, reliably decoded signals are used for cancellation and there would be much less performance degradation that results from error propagation.

Information-theoretic optimality of SIC is shown in (Gallager 1994) and also in (Cover & Thomas 1991). The combination of layered transmission and SIC decoding plays a crucial role in MIMO systems: (Foschini 1996), (Foschini *et al.* 1999), and (Foschini *et al.* 2003). More in-depth discussion for the combination of layered transmission and SIC decoding can be found in (Tse & Viswanath 2005).

Problems

Problem 8.1 Suppose that a 2×2 MIMO channel has the following channel matrix:

$$\mathbf{H} = \frac{1}{\sqrt{2}} \begin{bmatrix} 1 & 1 \\ 1 & 0.5 \end{bmatrix}.$$

Decide the order of the ZF-SIC detection that provides lower error probability.

Problem 8.2 Suppose that a 2×2 MIMO channel has the following channel matrix:

$$\mathbf{H} = \frac{1}{\sqrt{2}} \begin{bmatrix} 1 & 1 \\ 1 & 0.5 \end{bmatrix}.$$

It is assumed that the background noise vector is $\mathbf{n} \sim \mathcal{N}(0, \mathbf{I}_{2\times 2})$. The receiver employs ZF-SIC detection, where a soft-decision, which is the MMSE estimate, is used for cancellation.

(i) Find the signals' powers, $E_k = \mathcal{E}[|s_k|^2]$, that make the normalized MMSE of 10^{-1} for all k. The normalized MMSE is MMSE_k / E_k, where MMSE_k denotes the MMSE for the kth signal. Note that due to soft cancellation, there is cancellation error or residue.

(ii) Change the detection order and repeat (i).

Problem 8.3 Let $N = 2$ and $K = 3$. Suppose that $s_1, s_2, s_3 \in \{-1, 1\}$ and the channel matrix is given by

$$\mathbf{H} = \frac{1}{\sqrt{2}} \begin{bmatrix} 1 & 1 & 1 \\ 1 & 0.5 & -1 \end{bmatrix}.$$

In addition, assume that $\mathbf{n} \sim \mathcal{N}(0, \mathbf{I}_{2\times 2})$ and $\mathbf{r} = [r_1\ r_2]^{\mathrm{T}} = [3\ -1]^{\mathrm{T}}$.

(i) Perform the ZF-SIC detection when the last two data symbols are jointly detected first. (Since $K > N$, the last $K - N + 1$ data symbols should be jointly detected in the SIC detection.)

(ii) Suppose that the above ZF-SIC detector is used. Find the best detection order that maximizes the posteriori probability of the two data symbols detected first.

Problem 8.4 Find an exact expression for the average error probability of the ZF-SIC for a 2×2 MIMO channel when $[\mathbf{H}]_{n,k} \sim \mathcal{CN}(0, 1)$ and $s_k \in \{\pm\sqrt{E_k}\}$ and $\mathbf{n} \sim \mathcal{CN}(0, N_0\mathbf{I})$.

Problem 8.5 Derive (8.52).

Problem 8.6 Consider the channel in Problem 8.2. Find the achievable rate, $\mathsf{I}(\mathbf{r}; \mathbf{s})$, when $\mathbf{s} \sim \mathcal{CN}(0, \mathbf{I})$ and $\mathbf{n} \sim \mathcal{CN}(0, \mathbf{I})$ using the MMSE-SIC receiver. Find also the sum rate when the ZF-SIC receiver is used.

9 Lattice-reduction-aided MIMO detection

A lattice is a set of vectors that are generated by the integer linear combination of a certain set of vectors, which are called basis vectors. For example, the set of integer numbers is a lattice. Various optimization problems can be formulated in lattices and there are a number of applications related to these optimization problems. Cryptography is an example. Computationally efficient algorithms are developed to solve or approximately solve these optimization problems by exploiting structures and properties of lattices.

If a signal constellation is a subset of a lattice, the signal detection problem can be translated into an optimization problem in the lattice. From this, various lattice algorithms can help to solve the signal detection problem efficiently, in particular, MIMO detection problem. In this chapter, we mainly discuss suboptimal, but low complexity, MIMO detection methods that exploit the properties of lattices. MIMO precoding techniques are also briefly presented. In addition to efficient MIMO detection methods, we describe well-known lattice-reduction (LR) methods that play a crucial role in finding approximate solutions with low computational complexity for various lattice optimization problems, including LR-based MIMO detection.

9.1 Lattices and signal constellations

In this section, we briefly review the definition of a lattice with its properties and relation to signal constellations. For a detailed account, the reader is referred to Appendix 4 or (Cohen 1993), (Conway & Sloane 1991), (Forney, Jr. 1988), and (Hoffstein, Pipher & Silverman 2002).

9.1.1 Lattices

A lattice, which is a set of vectors, can be defined with its associated basis that is a set of linearly independent vectors in a vector space. Suppose that there are K real-valued linearly independent basis vectors, $\{\mathbf{b}_1, \mathbf{b}_2, \ldots, \mathbf{b}_K\}$, where $\mathbf{b}_k \in \mathbb{R}^N$ and $N \geq K$. The lattice associated with the basis $\{\mathbf{b}_1, \mathbf{b}_2, \ldots, \mathbf{b}_K\}$ can be defined as

$$\boldsymbol{\Lambda} = \left\{ \mathbf{a} \;\middle|\; \mathbf{a} = \sum_{k=1}^{K} \mathbf{b}_k u_k, \ u_k \in \mathbb{Z} \right\}, \tag{9.1}$$

where $\mathbb{Z}$ is the set of integer numbers. Thus, the lattice, $\mathbf{\Lambda}$, is generated by the *integer* linear combination of $\mathbf{b}_1, \mathbf{b}_2, \ldots, \mathbf{b}_K$. Let

$$\mathbf{B} = [\mathbf{b}_1 \ \mathbf{b}_2 \ \ldots \ \mathbf{b}_K].$$

The column vectors of $\mathbf{B}$ are referred to as the basis vectors (for a lattice) or generators that generate the corresponding lattice. Then, any lattice vector in $\mathbf{\Lambda}$ is expressed as $\mathbf{b} = \mathbf{B}\mathbf{u}$, where $\mathbf{u} \in \mathbb{Z}^K$. From this, we can also see that a linear transformation of a lattice becomes another lattice. That is,

$$\begin{aligned} \mathbf{\Lambda}_2 &= \{\mathbf{a} \mid \mathbf{a} = \mathbf{A}\mathbf{b}, \mathbf{b} \in \mathbf{\Lambda}_1\} \\ &= \{\mathbf{a} \mid \mathbf{a} = \mathbf{A}\mathbf{B}\mathbf{u}, \mathbf{u} \in \mathbb{Z}^K\}, \end{aligned} \tag{9.2}$$

where $\mathbf{\Lambda}_1$ is a lattice generated by $\mathbf{B}$ and $\mathbf{\Lambda}_2$ is another lattice that is obtained by a linear transformation, $\mathbf{A}$, from $\mathbf{\Lambda}_1$. According to (9.2), $\mathbf{\Lambda}_2$ is the lattice generated by $\mathbf{AB}$.

A lattice can be generated by different bases. Suppose that the lattices generated by $\mathbf{B}_1$ and $\mathbf{B}_2$ are the same. Then, each column vector of $\mathbf{B}_1$ should be represented by an integer linear combination of the column vectors of $\mathbf{B}_2$, and vice versa. This implies that there exists an integer $K \times K$ matrix, $\mathbf{U}$, such as

$$\mathbf{B}_1 = \mathbf{B}_2\mathbf{U}. \tag{9.3}$$

Furthermore, since $\mathbf{B}_2 = \mathbf{B}_1\mathbf{U}^{-1}$, $\mathbf{U}^{-1}$ should also be a $K \times K$ matrix whose elements are *integers*. Since $\det(\mathbf{U}^{-1}) = 1/\det(\mathbf{U})$, this implies that

$$\det(\mathbf{U}) = \pm 1.$$

This shows that $\mathbf{U}$ is *unimodular*. A unimodular matrix whose elements are integer is called an integer unimodular matrix.

For example, consider

$$\mathbf{B}_1 = \begin{bmatrix} 1 & 0 \\ 0 & 1 \end{bmatrix}, \ \mathbf{B}_2 = \begin{bmatrix} 1 & 3 \\ 0 & 1 \end{bmatrix}.$$

We can easily verify that they have the same lattice as $[3\ 1]^{\mathrm{T}} = 3 \times [1\ 0]^{\mathrm{T}} + [0\ 1]^{\mathrm{T}}$. We can also show that

$$\mathbf{U} = \mathbf{B}_2^{-1}\mathbf{B}_1 = \begin{bmatrix} 1 & -3 \\ 0 & 1 \end{bmatrix}, \ \mathbf{U}^{-1} = \begin{bmatrix} 1 & 3 \\ 0 & 1 \end{bmatrix},$$

whose elements are integers and their determinants are 1. Among the bases that generate the same lattice, there are some bases that may have shorter basis vectors. For instance, we can see that the column vectors of $\mathbf{B}_1$ are shorter than those of $\mathbf{B}_2$. We often need to find the basis whose basis vectors are shorter for a given lattice in many applications, including cryptography systems based on lattices (Hoffstein *et al.* 2002). In particular, in the LR-based MIMO detection, we need to find the basis whose basis vectors are the shortest as will be shown later.

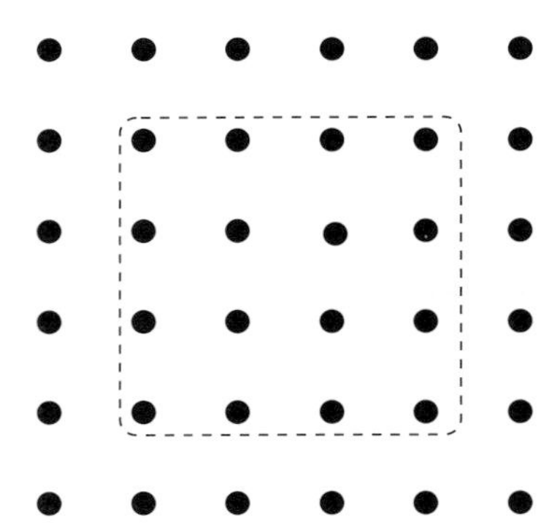

Figure 9.1 A subset of a lattice for a signal alphabet: the signals in the dashed line are for 16-QAM.

9.1.2 Signal constellation as a subset of lattice

A signal alphabet can be considered as a subset of a lattice. Consider the M-ary PAM. The signal alphabet of M-ary PAM is given by

$$\mathcal{S}_{\text{M-PAM}} = \{-M+1, -M+3, \ldots, -1, +1, \ldots, M-3, M-1\}$$

and the symbol energy is given by

$$E_{\text{s}} = \frac{M^2-1}{3}.$$

We can easily see that

$$\mathcal{S}_{\text{M-PAM}} \subset \mathbb{Z},$$

where $\mathbb{Z}$ is a lattice whose basis is $\mathbf{u} = 1$. In addition,

$$\mathcal{S}_{\text{M-PAM}} \subset 2\mathbb{Z}+1 = \{u \mid u = 2i+1, \ i \in \mathbb{Z}\}.$$

Scaling and shifting operations can be applied to a signal alphabet to show that the signal alphabet is a subset of a lattice. This is more appropriate for the signal detection. For example, suppose that $s_k \in \mathcal{S} = \{-3, -1, 1, 3\}$ for 4-ary PAM. Let $x_k = \frac{1}{2}s_k + \frac{3}{2}$. Then, the x_k's, are in $\mathbb{Z}_4$, where $\mathbb{Z}_Q \triangleq \{0, 1, \ldots, Q-1\}$.

Letting $m = \sqrt{M}$ be a power of 2, the signal alphabet for M-ary square QAM can be represented by

$$\mathcal{S}_{\text{M-QAM}} = \left\{a + jb \mid a, b \in \{-m+1, -m+3, \ldots, -1, +1, \ldots, m-3, m-1\}\right\},$$

where its symbol energy is given by

$$E_{\text{s}} = \frac{2(m^2-1)}{3} = \frac{2(M-1)}{3}.$$

For an M-ary square QAM, after applying scaling and shifting operations to the signal constellation, we can show that it becomes a subset of $\mathbb{Z} + j\mathbb{Z}$. As shown in Fig. 9.1, a 16-QAM constellation can be seen as a subset of the lattice, $\mathbb{Z} + i\mathbb{Z}$.

In general, for a certain signal alphabet, $\mathcal{S}$, there exist α and β such as

$$\{\alpha s + \beta \mid s \in \mathcal{S}\} \subseteq \mathbb{Z} \text{ or } \mathbb{Z} + j\mathbb{Z}. \tag{9.4}$$

Here, α and β are called the scaling and shifting coefficients, respectively. Throughout this chapter, we focus the signal detection problem for the signal alphabet that becomes a subset of a lattice after proper scaling and shifting.

9.2 MIMO detection over lattices

In this section, we formulate the MIMO detection problem as a lattice decoding problem. We assume that the number of signals or transmit antennas is K and the number of receive antennas is N throughout this chapter for the MIMO detection problem. Thus, the size of the channel matrix, $\mathbf{H}$, is $N \times K$. We also assume that $N \geq K$ throughout this chapter.

Consider the received signal vector through an MIMO channel:

$$\mathbf{r} = \mathbf{Hs} + \mathbf{n}, \tag{9.5}$$

where $\mathbf{s} \in \mathcal{S}^K$ is the signal vector and $\mathbf{n} \sim \mathcal{CN}(\mathbf{0}, N_0\mathbf{I})$ is a CSCG noise vector. Applying scaling and shifting operations to $\mathbf{s}$, we define

$$\underline{\mathbf{s}} = \alpha\mathbf{s} + \beta\mathbf{1}.$$

Then, we have

$$\begin{aligned}
\underline{\mathbf{r}} &= \alpha\mathbf{r} + \beta\mathbf{H1} \\
&= \mathbf{H}(\alpha\mathbf{s} + \beta\mathbf{1}) + \underline{\mathbf{n}} \\
&= \mathbf{H}\underline{\mathbf{s}} + \underline{\mathbf{n}},
\end{aligned} \tag{9.6}$$

where $\underline{\mathbf{n}} = \alpha\mathbf{n}$. With $\underline{\mathbf{r}}$, we can see that $\mathbf{H}\underline{\mathbf{s}}$ is a vector in the lattice generated by $\mathbf{H}$ if $\underline{\mathbf{s}} \in \mathbb{Z}^K$. Therefore, the detection problem becomes a search problem to find a vector in the lattice.

To illustrate the MIMO detection problem with $\underline{\mathbf{r}}$ and $\underline{\mathbf{s}}$, consider an example. If the original signal constellation is $\mathcal{S} = \{-3, -1, 1, 3\}$ and $\alpha = 1/2$ and $\beta = 3/2$, the ML detection becomes

$$\hat{\underline{\mathbf{s}}}_{\text{ml}} = \arg\min_{\underline{\mathbf{s}} \in \mathbb{Z}_4^K} ||\underline{\mathbf{r}} - \mathbf{H}\underline{\mathbf{s}}||^2.$$

The constraint on $\underline{\mathbf{s}}$, $\underline{\mathbf{s}} \in \mathbb{Z}_4^K$, could be relaxed such as $\underline{\mathbf{s}} \in \mathbb{Z}^K$. That is, the signal vector could be any vector in the lattice, $\mathbb{Z}^K$. With this relaxed constraint, an extended ML detection can be formulated as

$$\hat{\underline{\mathbf{s}}}_{\text{eml}} = \arg\min_{\underline{\mathbf{s}} \in \mathbb{Z}^K} ||\underline{\mathbf{r}} - \mathbf{H}\underline{\mathbf{s}}||^2. \tag{9.7}$$

Due to the relaxed constraint on $\underline{\mathbf{s}}$, the optimal solution of the extended ML detection may not belong to $\mathbb{Z}_4^K$. If $\hat{\underline{\mathbf{s}}}_{\text{eml}} \notin \mathbb{Z}_4^K$, a mapping of the solution into $\mathbb{Z}_4^K$ can be imposed.

For example, suppose that the optimal solution of the extended ML detection problem is $\underline{\mathbf{s}} = [5\ 1]^{\mathrm{T}} \notin \mathbb{Z}_4^2$. By a clipping operation, the solution can be mapped to $\underline{\mathbf{s}} = [3\ 1]^{\mathrm{T}} \in \mathbb{Z}_4^2$.

Note that the scaling and shifting operations can be extended to an M-ary square QAM. Since the elements of $\mathbf{r}$, $\mathbf{s}$, and $\mathbf{n}$ are complex-valued, it would be convenient to convert them to real-valued quantities as follows:

$$\begin{bmatrix} \Re(\mathbf{r}) \\ \Im(\mathbf{r}) \end{bmatrix} = \begin{bmatrix} \Re(\mathbf{H}) & -\Im(\mathbf{H}) \\ \Im(\mathbf{H}) & \Re(\mathbf{H}) \end{bmatrix} \begin{bmatrix} \Re(\mathbf{s}) \\ \Im(\mathbf{s}) \end{bmatrix} + \begin{bmatrix} \Re(\mathbf{n}) \\ \Im(\mathbf{n}) \end{bmatrix}.$$

Thus, throughout this chapter, we assume that signals and channels are real-valued unless stated otherwise and $\underline{\mathbf{s}}$ is properly scaled and shifted so that $\underline{\mathbf{s}} \in \mathbb{Z}_Q^K$, where Q is a positive integer.

For convenience, if there is no significant risk of confusion, hereafter, $\mathbf{r}$, $\mathbf{s}$, and $\mathbf{n}$ will be used to denote $\underline{\mathbf{r}}$, $\underline{\mathbf{s}}$ and $\underline{\mathbf{n}}$, respectively.

9.2.1 LR-based extended ML detection

Suppose that the lattices generated by the column vectors of $\mathbf{H}$ and $\mathbf{G}$ are the same. This implies there exists an integer unimodular matrix $\mathbf{U}$ that satisfies

$$\mathbf{H} = \mathbf{G}\mathbf{U}. \tag{9.8}$$

Then, the received signal vector can be rewritten as

$$\begin{aligned} \mathbf{r} &= \mathbf{H}\mathbf{s} + \mathbf{n} \\ &= \mathbf{G}\mathbf{U}\mathbf{s} + \mathbf{n} \\ &= \mathbf{G}\mathbf{c} + \mathbf{n}, \end{aligned} \tag{9.9}$$

where $\mathbf{c} = \mathbf{U}\mathbf{s}$. Since the elements of $\mathbf{U}$ are integers, $\mathbf{c} \in \mathbb{Z}^K$ and the extended ML detection problem to detect $\mathbf{c}$ rather than $\mathbf{s}$ can be given by

$$\hat{\mathbf{c}}_{\mathrm{eml}} = \arg \min_{\mathbf{c} \in \mathbb{Z}^K} ||\mathbf{r} - \mathbf{G}\mathbf{c}||^2. \tag{9.10}$$

Once $\hat{\mathbf{c}}_{\mathrm{eml}}$ is available, the solution of the extended ML problem in (9.7) can be found as

$$\hat{\mathbf{s}}_{\mathrm{eml}} = \mathbf{U}^{-1}\hat{\mathbf{c}}_{\mathrm{eml}} \in \mathbb{Z}^K.$$

If the column vectors of $\mathbf{G}$ are more orthogonal than those of $\mathbf{H}$, the optimal solution in (9.10) can be found by a computationally efficient search algorithm exploiting (near) orthogonality of $\mathbf{G}$. For example, if the column vectors of $\mathbf{G}$ are orthogonal, the ML detection problem in (9.10) can be decomposed into K one-dimensional search problems.

Example 9.2.1 Suppose that the channel matrix, $\mathbf{H}$, is given by

$$\mathbf{H} = \begin{bmatrix} 1 & 1 \\ 1 & 2 \end{bmatrix}.$$

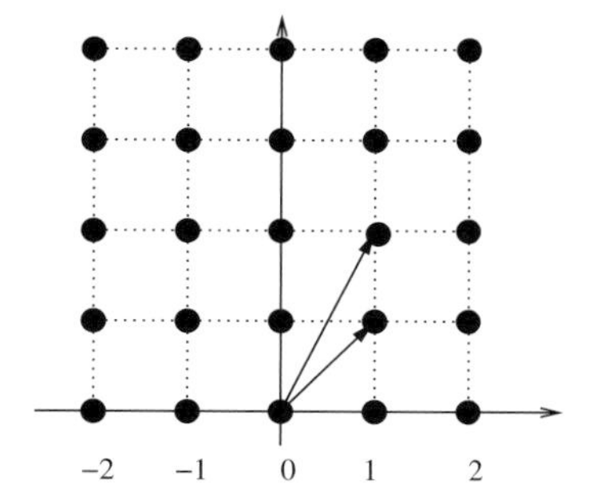

Figure 9.2 Lattice generated by $\mathbf{h}_1 = [1\ 1]^{\mathrm{T}}$ and $\mathbf{h}_2 = [1\ 2]^{\mathrm{T}}$.

This channel matrix can be factorized as follows:

$$\mathbf{H} = \underbrace{\begin{bmatrix} 1 & 0 \\ 0 & 1 \end{bmatrix}}_{=\mathbf{G}} \underbrace{\begin{bmatrix} 1 & 1 \\ 1 & 2 \end{bmatrix}}_{=\mathbf{U}},$$

where $\mathbf{U}$ is integer unimodular. Then, the extended ML detection can be carried out as follows:

$$\begin{aligned} \hat{\mathbf{c}} &= \arg\min_{\mathbf{c}\in\mathbb{Z}^2} ||\mathbf{r} - \mathbf{G}\mathbf{c}||^2 \\ &= \arg\min_{\mathbf{c}\in\mathbb{Z}^2} ||\mathbf{r} - \mathbf{c}||^2 \\ &\Rightarrow \begin{cases} \hat{c}_1 &= \arg\min_{c_1\in\mathbb{Z}} |r_1 - c_1|^2; \\ \hat{c}_2 &= \arg\min_{c_2\in\mathbb{Z}} |r_2 - c_2|^2. \end{cases} \end{aligned}$$

This shows that a new basis can reduce the computational complexity of the detection.

Note that once $\hat{\mathbf{s}}_{\mathrm{eml}}$ is found, the constraint that $\hat{\mathbf{s}} \in \mathbb{Z}_{\mathcal{S}}^K$ can be imposed, where $\mathbb{Z}_{\mathcal{S}}^K$ denotes a subset of $\mathbb{Z}^K$ associated with the signal alphabet $\mathcal{S}$. For example, if $\mathbb{Z}_{\mathcal{S}}^K = \mathbb{Z}_Q^K$, each coefficient of $\hat{\mathbf{s}}_{\mathrm{eml}}$ can be clipped between 0 and $Q - 1$. The clipped vector of $\hat{\mathbf{s}}_{\mathrm{eml}}$ may not be the optimal ML solution. However, if the SNR is sufficiently high, the clipped vector of $\hat{\mathbf{s}}_{\mathrm{eml}}$ usually becomes the optimal ML solution.

Through Example 9.2.1, we observe that new basis vectors that are orthogonal or nearly orthogonal can lead to a computationally efficient detection approach. LR algorithms are closely related to finding a new basis of nearly orthogonal basis vectors. LR algorithms attempt to find linearly independent K shortest vectors in the lattice $\mathbf{\Lambda}$ generated by a given $\mathbf{H}$. In general, the K shortest vectors are nearly orthogonal to each other. In order to see the relation between the K shortest vectors and (nearly) orthogonal new basis vectors, consider the lattice shown in Fig. 9.2. The lattice is generated by two basis vectors, $\mathbf{h}_1 = [1\ 1]^{\mathrm{T}}$ and $\mathbf{h}_2 = [1\ 2]^{\mathrm{T}}$. There are the two shortest vectors in the lattice which are given by

$$\mathbf{g}_1 = \begin{bmatrix} (\pm)1 \\ 0 \end{bmatrix} \text{ and } \mathbf{g}_2 = \begin{bmatrix} 0 \\ (\pm)1 \end{bmatrix}.$$

The two shortest vectors, $\mathbf{g}_1$ and $\mathbf{g}_2$, which are orthogonal in this example, form a new basis that can also generate the same lattice $\mathbf{\Lambda}$.

The outputs of LR algorithms are a new basis, $\mathbf{G}$, which is nearly orthogonal, and the associated integer unimodular matrix, $\mathbf{U}$. In Sections 9.4 and 9.5, we will discuss LR algorithms.

In the rest of this section, we assume that $\mathbf{H}$ is factorized as in (9.8) and the column vectors of $\mathbf{G}$ are nearly orthogonal.

9.2.2 LR-based ZF and MMSE detection

The signal detection can be reformulated with a new basis obtained by the LR. This signal detection is called LR-based detection. As the basis is nearly orthogonal, some low complexity suboptimal detection methods can provide reasonably good performance.

Suppose that $\mathbf{c}$ is the vector to be detected. Then, the output of the ZF detector is given by

$$\begin{aligned}\hat{\mathbf{c}}_{\text{zf}} &= \mathbf{G}^{\dagger}\mathbf{r}\\ &= \mathbf{G}^{\dagger}(\mathbf{Gc}+\mathbf{n})\\ &= \mathbf{c}+\mathbf{G}^{\dagger}\mathbf{n},\end{aligned} \tag{9.11}$$

where $\mathbf{G}^{\dagger}$ represents the pseudo-inverse of $\mathbf{G}$. A hard-decision of $\hat{\mathbf{c}}_{\text{zf}}$ can be obtained by rounding as $\lceil \hat{\mathbf{c}}_{\text{zf}} \rfloor$, where $\lceil \cdot \rfloor$ denotes the element-wise rounding operation. Then, the hard-decision of $\mathbf{s}$ is given by

$$\hat{\mathbf{s}}_{\text{zf}} = \mathbf{U}^{-1}\lceil \hat{\mathbf{c}}_{\text{zf}} \rfloor. \tag{9.12}$$

If the column vectors of $\mathbf{G}$ are orthogonal or near orthogonal, then the ZF detector to detect $\mathbf{c}$ can have reasonably good performance as the noise enhancement becomes insignificant. As the detection of $\mathbf{c}$ can be done with less errors, the detection of $\mathbf{s}$ can also be reliable.

Example 9.2.2 Suppose that the channel matrix, $\mathbf{H}$, is given by

$$\mathbf{H} = \begin{bmatrix} 1 & 1 \\ 0.5 & 2 \end{bmatrix}.$$

This channel matrix can be factorized as follows:

$$\mathbf{H} = \underbrace{\begin{bmatrix} 1 & -1 \\ 0.5 & 1 \end{bmatrix}}_{=\mathbf{G}} \underbrace{\begin{bmatrix} 1 & 2 \\ 0 & 1 \end{bmatrix}}_{=\mathbf{U}},$$

where $\mathbf{U}$ is integer unimodular. In Fig. 9.3, we show the basis vectors of $\mathbf{H}$ and $\mathbf{G}$. The two column vectors of $\mathbf{G}$ are more orthogonal than those of $\mathbf{H}$. The correlation between the two column vectors can be used to measure the orthogonality of a basis.

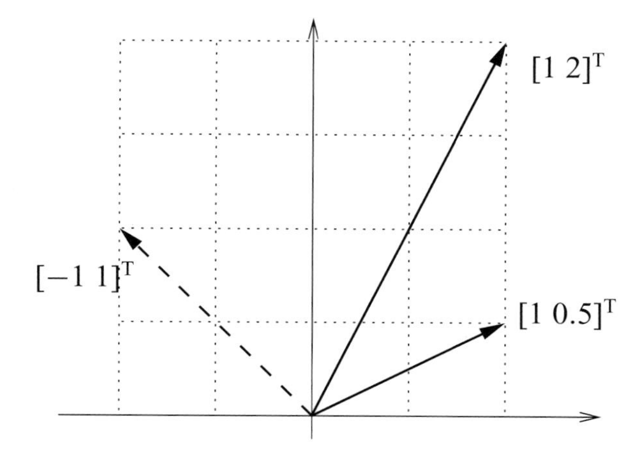

Figure 9.3 Basis vectors of **H** and **G**.

The normalized correlation of **H** is

$$\frac{\mathbf{h}_1^{\mathrm{T}}\mathbf{h}_2}{||\mathbf{h}_1||\ ||\mathbf{h}_2||} = 0.8,$$

while that of **G** is

$$\frac{\mathbf{g}_1^{\mathrm{T}}\mathbf{g}_2}{||\mathbf{g}_1||\ ||\mathbf{g}_2||} = -0.3162.$$

These correlations confirm that the two column vectors of **G** are more orthogonal than those of **H**.

Let $\mathbf{x} = [1\ -1]^{\mathrm{T}}$ and $\mathbf{n} = [-0.3\ 0.4]^{\mathrm{T}}$. In this case, the received signal vector becomes

$$\mathbf{r} = \mathbf{Hx} + \mathbf{n} = [-0.3\ -1.1]^{\mathrm{T}}.$$

The conventional ZF solution of $\mathbf{x}$ becomes

$$\mathbf{H}^{\dagger}\mathbf{r} = \begin{bmatrix} \frac{4}{3} & -\frac{2}{3} \\ -\frac{1}{3} & \frac{2}{3} \end{bmatrix} \mathbf{r} = \begin{bmatrix} 0.3333 \\ -0.6333 \end{bmatrix}.$$

After rounding, the hard-decision becomes $[0\ -1]^{\mathrm{T}}$, which is not correct.

On the other hand, with **G**, the ZF solution of **c** becomes

$$\mathbf{c}_{\mathrm{zf}} = \mathbf{G}^{\dagger}\mathbf{r} = \begin{bmatrix} \frac{2}{3} & -\frac{2}{3} \\ -\frac{1}{3} & \frac{2}{3} \end{bmatrix} \mathbf{r} = \begin{bmatrix} -0.9333 \\ -0.6333 \end{bmatrix}.$$

The hard-decision of $\mathbf{c}_{\mathrm{zf}}$ is $[-1\ -1]^{\mathrm{T}}$, and from this, the hard-decision of $\mathbf{x}$ is given by

$$\mathbf{U}^{-1}\begin{bmatrix} -1 \\ -1 \end{bmatrix} = \begin{bmatrix} 1 \\ -1 \end{bmatrix}.$$

This provides the correct answer.

MMSE detection is also available with the LR. Consider the MMSE filter matrix to estimate $\mathbf{c} - \mathcal{E}[\mathbf{c}]$ (we need to offset the mean in the MMSE combining):

$$\mathbf{W}_{\mathrm{mmse}} = \arg\min_{\mathbf{W}} \mathcal{E}\left[||(\mathbf{c} - \mathcal{E}[\mathbf{c}]) - \mathbf{W}^{\mathrm{T}}(\mathbf{r} - \mathcal{E}[\mathbf{r}])||^2\right]. \tag{9.13}$$

Then, the MMSE estimate becomes

$$\hat{\mathbf{c}}_{\mathrm{mmse}} = \mathbf{W}_{\mathrm{mmse}}^{\mathrm{T}}(\mathbf{r} - \mathcal{E}[\mathbf{r}]) + \mathcal{E}[\mathbf{c}].$$

After rounding, the MMSE estimate of $\mathbf{s}$ is given by

$$\hat{\mathbf{s}}_{\mathrm{mmse}} = \mathbf{U}^{-1}\lceil \hat{\mathbf{c}}_{\mathrm{mmse}} \rfloor.$$

9.2.3 LR-based ZF-SIC and LR-based MMSE-SIC detection

Using the QR factorization, the LR-based ZF-SIC can be derived. Consider the QR factorization of $\mathbf{G} = \mathbf{H}\mathbf{U}^{-1}$:

$$\mathbf{G} = \mathbf{Q}\mathbf{R}, \tag{9.14}$$

where $\mathbf{Q}$ is unitary and $\mathbf{R}$ is upper triangular. Pre-multiplying $\mathbf{Q}^{\mathrm{T}}$ to $\mathbf{r}$ in (9.9), we have

$$\begin{aligned}\mathbf{Q}^{\mathrm{T}}\mathbf{r} &= \mathbf{Q}^{\mathrm{T}}(\mathbf{H}\mathbf{s} + \mathbf{n}) \\ &= \mathbf{Q}^{\mathrm{T}}(\mathbf{G}\mathbf{c} + \mathbf{n}) \\ &= \mathbf{R}\mathbf{c} + \mathbf{Q}^{\mathrm{T}}\mathbf{n}.\end{aligned} \tag{9.15}$$

As shown in (9.15), the SIC can be carried out with $\mathbf{Q}^{\mathrm{T}}\mathbf{r}$ to detect $\mathbf{c}$.

Example 9.2.3 (Continued from Example 9.2.2) Suppose that $\mathbf{H}$ can be factorized as follows:

$$\mathbf{H} = \underbrace{\begin{bmatrix} 1 & -1 \\ 0.5 & 1 \end{bmatrix}}_{=\mathbf{G}} \underbrace{\begin{bmatrix} 1 & 2 \\ 0 & 1 \end{bmatrix}}_{=\mathbf{U}}.$$

To perform the ZF-SIC detection, the QR factorization of $\mathbf{G}$ can be considered:

$$\mathbf{G} = \underbrace{\begin{bmatrix} -\sqrt{0.8} & -\sqrt{0.2} \\ -\sqrt{0.2} & \sqrt{0.8} \end{bmatrix}}_{=\mathbf{Q}} \underbrace{\begin{bmatrix} -\sqrt{1.25} & \sqrt{0.2} \\ 0 & \sqrt{1.8} \end{bmatrix}}_{=\mathbf{R}}.$$

Since the column vectors of $\mathbf{G}$ are nearly orthogonal, the off-diagonal element of $\mathbf{R}$, which is $\sqrt{0.2}$, is smaller than the diagonal elements, which are $\sqrt{1.25}$ and $\sqrt{1.8}$. For comparison purposes, the QR factorization of the original channel matrix $\mathbf{H}$ can be considered as follows:

$$\mathbf{H} = \underbrace{\begin{bmatrix} -\sqrt{0.8} & -\sqrt{0.2} \\ -\sqrt{0.2} & \sqrt{0.8} \end{bmatrix}}_{=\mathbf{Q}} \underbrace{\begin{bmatrix} -\sqrt{1.25} & \sqrt{3.2} \\ 0 & \sqrt{1.8} \end{bmatrix}}_{=\mathbf{R}}.$$

In this case, the off-diagonal element of $\mathbf{R}$, which is $\sqrt{3.2}$, is larger than the diagonal elements. Since the off-diagonal element of $\mathbf{R}$ from $\mathbf{G}$ is much smaller than that from $\mathbf{H}$, we can see that the impact of error propagation is less significant if the SIC is performed with the LR, which could result in a better detection performance.

The LR-based MMSE-SIC detector can be derived with the extended channel matrix. With $\mathbf{H}_{\mathrm{ex}}$, the LR can be performed as follow:

$$\mathbf{H}_{\mathrm{ex}} = \begin{bmatrix} \mathbf{H} \\ c\mathbf{I} \end{bmatrix} = \mathbf{G}_{\mathrm{ex}} \mathbf{U}_{\mathrm{ex}},$$

where $c = \sqrt{N_0/E_s}$. With $\mathbf{G}_{\mathrm{ex}}$, the QR factorization is given by

$$\mathbf{G}_{\mathrm{ex}} = \mathbf{Q}_{\mathrm{ex}} \mathbf{R}_{\mathrm{ex}}.$$

The extended received signal vector can be now expressed as

$$\begin{aligned} \mathbf{r}_{\mathrm{ex}} &= [\mathbf{r}^{\mathrm{T}}\ \mathbf{0}^{\mathrm{T}}]^{\mathrm{T}} \\ &= \mathbf{H}_{\mathrm{ex}} \mathbf{s} + \mathbf{n}_{\mathrm{ex}} \\ &= \mathbf{G}_{\mathrm{ex}} \mathbf{c} + \mathbf{n}_{\mathrm{ex}} \\ &= \mathbf{Q}_{\mathrm{ex}} \mathbf{R}_{\mathrm{ex}} \mathbf{c} + \mathbf{n}_{\mathrm{ex}}, \end{aligned} \tag{9.16}$$

where $\mathbf{c} = \mathbf{U}_{\mathrm{ex}} \mathbf{s}$ and $\mathbf{n}_{\mathrm{ex}}^{\mathrm{T}} = [\mathbf{n}^{\mathrm{T}}\ -c\mathbf{s}^{\mathrm{T}}]^{\mathrm{T}}$. Then, by pre-multiplying $\mathbf{Q}_{\mathrm{ex}}$ to $\mathbf{r}_{\mathrm{ex}}$, we have

$$\mathbf{Q}_{\mathrm{ex}}^{\mathrm{T}} \mathbf{r}_{\mathrm{ex}} = \mathbf{R}_{\mathrm{ex}} \mathbf{c} + \mathbf{Q}_{\mathrm{ex}}^{\mathrm{T}} \mathbf{n}_{\mathrm{ex}}. \tag{9.17}$$

The SIC can be performed with (9.17) to obtain a hard-decision of $\mathbf{c}$ from $\mathbf{Q}_{\mathrm{ex}}^{\mathrm{T}} \mathbf{r}_{\mathrm{ex}}$. Once a hard-decision of $\mathbf{c}$ is available, a hard-decision of $\mathbf{s}$ can be found using the relation of $\mathbf{s} = \mathbf{U}^{-1}\mathbf{c}$.

Figure 9.4 shows simulation results for the BERs of various MIMO detectors when $N = K = 4$ and 16-QAM is used for signaling. The channel matrix is randomly generated for each run and each element of $\mathbf{H}$ is assumed to be an independent CSCG random variable with mean zero and unit variance. It is shown that the performance of the LR-based MIMO detectors can approach the optimal ML performance. In addition, it seems a full diversity order is achieved as the ML detector. Indeed, the LR-based detectors can achieve a full receive antenna diversity. This will be explained in Chapter 10.

As shown in Fig. 9.4, the LR-based linear or SIC detectors can provide excellent performance. Since linear and SIC detectors have low computational complexity that grows *linearly* with the number of transmit antennas, they could be good candidates for hardware implementation. A low complexity, but excellent performance results from the transformed channel matrix, $\mathbf{G}$, in (9.8), which has nearly orthogonal column vectors. Therefore, it is crucial to find nearly orthogonal $\mathbf{G}$ via the LR.

9.2.4 LR-based ML detection

So far, we can observe that LR can help suboptimal MIMO detectors to improve their performance. For ML detection, LR can help to find the ML solution with low

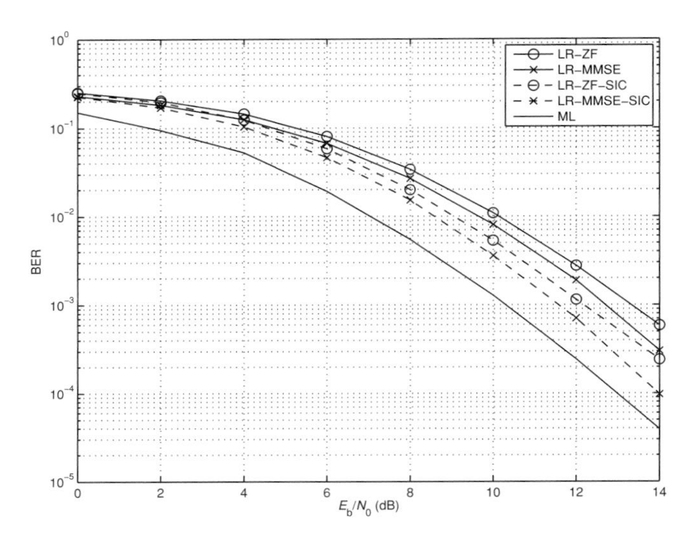

Figure 9.4 BER of various MIMO detectors for 4 × 4 MIMO channels (16-QAM is used for signaling).

computational complexity. In the context of lattice theory, the ML detection problem is a closest vector problem (CVP) ((Mow 2003), (Hoffstein *et al.* 2002)), which is known to be NP-hard. In (Viterbo & Boutros 1999), in order to solve an ML decoding problem, a computationally efficient method proposed in (Fincke & Pohst 1985), which was originally considered to find shortest vectors in a lattice, is applied. The resulting ML decoding algorithm is called the sphere decoding algorithm. For MIMO detection, the sphere decoding algorithm can also be applied to find the ML solution (e.g. (Vikalo, Hassibi & Kailath 2004), (Hochwald & ten Brink 2003), and (Choi 2006)).

While the sphere decoding algorithm can find the ML solution, there are also other approaches ((Agrell, Eriksson, Vardy & Zeger 2002) and (Damen, El Gamal & Caire 2003)) to approximate the ML solution with lower complexity than the sphere decoding algorithm. Those approaches are based on the suboptimal methods that yield an approximate CVP. In this subsection, we mainly focus on the approaches that are based on (Babai 1986).

Assuming that $\mathbf{s} \in \mathbb{Z}^K$, the ML detection problem can be given by

$$\begin{aligned}\hat{\mathbf{s}}_{\rm ml} &= \arg\min_{\mathbf{s}\in\mathbb{Z}^K} ||\mathbf{r} - \mathbf{H}\mathbf{s}||^2 \\ \Leftrightarrow \hat{\mathbf{b}} &= \arg\min_{\mathbf{b}\in\mathbf{\Lambda}} ||\mathbf{r} - \mathbf{b}||^2,\end{aligned} \tag{9.18}$$

where $\hat{\mathbf{b}} = \mathbf{H}\hat{\mathbf{s}}_{\rm ml}$ and

$$\mathbf{\Lambda} = \left\{ \mathbf{b} \;\middle|\; \mathbf{b} = \sum_{k=1}^{K} \mathbf{h}_k u_k,\ u_k \in \mathbb{Z} \right\}.$$

The second minimization problem is a CVP – for a given $\mathbf{r}$, the vector in a lattice $\mathbf{\Lambda}$ that minimizes the Euclidean distance is to be found. Suppose that the channel matrix is factorized as

$$\mathbf{H} = \mathbf{G}\mathbf{U}, \tag{9.19}$$

where $\mathbf{G}$ has the column vectors that are nearly orthogonal to each other and $\mathbf{U}$ is integer unimodular. Then, the ML detection is modified as follows:

$$\arg\min_{\mathbf{s}\in\mathbb{Z}^K} ||\mathbf{r} - \mathbf{G}\mathbf{U}\mathbf{s}||^2 \Rightarrow \arg\min_{\mathbf{c}\in\mathbb{Z}^K} ||\mathbf{r} - \mathbf{G}\mathbf{c}||^2, \tag{9.20}$$

where $\mathbf{c} = \mathbf{U}\mathbf{s} \in \mathbb{Z}^K$. Let

$$\begin{aligned}\hat{\mathbf{c}} &= \arg\min_{\mathbf{x}\in\mathbb{R}^K} ||\mathbf{r} - \mathbf{G}\mathbf{x}||^2 \\ &= \mathbf{G}^\dagger \mathbf{r}.\end{aligned} \tag{9.21}$$

Then, the approximate ML solution of (9.20) becomes

$$\begin{aligned}\hat{\mathbf{c}} &= \lceil \hat{\mathbf{c}} \rfloor \\ &= \lceil \mathbf{G}^\dagger \mathbf{r} \rfloor.\end{aligned} \tag{9.22}$$

This approximate ML solution becomes the exact one if the column vectors of $\mathbf{G}$ are orthogonal or the lattice generated by $\mathbf{G}$ is rectangular as explained in Example 9.2.1 (see also Problem 9.2 to know how this approach can provide an approximate closest vector in a lattice). Furthermore, we can see that it is identical to the LR-based ZF detector described in Subsection 9.2.2.

The approach in (9.22) can be further extended with the notion of the SIC if the column vectors of $\mathbf{G}$ are not sufficiently orthogonal to each other (see Problem 9.1), which is now identical to the LR-based ZF-SIC detector in Subsection 9.2.3. This approach, which is called Babai's approach, was originally proposed in (Babai 1986) to approximately solve the CVP.

9.3 Application to MIMO precoding

There could be various efficient transmission methods if the channel matrix is known at the transmitter in MIMO communication systems. Among them, in this section, we focus on a particular case where a transmitter is equipped with multiple transmit antennas and communicates with multiple receivers simultaneously. This is a typical example for downlink transmissions in cellular systems (Rappaport 1996), (Lee 1982). We introduce linear precoding with known channel matrices that allows simultaneous interference-free transmissions to multiple receivers.

9.3.1 System model

Suppose that a transmitter is equipped with N transmit antennas. For simplicity, assume that each receiver is equipped with a single receive antenna and there are K receivers.

The received signal at the kth receiver is given by

$$r_k = \mathbf{m}_k^{\mathrm{T}} \mathbf{x} + n_k, \tag{9.23}$$

where $[\mathbf{m}_k]_n, n = 1, 2, \ldots, N$, represents the channel gain from the nth transmit antenna to the kth receiver and n_k denotes the background noise at the kth receiver. Here, $\mathbf{x}$ denotes the $N \times 1$ transmitted signal vector that is a sum of the signals to be transmitted to all the K receivers. That is,

$$\begin{aligned}\mathbf{x} &= \sum_{k=1}^{K} \mathbf{v}_k s_k \\ &= \mathbf{V}\mathbf{s},\end{aligned} \tag{9.24}$$

where $\mathbf{v}_k$ is the precoding vector for the kth signal, s_k, to the kth receiver. Substituting (9.24) into (9.23), we have

$$\begin{aligned} r_k &= \mathbf{m}_k^{\mathrm{T}} \mathbf{V}\mathbf{s} + n_k \\ &= \mathbf{m}_k^{\mathrm{T}} \mathbf{v}_k s_k + \sum_{q \neq k} \mathbf{m}_k^{\mathrm{T}} \mathbf{v}_q s_q + n_k. \end{aligned}$$

The second term on the right-hand side becomes the interference. It is desirable to design $\mathbf{V}$ to suppress or minimize the interference to the other receivers. This certainly requires a joint optimization for $\mathbf{V}$.

9.3.2 Precoders

Stacking the K received signals, we have

$$\begin{aligned} \mathbf{r} &= [r_1 \; r_2 \; \ldots \; r_K]^{\mathrm{T}} \\ &= \underbrace{\begin{bmatrix} \mathbf{m}_1^{\mathrm{T}} \\ \mathbf{m}_2^{\mathrm{T}} \\ \vdots \\ \mathbf{m}_K^{\mathrm{T}} \end{bmatrix}}_{=\mathbf{A}} \mathbf{x} + \mathbf{n} \\ &= \mathbf{A}\mathbf{V}\mathbf{s} + \mathbf{n}, \end{aligned} \tag{9.25}$$

where $\mathbf{n} = [n_1 \; n_2 \; \ldots \; n_K]^{\mathrm{T}}$. Now, $\mathbf{A}$ is a $K \times N$ MIMO channel from a transmitter to multiple receivers. If $\mathbf{A}$ is known at the transmitter, a simple approach to suppress the interference is the ZF precoding where the precoding matrix, $\mathbf{V}$, becomes a pseudo-inverse of $\mathbf{A}$. That is,

$$\begin{aligned} \mathbf{V}_{\mathrm{zf}} &= \mathbf{A}^{\dagger} \\ &= \mathbf{A}^{\mathrm{H}}(\mathbf{A}\mathbf{A}^{\mathrm{H}})^{-1} \end{aligned} \tag{9.26}$$

if $N \geq K$ and the rank of $\mathbf{A}$ is K. Unfortunately, this approach could require unrealistically high transmission power. To see this, consider the total transmission power as

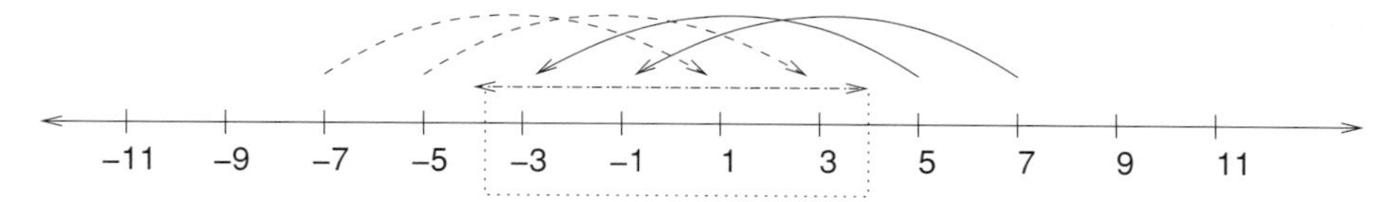

Figure 9.5 Modulo operation for 4-PAM.

follows:

$$\begin{aligned}\mathcal{E}[||\mathbf{x}||^2] &= \mathrm{tr}\left(\mathcal{E}\left[\mathbf{V}_{\mathrm{zf}}\mathbf{s}\mathbf{s}^{\mathrm{H}}\mathbf{V}_{\mathrm{zf}}^{\mathrm{H}}\right]\right) \\ &= E_{\mathrm{s}}\mathrm{tr}\left(\mathbf{V}_{\mathrm{zf}}\mathbf{V}_{\mathrm{zf}}^{\mathrm{H}}\right) \\ &= E_{\mathrm{s}}\mathrm{tr}\left((\mathbf{A}\mathbf{A}^{\mathrm{H}})^{-1}\right) \\ &= E_{\mathrm{s}}\sum_{k=1}^{K}\frac{1}{\lambda_k(\mathbf{A}\mathbf{A}^{\mathrm{H}})},\end{aligned} \tag{9.27}$$

where we assume that $\mathcal{E}[\mathbf{s}\mathbf{s}^{\mathrm{H}}] = E_{\mathrm{s}}\mathbf{I}$. This shows that if $\mathbf{A}$ becomes rank deficient,[1] the transmission power approaches infinity for the ZF precoding, which is impossible. To alleviate this problem, a regularization technique, which is called regularized channel inversion precoding, is used in (Peel, Hochwald & Swindlehurst 2005). The precoding matrix with a regularization parameter, ζ, is given by

$$\mathbf{V}_{\zeta} = \mathbf{A}^{\mathrm{H}}(\mathbf{A}\mathbf{A}^{\mathrm{H}} + \zeta\mathbf{I})^{-1}. \tag{9.28}$$

For a $\zeta > 0$, the transmission power could be reduced at the expense of increasing interference.

Another approach to avoid the problem is by employing the notion of lattices from (Hochwald, Peel & Swindlehurst 2005) and (Windpassinger, Fischer & Huber 2004). For convenience, assume that all the quantities are real-valued. Let $\mathbf{s} \in \mathcal{S}^K$, where $\mathcal{S}$ is a common signal alphabet for s_k. Furthermore, assume that $\mathcal{S}$ is a subset of a one-dimensional lattice. In the vector-perturbation technique (Hochwald *et al.* 2005), $\mathbf{s} + \bar{\mathbf{s}}$ rather than $\mathbf{s}$ is transmitted, where $\bar{\mathbf{s}}$ is an offset that will be decided later. To deal with the offset at the receiver, the following modulo operation is employed to the received signal:

$$x(\mathrm{mod})\Psi = x - \Psi\left\lfloor\frac{x}{\Psi} + \frac{1}{2}\right\rfloor, \tag{9.29}$$

where Ψ is decided to recover the transmitted symbol uniquely. For a symmetric signal constellation, Ψ can be determined as

$$\Psi = 2\left(\max_{s\in\mathcal{S}}|s| + \frac{1}{2}\min_{s_1,s_2\in\mathcal{S};\ s_1\neq s_2}|s_1 - s_2|\right).$$

For example, if $r_k = s_k \in \mathcal{S} = \{-3, -1, 1, 3\}$, $\Psi = 8$. Figure 9.5 illustrates this modulo operation for 4-PAM. Since an offset of an integer multiple of Ψ is suppressed by the

[1] This could happen if the $\mathbf{m}_k$'s are linearly dependent. In cellular systems, if some users are closely located, their $\mathbf{m}_k$'s can be linearly dependent.

modulo operation as follows

$$(s + \Psi u)(\text{mod})\Psi = s, \quad s \in \mathcal{S},\ u \in \mathbb{Z}, \tag{9.30}$$

we can add an offset to minimize the transmission power without causing any performance degradation at the receiver.

Suppose that the ZF precoder is used for precoding at the transmitter. For a given channel matrix $\mathbf{A}$, $\bar{\mathbf{s}}$ is decided to minimize the transmission power as follows:

$$\begin{aligned}\bar{\mathbf{s}} &= \arg\min_{\bar{\mathbf{s}}} ||\mathbf{V}_{\text{zf}}(\mathbf{s} + \bar{\mathbf{s}})||^2 \\ &= \arg\min_{\bar{\mathbf{s}}} (\mathbf{s} + \bar{\mathbf{s}})^{\text{T}}(\mathbf{A}\mathbf{A}^{\text{T}})^{-1}(\mathbf{s} + \bar{\mathbf{s}}) \\ &\text{subject to } [\bar{\mathbf{s}}]_k = \bar{s}_k \in \Psi\mathbb{Z} = \{\Psi u \mid u \in \mathbb{Z}\}, k = 1, 2, \ldots, K.\end{aligned} \tag{9.31}$$

If we look at the constraint carefully, it can be shown that searching for the optimal offset vector $\bar{\mathbf{s}}$ is equivalent to searching for the optimal signal alphabet that minimizes the transmission power. In order to see this clearly, define

$$\mathcal{S}_u = \{s \mid s + \Psi u,\ s \in \mathcal{S}\},\ u \in \mathbb{Z}.$$

It can be shown that

$$\mathcal{S}_u \cap \mathcal{S}_v = \emptyset,\ u \neq v \in \mathbb{Z}.$$

Since the signal alphabet for $\mathbf{s}$ is $\mathcal{S}^K$, the signal alphabet for $\mathbf{s} + \bar{\mathbf{s}}$ can be given by

$$\mathbf{s} + \bar{\mathbf{s}} \in \mathcal{S}_{u_1} \times \mathcal{S}_{u_2} \times \cdots \times \mathcal{S}_{u_K},$$

where $u_k = \bar{s}_k/\Psi$ and $\times$ represents the Cartesian product. Thus, for a given $\bar{\mathbf{s}}$, we can have a unique signal alphabet for $\mathbf{s} + \bar{\mathbf{s}}$. This implies that searching for the best offset vector is equivalent to searching for the best signal alphabet.

According to (9.30), the received signal at the kth receiver after the modulo operation becomes

$$\begin{aligned}\tilde{r}_k &= r_k(\text{mod})\Psi \\ &= (s_k + \bar{s}_k + n_k)(\text{mod})\Psi.\end{aligned} \tag{9.32}$$

If n_k is sufficiently small,

$$\begin{aligned}\tilde{r}_k &= ((s_k + n_k) + \bar{s}_k)(\text{mod})\Psi \\ &\simeq (s_k + n_k)(\text{mod})\Psi.\end{aligned} \tag{9.33}$$

In general, the modulo operation does not change the statistical properties of the received signal significantly. Thus, the detection performance may not be changed.

Noting that the offset vector can be given by

$$\bar{\mathbf{s}} = \Psi \mathbf{u},$$

where $\mathbf{u} \in \mathbb{Z}^K$, the optimization problem in (9.31) becomes a CVP. Consider a square-root factorization of $(\mathbf{A}\mathbf{A}^{\mathrm{T}})^{-1}$ as follows:

$$(\mathbf{A}\mathbf{A}^{\mathrm{T}})^{-1} = \mathbf{B}^{\mathrm{T}}\mathbf{B}.$$

Then, since

$$(\mathbf{s} + \bar{\mathbf{s}})^{\mathrm{T}}(\mathbf{A}\mathbf{A}^{\mathrm{T}})^{-1}(\mathbf{s} + \bar{\mathbf{s}}) = ||\mathbf{B}\mathbf{s} + \mathbf{B}\Psi\mathbf{u}||^2,$$

the corresponding CVP is to find

$$\hat{\mathbf{b}} = \arg \min_{\mathbf{b} \in \mathbf{\Lambda}} ||\mathbf{B}\mathbf{s} - \mathbf{b}||^2, \tag{9.34}$$

where $\mathbf{\Lambda}$ is the lattice generated by $\Psi\mathbf{B}$. Once $\hat{\mathbf{b}}$ is found, the offset vector is given by $\bar{\mathbf{s}} = -\mathbf{B}^{-1}\hat{\mathbf{b}}$.

The approaches discussed in Subsection 9.2.4 to solve the CVP can be applied to (9.34). In (Hochwald *et al.* 2005), the sphere-decoding algorithm is applied, while Babai's approach is applied in (Windpassinger *et al.* 2004). Thus, the precoding solution in (Windpassinger *et al.* 2004) is a low complexity suboptimal solution.

It is noteworthy that the precoding methods require the channel matrix, which is generally estimated. Thus, the precoding performance could be degraded by channel estimation errors.

9.4 Lattice reduction for two-dimensional lattices

So far, we assumed that a channel matrix can be factorized as $\mathbf{H} = \mathbf{G}\mathbf{U}$, where $\mathbf{G}$ has nearly orthogonal column vectors and $\mathbf{U}$ is integer unimodular. In this and the next section, we will focus on this factorization, which is called lattice basis reduction or LR for short. The LR for two-dimensional lattices, in which two shortest vectors are to be found, is relatively simple and explained in this section. A general case for lattices of dimension $K > 2$ is discussed in Section 9.5.

9.4.1 QR factorization

Since the QR factorization is closely related to the LR, we review the QR factorization.

Consider a matrix of two column vectors, say $\mathbf{H} = [\mathbf{h}_1 \; \mathbf{h}_2]$. The subspace spanned by $\mathbf{h}_1$ and $\mathbf{h}_2$ is denoted by $\mathrm{Span}\{\mathbf{h}_1, \mathbf{h}_2\}$. That is,

$$\mathrm{Span}\{\mathbf{h}_1, \mathbf{h}_2\} = \left\{\mathbf{h} \;\middle|\; \mathbf{h} = a_1\mathbf{h}_1 + a_2\mathbf{h}_2, \; a_1, a_2 \in \mathbb{R}\right\}.$$

We want to find another set of two orthonormal basis vectors that spans the same subspace. To this end, we can consider the following two orthogonal vectors:

$$\begin{aligned} \tilde{\mathbf{h}}_1 &= \tilde{\mathbf{h}}_1; \\ \tilde{\mathbf{h}}_2 &= \mathbf{h}_2 - \mu \mathbf{h}_1, \end{aligned} \tag{9.35}$$

where μ is determined to make sure that $< \tilde{\mathbf{h}}_1, \tilde{\mathbf{h}}_2 >= \tilde{\mathbf{h}}_1^{\mathrm{H}} \tilde{\mathbf{h}}_2 = 0$:

$$\mu = \frac{< \mathbf{h}_2, \tilde{\mathbf{h}}_1 >}{||\tilde{\mathbf{h}}_1||^2} = \frac{< \mathbf{h}_2, \mathbf{h}_1 >}{||\mathbf{h}_1||^2}. \tag{9.36}$$

It can be easily verified that

$$\mathrm{Span}\{\tilde{\mathbf{h}}_1, \tilde{\mathbf{h}}_2\} = \mathrm{Span}\{\mathbf{h}_1, \mathbf{h}_2\}$$

using the linear relationship in (9.35). From (9.35), we can also show that

$$\begin{aligned} \mathbf{h}_1 &= \tilde{\mathbf{h}}_1; \\ \mathbf{h}_2 &= \tilde{\mathbf{h}}_2 + \mu \tilde{\mathbf{h}}_1 \end{aligned}$$

or

$$[\mathbf{h}_1 \ \mathbf{h}_2] = [\tilde{\mathbf{h}}_1 \ \tilde{\mathbf{h}}_2] \begin{bmatrix} 1 & \mu \\ 0 & 1 \end{bmatrix}. \tag{9.37}$$

Define $\mathbf{q}_l = \tilde{\mathbf{h}}_l / ||\tilde{\mathbf{h}}_l||$ (provided that $\tilde{\mathbf{h}}_l$ is a non-zero vector). Then, Eq. (9.37) can be rewritten as

$$\begin{aligned} [\mathbf{h}_1 \ \mathbf{h}_2] &= [\mathbf{q}_1 \ \mathbf{q}_2] \begin{bmatrix} ||\tilde{\mathbf{h}}_1|| & 0 \\ 0 & ||\tilde{\mathbf{h}}_2|| \end{bmatrix} \begin{bmatrix} 1 & \mu \\ 0 & 1 \end{bmatrix} \\ &= [\mathbf{q}_1 \ \mathbf{q}_2] \begin{bmatrix} ||\tilde{\mathbf{h}}_1|| & \mu ||\tilde{\mathbf{h}}_1|| \\ 0 & ||\tilde{\mathbf{h}}_2|| \end{bmatrix}. \end{aligned} \tag{9.38}$$

Letting $\mathbf{Q} = [\mathbf{q}_1 \ \mathbf{q}_2]$ and

$$\mathbf{R} = \begin{bmatrix} ||\tilde{\mathbf{h}}_1|| & \mu ||\tilde{\mathbf{h}}_1|| \\ 0 & ||\tilde{\mathbf{h}}_2|| \end{bmatrix}, \tag{9.39}$$

we can show that $\mathbf{H} = \mathbf{QR}$, where $\mathbf{Q}$ is a unitary matrix whose column vectors are orthogonal (i.e. $\mathbf{Q}^{\mathrm{H}}\mathbf{Q} = \mathbf{I}$) and $\mathbf{R}$ is upper triangular. So far, what we have done is a QR factorization. A different QR factorization of $\mathbf{H}$ is available by letting $\tilde{\mathbf{h}}_2 = \mathbf{h}_2$ and $\tilde{\mathbf{h}}_1 = \mathbf{h}_1 - \mu' \mathbf{h}_2$.

In SIC detection, the detection order becomes important to mitigate error propagation as explained in Chapter 8. In addition, the overall detection performance depends on the two diagonal elements of $\mathbf{R}$, namely $r_{11} = ||\tilde{\mathbf{h}}_1||$ and $r_{22} = ||\tilde{\mathbf{h}}_2||$. The magnitude of r_{22} in $\mathbf{R}$ or the length of $\tilde{\mathbf{h}}_2$ (as shown in (9.39)) should be sufficiently large to reduce the chance of error propagation. The squared norm of $\tilde{\mathbf{h}}_2$ can be found as

$$\begin{aligned}
||\tilde{\mathbf{h}}_2||^2 &= ||\mathbf{h}_2 - \mu\mathbf{h}_1||^2 \\
&= < \mathbf{h}_2 - \mu\mathbf{h}_1, \mathbf{h}_2 - \mu\mathbf{h}_1 > \\
&= < \mathbf{h}_2, \mathbf{h}_2 - \mu\mathbf{h}_1 > \\
&= ||\mathbf{h}_2||^2 - \mu < \mathbf{h}_2, \mathbf{h}_1 > \\
&= ||\mathbf{h}_2||^2 - \frac{| < \mathbf{h}_2, \mathbf{h}_1 > |^2}{||\mathbf{h}_1||^2} \\
&\geq 0.
\end{aligned}$$

The third equality results from that $\mathbf{h}_1 \perp \mathbf{h}_2 - \mu\mathbf{h}_1$. From the above, we can see that the performance can be better if the correlation between $\mathbf{h}_1$ and $\mathbf{h}_2$ is lower, because $||\tilde{\mathbf{h}}_2||$ increases as the correlation decreases. Therefore, $\min\{r_{11}, r_{22}\} = \min\{||\mathbf{h}_1||, ||\tilde{\mathbf{h}}_2||\}$ can be a good performance indicator for the SIC detector. For a better performance of SIC detection, we need to maximize $\min\{||\mathbf{h}_1||, ||\tilde{\mathbf{h}}_2||\}$ by changing detection order.

Example 9.4.1 Consider a 2×2 matrix, say $\mathbf{H}_1$, and a modified one obtained by column swaps, say $\mathbf{H}_2$:

$$\mathbf{H}_1 = \begin{bmatrix} 1 & 2 \\ 4 & -1 \end{bmatrix} \quad \text{and} \quad \mathbf{H}_2 = \begin{bmatrix} 2 & 1 \\ -1 & 4 \end{bmatrix}.$$

In SIC detection, the detection order with $\mathbf{H}_1$ is $\{2 \rightarrow 1\}$, while that with $\mathbf{H}_2$ is $\{1 \rightarrow 2\}$. For $\mathbf{H}_1$,

$$\{r_{11}, r_{22}\} = \{||\mathbf{h}_1||, ||\tilde{\mathbf{h}}_2||\} = \{4.1231, 2.1828\},$$

while for $\mathbf{H}_2$,

$$\{r_{11}, r_{22}\} = \{||\mathbf{h}_1||, ||\tilde{\mathbf{h}}_2||\} = \{4.0249, 2.2361\}.$$

This shows that the optimal detection order is $\{1 \rightarrow 2\}$ (corresponding to $\mathbf{H}_2$) rather than $\{2 \rightarrow 1\}$ (corresponding to $\mathbf{H}_1$, which is the original channel matrix).

For convenience, for a given $\mathbf{A} = [\mathbf{a}_1 \ \mathbf{a}_2]$ of two column vectors, according to (9.39), let $\{r_{11}, r_{22}\}$ denote the diagonal elements of the upper triangular matrix of $\mathbf{A} = \mathbf{QR}$, where

$$\{r_{11}, r_{22}\} = \{||\mathbf{a}_1||, ||\tilde{\mathbf{a}}_2||\}.$$

Here, $\tilde{\mathbf{a}}_2 = \mathbf{a}_2 - \mu\mathbf{a}_1$, where $\mu =< \mathbf{a}_2, \mathbf{a}_1 > /||\mathbf{a}_1||^2$. From this representation, we have the following result.

Lemma 9.4.1 *The determinant of* $\mathbf{A}$ *of two column vectors is given by*

$$\det(\mathbf{A}) = r_{11}r_{22} = ||\mathbf{a}_1|| \cdot ||\tilde{\mathbf{a}}_2||. \tag{9.40}$$

Furthermore, if $\mathbf{B} = [\mathbf{b}_1\ \mathbf{b}_2] = \mathbf{AU}$, *where* $\mathbf{U}$ *of size* 2×2 *is unimodular,*

$$|\det(\mathbf{B})| = ||\mathbf{b}_1|| \cdot ||\tilde{\mathbf{b}}_2|| = |\det(\mathbf{A})| = ||\mathbf{a}_1|| \cdot ||\tilde{\mathbf{a}}_2||. \tag{9.41}$$

Proof: See Problem 9.5. □

9.4.2 Lattice basis reduction method

As mentioned earlier, LR is closely related to QR factorization. While a new (orthogonal) basis that has the same subspace of a given matrix, say $\mathbf{H}$, is to be found in QR factorization, a new (shortest) basis that generates the same lattice is to be found in LR. As shown in Section 9.2, LR plays a crucial role in improving the performance of suboptimal MIMO detectors, including MMSE or SIC detectors, by finding a new basis that is nearly orthogonal.

To see the impact of LR clearly, consider SIC detection with $K = 2$. As discussed in Subsection 9.4.1, for SIC detection, the best matrix, $\mathbf{G}$, that maximizes $\min\{||\mathbf{g}_1||, ||\tilde{\mathbf{g}}_2||\}$ can be found as follows:

$$\begin{aligned} &\max_{\mathbf{U}} \min\{||\mathbf{g}_1||, ||\tilde{\mathbf{g}}_2||\} \\ &\text{subject to } \mathbf{H} = \mathbf{GU}, \text{ where } \mathbf{U} \text{ is integer unimodular.} \end{aligned} \tag{9.42}$$

The following result is given in (Yao & Wornell 2002).

Lemma 9.4.2 *Consider a channel matrix,* $\mathbf{H} = [\mathbf{h}_1\ \mathbf{h}_2]$, *where* $K = 2$. *Let* $\mathbf{\Lambda}$ *denote the lattice generated by* $\{\mathbf{h}_1, \mathbf{h}_2\}$. *Let* $\mathbf{u}$ *be a shortest vector in* $\mathbf{\Lambda}$ *and* $\mathbf{v}$ *is a shortest vector that is not proportional to* $\mathbf{u}$ *in* $\mathbf{\Lambda}$. *Then,*

$$\min\{||\mathbf{u}||, ||\tilde{\mathbf{v}}||\} \geq \min\{||\mathbf{g}_1||, ||\tilde{\mathbf{g}}_2||\},$$

where $\mathbf{H} = \mathbf{GU}$ *and* $\mathbf{U}$ *is any integer unimodular.*

Proof: This can be proved by the following two parts:

$$(1)\ ||\mathbf{u}|| \geq \min\{||\mathbf{g}_1||, ||\tilde{\mathbf{g}}_2||\};$$

$$(2)\ ||\mathbf{v}|| \geq \min\{||\mathbf{g}_1||, ||\tilde{\mathbf{g}}_2||\}.$$

(1) Since the same lattice, $\mathbf{\Lambda}$, can be generated by $\{\mathbf{g}_1, \mathbf{g}_2\}$, we have

$$\mathbf{u} = c_1\mathbf{g}_1 + c_2\mathbf{g}_2,$$

where $c_1, c_2 \in \mathbb{Z}$. From this, it follows that

$$\begin{aligned} ||\mathbf{u}|| &= ||c_1\mathbf{g}_1 + c_2\mathbf{g}_2|| \\ &= ||c_1\mathbf{g}_1 + c_2(\tilde{\mathbf{g}}_2 + \hat{\mathbf{g}}_2)|| \\ &= ||(c_1\mathbf{g}_1 + c_2\hat{\mathbf{g}}_2) + c_2\tilde{\mathbf{g}}_2|| \\ &\geq ||c_2\tilde{\mathbf{g}}_2||, \end{aligned}$$

where $\hat{\mathbf{g}}_2 = \mathbf{g}_2 - \tilde{\mathbf{g}}_2 \perp \tilde{\mathbf{g}}_2$. The inequality results from that $\tilde{\mathbf{g}}_2$ is orthogonal to $\mathbf{g}_1$ and $\hat{\mathbf{g}}_2$. If $|c_2| \ge 1$, it can be shown that

$$||\mathbf{u}|| \ge ||c_2\tilde{\mathbf{g}}_2|| \ge ||\tilde{\mathbf{g}}_2|| \ge \min\{||\mathbf{g}_1||, ||\tilde{\mathbf{g}}_2||\}.$$

If $c_2 = 0$ and $c_1 \neq 0$, $\mathbf{u} = c_1\mathbf{g}_1$. Thus, $||\mathbf{u}|| = |c_1|||\mathbf{g}_1|| \ge ||\mathbf{g}_1||$. In addition, since $\mathbf{u}$ is a shortest vector, $||\mathbf{u}|| \le ||\mathbf{g}_1||$. Consequently, in this case, we can show that

$$||\mathbf{u}|| = ||\mathbf{g}_1|| \ge \min\{||\mathbf{g}_1||, ||\tilde{\mathbf{g}}_2||\}.$$

This completes Part (1).

(2) We assumed that $\{\mathbf{u}, \mathbf{v}\}$ and $\{\mathbf{g}_1, \mathbf{g}_2\}$ build the same lattice, i.e. $[\mathbf{u}\ \mathbf{v}] = [\mathbf{g}_1\ \mathbf{g}_2]\mathbf{U}$ for a certain (integer) unimodular matrix $\mathbf{U}$. Thus, according to Lemma 9.4.1, we have $||\mathbf{u}|| \cdot ||\tilde{\mathbf{v}}|| = ||\mathbf{g}_1|| \cdot ||\tilde{\mathbf{g}}_2||$. Since $||\mathbf{u}|| \le ||\mathbf{g}_1||$ (as $\mathbf{u}$ is a shortest vector in $\mathbf{\Lambda}$),

$$||\tilde{\mathbf{v}}|| \ge ||\tilde{\mathbf{g}}_2|| \ge \min\{||\mathbf{g}_1||, ||\tilde{\mathbf{g}}_2||\}.$$

This completes Part (2). □

According to Lemma 9.4.2, we can see that the best $\mathbf{G}$ for the optimization problem in (9.42) is $\mathbf{G} = [\mathbf{u}\ \mathbf{v}]$. In other words, the basis consisting of two shortest vectors in a lattice can provide the best performance for the LR based SIC detector. This basis can be found by LR.

Since the performance of SIC detection depends on the detection order, we need to find the best order. Since $K = 2$, there are two possible orders or there are two possible $\mathbf{G}$'s: $\mathbf{G} = [\mathbf{u}\ \mathbf{v}]$ or $\mathbf{G} = [\mathbf{v}\ \mathbf{u}]$. The following result shows that $\mathbf{G} = [\mathbf{u}\ \mathbf{v}]$ is better.

Lemma 9.4.3 *Consider a two-dimensional lattice,* $\mathbf{\Lambda}$. *Let* $\mathbf{u}$ *be the shortest vector in* $\mathbf{\Lambda}$ *and* $\mathbf{v}$ *is the shortest vector that is not proportional to* $\mathbf{u}$ *in* $\mathbf{\Lambda}$. *Then,*

$$\min\{||\mathbf{u}||, ||\tilde{\mathbf{v}}||\} \ge \min\{||\mathbf{v}||, ||\tilde{\mathbf{u}}||\}.$$

Proof: We can easily show that $\min\{||\mathbf{v}||, ||\tilde{\mathbf{u}}||\} = ||\tilde{\mathbf{u}}||$ because $||\tilde{\mathbf{u}}|| \le ||\mathbf{u}|| \le ||\mathbf{v}||$ (by definitions). In addition, since $||\mathbf{u}|| \ge ||\tilde{\mathbf{u}}||$, we only need to show $||\tilde{\mathbf{u}}|| \le ||\tilde{\mathbf{v}}||$. From

$$||\tilde{\mathbf{u}}||^2 = ||\mathbf{u}||^2 - \frac{|<\mathbf{u}, \mathbf{v}>|^2}{||\mathbf{v}||^2};$$

$$||\tilde{\mathbf{v}}||^2 = ||\mathbf{v}||^2 - \frac{|<\mathbf{v}, \mathbf{u}>|^2}{||\mathbf{u}||^2},$$

we have

$$\begin{aligned} |<\mathbf{v}, \mathbf{u}>|^2 &= ||\mathbf{v}||^2(||\mathbf{u}||^2 - ||\tilde{\mathbf{u}}||^2) \\ &= ||\mathbf{u}||^2(||\mathbf{v}||^2 - ||\tilde{\mathbf{v}}||^2). \end{aligned}$$

From this, it follows

$$\frac{||\tilde{\mathbf{u}}||^2}{||\mathbf{u}||^2} = \frac{||\tilde{\mathbf{v}}||^2}{||\mathbf{v}||^2}$$

$$\text{or} \quad ||\tilde{\mathbf{u}}||^2 = \frac{||\mathbf{u}||^2}{||\mathbf{v}||^2}||\tilde{\mathbf{v}}||^2 \le ||\tilde{\mathbf{v}}||^2.$$

This completes the proof. □

Using the results above, we can derive the Gaussian lattice reduction algorithm, which is an LR algorithm for two-dimensional lattices. Consider a matrix of two column vectors, $\mathbf{B} = [\mathbf{b}_1\ \mathbf{b}_2]$, and suppose that $||\mathbf{b}_1|| \leq ||\mathbf{b}_2||$. Let $\mathbf{v}_1 = \mathbf{b}_1$ and

$$\mathbf{v}_2 = \mathbf{b}_2 - t\mathbf{b}_1,$$

where $t \in \mathbb{Z}$. Then, the following relation holds

$$[\mathbf{v}_1\ \mathbf{v}_2] = [\mathbf{b}_1\ \mathbf{b}_2] \underbrace{\begin{bmatrix} 1 & -t \\ 0 & 1 \end{bmatrix}}_{=\mathbf{U}}, \tag{9.43}$$

where $\mathbf{U}$ is integer unimodular. This means that the basis $\{\mathbf{v}_1, \mathbf{v}_2\}$ has the same lattice as the basis $\{\mathbf{b}_1, \mathbf{b}_2\}$. We want to determine t that minimizes the length of $\mathbf{v}_2$. That is,

$$\begin{aligned} \hat{t} &= \arg\min_{t \in \mathbb{Z}} ||\mathbf{v}_2||^2 \\ &= \arg\min_{t \in \mathbb{Z}} ||\mathbf{b}_2 - t\mathbf{b}_1||^2. \end{aligned}$$

Since the objective function is quadratic, the solution can be easily found as

$$\hat{t} = \lceil \frac{< \mathbf{b}_2, \mathbf{b}_1 >}{||\mathbf{b}_1||^2} \rfloor. \tag{9.44}$$

The minimization of the length of $\mathbf{v}_2$ is equivalent to the minimization of the absolute correlation between $\mathbf{b}_1$ and $\mathbf{b}_2 - t\mathbf{b}_1$ or $| < \mathbf{b}_2 - t\mathbf{b}_1, \mathbf{b}_1 > |^2$:

$$\begin{aligned} \hat{t} &= \arg\min_{t \in \mathbb{Z}} | < \mathbf{b}_2 - t\mathbf{b}_1, \mathbf{b}_1 > |^2 \\ &= \arg\min_{t \in \mathbb{Z}} | < \mathbf{b}_2, \mathbf{b}_1 > - t||\mathbf{b}_1||^2|^2 \\ &= \lceil \frac{< \mathbf{b}_2, \mathbf{b}_1 >}{||\mathbf{b}_1||^2} \rfloor. \end{aligned}$$

If t is not restricted to $\mathbb{Z}$, t becomes μ in (9.36) and the two vectors, $\mathbf{v}_1$ and $\mathbf{v}_2$, can be orthogonal. However, since $t \in \mathbb{Z}$, the two vectors, $\mathbf{v}_1$ and $\mathbf{v}_2$, may not be orthogonal and their correlation is interesting. The correlation between $\mathbf{v}_1$ and $\mathbf{v}_2$ is given by

$$\begin{aligned} < \mathbf{v}_2, \mathbf{v}_1 > &= < (\mathbf{b}_2 - \hat{t}\mathbf{b}_1), \mathbf{b}_1 > \\ &= < \mathbf{b}_2, \mathbf{b}_1 > - \hat{t}||\mathbf{b}_1||^2 \\ &= \left(\frac{< \mathbf{b}_2, \mathbf{b}_1 >}{||\mathbf{b}_1||^2} - \hat{t} \right) ||\mathbf{b}_1||^2. \end{aligned} \tag{9.45}$$

From (9.44), we can show that

$$\left| \frac{< \mathbf{b}_2, \mathbf{b}_1 >}{||\mathbf{b}_1||^2} - \hat{t} \right| \leq \frac{1}{2}.$$

Lemma 9.4.4 *Suppose that a basis $\{\mathbf{u}, \mathbf{v}\}$, which generates a lattice $\mathbf{\Lambda}$, is given. If $||\mathbf{u}|| < ||\mathbf{v}||$ and $| < \mathbf{u}, \mathbf{v} > | \leq \frac{1}{2}||\mathbf{u}||^2$, then (i) $\mathbf{u}$ is the shortest vector in $\mathbf{\Lambda}$ and (ii) $\mathbf{v}$ is the shortest vector in $\mathbf{\Lambda} \backslash \{ \alpha\mathbf{u},\ \alpha \in \mathbb{Z}\}$.*

Proof: (i) Consider a vector in $\mathbf{\Lambda}$, $\mathbf{s} = a\mathbf{u} + b\mathbf{v}$, $a, b \in \mathbb{Z}$. We have

$$\begin{aligned} ||\mathbf{s}||^2 &= |a|^2||\mathbf{u}||^2 + |b|^2||\mathbf{v}||^2 + 2 < a\mathbf{u}, b\mathbf{v} > \\ &\geq |a|^2||\mathbf{u}||^2 + |b|^2||\mathbf{v}||^2 + 2|ab| \, |< \mathbf{u}, \mathbf{v} >| \\ &\geq (|a|^2 + |b|^2 + |ab|)||\mathbf{u}||^2. \end{aligned} \tag{9.46}$$

As $a, b \in \mathbb{Z}$, $|a|^2 + |b|^2 + |ab| \geq 1$ unless $a = b = 0$. Therefore, $||\mathbf{s}||^2 \geq ||\mathbf{u}||^2$ for any $\mathbf{s} \in \mathbf{\Lambda}$, $\mathbf{s} \neq \mathbf{0}$. This shows that $\mathbf{u}$ is the shortest vector in $\mathbf{\Lambda}$.

(ii) For a $\mathbf{s} \in \mathbf{\Lambda} \backslash \{ \alpha\mathbf{u},\ \alpha \in \mathbb{Z}\}$, we have $\mathbf{s} = a\mathbf{u} + b\mathbf{v}$, where $a, b \in \mathbb{Z}$ and $b \neq 0$. Then, we can show that

$$\begin{aligned} ||\mathbf{s}||^2 &= |a|^2||\mathbf{u}||^2 + |b|^2||\mathbf{v}||^2 + 2 < a\mathbf{u}, b\mathbf{v} > +(|b|^2||\mathbf{u}||^2 - |b|^2||\mathbf{u}||^2) \\ &= |b|^2(||\mathbf{v}||^2 - ||\mathbf{u}||^2) + \underbrace{|a|^2||\mathbf{u}||^2 + |b|^2||\mathbf{u}||^2 + 2 < a\mathbf{u}, b\mathbf{v} >}_{\geq ||\mathbf{u}||^2} \\ &\geq (|b|^2 - 1)(||\mathbf{v}||^2 - ||\mathbf{u}||^2) + ||\mathbf{v}||^2. \end{aligned} \tag{9.47}$$

Since $|b|^2 - 1 \geq 0$ and $||\mathbf{v}||^2 - ||\mathbf{u}||^2 \geq 0$,

$$||\mathbf{s}||^2 \geq ||\mathbf{v}||^2.$$

This completes the proof. □

If the basis, $\{\mathbf{b}_1, \mathbf{b}_2\}$, satisfies the conditions in Lemma 9.4.4, i.e.

$$\begin{aligned} \frac{|\mathbf{b}_1^{\mathrm{T}}\mathbf{b}_2|}{||\mathbf{b}_1||^2} &\leq \frac{1}{2}; \\ ||\mathbf{b}_1|| &\leq ||\mathbf{b}_2||, \end{aligned} \tag{9.48}$$

they are called Gaussian reduced. The second condition in (9.48) is often relaxed as follows:

$$||\mathbf{b}_1|| \leq \omega||\mathbf{b}_2||, \tag{9.49}$$

where $\omega \geq 1$ to find the Gaussian reduced basis quickly. The Gaussian lattice reduction algorithm updates the basis iteratively by the linear transformation in (9.43) and column swaps to make the basis Gaussian reduced. A pseudo code for the Gaussian lattice reduction is given below.

S1: Input $(\mathbf{h}_1, \mathbf{h}_2)$
S2: Set $\mathbf{b}_1 = \mathbf{h}_1$ and $\mathbf{b}_2 = \mathbf{h}_2$
S3: Set $\mathbf{J} = \begin{bmatrix} 0 & 1 \\ 1 & 0 \end{bmatrix}$ and $\mathbf{Q} = \begin{bmatrix} 1 & 0 \\ 0 & 1 \end{bmatrix}$
S4: do
S5:   if $||\mathbf{b}_1|| > ||\mathbf{b}_2||$
S6:   swap $\mathbf{b}_1$ and $\mathbf{b}_2$, and $\mathbf{Q} = \mathbf{QJ}$
S7:   end if
S8:   if $|< \mathbf{b}_2, \mathbf{b}_1 > /||\mathbf{b}_1||^2| > 1/2$
S9: $t = \lceil \frac{<\mathbf{b}_2, \mathbf{b}_1>}{||\mathbf{b}_1||^2} \rfloor$

S10: $\quad \mathbf{b}_2 = \mathbf{b}_2 - t\mathbf{b}_1$ and $\mathbf{Q} = \mathbf{Q}\begin{bmatrix} 1 & -t \\ 0 & 1 \end{bmatrix}$

S11: `end if`

S12: `while` $||\mathbf{b}_1|| \leq ||\mathbf{b}_2||$

S13: `return` $(\mathbf{b}_1, \mathbf{b}_2, \mathbf{Q})$

If the two column vectors of $\mathbf{H} = [\mathbf{h}_1 \ \mathbf{h}_2]$ are the input vectors to the Gaussian lattice reduction, we have

$$\mathbf{HQ} = \mathbf{G},$$

where the two column vectors of $\mathbf{G} = [\mathbf{g}_1 \ \mathbf{g}_2]$ are the output vectors. Thus, we also have

$$\mathbf{H} = \mathbf{G}\mathbf{Q}^{-1},$$

where $\mathbf{Q}^{-1} = \mathbf{U}$. Note that, in the pseudo code, the matrix $\mathbf{Q}$ is used to trace the update of the integer unimodular matrix $\mathbf{U}$.

Example 9.4.2 Suppose that the input matrix is given by

$$\mathbf{H} = \begin{bmatrix} 60 & 49 \\ 52 & 42 \end{bmatrix}.$$

Let $\mathbf{b}_1 = \mathbf{h}_1$ and $\mathbf{b}_2 = \mathbf{h}_2$. Since $||\mathbf{b}_1|| > ||\mathbf{b}_2||$, a column swap takes place. Now, we have

$$\mathbf{b}_1 = \begin{bmatrix} 49 \\ 42 \end{bmatrix}, \ \mathbf{b}_2 = \begin{bmatrix} 60 \\ 52 \end{bmatrix}.$$

We have

$$\frac{< \mathbf{b}_2, \mathbf{b}_1 >}{||\mathbf{b}_1||^2} = 1.2303 > 1/2.$$

For the linear transformation, t becomes $t = \lceil \frac{<\mathbf{b}_2, \mathbf{b}_1>}{||\mathbf{b}_1||^2} \rfloor = 1$. The updated basis becomes

$$[\mathbf{b}_1 \ \mathbf{b}_2] \Leftarrow [\mathbf{b}_1 \ \mathbf{b}_2]\begin{bmatrix} 1 & -t \\ 0 & 1 \end{bmatrix} = [\mathbf{b}_1 \ \mathbf{b}_2]\begin{bmatrix} 1 & -1 \\ 0 & 1 \end{bmatrix} = \begin{bmatrix} 49 & 11 \\ 42 & 10 \end{bmatrix}.$$

We can continue the Gaussian lattice reduction further. In summary, the sequence of the updated bases by the Gaussian lattice reduction is given by

$$[\mathbf{b}_1 \ \mathbf{b}_2] = \begin{bmatrix} 60 & 49 \\ 52 & 42 \end{bmatrix} \rightarrow \begin{bmatrix} 49 & 11 \\ 42 & 10 \end{bmatrix} \rightarrow \begin{bmatrix} 11 & 5 \\ 10 & 2 \end{bmatrix} \rightarrow \begin{bmatrix} 5 & -4 \\ 2 & 4 \end{bmatrix}.$$

Furthermore,

$$\mathbf{Q} = \begin{bmatrix} 5 & -4 \\ 2 & 4 \end{bmatrix}$$

or

$$\mathbf{U} = \mathbf{Q}^{-1} = \begin{bmatrix} 16 & 13 \\ 5 & 4 \end{bmatrix}.$$

The final Gaussian reduced basis is small and has low correlation. For comparison, we can see the following table:

Bases	norms $(\{||\mathbf{b}_1||, ||\mathbf{b}_2||\})$	normalized correlation $(\frac{\mathbf{b}_1^T\mathbf{b}_2}{||\mathbf{b}_1||\cdot||\mathbf{b}_2||})$
Original	$\{79.39,\ 64.53\}$	$\simeq 1$
Reduced	$\{5.38, 5.65\}$	-0.3939

The Gaussian lattice reduction algorithm can be further generalized with a complex-valued lattice and the result in Lemma 9.4.4 can also be generalized (see (Yao & Wornell 2002)). Consider the set of complex integers, $\mathbb{Z} + j\mathbb{Z}$. Suppose that $||\mathbf{b}_1|| < ||\mathbf{b}_2||$. We can find a new basis, $\{\mathbf{v}_1, \mathbf{v}_2\}$, as in (9.43). The difference would be that $t \in \mathbb{Z} + j\mathbb{Z}$. Thus, we have

$$\begin{aligned}
\hat{t} &= \arg \min_{t \in \mathbb{Z}+j\mathbb{Z}} | < \mathbf{b}_2 - t\mathbf{b}_1, \mathbf{b}_1 > |^2 \\
&= \arg \min_{t \in \mathbb{Z}+j\mathbb{Z}} | < \mathbf{b}_2, \mathbf{b}_1 > -t||\mathbf{b}_1||^2|^2 \\
&= \lceil \frac{< \mathbf{b}_2, \mathbf{b}_1 >}{||\mathbf{b}_1||^2} \rfloor,
\end{aligned}$$

where $\lceil x \rfloor = \lceil \Re(x) \rfloor + j \lceil \Im(x) \rfloor$ for a complex number x. From (9.45), we have

$$< \mathbf{v}_2, \mathbf{v}_1 >= \left(\frac{< \mathbf{b}_2, \mathbf{b}_1 >}{||\mathbf{b}_1||^2} - \hat{t} \right) ||\mathbf{b}_1||^2,$$

and it implies that

$$\begin{aligned}
\left| \Re \left(\frac{< \mathbf{b}_2, \mathbf{b}_1 >}{||\mathbf{b}_1||^2} - \hat{t} \right) \right| &\le 1/2; \\
\left| \Im \left(\frac{< \mathbf{b}_2, \mathbf{b}_1 >}{||\mathbf{b}_1||^2} - \hat{t} \right) \right| &\le 1/2.
\end{aligned} \tag{9.50}$$

Thus, for complex basis vectors, Lemma 9.4.4 can be modified as follows.

Lemma 9.4.5 *Suppose that a basis $\{\mathbf{u}, \mathbf{v}\}$ generates a lattice $\mathbf{\Lambda}$. If $||\mathbf{u}|| < ||\mathbf{v}||$, $|\Re(< \mathbf{u}, \mathbf{v} >)| \le \frac{1}{2}||\mathbf{u}||^2$ and $|\Im(< \mathbf{u}, \mathbf{v} >)| \le \frac{1}{2}||\mathbf{u}||^2$, then (i) $\mathbf{u}$ is the shortest vector in $\mathbf{\Lambda}$ and (ii) $\mathbf{v}$ is the shortest vector in $\mathbf{\Lambda} \backslash \{ \alpha\mathbf{u},\ \alpha \in \mathbb{Z} + j\mathbb{Z}\}$.*

The same pseudo code for real-valued basis vectors can be used for complex-valued basis vectors by replacing the rounding operation with that for a complex number and replacing Step 8 by the following step:

S8′: `if` $|\Re(< \mathbf{b}_2, \mathbf{b}_1 >)|/||\mathbf{b}_1||^2 > 1/2$ `or` $|\Im(< \mathbf{b}_2, \mathbf{b}_1 >)|/||\mathbf{b}_1||^2 > 1/2$

9.5 LLL algorithm: a lattice reduction algorithm

LR is to find a basis whose basis vectors are shortest. As shown in Section 9.4, for two-dimensional lattices, Gaussian lattice reduction provides two shortest basis vectors. The resulting basis is called a reduced basis. For higher-dimensional lattices, we can also define a reduced basis. Unfortunately, unlike the orthogonalized basis, there is more than one definition for a reduced basis. A strict definition is known to be the Minkowski reduced form (Pohst 1993) (see also (Mow 2003)). It is a straightforward extension from the case of $K = 2$. For a given lattice $\mathbf{\Lambda}$ generated by $\mathbf{B}$, the basis $\mathbf{B}$ is *Minkowski reduced* if $\mathbf{b}_1$ is a shortest vector in $\mathbf{\Lambda}$ and $\mathbf{b}_k$ is a shortest vector independent from $\mathbf{b}_1, \mathbf{b}_2, \ldots, \mathbf{b}_{k-1}$, for $k = 2, 3, \ldots, K$. While Gaussian lattice reduction can find a Minkowski reduced basis for the case of $K = 2$ with a small number of iterations, it is known that LR for the Minkowski reduced form is NP-hard. This implies that the complexity of LR becomes prohibitively high as the dimension increases. Thus, it is desirable to have a relaxed definition for a reduced basis to derive computationally efficient LR algorithms for lattices of dimension $K > 2$.

As a generalized Gaussian reduction algorithm for higher-dimensional lattices, the LLL algorithm is proposed in (Lenstra, Lenstra & Lovasz 1982) to find a certain reduced basis, which is called the LLL reduced basis. The significance of the LLL algorithm is that it has a polynomial time complexity. This low complexity becomes possible by relaxing basis reduction conditions.

In this section, we explain the conditions for LLL reduction. These conditions are similar to those for Gaussian reduction for lattices of dimension 2. Then, we describe two different versions of the LLL algorithm.

9.5.1 LLL reduced basis

Given a basis $\{\mathbf{b}_1, \mathbf{b}_2, \ldots, \mathbf{b}_K\}$, we can find an orthogonal basis using the Gram–Schmidt procedure. Let $\tilde{\mathbf{b}}_1 = \mathbf{b}_1$. Then, the other orthogonal basis vectors are given by

$$\tilde{\mathbf{b}}_k = \mathbf{b}_k - \sum_{q=1}^{k-1} \mu_{k,q} \tilde{\mathbf{b}}_q, \; k = 1, 2, \ldots, K, \tag{9.51}$$

where

$$\mu_{k,q} = \frac{\tilde{\mathbf{b}}_q^{\mathrm{T}} \mathbf{b}_k}{||\tilde{\mathbf{b}}_q||^2}, \; q = 1, 2, \ldots, k-1.$$

The coefficients $\mu_{k,q}$ are determined to ensure the following orthogonality:

$$\tilde{\mathbf{b}}_l^{\mathrm{T}} \tilde{\mathbf{b}}_k = 0, \; l = 1, 2, \ldots, k-1; \; k = 1, 2, \ldots, K.$$

As a result, we can see that the column vectors $\tilde{\mathbf{b}}_k$'s are mutually orthogonal. From the $\tilde{\mathbf{b}}_k$'s, the original vectors $\mathbf{b}_k$'s can also be found. From (9.51), it follows that

$$\begin{aligned} \mathbf{b}_k &= \tilde{\mathbf{b}}_k + \sum_{q=1}^{k-1} \mu_{k,q} \tilde{\mathbf{b}}_q \\ &= \sum_{q=1}^{k} \mu_{k,q} \tilde{\mathbf{b}}_q, \ k = 1, 2, \ldots, K, \end{aligned} \tag{9.52}$$

if we let $\mu_{k,k} = 1$ for all k. For convenience, we can adopt a matrix representation as follows:

$$\begin{aligned} \mathbf{B} &= [\mathbf{b}_1 \ \mathbf{b}_2 \ \ldots \ \mathbf{b}_K] \\ &= \tilde{\mathbf{B}}[\boldsymbol{\mu}], \end{aligned} \tag{9.53}$$

where

$$\tilde{\mathbf{B}} = [\tilde{\mathbf{b}}_1 \ \tilde{\mathbf{b}}_2 \ \ldots \ \tilde{\mathbf{b}}_K];$$

$$[\boldsymbol{\mu}] = \begin{bmatrix} \mu_{1,1} & \mu_{2,1} & \cdots & \mu_{K,1} \\ 0 & \mu_{2,2} & \cdots & \mu_{K,2} \\ \vdots & \vdots & \ddots & \vdots \\ 0 & 0 & \cdots & \mu_{K,K} \end{bmatrix}. \tag{9.54}$$

Consider a lattice generated by a pair of basis vectors $\mathbf{b}_k$ and $\mathbf{b}_{k+1}$ projected orthogonally to $\mathbf{b}_1, \mathbf{b}_2, \ldots, \mathbf{b}_{k-1}$ (or $\tilde{\mathbf{b}}_1, \tilde{\mathbf{b}}_2, \ldots, \tilde{\mathbf{b}}_{k-1}$). We can show that this lattice is generated by the following basis vectors:

$$\begin{aligned} \mathbf{b}_k(k) &= \tilde{\mathbf{b}}_k; \\ \mathbf{b}_{k+1}(k) &= \mu_{k+1,k} \tilde{\mathbf{b}}_k + \tilde{\mathbf{b}}_{k+1}, \end{aligned}$$

because, from (9.52),

$$\mathbf{b}_k = \tilde{\mathbf{b}}_k + \mu_{k,k-1} \tilde{\mathbf{b}}_{k-1} + \cdots + \mu_{k,1} \tilde{\mathbf{b}}_1.$$

The Gaussian reduction conditions in (9.48) can be applied to the basis $\{\mathbf{b}_k(k), \mathbf{b}_{k+1}(k)\}$. The LLL basis reduction conditions are as follows (Lenstra *et al.* 1982):

$$|\mu_{k,l}| \le \frac{1}{2}, \quad 1 \le l < k \le K; \tag{9.55}$$

$$||\tilde{\mathbf{b}}_k||^2 \le \omega ||\mu_{k+1,k} \tilde{\mathbf{b}}_k + \tilde{\mathbf{b}}_{k+1}||^2, \ k = 1, 2, \ldots, K-1, \tag{9.56}$$

where ω in (9.49) becomes $4/3$ in (9.56). If a basis satisfies the conditions in (9.55) and (9.56), this basis is called LLL reduced. It is noteworthy that ω can be any constant in a range between 1 and 4, i.e. $1 < \omega < 4$.

For a given basis or matrix, $\mathbf{B}$, the LLL algorithm is to find a basis that satisfies the conditions in (9.55) and (9.56) and more importantly has the same lattice generated by $\mathbf{B}$. Thus, if $\{\mathbf{b}_1, \mathbf{b}_2, \ldots, \mathbf{b}_k\}$ is an LLL reduced basis from $\{\mathbf{h}_1, \mathbf{h}_2, \ldots, \mathbf{h}_K\}$, we should have the following relationship:

$$\begin{aligned}\mathbf{B} &= \mathbf{H}\mathbf{U} \\ &= \tilde{\mathbf{H}}[\mu]\mathbf{U},\end{aligned} \tag{9.57}$$

where $\mathbf{U}$ is integer unimodular.

9.5.2 LLL algorithm

The LLL algorithm is an algorithm that can generate an LLL reduced basis from a given basis in a polynomial time (Lenstra *et al.* 1982). The LLL algorithm consists of the following two steps: (i) size reduction and (ii) column swap. The size reduction is to reduce the correlation using a linear transformation and the column swap is to arrange smaller indices for the shorter basis vectors (in the Gaussian lattice reduction, the first column vector is smaller than or equal to the second one as shown in (9.48)).

The size reduction can be carried out by the following transformation:

$$\mathbf{M}_{k,l} = \mathbf{I} - \gamma \mathbf{e}_k \mathbf{e}_l^{\mathrm{T}}, \ k < l,$$

where γ is an integer and $\mathbf{e}_k$ is the unit vector whose elements are all zeros except the kth element which is one. It can be easily verified that $\mathbf{M}_{k,l}$ is integer unimodular. In addition, for convenience, let

$$f_{k,l} = \begin{cases} \mu_{l,k}, & l \le k; \\ 0, & k > l. \end{cases}$$

From (9.57), suppose that $\mathbf{H}$ is the input matrix. To find the LLL reduced basis, $\mathbf{B}$, let

$$\begin{aligned}\mathbf{B} &\leftarrow \mathbf{H}; \\ \mathbf{U} &\leftarrow \mathbf{I}.\end{aligned}$$

After the Gram–Schmidt orthogonalization, we have

$$\mathbf{B} = \tilde{\mathbf{B}}[\mu] = \tilde{\mathbf{B}}\mathbf{F},$$

where $[\mathbf{F}]_{k,l} = f_{k,l}$. If $|f_{k,l}| > 0.5$, the transformation with $\mathbf{M}_{k,l}$ can be used to convert $|f_{k,l}| < 0.5$ as follows:

$$\begin{aligned}\mathbf{B} &\leftarrow \mathbf{B}\mathbf{M}_{k,l}; \\ \mathbf{F} &\leftarrow \mathbf{F}\mathbf{M}_{k,l}; \\ \mathbf{U} &\leftarrow \mathbf{U}\mathbf{M}_{k,l};\end{aligned} \tag{9.58}$$

with $\gamma = \lceil f_{k,l} \rfloor$. Thus, for any (k, l) with $|f_{k,l}| > 0.5$, the transformation in (9.58) is applied for the size reduction. Note that

$$\text{the } i\text{th column of } \mathbf{F}\mathbf{M}_{k,l} = \begin{cases} \mathbf{f}_l - \gamma \mathbf{f}_k, & \text{if } i = l; \\ \mathbf{f}_i, & \text{if } i \ne l. \end{cases}$$

To perform the size reduction, $\mathbf{M}_{k,l}$ can be applied sequentially. A pseudo code for the size reduction is given below.

```
S1: Input (B, F) % F: upper triangular matrix
S2: Set U = I
S3: for l = 2 : 1 : K
S4:    for k = l − 1 : −1 : 1
S5:       if |f_{k,l}| > 0.5
S6:          B ← BM_{k,l}
S7:          F ← FM_{k,l}
S8:          U ← UM_{k,l}
S9:       end if
S10:   end for
S11: end for
S12: Return (B, F, U)
```

Example 9.5.1 Let

$$\mathbf{F} = \begin{bmatrix} 1 & 0.7 & 0.1 \\ 0 & 1 & -2.6 \\ 0 & 0 & 1 \end{bmatrix}.$$

Since $|f_{1,2}| > 0.5$, the $M_{k,l}$-transformation can be applied as follows:

$$\mathbf{F} \leftarrow \mathbf{F}\mathbf{M}_{1,2} = \begin{bmatrix} 1 & 0.7 & 0.1 \\ 0 & 1 & -2.6 \\ 0 & 0 & 1 \end{bmatrix} \left(\mathbf{I} - 1 \begin{bmatrix} 0 & 1 & 0 \\ 0 & 0 & 0 \\ 0 & 0 & 0 \end{bmatrix} \right)$$

$$= \begin{bmatrix} 1 & -0.3 & 0.1 \\ 0 & 1 & -2.6 \\ 0 & 0 & 1 \end{bmatrix}.$$

The same process can be repeated for $f_{2,3}$:

$$\mathbf{F} \leftarrow \mathbf{F}\mathbf{M}_{2,3} = \begin{bmatrix} 1 & -0.3 & 0.1 \\ 0 & 1 & -2.6 \\ 0 & 0 & 1 \end{bmatrix} \left(\mathbf{I} - (-3) \begin{bmatrix} 0 & 0 & 0 \\ 0 & 0 & 1 \\ 0 & 0 & 0 \end{bmatrix} \right)$$

$$= \begin{bmatrix} 1 & -0.3 & -0.8 \\ 0 & 1 & 0.4 \\ 0 & 0 & 1 \end{bmatrix}.$$

As $|f_{1,3}| > 0.5$, $\mathbf{M}_{1,3}$ has to be applied. Then, the resulting integer unimodular matrix, $\mathbf{U}$, is given by

$$\mathbf{U} = \mathbf{M}_{12}\mathbf{M}_{23}\mathbf{M}_{13} = \begin{bmatrix} 1 & -1 & -2 \\ 0 & 1 & 3 \\ 0 & 0 & 1 \end{bmatrix}.$$

To satisfy the condition in (9.56), we need column swaps. If the condition in (9.56) is not satisfied for some k, we swap the two basis vectors $\mathbf{b}_k$ and $\mathbf{b}_{k+1}$. This operation can be done by the following transformation:

$$\begin{aligned}\mathbf{B} &\leftarrow \mathbf{B}\boldsymbol{\Phi}_k;\\ \mathbf{F} &\leftarrow \mathbf{F}\boldsymbol{\Phi}_k,\end{aligned} \tag{9.59}$$

where

$$\boldsymbol{\Phi}_k = \begin{bmatrix} \mathbf{I}_{k-1} & & & \\ & 0 & 1 & \\ & 1 & 0 & \\ & & & \mathbf{I}_{K-k-1} \end{bmatrix}.$$

Since $\det(\boldsymbol{\Phi}_k) = -1$, $\boldsymbol{\Phi}_k$ is integer unimodular. Note that $\mathbf{F}\boldsymbol{\Phi}_k$ is not upper triangular anymore. Thus, it is required to update $\tilde{\mathbf{B}}$ and $\mathbf{F}$ from $\mathbf{B}\boldsymbol{\Phi}_k$ (another Gram–Schmidt orthogonalization may be required).

A brief (but not computationally efficient yet) pseudo code for the LLL algorithm can be given as below.

S1: `Input` $(\mathbf{B}, \tilde{\mathbf{B}}, \mathbf{F})$
S2: $(\mathbf{B}, \mathbf{F}, \mathbf{U})$ = `SizeReduction`$(\mathbf{B}, \mathbf{F})$
S3: `while` (there exists any k violating (9.56))
S4: `swap` $\mathbf{b}_k$ `and` $\mathbf{b}_{k+1}$ `and update` $\mathbf{U}$
S5: `update` $\tilde{\mathbf{B}}$ `and` $\mathbf{F}$
S6: $(\mathbf{B}, \mathbf{F}, \mathbf{U})$ = `SizeReduction`$(\mathbf{B}, \mathbf{F})$
S7: `end while`
S8: `Return` $(\mathbf{B}, \mathbf{F}, \mathbf{U})$

In (Lenstra *et al.* 1982), it is shown that the above algorithm can be carried out in a polynomial time.

9.5.3 Another version of LLL algorithm

As mentioned earlier, after column swaps, we notice that $\mathbf{F}\boldsymbol{\Phi}_k$ is not upper triangular. A straightforward, but not computationally efficient approach is to perform the Gram–Schmidt orthogonalization. Fortunately, there are other computationally efficient approaches that replace the Gram–Schmidt orthogonalization.

In (Wubben, Bohnke, Kuhn & Kammeyer 2004) and (Luk & Tracy 2008) Givens rotations (Golub & Loan 1983) are used to transform $\mathbf{F}\boldsymbol{\Phi}_k$ into an upper triangular matrix. The approach in (Wubben *et al.* 2004) and (Luk & Tracy 2008) can be explained better with the QR factorization of $\mathbf{B}$ than with the Gram–Schmidt orthogonalization. Let

$$\mathbf{B} = \mathbf{QR}.$$

The LLL basis reduction conditions in (9.55) and (9.56) can be modified with the coefficients in $\mathbf{R}$ as follows:

$$|r_{l,k}| \leq \frac{r_{l,l}}{2}, \quad 1 \leq l < k \leq K; \tag{9.60}$$

$$r_{k,k}^2 \leq \omega\left(r_{k,k+1}^2 + r_{k+1,k+1}^2\right), \quad k = 1, 2, \ldots, K-1. \tag{9.61}$$

For the size reduction in (9.60), we modify (9.58) as follows:

$$\begin{aligned} \mathbf{B} &\leftarrow \mathbf{B}\mathbf{M}_{k,l}; \\ \mathbf{R} &\leftarrow \mathbf{R}\mathbf{M}_{k,l}; \\ \mathbf{U} &\leftarrow \mathbf{U}\mathbf{M}_{k,l}, \end{aligned} \tag{9.62}$$

where

$$\mathbf{M}_{k,l} = \mathbf{I} - \lceil r_{k,l}/r_{k,k} \rfloor \mathbf{e}_k \mathbf{e}_l^{\mathrm{T}}.$$

This reduction operation is denoted by `Reduction(`k, l`)`.

For column swaps, define

$$\mathbf{J}_k = \begin{bmatrix} \mathbf{I}_{k-1} & & & \\ & c & s & \\ & -s & c & \\ & & & \mathbf{I}_{K-k-1} \end{bmatrix},$$

where $c = \cos(\theta)$ and $s = \sin(\theta)$. We can show that

$$\mathbf{J}_k \mathbf{J}_k^{\mathrm{T}} = \mathbf{J}_k^{\mathrm{T}} \mathbf{J}_k = \mathbf{I}$$

for any θ. After a column swap, we have

$$\mathbf{R}\boldsymbol{\Phi}_k = \begin{bmatrix} \mathbf{X}_1 & \times & \times \\ \mathbf{0} & \mathbf{A} & \times \\ \mathbf{0} & \mathbf{0} & \mathbf{X}_2 \end{bmatrix},$$

where $\mathbf{X}_1$ and $\mathbf{X}_2$ denote upper triangular matrices, $\times$ represents irrelevant terms, and

$$\mathbf{A} = \begin{bmatrix} r_{k,k+1} & r_{k,k} \\ r_{k+1,k+1} & 0 \end{bmatrix}.$$

For convenience, let

$$\boldsymbol{\Theta} = \begin{bmatrix} c & s \\ -s & c \end{bmatrix}.$$

Then, we can show that

$$\mathbf{J}_k \mathbf{F} \mathbf{\Phi}_k = \begin{bmatrix} \mathbf{X}_1 & \times & \times \\ \mathbf{0} & \mathbf{\Theta A} & \times \\ \mathbf{0} & \mathbf{0} & \mathbf{X}_2 \end{bmatrix}.$$

As $\mathbf{\Theta}$ is a rotation operation, there exists θ that makes $\mathbf{\Theta A}$ upper triangular

$$\mathbf{\Theta A} = \begin{bmatrix} \times & \times \\ 0 & \times \end{bmatrix}$$

by letting

$$\theta = \tan^{-1} \frac{r_{k+1,k+1}}{r_{k,k+1}}.$$

Thus, we can see that $\mathbf{J}_k \mathbf{F} \mathbf{\Phi}_k$ is upper triangular even after a column swap.

Example 9.5.2 Let

$$\mathbf{A} = \begin{bmatrix} 4 & -3 \\ 1 & 0 \end{bmatrix}.$$

If $\theta = \tan^{-1}(1/4) = 0.2450$,

$$\mathbf{\Theta A} = \begin{bmatrix} 4.1231 & -2.9104 \\ 0 & 0.7276 \end{bmatrix}.$$

Consequently, the column swap operation, denoted by `Swap`$(k, k+1)$, can be carried out as follows:

$$\begin{aligned} \mathbf{B} &\leftarrow \mathbf{B}\mathbf{\Phi}_k; \\ \mathbf{Q} &\leftarrow \mathbf{Q}\mathbf{J}_k^{\mathrm{T}}; \\ \mathbf{R} &\leftarrow \mathbf{J}_k \mathbf{R} \mathbf{\Phi}_k; \\ \mathbf{U} &\leftarrow \mathbf{U}\mathbf{\Phi}_k. \end{aligned}$$

We can confirm that

$$\begin{aligned} \mathbf{B}\mathbf{\Phi}_k &= \tilde{\mathbf{B}} \mathbf{F} \mathbf{\Phi}_k \\ &= \tilde{\mathbf{B}} \underbrace{(\mathbf{J}_k^{\mathrm{T}} \mathbf{J}_k)}_{=\mathbf{I}} \mathbf{F} \mathbf{\Phi}_k \\ &= (\tilde{\mathbf{B}} \mathbf{J}_k^{\mathrm{T}})(\mathbf{J}_k \mathbf{F} \mathbf{\Phi}_k). \end{aligned}$$

A pseudo code for the LLL algorithm with QR factorization and Givens rotations is given below.

S1: `Input` $(\mathbf{B}, \mathbf{Q}, \mathbf{R})$
S2: `Set` $k = 1$ `and` $\mathbf{U} = \mathbf{I}$

```
S3:  while k ≤ K − 1
S4:    if r_{k,k} < 2|r_{k,k+1}|
S5:      Reduction(k, k + 1)
S6:    end if
S7:    if r_{k,k}^2 > ω (r_{k,k+1}^2 + r_{k+1,k+1}^2)
S8:      Swap(k, k + 1)
S9:      k ← max(k − 1, 1)
S10:   else
S11:     for i = k − 1 : −1 : 1
S12:       if r_{i,i} < 2|r_{i,k+1}|
S13:         Reduction(i, k + 1)
S14:       end if
S15:     end for
S16:     k ← k + 1
S17:   end if
S18: end while
```

This pseudo code can be found in (Wubben *et al.* 2004) and (Luk & Tracy 2008). In the original LLL algorithm in (Lenstra *et al.* 1982), a different transformation, which does not require the computation of square roots, is used to make $\mathbf{F}\mathbf{\Phi}_k$ upper triangular. Thus, compared with the original LLL algorithm, the above algorithm requires the computation of square roots due to Givens rotations. However, in (Luk & Tracy 2008), it is claimed that this algorithm is more numerically stable than the original one. The LLL algorithm can also be modified for a complex-valued matrix directly (Gan & Mow 2005). In Chapter 10, we consider LR for complex-valued MIMO channels for analysis purposes.

9.6 Summary and notes

We discussed the LR-based MIMO detectors and well-known LR algorithms in this chapter. The LR-based MIMO detection was considered in (Yao & Wornell 2002) and extended in (Wubben *et al.* 2004) with suboptimal MIMO detectors such as ZF, MMSE, and SIC detectors. To solve ML decoding or detection problems efficiently, LR was also considered in (Agrell *et al.* 2002) and (Damen *et al.* 2003). As shown in Subsection 9.2.4, they are closely related. For instance, it was shown that approximate ML solutions using Babai's approach based on LR are identical to the LR-based ZF and SIC detectors.

We presented the Gaussian lattice reduction and LLL algorithms in this chapter with emphasis on algorithm implementation. For a detailed account of the LLL algorithm, the reader is referred to the original paper of the LLL algorithm (Lenstra *et al.* 1982) or the textbooks (Hoffstein *et al.* 2002) and (Cohen 1993). A general overview of the LLL and other LR algorithms and their applications can also be found in (Cary 2002) and (Hoffstein *et al.* 2002). In particular, (Hoffstein *et al.* 2002) is easy to follow. An extensive overview of lattice decoding with major LR algorithms is given in (Mow 2003).

Problems

Problem 9.1

(i) Explain how SIC can be applied to the ML detection problem in (9.20).
Hint: Use QR factorization of $\mathbf{G} = \mathbf{QR}$ and show that

$$||\mathbf{r} - \mathbf{Gc}||^2 = ||\mathbf{Q}^{\mathrm{T}}\mathbf{r} - \mathbf{Rc}||^2.$$

(ii) Using (9.22) and the above SIC-based approach, find approximate ML solutions to the following problem:

$$\hat{\mathbf{s}} = \arg\min_{\mathbf{s}\in\mathbb{Z}^2} ||\mathbf{r} - \mathbf{Hs}||^2,$$

when $\mathbf{r} = [2.3 \quad -1.4]^{\mathrm{T}}$ and

$$\mathbf{H} = \begin{bmatrix} 1 & 1 \\ 0.5 & 2 \end{bmatrix} = \underbrace{\begin{bmatrix} 1 & -1 \\ 0.5 & 1 \end{bmatrix}}_{=\mathbf{G}} \underbrace{\begin{bmatrix} 1 & 2 \\ 0 & 1 \end{bmatrix}}_{=\mathbf{U}}.$$

Problem 9.2

(i) Consider the following CVP:

$$\hat{\mathbf{x}} = \arg\min_{\mathbf{x}\in\mathbf{\Lambda}} ||\mathbf{r} - \mathbf{x}||^2,$$

where $\mathbf{\Lambda} = \{\mathbf{x} \mid \mathbf{x} = \sum_{k=1}^{K} \mathbf{h}_k u_k,\ u_k \in \mathbb{Z}\}$ and $\mathbf{h}_1, \mathbf{h}_2, \ldots, \mathbf{h}_K$ are linearly independent. Let $\mathbf{H} = [\mathbf{h}_1 \ \mathbf{h}_2 \ \ldots \ \mathbf{h}_K] = \mathbf{GU}$, where $\mathbf{U}$ is integer unimodular. Suppose that

$$\mathbf{r} = \sum_{k=1}^{K} t_k \mathbf{g}_k,$$

where $\mathbf{g}_k$ is the kth column vector of $\mathbf{G}$. Show that the solution vector from Babai's approach, which is $\hat{\mathbf{x}} = \sum_{k=1}^{K} \lceil t_k \rfloor \mathbf{g}_k$, is the closest vector among the vertices of the following parallelepiped

$$\mathcal{F} = \left\{\mathbf{g} \mid \mathbf{g} = w_1\mathbf{g}_1 + w_2\mathbf{g}_2 + \cdots + w_K\mathbf{g}_K,\ \lfloor t_k \rfloor \le w_k \le \lceil t_k \rceil\right\}.$$

if the column vectors of $\mathbf{G}$ are orthogonal.

(ii) Consider the following two bases that have the same lattice:

$$\mathbf{G}_1 = \begin{bmatrix} 60 & 49 \\ 52 & 42 \end{bmatrix} \text{ and } \mathbf{G}_2 = \begin{bmatrix} 5 & -4 \\ 2 & 4 \end{bmatrix}.$$

When $\mathbf{r} = [54 \ 46]^{\mathrm{T}}$, find the approximate closest vectors with $\mathbf{G}_1$ and $\mathbf{G}_2$ using Babai's approach and compute the distances between $\mathbf{r}$ and $\hat{\mathbf{x}}$ for each case.

Problem 9.3 Find the transmission power of the regularized precoding in (9.28) in terms of ζ and the eigenvalues of $\mathbf{AA}^{\mathrm{H}}$, $\lambda_k(\mathbf{AA}^{\mathrm{H}})$, $k = 1, 2, \ldots, K$.

Problem 9.4 Show that (9.30) is true.

Problem 9.5 Prove Lemma 9.4.1 using $\det(\mathbf{AB}) = \det(\mathbf{A})\det(\mathbf{B})$.

Problem 9.6 Suppose that the channel matrix is given as in Example 9.4.2. The received signal is given by

$$\mathbf{r} = \mathbf{Hs} + \mathbf{n},$$

where $\mathbf{s}$ is the desired signal with $\mathcal{E}[\mathbf{s}] = \mathbf{0}$ and $\mathcal{E}[\mathbf{ss}^{\mathrm{T}}] = \mathbf{I}$ and $\mathbf{n} \sim \mathcal{N}(0, N_0\mathbf{I})$ is the background noise. With $N_0 = 0.1$, find the MMSE combining matrix:

$$\mathbf{W}_{\mathrm{mmse}} = \arg\min_{\mathbf{W}} \mathcal{E}[(\mathbf{s} - \mathbf{W}^{\mathrm{T}}\mathbf{r})^{\mathrm{T}}(\mathbf{s} - \mathbf{W}^{\mathrm{T}}\mathbf{r})]$$

and its MMSE. Repeat with the Gaussian lattice reduced channel matrix found in Example 9.4.2. Compare the MMSEs.

Problem 9.7 Show that the LLL reduction conditions in (9.60) and (9.61) are identical to the original conditions in (9.55) and (9.56), respectively.

10 Analysis of LR-based MIMO detection

Performance analysis of LR-based MIMO detectors for wireless communication systems are not relatively well studied yet. However, some important results are available in the literature for the performance over wireless MIMO channels. For instance, the relationship between the error probability and the length of a shortest basis vector is studied in (Yao & Wornell 2002) when the receiver is equipped with two receive antennas. The diversity order of LR-based detectors when the receiver has more than two receive antennas is discussed in (Taherzadeh, Mobasher & Khandani 2007*b*). A more complete and comprehensive analysis is presented in (Ling 2006).

In this chapter, we mainly focus on the performance of the LR-based SIC detector that can provide a good performance with relatively low complexity.

10.1 Assumptions for analysis

In this section, we present fundamental assumptions to carry out the analysis for LR-based MIMO detection over wireless channels. It is assumed that there are K transmit antennas at a transmitter and N receive antennas at a receiver. Furthermore, we assume that $N \geq K$.

For analysis, we consider the following received signal over an MIMO channel:

$$\mathbf{r} = \mathbf{H}\mathbf{s} + \mathbf{n}, \tag{10.1}$$

where $\mathbf{H}$ is the $N \times K$ channel matrix, $\mathbf{s}$ is the $K \times 1$ symbol vector, and $\mathbf{n}$ is the $N \times 1$ noise vector. It is assumed that $\mathbf{n} \sim \mathcal{CN}(\mathbf{0}, N_0\mathbf{I})$.

In general, the performance of MIMO detectors depends on statistical properties of MIMO channels and signal constellation. For them, we have the following assumptions.

(A1) $s_k \in \mathbb{Z} + j\mathbb{Z}$, where s_k denotes the kth signal in $\mathbf{s}$ and $\mathbb{Z} + j\mathbb{Z}$ denotes the set of complex integers. This implies that the signal constellation or alphabet is a subset of $\mathbb{Z} + j\mathbb{Z}$. For example, the signal alphabet of 16-QAM is a subset of $\mathbb{Z} + j\mathbb{Z}$ (after shifting and scaling operations), where there are 16 lattice points. It is also assumed that the s_k's have the same alphabet, denoted by $\mathcal{S} \subset \mathbb{Z} + j\mathbb{Z}$.

(A2) The elements of the channel matrix $\mathbf{H}$ are independent zero-mean CSCG random variables with variance σ_h^2 (in this variance term, the signal power is absorbed for convenience).

Due to **(A1)**, we need to generalize the definition of a lattice for a given basis $\mathbf{B} = [\mathbf{b}_1\ \mathbf{b}_2\ \ldots\ \mathbf{b}_K]$, where the $\mathbf{b}_k$'s are linearly independent complex-valued basis vectors, as follows:

$$\Lambda = \left\{ \mathbf{a} \mid \mathbf{a} = \sum_{k=1}^{K} \mathbf{b}_k u_k,\ u_k \in \mathbb{Z} + j\mathbb{Z} \right\}. \tag{10.2}$$

For LR-based MIMO detection, it is assumed that the channel matrix $\mathbf{H}$ is factorized using LR as follows

$$\mathbf{H} = \mathbf{G}\mathbf{U},$$

where $\mathbf{G}$ has nearly orthogonal column vectors and $\mathbf{U}$ is complex integer unimodular (a matrix is called complex integer unimodular if its determinant is ± 1 and each element is a complex integer). Then, the received signal is re-written as

$$\begin{aligned} \mathbf{r} &= \mathbf{H}\mathbf{s} + \mathbf{n} \\ &= \mathbf{G}\mathbf{c} + \mathbf{n}, \end{aligned} \tag{10.3}$$

where $\mathbf{c} = \mathbf{U}\mathbf{s} \in \mathbb{Z}^K + j\mathbb{Z}^K$. In LR-based detection, it is attempted to detect $\mathbf{c}$ rather than $\mathbf{s}$. For analysis purposes, we assume that a complex LR (Gan & Mow 2005) is performed with the complex-valued channel matrix $\mathbf{H}$. Thus, all the quantities are complex-valued in this chapter.

10.2 Analysis of two-dimensional systems

In (Yao 2003), LR-based MIMO detectors are extensively analyzed in terms of the complexity and performance when $K = 2$ (two transmit antennas). As a special case, the analysis becomes simpler. In this section, based on (Yao 2003), we discuss the complexity and performance issues of LR-based MIMO detectors that use Gaussian lattice reduction.

10.2.1 Performance analysis

From Subsection 7.5.2, for $\mathbf{s}_1, \mathbf{s}_2 \in \mathbb{Z}^K + j\mathbb{Z}^K$, the PEP is given by

$$\begin{aligned} \Pr(\mathbf{s}_1 \to \mathbf{s}_2) &= \Pr(||\mathbf{r} - \mathbf{H}\mathbf{s}_1|| > ||\mathbf{r} - \mathbf{H}\mathbf{s}_2||) \\ &= \Pr(||\mathbf{r} - \mathbf{G}\mathbf{c}_1|| > ||\mathbf{r} - \mathbf{G}\mathbf{c}_2||) \\ &= \mathcal{Q}\left(\sqrt{\frac{||\mathbf{G}(\mathbf{c}_1 - \mathbf{c}_2)||^2}{2N_0}} \right), \end{aligned} \tag{10.4}$$

where $\mathcal{Q}(x) = \int_x^\infty \frac{1}{\sqrt{2\pi}} e^{-z^2/2} dz$ and $\mathbf{c}_i = \mathbf{U}\mathbf{s}_i$, $i = 1, 2$. Since $\mathbf{c}_1 - \mathbf{c}_2 \in \mathbb{Z}^K + j\mathbb{Z}^K$, $\mathbf{G}(\mathbf{c}_1 - \mathbf{c}_2)$ is a vector in the lattice generated by $\mathbf{G}$ or $\mathbf{H}$. For convenience, let Λ

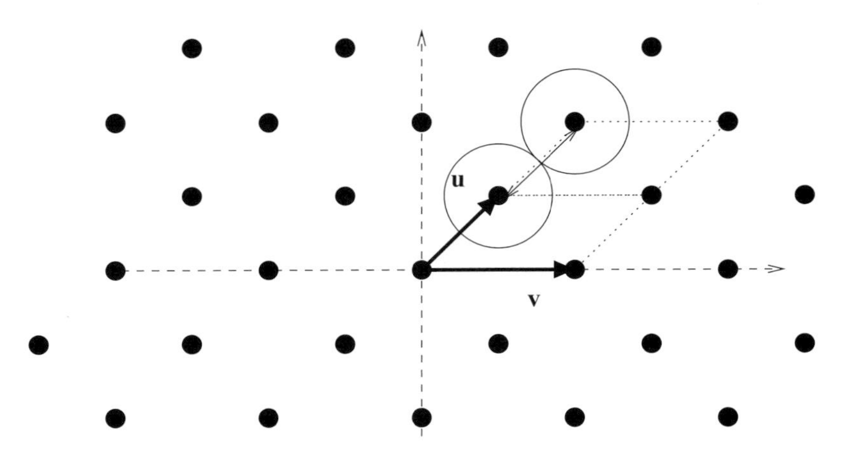

Figure 10.1 The minimum distance of a two-dimensional lattice. Each circle in the figure can be considered as a decision boundary.

denote this lattice. Using Gaussian lattice reduction, we can have

$$\mathbf{G} = [\mathbf{u}\ \mathbf{v}], \tag{10.5}$$

where $\mathbf{u}$ is a shortest vector in $\boldsymbol{\Lambda}$ and $\mathbf{v}$ is a shortest vector that is not proportional to $\mathbf{u}$ in $\boldsymbol{\Lambda}$. Then, it can be shown that

$$||\mathbf{G}(\mathbf{c}_1 - \mathbf{c}_2)|| \geq ||\mathbf{u}||, \ \mathbf{c}_1 \neq \mathbf{c}_2. \tag{10.6}$$

Thus, the minimum distance becomes $||\mathbf{u}||$, the length of a shortest vector in $\boldsymbol{\Lambda}$, as shown in Fig. 10.1. Using (10.6), an upper bound on the PEP in (10.4) can be found as

$$\Pr(\mathbf{s}_1 \to \mathbf{s}_2) \leq \mathcal{Q}\left(\frac{||\mathbf{u}||}{\sqrt{2N_0}}\right). \tag{10.7}$$

This shows that the performance of the ML detector depends on a shortest vector in the lattice generated by $\mathbf{H}$. We can obtain a similar result with the error probability.

According to Fig. 10.1, for each lattice point, a circle of radius $||\mathbf{u}||/2$ can be considered as a decision boundary. Thus, a necessary condition for an error event is

$$|n| \geq d_{\mathrm{ml,min}} = \frac{||\mathbf{u}||}{2},$$

where $n \sim \mathcal{N}(0, N_0/2)$ denotes a noise term per dimension. For a high SNR, the error probability becomes

$$\begin{aligned} P_{\mathrm{err}}(\mathbf{H}) &\leq \kappa \mathcal{Q}\left(\frac{d_{\mathrm{ml,min}}}{\sqrt{N_0/2}}\right) \\ &= \kappa \mathcal{Q}\left(\frac{||\mathbf{u}||}{\sqrt{2N_0}}\right), \end{aligned} \tag{10.8}$$

where κ is the number of the nearest lattice points within a range of the minimum distance from a lattice point. From this result, we can see that the analyses with the

PEP in (10.7) and error probability for a high SNR provide the same conclusion that the performance depends on $||\mathbf{u}||$.

Since $\mathbf{u}$ is a (non-zero) shortest vector in the lattice generated by $\mathbf{H}$, we can show that

$$||\mathbf{u}||^2 = \min_{\mathbf{d} \in \mathbb{Z}^2 + j\mathbb{Z}^2, \mathbf{d} \neq \mathbf{0}} ||\mathbf{H}\mathbf{d}||^2.$$

Thus, we can have the following expression for the average error probability:

$$\begin{aligned}\bar{P}_{\text{err}} &= \mathcal{E}[P_{\text{err}}(\mathbf{H})] \\ &\leq \kappa \mathcal{E}\left[\mathcal{Q}\left(\frac{||\mathbf{H}\bar{\mathbf{d}}||}{\sqrt{2N_0}}\right)\right] \\ &\leq \kappa \mathcal{E}\left[\exp\left(-\frac{||\mathbf{H}\bar{\mathbf{d}}||^2}{4N_0}\right)\right],\end{aligned} \tag{10.9}$$

where $\bar{\mathbf{d}}$ is a non-zero vector that satisfies $||\mathbf{u}|| = ||\mathbf{H}\bar{\mathbf{d}}||$. For given $\bar{\mathbf{d}}$, $\mathbf{H}\bar{\mathbf{d}} \sim \mathcal{CN}(\mathbf{0}, \sigma_h^2||\bar{\mathbf{d}}||^2\mathbf{I})$ under **(A2)** (see Problem 10.3). Using the result in Lemma 7.5.1, we can show that

$$\begin{aligned}\bar{P}_{\text{err}} &\leq \kappa \mathcal{E}\left[\exp\left(-\frac{||\mathbf{H}\bar{\mathbf{d}}||^2}{4N_0}\right)\right] \\ &= \kappa \det\left(\mathbf{I} + \frac{\sigma_h^2||\bar{\mathbf{d}}||^2}{4N_0}\mathbf{I}\right)^{-1} \\ &= \kappa\left(1 + \frac{\sigma_h^2||\bar{\mathbf{d}}||^2}{4N_0}\right)^{-N}.\end{aligned} \tag{10.10}$$

This shows that the diversity order is N, which is the number of receive antennas. Although this result is just a different version of the well-known result that the ML detector has a full receive antenna diversity, it is useful to show that the LR-based SIC detector also has a full receive antenna diversity.

The performance of the SIC detector depends on $\min\{||\mathbf{u}||, ||\tilde{\mathbf{v}}||\}$. Since $\min\{||\mathbf{u}||, ||\tilde{\mathbf{v}}||\} \geq ||\tilde{\mathbf{u}}||$ and

$$||\tilde{\mathbf{u}}||^2 = ||\mathbf{u}||^2 - \frac{|<\mathbf{u}, \mathbf{v}>|^2}{||\mathbf{v}||^2} \geq \frac{1}{2}||\mathbf{u}||^2,$$

the minimum distance for SIC detection can be low-bounded as

$$d_{\text{sic,min}} = \frac{1}{2}\min\{||\mathbf{u}||, ||\tilde{\mathbf{v}}||\} \geq \frac{1}{2}||\tilde{\mathbf{u}}|| \geq \frac{1}{2}\frac{||\mathbf{u}||}{\sqrt{2}}. \tag{10.11}$$

It can be concluded that

$$\frac{1}{\sqrt{2}}d_{\text{ml,min}} \leq d_{\text{sic,min}} \leq d_{\text{ml,min}}. \tag{10.12}$$

This implies that the performance of the LR-based SIC detector is at most 3 dB worse than that of the ML detector for all range of SNR. Thus, we can easily conclude that the diversity order of the LR-based SIC detector is identical to that of the ML detector. In order words, the LR-based SIC detector can achieve a full receive antenna diversity.

Furthermore, the LR-based ZF detector can have a full receive antenna diversity (see Problem 10.2). A general analysis is given in Section 10.3.

10.2.2 Computational complexity

The computational complexity of Gaussian lattice reduction depends on the number of iterations. For a given channel matrix $\mathbf{H}$, if $\{\mathbf{h}_1, \mathbf{h}_2\}$ or $\{\mathbf{h}_2, \mathbf{h}_1\}$ is the shortest vector sequence, any iteration is not required. Thus, if we can find the probability that $\{\mathbf{h}_1, \mathbf{h}_2\}$ or $\{\mathbf{h}_2, \mathbf{h}_1\}$ is the shortest vector sequence, it can provide some idea of the computational complexity.

Consider the QR factorization of $\mathbf{H} = \mathbf{QR}$. Since $\mathbf{Q}$ is unitary, the geometrical properties of $\mathbf{H}$ and $\mathbf{R}$ are equivalent. Let

$$\mathbf{R} = [\mathbf{r}_1 \ \mathbf{r}_2] = \begin{bmatrix} r_{1,1} & r_{1,2} \\ 0 & r_{2,2} \end{bmatrix}. \tag{10.13}$$

Lemma 10.2.1 *Suppose that $\mathbf{r}_1$ and $\mathbf{r}_2$ are linearly independent. The probability that $\{\mathbf{r}_1, \mathbf{r}_2\}$ or $\{\mathbf{r}_2, \mathbf{r}_1\}$ is the shortest vector sequence, denoted by $P_{\text{short}}(\mathbf{r}_1, \mathbf{r}_2)$, is bounded as follows:*

$$P_{\text{short}}(\mathbf{r}_1, \mathbf{r}_2) \le \min\left\{ \Pr\left(|r_{1,2}| \le \frac{|r_{1,1}|}{2} \right), \Pr\left(|r_{1,1}| \le \frac{1}{2}\left(|r_{1,2}| + \frac{|r_{2,2}|^2}{|r_{1,2}|} \right) \right) \right\}. \tag{10.14}$$

Proof: According to Lemma 9.4.4, sufficient conditions that $\{\mathbf{r}_1, \mathbf{r}_2\}$ or $\{\mathbf{r}_2, \mathbf{r}_1\}$ is the shortest vector sequence are

$$\begin{aligned} &(1) \quad |r_{1,1}||r_{1,2}| \le \tfrac{1}{2}|r_{1,1}|^2, && \text{if } ||\mathbf{r}_1||^2 \le ||\mathbf{r}_2||^2; \\ &(2) \ |r_{1,1}||r_{1,2}| \le \tfrac{1}{2}(|r_{2,2}|^2 + |r_{1,2}|^2), && \text{if } ||\mathbf{r}_2||^2 \le ||\mathbf{r}_1||^2. \end{aligned} \tag{10.15}$$

This leads to

$$\begin{aligned} P_{\text{short}}(\mathbf{r}_1, \mathbf{r}_2) &= \Pr\left(|r_{1,1}||r_{1,2}| \le \frac{1}{2} \min\{|r_{1,1}|^2, (|r_{2,2}|^2 + |r_{1,2}|^2)\} \right) \\ &\le \min\left\{ \Pr\left(|r_{1,2}| \le \frac{|r_{1,1}|}{2} \right), \Pr\left(|r_{1,1}| \le \frac{1}{2}\left(|r_{1,2}| + \frac{|r_{2,2}|^2}{|r_{1,2}|} \right) \right) \right\}. \end{aligned}$$

This completes the proof. □

We can obtain a closed-form expression for the following upper bound:

$$P_{\text{short}}(\mathbf{r}_1, \mathbf{r}_2) \le \Pr\left(|r_{1,2}| \le \frac{|r_{1,1}|}{2} \right) = \Pr\left(|r_{1,2}|^2 \le \frac{|r_{1,1}|^2}{4} \right). \tag{10.16}$$

The pdf of a chi-square random variable, X, with $2n$ degrees of freedom is given by

$$f_n(x) = \frac{1}{(n-1)!} x^{n-1} \mathrm{e}^{-x}.$$

Suppose that $X = |r_{1,2}|^2$ $Y = |r_{1,1}|^2$ are chi-square random variables with 2 and $2n$ degrees of freedom, respectively. Then, we can find a closed-form expression for the

Table 10.1 Empirical results (from 10^6 runs) of the probability of the number of iterations in the Gaussian lattice reduction.

Number of iterations	$N = 2$	$N = 4$	$N = 6$	$N = 8$	$N = 10$
1	0.5989	0.8534	0.9358	0.9703	0.9859
2	0.3785	0.1464	0.0642	0.0297	0.0141
3	0.0218	0.0001	$< 10^{-5}$	0	0
4	0.0007	0	0	0	0
5	$< 10^{-5}$	0	0	0	0

upper bound as follows:

$$\begin{aligned}\Pr\left(X \le \frac{Y}{4}\right) &= \int_0^\infty \int_0^{y/4} f_X(x)\mathrm{d}x f_Y(y)\mathrm{d}y \\ &= \int_0^\infty \left[\int_0^{y/4} \mathrm{e}^{-x}\mathrm{d}x\right] f_Y(y)\mathrm{d}y \\ &= 1 - \mathcal{E}[\mathrm{e}^{-Y/4}] \\ &= 1 - \left(\frac{4}{5}\right)^n. \end{aligned} \tag{10.17}$$

It is shown in (Edelman 1989) that if each element of $\mathbf{H}$ is an independent CSCG random variable as in **(A2)**, X is a chi-square random variable with 2 degrees of freedom and Y is a chi-square random variable with $2(N - K + 1)$ degrees of freedom. Thus, it can be shown that

$$P_{\text{short}}(\mathbf{r}_1, \mathbf{r}_2) \le 1 - \left(\frac{4}{5}\right)^{N-K+1}. \tag{10.18}$$

From this result, we can see that as $N - K$ increases, the complexity for the Gaussian lattice reduction decreases.

While the result in (10.18) provides a rough idea of the computational complexity, it may be necessary to have a distribution of the number of iterations for random channels. As analytical approaches are difficult, we consider simulations. For channel matrices whose elements are independent CSCG random variables, simulations are carried out to obtain empirical results for the probability of the number of iterations in the Gaussian lattice reduction. In Table 10.2.2, the results are shown. As expected, the number of iterations decreases with $N - K$ in general. Furthermore, it can also be confirmed that one or two iterations are usually required to perform the LR.

10.3 Performance of LR-based SIC detector for higher-dimensional lattices

The performance of LR-based detectors when $K > 2$ is considered in (Taherzadeh, Mobasher & Khandani 2007*a*), (Taherzadeh *et al.* 2007*b*), and (Ling 2006). It is known that LR-based detectors can achieve a full receive antenna diversity as the ML detector when the LLL algorithm is used for LR. This is an important finding since most suboptimal MIMO detectors including ZF and MMSE detectors cannot fully exploit the receive antenna diversity without exploiting LR. In this section, we focus on the LR-based SIC detector and show that it can achieve a full receive antenna diversity when the LLL algorithm is used for LR based on a relatively simple approach.

10.3.1 Performance of the ML detector

Prior to the performance analysis of the LR-based SIC detector, it would be useful to consider the performance of the ML detector over a lattice.

To find the PEP, suppose that $\mathbf{s}_1$ is transmitted, while $\mathbf{s}_2$ is erroneously detected, where $\mathbf{s}_i \in \mathcal{S}^K \subseteq \mathbb{Z}^K + j\mathbb{Z}^K$, $i = 1, 2$. Then, the PEP is given by

$$\begin{aligned} P(\mathbf{s}_1 \to \mathbf{s}_2) &= \Pr(||\mathbf{r} - \mathbf{H}\mathbf{s}_2||^2 \le ||\mathbf{r} - \mathbf{H}\mathbf{s}_1||^2) \\ &= \mathcal{Q}\left(\sqrt{\frac{||\mathbf{H}\mathbf{d}||^2}{2N_0}}\right) \\ &\le \exp\left(-\frac{||\mathbf{H}\mathbf{d}||^2}{4N_0}\right), \end{aligned} \tag{10.19}$$

where $\mathbf{d} = \mathbf{s}_1 - \mathbf{s}_2$. The last inequality is due to the Chernoff bound. We can easily show that $\mathbf{d} \in \mathbb{Z}^K + j\mathbb{Z}^K$. Denote by $S(\mathbf{H})$ a (non-zero) shortest vector in the lattice generated by $\mathbf{H}$. Then, we have

$$S(\mathbf{H}) = \min_{\mathbf{d} \in \mathbb{Z}^K + j\mathbb{Z}^K, \mathbf{d} \neq \mathbf{0}} ||\mathbf{H}\mathbf{d}||,$$

an upper bound on the PEP can be given by

$$P(\mathbf{s}_1 \to \mathbf{s}_2) \le \exp\left(-\frac{S^2(\mathbf{H})}{4N_0}\right). \tag{10.20}$$

This upper bound is valid when the signal alphabet is any subset of $\mathbb{Z}^K + j\mathbb{Z}^K$.

The average PEP can be obtained by taking the expectation with respect to $\mathbf{H}$ under **(A2)**. Define

$$\mathcal{D} = \{\mathbf{d} = \mathbf{s}_1 - \mathbf{s}_2 \mid \mathbf{s}_1, \mathbf{s}_2 \in \mathbb{Z}^K + j\mathbb{Z}^K\}.$$

Since

$$S^2(\mathbf{H}) = ||\mathbf{H}\bar{\mathbf{d}}||^2,$$

where

$$\bar{\mathbf{d}} = \arg \min_{\mathbf{d} \in \mathcal{D}, \mathbf{d} \neq \mathbf{0}} ||\mathbf{H}\mathbf{d}||^2,$$

we have

$$\exp\left(-\frac{S^2(\mathbf{H})}{4N_0}\right) \le \sum_{\mathbf{d} \in \mathcal{D}, \mathbf{d} \neq \mathbf{0}} \exp\left(-\frac{||\mathbf{H}\mathbf{d}||^2}{4N_0}\right). \tag{10.21}$$

It can be shown that $\mathbf{Hd} \sim \mathcal{CN}(\mathbf{0}, \sigma_h^2 ||\mathbf{d}||^2 \mathbf{I})$ under **(A2)**. Taking the expectation with respect to $\mathbf{H}$, from Lemma 7.5.1, we can show that

$$\mathcal{E}\left[\exp\left(-\frac{||\mathbf{Hd}||^2}{4N_0}\right)\right] = \det\left(\mathbf{I} + \frac{\sigma_h^2 ||\mathbf{d}||^2}{4N_0}\mathbf{I}\right)^{-1}. \tag{10.22}$$

Finally, using (10.21) and (10.22), an upper bound on the PEP over a lattice is given by

$$\begin{aligned} \mathcal{E}[P(\mathbf{s}_{(1)} \to \mathbf{s}_{(2)})] &\leq \mathcal{E}\left[\exp\left(-\frac{S^2(\mathbf{H})}{4N_0}\right)\right] \\ &= \mathcal{E}\left[\exp\left(-\frac{||\mathbf{H}\bar{\mathbf{d}}||^2}{4N_0}\right)\right] \\ &\leq \sum_{\mathbf{d} \in \mathcal{D}, \mathbf{d} \neq \mathbf{0}} \left(1 + \frac{\sigma_h^2 ||\mathbf{d}||^2}{4N_0}\right)^{-N}, \end{aligned} \tag{10.23}$$

which shows that the diversity order is N, i.e. a full receive antenna diversity gain is achieved. This average PEP is different from the PEP derived in Chapter 7 and would be loose. However, this result would be useful to study the diversity gain of LR-based detectors. Note that the size of $\mathcal{D}$ can be finite as the size of $\mathcal{S}$ is finite.

10.3.2 Properties of LLL reduced basis

We focus on key properties of an LLL reduced basis in this subsection. Although the proofs of some key properties of an LLL reduced basis are a bit straightforward, we present them as they help to understand the properties of LR-based detectors.

Consider the following LLL reduced basis:

$$\mathbf{B} = [\mathbf{b}_1 \ \mathbf{b}_2 \ \ldots \ \mathbf{b}_K]$$

from a given basis $\mathbf{H}$, which is a complex-valued channel matrix in this chapter. Let $\mathbf{\Lambda}$ denote the lattice generated by $\mathbf{B}$. Define the determinant of the lattice $\mathbf{\Lambda}$, $V(\mathbf{\Lambda})$, as

$$V(\mathbf{\Lambda}) = \sqrt{\det(\mathbf{B}^{\mathrm{H}}\mathbf{B})}. \tag{10.24}$$

We can show that this determinant does not depend on the choice of the basis. Let $\mathbf{A} = \mathbf{BU}$, where $\mathbf{U}$ is complex integer unimodular. In this case, $\mathbf{A}$ also generates the lattice $\mathbf{\Lambda}$. Since $\mathbf{U}$ is unimodular, we have

$$\det(\mathbf{B}^{\mathrm{H}}\mathbf{B}) = \det(\mathbf{A}^{\mathrm{H}}\mathbf{A}) \tag{10.25}$$

and it implies that the determinant of $\mathbf{\Lambda}$,

$$V(\mathbf{\Lambda}) = \sqrt{\det(\mathbf{B}^{\mathrm{H}}\mathbf{B})} = \sqrt{\det(\mathbf{A}^{\mathrm{H}}\mathbf{A})},$$

does not depend on the choice of the basis. However, for convenience, we also write $V(\mathbf{\Lambda}) = V(\mathbf{B})$.

With the Gram–Schmidt orthogonalization of $\mathbf{B}$ from (9.51)

$$\mathbf{B} = \tilde{\mathbf{B}}[\mu] \tag{10.26}$$

or

$$\begin{aligned}\mathbf{b}_k &= \tilde{\mathbf{b}}_k + \sum_{q=1}^{k-1} \mu_{k,q}\tilde{\mathbf{b}}_q \\ &= \sum_{q=1}^{k} \mu_{k,q}\tilde{\mathbf{b}}_q,\end{aligned} \tag{10.27}$$

the LLL basis reduction conditions for complex-valued matrices are given by (Gan, Ling & Mow 2006):

$$|\Re(\mu_{k,l})| \le \frac{1}{2} \text{ and } |\Im(\mu_{k,l})| \le \frac{1}{2}, \quad 1 \le l < k \le K; \tag{10.28}$$

$$||\tilde{\mathbf{b}}_k||^2 \le \omega||\mu_{k+1,k}\tilde{\mathbf{b}}_k + \tilde{\mathbf{b}}_{k+1}||^2, \; k = 1, 2, \ldots, K-1, \tag{10.29}$$

where $1 < \omega < 4$ is a constant. In general, we set $\omega = 4/3$.

From (10.28) and (10.29), the following properties are derived in (Lenstra *et al.* 1982).

Theorem 10.3.1 *Suppose that* $\mathbf{B}$ *is an LLL reduced basis. Then, we have*

$$||\mathbf{b}_1|| \le \beta^{(K-1)/4} V^{1/K}(\mathbf{B}); \tag{10.30}$$

$$||\mathbf{b}_1|| \le \beta^{(K-1)/2} S(\mathbf{B}); \tag{10.31}$$

$$\prod_{k=1}^{K} ||\mathbf{b}_k|| \le \beta^{K(K-1)/4} V(\mathbf{B}), \tag{10.32}$$

where $\beta = (\omega^{-1} - \frac{1}{4})^{-1} > 4/3$. *If* $\omega = 4/3$, *we have* $\beta = 2$.

Proof of (10.30): From (10.29),

$$\begin{aligned}\omega^{-1}||\tilde{\mathbf{b}}_k||^2 &\le ||\mu_{k+1,k}\tilde{\mathbf{b}}_k + \mathbf{b}_{k+1}||^2 \\ &= |\mu_{k+1,k}|^2||\tilde{\mathbf{b}}_k||^2 + ||\mathbf{b}_{k+1}||^2.\end{aligned}$$

Since $|\mu_{k+1,k}|^2 \le 1/4$ from (10.29), it follows that

$$||\tilde{\mathbf{b}}_{k+1}||^2 \ge \beta^{-1}||\tilde{\mathbf{b}}_k||^2. \tag{10.33}$$

By induction, we have

$$||\tilde{\mathbf{b}}_k||^2 \ge \beta^{l-k}||\tilde{\mathbf{b}}_l||^2, \; 1 \le l \le k \le K. \tag{10.34}$$

This implies

$$\begin{aligned}||\tilde{\mathbf{b}}_1||^2||\tilde{\mathbf{b}}_2||^2 \cdots ||\tilde{\mathbf{b}}_K||^2 &\ge ||\tilde{\mathbf{b}}_1||^2(\beta^{-1}||\tilde{\mathbf{b}}_1||^2)\cdots(\beta^{1-K}||\tilde{\mathbf{b}}_1||^2) \\ &= ||\tilde{\mathbf{b}}_1||^{2K}\beta^{-(1+2+\ldots+(K-1))}.\end{aligned}$$

Since $\sum_{i=1}^{n} i = \frac{n(n+1)}{2}$,

$$||\tilde{\mathbf{b}}_1||^2||\tilde{\mathbf{b}}_2||^2 \cdots ||\tilde{\mathbf{b}}_K||^2 \geq ||\tilde{\mathbf{b}}_1||^{2K} \beta^{-K(K-1)/2}$$

or

$$\begin{aligned} ||\tilde{\mathbf{b}}_1|| &\leq \beta^{(K-1)/4} \left(||\tilde{\mathbf{b}}_1||^2||\tilde{\mathbf{b}}_2||^2 \cdots ||\tilde{\mathbf{b}}_K||^2\right)^{\frac{1}{2K}} \\ &= \beta^{(K-1)/4} V^{1/K}(\mathbf{B}). \end{aligned} \tag{10.35}$$

For the equality in (10.35), see Problem 10.5. Since $\tilde{\mathbf{b}}_1 = \mathbf{b}_1$, this completes the proof. □

Proof of (10.31): Firstly, we need to show that

$$S(\mathbf{B}) \geq \min_k ||\tilde{\mathbf{b}}_k||^2 \tag{10.36}$$

To show this, let $\mathbf{v}$ denote the shortest vector. Then, we have

$$\mathbf{v} = \sum_{l=1}^{k} v_l \mathbf{b}_l,$$

where v_l are integers and k is the largest index of the basis vector to represent $\mathbf{v}$. Thus, $v_k \neq 0$. Since $\mathbf{b}_l = \mu_{l,l}\tilde{\mathbf{b}}_l + \sum_{q=1}^{l-1} \mu_{l,q}\tilde{\mathbf{b}}_q$, we have

$$\mathbf{v} = v_k \mu_{k,k} \tilde{\mathbf{b}}_k + \text{a linear combination of } \tilde{\mathbf{b}}_{k-1}, \tilde{\mathbf{b}}_{k-2}, \ldots, \tilde{\mathbf{b}}_1.$$

As the $\tilde{\mathbf{b}}_l$'s are orthogonal and $\mu_{k,k} = 1$, we have

$$||\mathbf{v}||^2 \geq ||v_k \mu_{k,k} \tilde{\mathbf{b}}_k||^2 = |v_k|^2 ||\tilde{\mathbf{b}}_k||^2 \geq ||\tilde{\mathbf{b}}_k||^2.$$

Therefore,

$$S(\mathbf{B}) = ||\mathbf{v}||^2 \geq ||\tilde{\mathbf{b}}_k||^2 \geq \min_l ||\tilde{\mathbf{b}}_l||^2.$$

This completes the proof of (10.36).

From (10.34), we have

$$||\tilde{\mathbf{b}}_k|| \geq \beta^{(1-k)/2} ||\tilde{\mathbf{b}}_1||.$$

Then, it follows that

$$\min_k ||\tilde{\mathbf{b}}_k|| \geq \min_k \beta^{(1-k)/2} ||\tilde{\mathbf{b}}_1|| = \beta^{-(K-1)/2} ||\tilde{\mathbf{b}}_1||. \tag{10.37}$$

Substituting (10.37) into (10.36), we can show that

$$S(\mathbf{B}) \geq \min_k ||\tilde{\mathbf{b}}_k|| \geq \beta^{-(K-1)/2} ||\tilde{\mathbf{b}}_1||.$$

This completes the proof of (10.31). □

Proof of (10.32): From (10.27) and using (10.28), it can be shown that

$$\begin{aligned}||\mathbf{b}_k||^2 &= ||\tilde{\mathbf{b}}_k + \sum_{q=1}^{k-1} \mu_{k,q}\tilde{\mathbf{b}}_q||^2 \\ &\le ||\tilde{\mathbf{b}}_k||^2 + \frac{1}{4}\sum_{q=1}^{k-1} ||\tilde{\mathbf{b}}_q||^2. \end{aligned} \tag{10.38}$$

Applying (10.34) to (10.38), we have

$$||\mathbf{b}_k||^2 \le \left(1 + \frac{1}{4}\sum_{q=1}^{k-1} \beta^{k-q}\right) ||\tilde{\mathbf{b}}_k||^2. \tag{10.39}$$

Since

$$\sum_{q=1}^{k-1} \beta^{k-q} = \beta^k \sum_{q=1}^{k-1} \beta^{-q} = \frac{\beta^k - \beta}{\beta - 1},$$

we have

$$1 + \frac{1}{4}\sum_{q=1}^{k-1} \beta^{q-k} = 1 + \frac{\beta^k - \beta}{4(\beta - 1)}.$$

Furthermore, we can show that

$$1 + \frac{\beta^k - \beta}{4(\beta - 1)} \le \beta^{k-1}. \tag{10.40}$$

This implies that

$$1 + \frac{1}{4}\sum_{q=1}^{k-1} \beta^{k-q} \le \beta^{k-1}. \tag{10.41}$$

Substituting (10.41) into (10.39), we have

$$||\mathbf{b}_k||^2 \le \beta^{k-1}||\tilde{\mathbf{b}}_k||^2.$$

Then, it follows that

$$\begin{aligned}\prod_{k=1}^{K} ||\mathbf{b}_k||^2 &\le \prod_{k=1}^{K} \beta^{k-1}||\tilde{\mathbf{b}}_k||^2 \\ &= \beta^{\sum_{k=1}^{K}(k-1)} V^2(\mathbf{B}). \end{aligned} \tag{10.42}$$

Noting that

$$\sum_{k=1}^{K}(k-1) = \frac{K(K+1)}{2} - K = \frac{K(K-1)}{2},$$

we have

$$\prod_{k=1}^{K} ||\mathbf{b}_k||^2 \leq \beta^{K(K-1)/2} V^2(\mathbf{B}).$$

This completes the proof of (10.32). □

10.3.3 Diversity gain of LR-based SIC detector

Based on the properties of an LLL reduced basis, we can show that the LR-based SIC detector can have a full receive antenna diversity under **(A2)**.

For SIC detection, we consider the QR factorization of $\mathbf{G}$. Then, from (10.3), the received signal becomes

$$\mathbf{r} = \mathbf{QRc} + \mathbf{n}.$$

Pre-multiplying $\mathbf{Q}^{\mathrm{H}}$ to $\mathbf{r}$, we have

$$\begin{aligned} \mathbf{x} &= \mathbf{Q}^{\mathrm{H}}\mathbf{r} \\ &= \mathbf{Rc} + \mathbf{Q}^{\mathrm{H}}\mathbf{n}. \end{aligned}$$

As the statistical properties of $\mathbf{Q}^{\mathrm{H}}\mathbf{n}$ are the same as those of $\mathbf{n}$, we simply write $\mathbf{n}$ to denote $\mathbf{Q}^{\mathrm{H}}\mathbf{n}$. To detect c_K, we can use x_K:

$$x_K = r_{K,K} c_K + n_K.$$

Since $c_k \in \mathbb{Z} + j\mathbb{Z}$, a sufficient condition for the correct detection of c_K is

$$|n_K| < \frac{|r_{K,K}|}{2}.$$

Once c_K is correctly detected, its cancellation becomes perfect in detecting c_{K-1}. Thus, we can conclude that a sufficient condition for the correct detection of $\mathbf{c}$ (i.e., all the K symbols) is

$$|n_k| < \frac{|r_{k,k}|}{2} \text{ for all } k. \tag{10.43}$$

From this, we can find a lower bound on the probability that there is no error in detecting $\mathbf{c}$ as follows:

$$\Pr(\text{no error}) \geq \prod_{k=1}^{K} \Pr\left(|n_k| < \frac{|r_{k,k}|}{2} \right). \tag{10.44}$$

Since the n_k's are independent, the joint probability that $|n_k| < |r_{k,k}|/2$, for all $k = 1, 2, \ldots, K$, becomes a product of all the individual probabilities. As $|n_k|^2$ is a chi-square random variable with two degrees of freedom (which is an exponential random variable), we can show that

$$\Pr\left(|n_k| < \frac{|r_{k,k}|}{2}\right) = \Pr\left(|n_k|^2 < \frac{|r_{k,k}|^2}{4}\right)$$
$$= 1 - \mathrm{e}^{-\frac{|r_{k,k}|^2}{4N_0}}. \tag{10.45}$$

Substituting (10.45) into (10.44), we have

$$\Pr(\text{error}) = 1 - \Pr(\text{no error})$$
$$\leq 1 - \prod_{k=1}^{K}\left(1 - \mathrm{e}^{-\frac{|r_{k,k}|^2}{4N_0}}\right). \tag{10.46}$$

As $N_0 \to 0$, we can show that

$$\Pr(\text{error}) \leq 1 - \prod_{k=1}^{K}\left(1 - \mathrm{e}^{-\frac{|r_{k,k}|^2}{4N_0}}\right)$$
$$\simeq \sum_{k=1}^{K} \mathrm{e}^{-\frac{|r_{k,k}|^2}{4N_0}}$$
$$\simeq \mathrm{e}^{-\min_k \frac{|r_{k,k}|^2}{4N_0}}. \tag{10.47}$$

Theorem 10.3.2

$$\min_k |r_{k,k}|^2 \geq \beta^{-K+1} S^2(\mathbf{H}). \tag{10.48}$$

Proof: From (10.33) (see also Problem 9.7), we have

$$|r_{k+1,k+1}|^2 \geq \beta^{-1}|r_{k,k}|^2. \tag{10.49}$$

Using (10.49), we can show that

$$\min\{|r_{1,1}|^2, |r_{2,2}|^2\} \geq \min\{|r_{1,1}|^2, \beta^{-1}|r_{1,1}|^2\} \geq \beta^{-1}|r_{1,1}|^2.$$

The second inequality is due to $\beta > 4/3$. By repeatedly applying this inequality with more $|r_{k,k}|^2$ terms, we can have

$$\min_k |r_{k,k}|^2 \geq \beta^{-K+1}|r_{1,1}|^2. \tag{10.50}$$

Since $\mathbf{G} = \mathbf{QR}$, we have

$$||\mathbf{g}_1||^2 = |r_{1,1}|^2.$$

We can also show that

$$||\mathbf{g}_1||^2 = |r_{1,1}|^2 \geq S^2(\mathbf{H}). \tag{10.51}$$

Substituting (10.51) into (10.50), we can show (10.48). This completes the proof. □

Theorem 10.3.3 *Under* **(A2)**, *the average probability of error is given by*

$$\mathcal{E}[\Pr(\text{error})] \le \sum_{\mathbf{d}\in\mathcal{D},\mathbf{d}\neq\mathbf{0}} \left(1 + \frac{\beta^{-K+1}\sigma_h^2||\mathbf{d}||^2}{4N_0}\right)^{-N}. \quad (10.52)$$

Proof: Substituting (10.48) into (10.47), we have

$$\mathcal{E}[\Pr(\text{error})] \le \mathcal{E}\left[\exp\left(-\frac{\beta^{-K+1}S^2(\mathbf{H})}{4N_0}\right)\right]. \quad (10.53)$$

Then, using the result in (10.23), the upper bound in (10.52) can be obtained. □

The result in (10.52) certainly shows that the LR-based SIC detector can achieve a full receiver diversity gain as the ML detector. Compared with (10.23), the PEP for the LR-based SIC detector in (10.52) has an extra term of β. Since $\beta = 4/3$ for an LLL reduced basis, we can see that the PEP (actually an upper bound) for the LR-based SIC detector is higher than that for the ML detector.

It is well-known that the LLL algorithm has a polynomial time complexity. However, since its complexity varies depending on a given input basis, it is difficult to derive an exact computational complexity expression for random MIMO channels. Since the complexity analysis of the LLL algorithm is beyond the scope of the book, we do not address this issue in this chapter. The reader is referred to the original paper (Lenstra *et al.* 1982). In (Sandell, Lillie, McNamara, Ponnampalam & Miforf 2007) and (Jalden, Seethaler & Matz 2008), the complexity of LR-based MIMO detectors is discussed with random MIMO channels.

10.4 Summary and notes

In this chapter, we mainly discussed the performance of the LR-based SIC detector. It was shown that it can achieve a full receive antenna diversity as the ML detector. This was also found in (Ling 2006).

Analysis of LR-based MIMO detectors usually requires reasonable understanding of lattice theory. The reader is referred to (Pohst 1993), (Hoffstein *et al.* 2002), and (Conway & Sloane 1991) for a detailed account of lattice theory.

Problems

Problem 10.1 Assume that all the quantities are real-valued. Suppose that the received signal is given by

$$\mathbf{r} = \mathbf{G}\mathbf{c} + \mathbf{n},$$

where $\mathbf{c} \in \mathbb{Z}^K$. Show that the decision area of the ZF detector is identical to the following parallelepiped:

$$\mathcal{F}(\mathbf{G}) = \left\{ \mathbf{b} \mid \mathbf{b} = w_1 \mathbf{g}_1 + w_2 \mathbf{g}_2 + \cdots + w_K \mathbf{g}_K, -\frac{1}{2} \leq w_k < \frac{1}{2} \right\}.$$

Problem 10.2 Suppose that $K = 2$ and Gaussian lattice reduction is applied as in Section 10.2. Show that the minimum distance for the LR-based ZF detector is given by

$$d_{\text{zf,min}} = \frac{1}{2} ||\tilde{\mathbf{u}}||.$$

Problem 10.3 Let $\mathbf{h}_1 \sim \mathcal{CN}(0, \mathbf{R}_1)$ and $\mathbf{h}_2 \sim \mathcal{CN}(0, \mathbf{R}_2)$ be independent CSCG random vectors. For given a_1 and a_2, show that $\mathbf{x} = a_1 \mathbf{h}_1 + a_2 \mathbf{h}_2 \sim \mathcal{CN}(\mathbf{0}, |a_1|^2 \mathbf{R}_1 + |a_2|^2 \mathbf{R}_2)$. (*This result is used to derive the average error probability under* **(A2)** *in* (10.10).)

Problem 10.4 Show that (10.25) is true.

Problem 10.5 Consider the Gram–Schmidt orthogonalization of **B** in (10.26). Show that

$$V^2(\mathbf{B}) = \det(\mathbf{B}^{\mathrm{H}} \mathbf{B}) = ||\tilde{\mathbf{b}}_1||^2 ||\tilde{\mathbf{b}}_2||^2 \cdots ||\tilde{\mathbf{b}}_K||^2.$$

In order to show this, it is necessary to use the fact that the determinant of an upper triangular matrix whose diagonal elements are all 1's is 1.

Problem 10.6 Show that the inequality in (10.40) is true.

Problem 10.7 Derive (10.45).

Appendix 1 Review of signals and systems

A1.1 Key definitions and operations

We consider discrete-time signals and linear systems in this appendix. Discrete-time signals are number sequences and systems are mappings of sequences. In general, discrete-time signals can be considered as sampled signal sequences of analog signals.

(i) Linear systems: Consider an operator $H[\cdot]$ to denote a system. The output of the system $H[\cdot]$ given input x_l is denoted by $H[x_l]$. Then, a system is called *linear* if

$$H[\alpha x_l + \beta y_l] = \alpha H[x_l] + \beta H[y_l], \tag{1.1}$$

where α and β are constants and x_l and y_l are input signals.

(ii) Impulse response: The *impulse response* of a linear system is the output of the linear system when the input is a unit impulse:

$$h_l = H[\delta_l], \ l = \ldots, -1, 0, 1, \ldots,$$

where δ_l denotes the unit impulse (also called the Kronecker delta) defined as

$$\delta_l = \begin{cases} 1, & \text{if } l = 0; \\ 0, & \text{otherwise.} \end{cases}$$

(iii) Causal: A linear system is called *causal* if

$$H[\delta_l] = 0, \ \text{for all } l < 0.$$

(iv) Convolution: The *convolution* of x_l and y_l is defined as follows:

$$\begin{aligned} x_l * y_l &= \sum_{m=-\infty}^{\infty} x_m y_{l-m} \\ &= \sum_{m=-\infty}^{\infty} x_{l-m} y_m, \end{aligned} \tag{1.2}$$

where $*$ denotes the convolution (operation).

(v) Output of linear systems: If a linear system has impulse response $\{h_m\}$, given input x_l, the output is given by

$$H[x_l] = x_l * h_l. \tag{1.3}$$

Noting that $x_l = \sum_{m=-\infty}^{\infty} x_m \delta_{l-m}$, it is straightforward to show that (1.3) is true as follows:

$$\begin{aligned} H[x_l] &= H\left[\sum_{m=-\infty}^{\infty} x_m \delta_{l-m}\right] \\ &= \sum_{m=-\infty}^{\infty} H\left[x_m \delta_{l-m}\right] \\ &= \sum_{m=-\infty}^{\infty} x_m H\left[\delta_{l-m}\right] \\ &= \sum_{m=-\infty}^{\infty} x_m h_{l-m} \\ &= \sum_{m=-\infty}^{\infty} h_m x_{l-m} \\ &= x_l * h_l. \end{aligned} \tag{1.4}$$

The second and third equalities are due to the properties of linear systems in (1.1). If the linear system is causal, we have

$$\begin{aligned} H[x_l] &= x_l * h_l \\ &= \sum_{m=0}^{\infty} h_m x_{l-m}. \end{aligned} \tag{1.5}$$

A1.2 Vector and matrix notations

Suppose that the length of discrete-time signals is finite. In addition, the length of impulse responses of linear systems is finite. Then, a discrete-time signal can be represented by a vector as follows:

$$\{x_l\} \Leftrightarrow \mathbf{x} = [x_0\ x_1\ \ldots\ x_{L-1}]^{\mathrm{T}}.$$

A linear system can be represented by a matrix as follows:

$$H[\alpha x_l + \beta y_l] = \alpha H[x_l] + \beta H[y_l] \Leftrightarrow \mathbf{H}(\alpha \mathbf{x} + \beta \mathbf{y}) = \alpha \mathbf{H}\mathbf{x} + \beta \mathbf{H}\mathbf{y}. \tag{1.6}$$

A1.3 $\mathcal{Z}$-transform

(i) Definition: The $\mathcal{Z}$-transform of signal x_l is defined as

$$\begin{aligned} X(z) &= \mathcal{Z}(x_l) \\ &= \sum_l x_l z^{-l}, \end{aligned}$$

where z is a complex variable.

(ii) Convolution property: If $y_l = x_l * h_l$, then

$$Y(z) = X(z)H(z),$$

where $Y(z)$, $X(z)$, and $H(z)$ stand for the $\mathcal{Z}$-transforms of y_l, x_l, and h_l, respectively.

(iii) Shift property: If $y_l = x_{l-m}$, then

$$Y(z) = z^{-m}X(z).$$

(iv) Parseval's theorem: Let $X(z)$ and $Y(z)$ be the $\mathcal{Z}$-transforms of x_l and y_l, respectively. Then, we have

$$\sum_l x_l y_l = \frac{1}{j2\pi} \oint X(z)Y(z^{-1})\frac{\mathrm{d}z}{z}.$$

A1.4 Sampling theorem

An analog signal can be converted into a sampled signal (i.e. a discrete-time sequence) by the sampling process. The impulse sampling is a conceptual sampling process that can help to determine the sampling rate to recover the original analog signal from a sampled signal: see (Haykin & Veen 1999) for other sampling processes.

Suppose that $x(t)$ is a bandlimited analog signal. The Fourier transform of $x(t)$, denoted by $X(f)$, is zero outside the interval, $-W \le f \le W$, where W is the bandwidth. The impulse sampling uses an impulse sequence to sample analog signals as follows:

$$\begin{aligned} x_\mathrm{s}(t) &= x(t)\left(\sum_{l=-\infty}^{\infty} \delta(t - lT_\mathrm{s})\right) \\ &= \sum_{l=-\infty}^{\infty} x(t)\delta(t - lT_\mathrm{s}) \\ &= \sum_{l=-\infty}^{\infty} x(lT_\mathrm{s})\delta(t - lT_\mathrm{s}), \end{aligned}$$

where $x_s(t)$ is the sampled signal, T_s is the sampling interval, and $\delta(t)$ is the Dirac delta function. The Dirac delta function is characterized as follows:

$$\int_{-\infty}^{\infty} \delta(t)\mathrm{d}t = 1;$$

$$\int_{-\infty}^{\infty} f(\tau)\delta(t-\tau)\mathrm{d}\tau = f(t),$$

where $f(t)$ is a smooth function. It is important to determine the sampling interval so that the original analog signal $x(t)$ can be recovered from the sampled signal $x_s(t)$.

Using the convolution property, we can find the Fourier transform of $x_s(t)$ as follows:

$$\begin{aligned} X_s(f) &= X(f) * \mathcal{F}\left(\sum_{l=-\infty}^{\infty} \delta(t - lT_s)\right) \\ &= X(f) * \left(\frac{1}{T_s}\sum_{n=-\infty}^{\infty} \delta(f - nf_s)\right) \\ &= \frac{1}{T_s}\sum_{n=-\infty}^{\infty} X(f - nf_s), \end{aligned}$$

where

$$\mathcal{F}\left(\sum_{l=-\infty}^{\infty} \delta(t - lT_s)\right) = \frac{1}{T_s}\sum_{n=-\infty}^{\infty} \delta(f - nf_s)$$

and $f_s = 1/T_s$. Thus, if $f_s \geq 2W$, $X(f)$ can be found from $X_s(f)$ without aliasing by taking an ideal low pass filtering. The sampling rate $f_s \geq 2W$ is called the Nyquist sampling rate.

Appendix 2 A brief review of entropy, mutual information, and channel capacity

We briefly introduce the concepts of the entropy, channel capacity, and mutual information. The reader is referred to (Cover & Thomas 1991) for a detailed account of information theory.

A2.1 Entropy

Consider a probabilistic experiment, the outcome of which is one of the alphabet $\{s_0, s_1, \ldots, s_{K-1}\}$. Let $\Pr(S = s_k) = p_k$ denote the probability that the symbol s_k is observed, where S is the random variable for the probabilistic experiment. It should be satisfied that $\sum_{k=0}^{K-1} p_k = 1$ and $p_k \geq 0$ for all k. Then, the information of s_k is defined as follows:

$$\begin{aligned} \mathsf{I}(s_k) &= \log_2 \left(\frac{1}{\Pr(S = s_k)} \right) \\ &= -\log_2 (\Pr(S = s_k)) \end{aligned} \tag{2.1}$$

and the *entropy* of random variable S is given by

$$\begin{aligned} \mathcal{H}(S) &= \mathcal{E}[\mathsf{I}(S)] \\ &= -\sum_{k=0}^{K-1} p_k \log_2 (p_k) . \end{aligned} \tag{2.2}$$

It is assumed that $0 \cdot \log(\infty) = 0$. The dimension of the entropy is bits per symbol, where "bit" is an acronym for binary information digit. The entropy can be understood in various ways; it can indicate randomness or uncertainty of random variables. As the entropy increases, a random variable can be more random or uncertain. A random variable of a higher entropy is regarded as a random variable of more information.

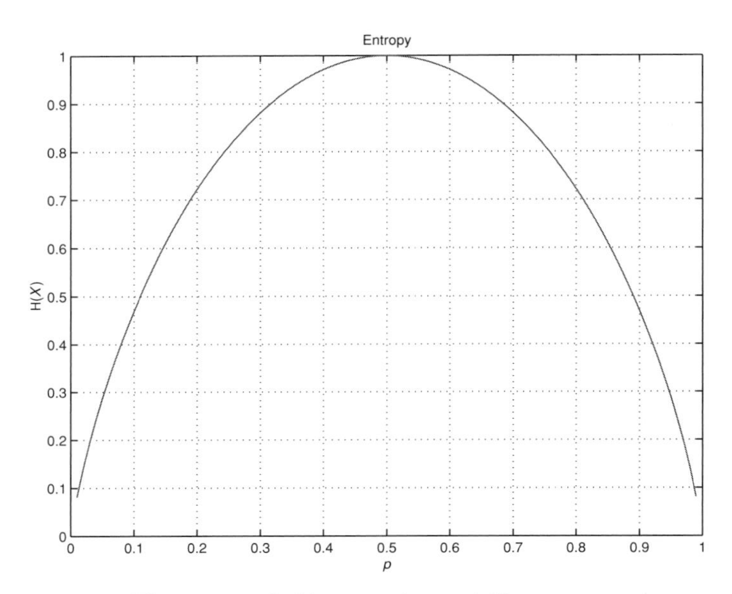

Figure A2.1 The entropy of a binary random variable.

For example, consider a binary random variable $X \in \{0, 1\}$, where $\Pr(X = 0) = p$ and $\Pr(X = 1) = 1 - p$. Then, the entropy of X is given by

$$\begin{aligned} \mathsf{H}(X) &= \mathcal{E}[-\log_2 \Pr(X)] \\ &= \Pr(X = 0) \times -\log_2 \Pr(X = 0) \\ &\quad + \Pr(X = 1) \times -\log_2 \Pr(X = 1) \\ &= -p \log_2 p - (1 - p) \log_2 (1 - p). \end{aligned}$$

Figure A2.1 shows the entropy curve in terms of the probability p. The entropy becomes the maximum when p is $1/2$, i.e. the maximum uncertainty is achieved if $p = 1/2$. If there is no uncertainty (i.e. $p = 0$ or $p = 1$), the entropy becomes zero.

A2.2 Mutual information and channel capacity

A digital communication system is often simplified as a memoryless system in which the current output is dependent only on the current input, not past or future inputs. If the input and output signals are discrete, the memoryless channel is called the discrete memoryless channel (DMC).

The binary symmetric channel (BSC) is a special case of the DMC, where the input and output are bits. A BSC is illustrated in Fig. A2.2. In this case, the BSC is fully

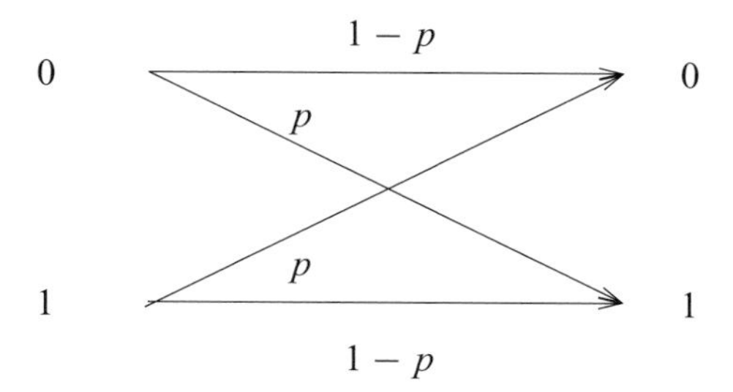

Figure A2.2 A binary symmetric channel (BSC).

characterized by the following conditional probability:

$$\Pr(y|x) = f_{Y|X}(y|x) = \begin{cases} 1-p, & \text{if } y = x; \\ p, & \text{if } y \neq x. \end{cases}$$

where x and y stand for the input and output, respectively. Note that the input distribution $\Pr(X = x) = f_X(x)$ has not yet been given.

Suppose that we observe $Y = y$ from the BSC. The entropy or uncertainty of X with observation $Y = y$ can be measured as

$$\mathsf{H}(X|Y = y) = \mathcal{E}[-\log_2 f_{X|Y}(X|y)].$$

Note that since $Y = y$, y is not a random variable and the expectation is carried out over X. That is,

$$\begin{aligned}\mathsf{H}(X|Y = y) = &-\log_2 f_{X|Y}(X = 0|y) f_{X|Y}(X = 0|y) \\ &-\log_2 f_{X|Y}(X = 1|y) f_{X|Y}(X = 1|y).\end{aligned}$$

If $\mathsf{H}(X|Y = y)$ is close to zero, the decision based on $Y = y$ will be correct with a high probability. Otherwise (i.e. if close to 1), incorrect decisions will happen with a higher probability.

In the above, we consider a particular case that $Y = y$. With a random variable Y, the *conditional entropy* is defined as

$$\begin{aligned}\mathsf{H}(X|Y) &= \mathcal{E}_{X,Y}[-\log_2 f_{X|Y}(X|Y)] \\ &= \sum_x \sum_y -\log_2 f_{X|Y}(x|y) f_{X,Y}(X = x, Y = y).\end{aligned}$$

If X and Y are independent, then

$$\begin{aligned}
\mathsf{H}(X|Y) &= \mathcal{E}[-\log_2 f_{X|Y}(X|Y)] \\
&= \mathcal{E}\left[-\log_2 \frac{f_{X,Y}(X,Y)}{f_Y(Y)}\right] \\
&= \mathcal{E}\left[-\log_2 \frac{f_X(X)f_Y(Y)}{f_Y(Y)}\right] \\
&= \mathcal{E}\left[-\log_2 f_X(X)\right] \\
&= \mathsf{H}(X).
\end{aligned}$$

This shows that there is no entropy or uncertainty reduction by observing Y. On the other hand, if $X = Y$, then

$$\begin{aligned}
\mathsf{H}(X|Y) &= \mathcal{E}[-\log_2 f_{X|Y}(X|Y)] \\
&= \mathcal{E}\left[-\log_2 \frac{f_Y(Y)}{f_Y(Y)}\right] \\
&= 0.
\end{aligned}$$

That is, the entropy becomes zero by observing Y. Thus, there is no uncertainty of the information of X when Y is given.

Define the *mutual information* as

$$\mathsf{I}(X;Y) = \mathsf{H}(X) - \mathsf{H}(X|Y)$$

which is the difference between the entropy of X (the original entropy) and the conditional entropy of X given Y. We can show that

$$\begin{aligned}
\mathsf{I}(X;Y) &= \mathsf{H}(X) - \mathsf{H}(X|Y) \\
&= \mathcal{E}\left[-\log_2 f_X(X)\right] - \mathcal{E}\left[-\log_2 \frac{f_{X,Y}(X,Y)}{f_Y(Y)}\right] \\
&= \mathcal{E}\left[-\log_2 \frac{f_X(X)f_Y(Y)}{f_{X,Y}(X,Y)}\right] \\
&= \mathsf{H}(X) + \mathsf{H}(Y) - \mathsf{H}(X,Y) \\
&= \mathsf{I}(Y;X) \\
&= \mathsf{H}(Y) - \mathsf{H}(Y|X).
\end{aligned}$$

The mutual information is related to the Kullback–Leibler distance in statistics. The Kullback–Leibler distance is used to measure the distance between two distributions and defined as

$$D(f(x)||g(x)) = \sum_x f(x) \log_2 \frac{f(x)}{g(x)},$$

where $f(x)$ and $g(x)$ are two distributions. It can be shown that

$$\mathsf{I}(X;Y) = D\left(f_{X,Y}(x,y) || f_X(x) f_Y(y)\right).$$

Note that if X and Y are independent, we have

$$\begin{aligned} \mathsf{H}(X,Y) &= \mathcal{E}[-\log_2 f_{X,Y}(X,Y)] \\ &= \mathcal{E}[-\log_2 (f_X(X) f_Y(Y))] \\ &= \mathcal{E}[-\log_2 f_X(X)] + \mathcal{E}[-\log_2 f_Y(Y)] \\ &= \mathsf{H}(X) + \mathsf{H}(Y) \end{aligned}$$

Hence, in this case, it follows that

$$\mathsf{I}(X;Y) = 0.$$

If $X = Y$, then

$$\mathsf{I}(X;Y) = \mathsf{H}(X).$$

Note that the dimension of $\mathsf{I}(X;Y)$ is the same as $\mathsf{H}(X)$, i.e. bits per symbol.

Suppose that X and Y denote the input and output of a channel, respectively. The mutual information shows the amount of information that can reliably be delivered over the channel. The entropy $\mathsf{H}(X)$ is the amount of information of X and the conditional entropy $\mathsf{H}(X|Y)$ is the amount of uncertainty of X given Y. Thus, the difference, $\mathsf{I}(X;Y)$, becomes the amount of information of X from Y when Y is observed.

We can show that

$$\mathsf{I}(X;Y) \leq \mathsf{H}(X).$$

This inequality is a consequence of the fact that the amount of the information reliably transmitted over the channel cannot exceed the original amount of information of X.

The mutual information can be maximized with respect to the input distribution, $f_X(x)$. The *channel capacity* is the maximum of mutual information and defined as

$$C = \max_{f_X(x)} \mathsf{I}(X;Y) \qquad \text{(bits per symbol)}.$$

It is important to note that for a given channel of capacity C, the channel input X whose entropy $\mathsf{H}(X) > C$ cannot be transmitted without any loss.

A2.3 Differential entropy and mutual information for continuous random variables

Previously we only considered the case of discrete random variables. For a continuous random variable, the entropy can be defined in a similar way as

$$\begin{aligned}\mathsf{h}(X) &= \mathcal{E}[-\log_2 f_X(X)] \\ &= -\int_{-\infty}^{\infty} f_X(x)\log_2 f_X(x)\mathrm{d}x. \end{aligned} \tag{2.3}$$

This is called the *differential entropy* of X. Consider an example. Let $X \sim \mathcal{N}(\mu, \sigma^2)$. The differential entropy of a Gaussian random variable becomes

$$\begin{aligned}\mathsf{h}(X) &= \int_{-\infty}^{\infty} \frac{1}{\sqrt{2\pi\sigma^2}} \exp\left(-\frac{(x-\mu)^2}{2\sigma^2}\right) \\ &\quad \times \left[\log_2(\sqrt{2\pi\sigma^2}) + (\log_2 \mathrm{e})\frac{(x-\mu)^2}{2\sigma^2}\right]\mathrm{d}x \\ &= \log_2(\sqrt{2\pi\sigma^2}) + (\log_2 \mathrm{e})\frac{1}{2} \\ &= \frac{1}{2}\log_2(2\pi \mathrm{e}\sigma^2). \end{aligned} \tag{2.4}$$

Interestingly, the mean μ does not affect the differential entropy, while the variance σ^2 does. The differential entropy increases with σ^2.

Although $\mathsf{H}(X) \geq 0$ for a discrete random variable X, the differential entropy $\mathsf{h}(X)$ for a continuous random variable X can have a negative value. For example, let $\sigma^2 = 1/4\pi\mathrm{e}$ in Eq. (2.4). Then, $\mathsf{h}(X) = -0.5$.

The mutual information between two continuous random variables X and Y can be defined as

$$\mathsf{I}(X;Y) = \int_{-\infty}^{\infty}\int_{-\infty}^{\infty} f_{X,Y}(x,y)\log_2 \frac{f_{X|Y}(x|y)}{f_X(x)}\mathrm{d}x\mathrm{d}y. \tag{2.5}$$

There are some important properties of the mutual information as follows.

(i) $\mathsf{I}(X;Y) = \mathsf{I}(Y;X)$. That is,

$$\begin{aligned}\mathsf{I}(X;Y) &= \mathsf{h}(X) - \mathsf{h}(X|Y) \\ &= \mathsf{h}(Y) - \mathsf{h}(Y|X), \end{aligned} \tag{2.6}$$

where

$$\mathsf{h}(X|Y) = \int_{-\infty}^{\infty}\int_{-\infty}^{\infty} f_{X,Y}(x,y)\log_2 \frac{1}{f_{X|Y}(x|y)}\mathrm{d}x\mathrm{d}y. \tag{2.7}$$

(ii) $\mathsf{I}(X;Y) \geq 0$.

(iii) Suppose that X and N are independent continuous random variables. We have

$$\mathsf{h}(X+N|X) = \mathsf{h}(N).$$

(iv) Consider two pdfs $f_X(x)$ and $f_Y(x)$. Using $\log_2 x \le x - 1$, $x \ge 0$, it can be shown that

$$\int_{-\infty}^{\infty} f_Y(x) \log_2 \frac{f_X(x)}{f_Y(x)} \mathrm{d}x \le 0. \tag{2.8}$$

Another important result is that the differential entropy of a Gaussian random variable X is always greater than or equal to that of any other continuous random variable Y which has the same mean and variance as X. That is,

$$\mathsf{h}(Y) \le \mathsf{h}(X),$$

where $\mathcal{E}[X] = \mathcal{E}[Y]$ and $\mathrm{Var}(X) = \mathrm{Var}(Y)$. To show this, we need to use Eq. (2.8). It can be shown that

$$-\int f_Y(x) \log_2 f_Y(x) \mathrm{d}x \le -\int f_Y(x) \log_2 f_X(x) \mathrm{d}x. \tag{2.9}$$

The term on the left-hand side in (2.9) is $\mathsf{h}(Y)$. Let μ and σ^2 be the mean and variance, respectively. Then, the term on the right-hand side in (2.9) becomes

$$\begin{aligned}
-\int f_Y(x) \log_2 f_X(x) \mathrm{d}x &= \int f_Y(x) \left[\frac{1}{2} \log_2(2\pi\sigma^2) + \log_2 \mathrm{e} \frac{(x-\mu)^2}{2\sigma^2} \right] \mathrm{d}x \\
&= \frac{1}{2} \log_2(2\pi\sigma^2) + \frac{\log_2 \mathrm{e}}{2\sigma^2} \underbrace{\int f_Y(x)(x-\mu)^2 \mathrm{d}x}_{=\mathcal{E}[(Y-\mu)^2]} \\
&= \frac{1}{2} \log_2(2\pi\sigma^2) + \frac{\log_2 \mathrm{e}}{2} \\
&= \frac{1}{2} \log_2(2\pi \mathrm{e} \sigma^2) \\
&= \mathsf{h}(X).
\end{aligned}$$

Thus, we can confirm that

$$\mathsf{h}(Y) \le \frac{1}{2} \log_2(2\pi \mathrm{e} \sigma^2). \tag{2.10}$$

A2.4 Information capacity theorem

Consider the additive white Gaussian noise (AWGN) channel. The received signal is given by

$$Y = X + N,$$

where X is the transmitted signal that is a continuous random variable and N is the white Gaussian noise with mean zero and variance σ^2. The mutual information becomes

$$\begin{aligned}
\mathsf{I}(X;Y) &= \mathsf{h}(Y) - \mathsf{h}(Y|X) \\
&= \mathsf{h}(Y) - \mathsf{h}(N),
\end{aligned} \tag{2.11}$$

where N and X are independent of each other. Suppose that X has mean zero and variance P. The capacity of the AWGN channel becomes

$$\begin{aligned} C_{\text{awgn}} &= \max_{f_X(x)} \mathsf{I}(X, Y) \\ &= \max_{f_X(x)} \mathsf{h}(Y) - \mathsf{h}(N). \end{aligned} \tag{2.12}$$

The variance of the random variable Y is $P + \sigma^2$ and the mean is zero. From (2.10), it can be shown that

$$\mathsf{h}(Y) \le \frac{1}{2} \log_2 \left(2\pi \mathrm{e}(P + \sigma^2)\right)$$

and the upper bound is achieved when Y is a Gaussian random variable. This means that X should be a Gaussian random variable, i.e. $X \sim \mathcal{N}(0, P)$, to achieve the channel capacity. Consequently, the channel capacity over the AWGN channel is given by

$$\begin{aligned} C_{\text{awgn}} &= \frac{1}{2} \log_2 \left(\frac{P + \sigma^2}{\sigma^2} \right) \\ &= \frac{1}{2} \log_2 \left(1 + \frac{P}{\sigma^2} \right) \quad \text{(bits per symbol)}. \end{aligned} \tag{2.13}$$

This is the maximum rate (of bits per symbol) that can be transmitted over the AWGN channel.

Consider a random variable $\mathbf{x} \sim \mathcal{CN}(\bar{\mathbf{x}}, \mathbf{R})$ whose pdf is given by

$$f(\mathbf{x}) = \frac{1}{\det(\pi \mathbf{R})} \exp\left(-(\mathbf{x} - \bar{\mathbf{x}})^{\mathrm{H}} \mathbf{R}^{-1} (\mathbf{x} - \bar{\mathbf{x}})\right).$$

The differential entropy of $\mathbf{x}$ is given by

$$\begin{aligned} \mathsf{h}(\mathbf{x}) &= \mathcal{E}[-\log_2(f(\mathbf{x}))] \\ &= \log_2 \det(\pi \mathbf{R}) - \int \log_2 \left(\exp\left(-(\mathbf{x} - \bar{\mathbf{x}})^{\mathrm{H}} \mathbf{R}^{-1} (\mathbf{x} - \bar{\mathbf{x}})\right)\right) f(\mathbf{x}) \mathrm{d}\mathbf{x} \\ &= \log_2 \det(\pi \mathbf{R}) + \int (\mathbf{x} - \bar{\mathbf{x}})^{\mathrm{H}} \mathbf{R}^{-1} (\mathbf{x} - \bar{\mathbf{x}}) f(\mathbf{x}) \mathrm{d}\mathbf{x} \log_2 \mathrm{e} \\ &= \log_2 \det(\pi \mathbf{R}) + \mathcal{E}\left[(\mathbf{x} - \bar{\mathbf{x}})^{\mathrm{H}} \mathbf{R}^{-1} (\mathbf{x} - \bar{\mathbf{x}})\right] \log_2 \mathrm{e} \\ &= \log_2 \det(\pi \mathbf{R}) + \mathcal{E}\left[\mathrm{Tr}\left(\mathbf{R}^{-1} (\mathbf{x} - \bar{\mathbf{x}})(\mathbf{x} - \bar{\mathbf{x}})^{\mathrm{H}}\right)\right] \log_2 \mathrm{e} \\ &= \log_2 \det(\pi \mathbf{R}) + \mathrm{Tr}\left(\mathbf{R}^{-1} \mathcal{E}\left[(\mathbf{x} - \bar{\mathbf{x}})(\mathbf{x} - \bar{\mathbf{x}})^{\mathrm{H}}\right]\right) \log_2 \mathrm{e} \\ &= \log_2 \det(\pi \mathbf{R}) + N \log_2 \mathrm{e} \\ &= \log_2 \det(\pi \mathrm{e} \mathbf{R}). \end{aligned}$$

With the differential entropy of a CSCG random vector, we can find the mutual information of a multiple input multiple output (MIMO) channel. Suppose that the received signal vector is given by

$$\mathbf{r} = \mathbf{H}\mathbf{s} + \mathbf{n},$$

where $\mathbf{s}$ is the transmitted signal vector, $\mathbf{H}$ is the MIMO channel matrix, and $\mathbf{n}$ is the background noise vector. Assume that $\mathbf{s} \sim \mathcal{CN}(\mathbf{0}, \mathbf{R_s})$ and $\mathbf{n} \sim \mathcal{CN}(\mathbf{0}, \mathbf{R_n})$. Then, we can show that $\mathbf{r} \sim \mathcal{CN}(\mathbf{0}, \mathbf{HR_sH}^{\mathrm{H}} + \mathbf{R_n})$. The mutual information between $\mathbf{s}$ and $\mathbf{r}$ or mutual information of MIMO channel is given by

$$\begin{aligned}
\mathsf{I}(\mathbf{r};\mathbf{s}) &= \mathsf{h}(\mathbf{r}) - \mathsf{h}(\mathbf{r}|\mathbf{s}) \\
&= \mathsf{h}(\mathbf{r}) - \mathsf{h}(\mathbf{n}) \\
&= \log_2 \det\left(\pi \mathrm{e}(\mathbf{HR_sH}^{\mathrm{H}} + \mathbf{R_n})\right) - \log_2 \det(\pi \mathrm{e}\mathbf{R_n}) \\
&= \log_2 \det\left(\mathbf{R_n}^{-1}(\mathbf{HR_sH}^{\mathrm{H}} + \mathbf{R_n})\right) \\
&= \log_2 \det\left(\mathbf{I} + \mathbf{R_n}^{-1}\mathbf{HR_sH}^{\mathrm{H}}\right).
\end{aligned}$$

Appendix 3 Important properties of matrices and vectors

A3.1 Vectors and matrices

An $N \times K$ matrix $\mathbf{A}$ is an array of numbers:

$$\mathbf{A} = \begin{bmatrix} a_{1,1} & a_{1,2} & \cdots & a_{1,K} \\ a_{2,1} & a_{2,2} & \cdots & a_{2,K} \\ \vdots & \vdots & \ddots & \vdots \\ a_{N,1} & a_{N,2} & \cdots & a_{N,K} \end{bmatrix},$$

where $a_{n,k}$ denotes the (n, k)th element. If $N = K$, the matrix $\mathbf{A}$ is called square. An $N \times 1$ vector $\mathbf{a}$ is a matrix with one column:

$$\mathbf{a} = \begin{bmatrix} a_1 \\ a_2 \\ \vdots \\ a_N \end{bmatrix},$$

where a_n denotes the nth element of $\mathbf{a}$.

Basic manipulations with matrices are shown below:

(i) Addition:

$$\mathbf{C} = \mathbf{A} + \mathbf{B} \Leftrightarrow c_{n,k} = a_{n,k} + b_{n,k},$$

where the sizes of $\mathbf{A}$, $\mathbf{B}$, and $\mathbf{C}$ are the same.

(ii) Multiplications:

- For $\mathbf{A}$ and $\mathbf{B}$ of size $N \times M$ and $M \times K$, respectively:

$$\mathbf{C} = \mathbf{AB} \Leftrightarrow c_{n,k} = \sum_{m=1}^{M} a_{n,m} b_{m,k},$$

where the size of $\mathbf{C}$ is $N \times K$.

- For a scalar α:

$$\mathbf{C} = \alpha\mathbf{A} \Leftrightarrow c_{n,k} = \alpha a_{n,k}.$$

(iii) Transpositions:

- The transpose of $\mathbf{A}$, denoted by $\mathbf{A}^{\mathrm{T}}$, is defined as

$$[\mathbf{A}^{\mathrm{T}}]_{n,k} = [\mathbf{A}]_{k,n},$$

where $[\mathbf{A}]_{n,k}$ stands for the (n, k)th element of $\mathbf{A}$. We also have

$$(\mathbf{A} + \mathbf{B})^{\mathrm{T}} = \mathbf{A}^{\mathrm{T}} + \mathbf{B}^{\mathrm{T}};$$
$$(\mathbf{AB})^{\mathrm{T}} = \mathbf{B}^{\mathrm{T}}\mathbf{A}^{\mathrm{T}}.$$

- The Hermitian transpose of $\mathbf{A}$, denoted by $\mathbf{A}^{\mathrm{H}}$, is defined as

$$[\mathbf{A}^{\mathrm{H}}]_{n,k} = [\mathbf{A}]^*_{k,n},$$

where the superscript $*$ denotes the complex conjugate. We also have

$$(\mathbf{A} + \mathbf{B})^{\mathrm{H}} = \mathbf{A}^{\mathrm{H}} + \mathbf{B}^{\mathrm{H}};$$
$$(\mathbf{AB})^{\mathrm{H}} = \mathbf{B}^{\mathrm{H}}\mathbf{A}^{\mathrm{H}}.$$

(iv) Triangular matrices: $\mathbf{A}$ is called lower triangular if

$$[\mathbf{A}]_{n,k} = \begin{cases} a_{n,k}, & \text{if } n \geq k; \\ 0, & \text{otherwise} \end{cases}$$

If the diagonal elements of $\mathbf{A}$ are all zeros, $\mathbf{A}$ is called strictly lower triangular.
$\mathbf{A}$ is called upper triangular if

$$[\mathbf{A}]_{n,k} = \begin{cases} a_{n,k}, & \text{if } n \leq k; \\ 0, & \text{otherwise} \end{cases}$$

If the diagonal elements of $\mathbf{A}$ are all zeros, $\mathbf{A}$ is called strictly upper triangular.

(v) Symmetric: A square matrix $\mathbf{A}$ is called symmetric if

$$\mathbf{A}^{\mathrm{H}} = \mathbf{A}.$$

(vi) Identity matrix: $\mathbf{I}$ is a square matrix which is defined as

$$\mathbf{I} = \mathrm{Diag}(1, 1, \ldots, 1),$$

where $\mathrm{Diag}(a_1, a_2, \ldots, a_N)$ denotes a diagonal matrix that is given by

$$\mathrm{Diag}(a_1, a_2, \ldots, a_N) = \begin{bmatrix} a_1 & 0 & \cdots & 0 \\ 0 & a_2 & \cdots & 0 \\ \vdots & \vdots & \ddots & \vdots \\ 0 & 0 & \cdots & a_N \end{bmatrix}.$$

Hence, it follows, if multiplications can be defined, that

$$\mathbf{AI} = \mathbf{A};$$
$$\mathbf{IA} = \mathbf{A}.$$

(vii) Inverse of $\mathbf{A}$: $\mathbf{B}$ is the inverse of a square matrix $\mathbf{A}$ if

$$\mathbf{AB} = \mathbf{I}.$$

$\mathbf{B}$ is denoted by $\mathbf{A}^{-1}$. We also have

$$\begin{aligned} \mathbf{A}^{-\mathrm{T}} &\triangleq (\mathbf{A}^{-1})^{\mathrm{T}} = (\mathbf{A}^{\mathrm{T}})^{-1}; \\ \mathbf{A}^{-\mathrm{H}} &\triangleq (\mathbf{A}^{-1})^{\mathrm{H}} = (\mathbf{A}^{\mathrm{H}})^{-1}. \end{aligned}$$

(viii) Rank: The rank of a matrix is the minimum of the numbers of linearly independent rows and columns. If the rank of an $N \times N$ matrix is N (i.e., of full rank), the matrix is called nonsingular. Otherwise, the matrix is called singular.

(ix) Trace: The trace of a square matrix $\mathbf{A}$ is defined as

$$\mathrm{tr}(\mathbf{A}) = \sum_{n=1}^{N} a_{n,n}.$$

Some properties of the trace are as follows.

- $\mathrm{tr}(\mathbf{A} + \mathbf{B}) = \mathrm{tr}(\mathbf{A}) + \mathrm{tr}(\mathbf{B})$;
- $\mathrm{tr}(\mathbf{AB}) = \mathrm{tr}(\mathbf{BA})$

(x) Determinant: For an $N \times N$ matrix $\mathbf{A}$, the determinant is given by

$$\begin{aligned} \det(\mathbf{A}) &= |\mathbf{A}| \\ &= \sum_{\text{permutation}:i_1,i_2,\ldots,i_N} \pm\, a_{1,i_1} a_{2,i_2} \cdots a_{N,i_N}, \end{aligned}$$

where the sum is taken over all the possible permutations and the sign is $+$ if the permutation is even and -1 if the permutation is odd.

Some examples are as follows.

- If

$$\mathbf{A} = \begin{bmatrix} \alpha & \beta \\ \gamma & \delta \end{bmatrix},$$

$$|\mathbf{A}| = \underbrace{a_{1,1}a_{2,2}}_{\text{permutation: } 1,2} - \underbrace{a_{1,2}a_{2,1}}_{\text{permutation: } 2,1} = \alpha\delta - \beta\gamma.$$

- If $\mathbf{A}$ is square and (upper or lower) triangle, we can also show that

$$|\mathbf{A}| = \prod_{n=1}^{N} a_{n,n}.$$

Some properties of the determinant are as follows.

- $|\mathbf{A}^{\mathrm{T}}| = |\mathbf{A}|$
- $|\mathbf{AB}| = |\mathbf{A}||\mathbf{B}|$
- $|\mathbf{A}^{-1}| = \frac{1}{|\mathbf{A}|}$

(xi) Norms: The 2-norm of a vector $\mathbf{x}$ is defined as

$$||\mathbf{x}|| = \sqrt{\mathbf{x}^{\mathrm{T}}\mathbf{x}}.$$

The Frobenius norm of an $N \times K$ matrix $\mathbf{A}$ is defined as

$$||\mathbf{A}||_{\mathrm{F}} = \sqrt{\sum_{n=1}^{N} \sum_{k=1}^{K} |a_{n,k}|^2}.$$

We can show that

$$\begin{aligned} ||\mathbf{A}||_{\mathrm{F}}^2 &= \sum_{k=1}^{K} ||\mathbf{a}_k||^2 \\ &= \mathrm{tr}(\mathbf{A}^{\mathrm{H}}\mathbf{A}), \end{aligned}$$

where $\mathbf{a}_k$ denotes the kth column vector of $\mathbf{A}$.

The 2-norm of a matrix $\mathbf{A}$ is defined as

$$||\mathbf{A}|| = \max_{\mathbf{x} \neq \mathbf{0}} \frac{||\mathbf{A}\mathbf{x}||}{||\mathbf{x}||}.$$

(xii) Positive definite: A square matrix $\mathbf{A}$ is called positive definite if

$$\mathbf{x}^{\mathrm{H}}\mathbf{A}\mathbf{x} > 0 \quad \text{for any } \mathbf{x} \neq \mathbf{0}.$$

If $\mathbf{x}^{\mathrm{H}}\mathbf{A}\mathbf{x} \geq 0$ for any $\mathbf{x} \neq \mathbf{0}$, $\mathbf{A}$ is called positive semi-definite or nonnegative definite.

(xiii) Eigenvalues and eigenvectors: A square matrix $\mathbf{A}$ has an eigenvector $\mathbf{e}$ and an eigenvalue λ if

$$\mathbf{A}\mathbf{e} = \lambda\mathbf{e}.$$

Generally, $\mathbf{e}$ is normalized to be $||\mathbf{e}|| = 1$. For a square and symmetric matrix $\mathbf{A}$ of size $N \times N$, there are N real-valued eigenvalues, $\lambda_1, \lambda_2, \ldots, \lambda_N$. Denote by $\mathbf{e}_n$ the eigenvector corresponding to λ_n. Furthermore, let

$$\begin{aligned} \mathbf{\Lambda} &= \mathrm{Diag}(\lambda_1, \lambda_2, \ldots, \lambda_N); \\ \mathbf{E} &= [\mathbf{e}_1\ \mathbf{e}_2\ \ldots\ \mathbf{e}_N]. \end{aligned}$$

Then, $\mathbf{A}$ can be expressed as

$$\mathbf{A} = \mathbf{E}\mathbf{\Lambda}\mathbf{E}^{\mathrm{H}},$$

where $\mathbf{E}$ is unitary:

$$\mathbf{E}\mathbf{E}^{\mathrm{H}} = \mathbf{E}^{\mathrm{H}}\mathbf{E} = \mathbf{I}.$$

A3.2 Subspaces, orthogonal projection, and pseudo-inverse

(i) Subspace: A subspace is a subset of a vector space. A subspace can be defined by a span of vectors as

$$\mathrm{Span}\,\{\mathbf{a}_1, \mathbf{a}_2, \ldots, \mathbf{a}_N\} = \left\{ \sum_{n=1}^{N} c_n \mathbf{a}_n \mid c_1, c_2, \ldots, c_N \in \mathbb{C} \right\}.$$

The range of a matrix is a subspace that is defined as

$$\begin{aligned}\text{Range}(\mathbf{A}) &= \{\mathbf{x} \mid \mathbf{x} = \mathbf{A}\mathbf{c} \ \text{ for any vector } \mathbf{c}\} \\ &= \text{Span}\{\mathbf{a}_1, \mathbf{a}_2, \ldots, \mathbf{a}_N\}.\end{aligned}$$

The null space of $\mathbf{A}$ is defined as

$$\text{Null}(\mathbf{A}) = \{\mathbf{x} \mid \mathbf{A}\mathbf{x} = \mathbf{0}\}.$$

(ii) Orthogonal projection: A matrix $\mathbf{P}$ is the orthogonal projection onto the subspace $\mathcal{S}$ if

(a) $\text{Range}(\mathbf{P}) = \mathcal{S}$;

(b) $\mathbf{P}^2 = \mathbf{P}\mathbf{P} = \mathbf{P}$;

(c) $\mathbf{P}^{\text{H}} = \mathbf{P}$.

For an $N \times K$ matrix $\mathbf{A}$, where $N \geq K$, the orthogonal projection matrix onto the subspace $\text{Range}(\mathbf{A})$ can be found as

$$\mathbf{P}(\mathbf{A}) = \mathbf{A}(\mathbf{A}^{\text{H}}\mathbf{A})^{-1}\mathbf{A}.$$

If the inverse does not exist, the pseudo-inverse can replace the inverse. We can easily verify that

(a) $\text{Range}(\mathbf{P}(\mathbf{A})) = \mathcal{S}$;

(b) $\mathbf{P}^2(\mathbf{A}) = \mathbf{P}(\mathbf{A})$;

(c) $\mathbf{P}^{\text{H}}(\mathbf{A}) = \mathbf{P}(\mathbf{A})$.

(iii) Pseudo-inverse: The pseudo-inverse of an $N \times M$ matrix $\mathbf{A}$, denoted by $\mathbf{A}^{\dagger}$ of size $M \times N$, is uniquely defined by the following equations:

$$\begin{aligned}\mathbf{A}^{\dagger}\mathbf{A}\mathbf{x} = \mathbf{x}, &\quad \text{for all } \mathbf{x} \in \text{Range}(\mathbf{A}^{\text{H}}); \\ \mathbf{A}^{\dagger}\mathbf{x} = \mathbf{0}, &\quad \text{for all } \mathbf{x} \in \text{Null}(\mathbf{A}^{\text{H}}).\end{aligned}$$

If the rank of $\mathbf{A}$ is M,

$$\mathbf{A}^{\dagger} = (\mathbf{A}^{\text{H}}\mathbf{A})^{-1}\mathbf{A}^{\text{H}}.$$

If the rank of $\mathbf{A}$ is N,

$$\mathbf{A}^{\dagger} = \mathbf{A}^{\text{H}}(\mathbf{A}\mathbf{A}^{\text{H}})^{-1}.$$

A3.3 Various forms of matrix inversion lemma

(i) A standard matrix inversion lemma is given by

$$(\mathbf{A} + \mathbf{B}\mathbf{C}\mathbf{D})^{-1} = \mathbf{A}^{-1} - \mathbf{A}^{-1}\mathbf{B}(\mathbf{C}^{-1} + \mathbf{D}\mathbf{A}^{-1}\mathbf{B})^{-1}\mathbf{D}\mathbf{A}^{-1}. \tag{3.1}$$

(ii) Woodbury's identity

$$(\mathbf{R} + \alpha\mathbf{u}\mathbf{u}^{\text{H}})^{-1} = \mathbf{R}^{-1} - \frac{\alpha}{1 + \alpha\mathbf{u}^{\text{H}}\mathbf{R}^{-1}\mathbf{u}}\mathbf{R}^{-1}\mathbf{u}\mathbf{u}^{\text{H}}\mathbf{R}^{-1} \tag{3.2}$$

Note that

$$\mathbf{u}^{\mathrm{H}}(\mathbf{R} + \alpha \mathbf{u}\mathbf{u}^{\mathrm{H}})^{-1}\mathbf{u} = \frac{\mathbf{u}^{\mathrm{H}}\mathbf{R}^{-1}\mathbf{u}}{1 + \alpha \mathbf{u}^{\mathrm{H}}\mathbf{R}^{-1}\mathbf{u}}. \tag{3.3}$$

(iii) Sherman–Morrison formula

$$\left(\mathbf{R} + \mathbf{u}\mathbf{v}^{\mathrm{H}}\right)^{-1} = \mathbf{R}^{-1} - \frac{1}{1 + \mathbf{v}^{\mathrm{H}}\mathbf{R}^{-1}\mathbf{u}}\mathbf{R}^{-1}\mathbf{u}\mathbf{v}^{\mathrm{H}}\mathbf{R}^{-1}. \tag{3.4}$$

(iv) Partitioned matrix inverse

$$\begin{aligned}\begin{bmatrix} \mathbf{A} & \mathbf{B} \\ \mathbf{C} & \mathbf{D} \end{bmatrix}^{-1} &= \begin{bmatrix} \mathbf{A}^{-1} & \mathbf{0} \\ \mathbf{0} & \mathbf{0} \end{bmatrix} + \begin{bmatrix} -\mathbf{A}^{-1}\mathbf{B} \\ \mathbf{1} \end{bmatrix} (\mathbf{D} - \mathbf{C}\mathbf{A}^{-1}\mathbf{B})^{-1} \left[-\mathbf{C}\mathbf{A}^{-1} \; \mathbf{1}\right] \\ &= \begin{bmatrix} \mathbf{0} & \mathbf{0} \\ \mathbf{0} & \mathbf{D}^{-1} \end{bmatrix} + \begin{bmatrix} \mathbf{1} \\ -\mathbf{D}^{-1}\mathbf{C} \end{bmatrix} (\mathbf{A} - \mathbf{B}\mathbf{D}^{-1}\mathbf{C})^{-1} \left[\mathbf{1} \; -\mathbf{B}\mathbf{D}^{-1}\right] \end{aligned} \tag{3.5}$$

A3.4 Vectorization and Kronecker product

Consider an $N \times K$ matrix $\mathbf{A}$:

$$\mathbf{A} = [\mathbf{a}_1 \; \mathbf{a}_2 \; \ldots \; \mathbf{a}_K].$$

The vectorization of $\mathbf{A}$ is given by

$$\mathrm{vec}(\mathbf{A}) = [\mathbf{a}_1^{\mathrm{T}} \; \mathbf{a}_2^{\mathrm{T}} \; \ldots \; \mathbf{a}_K^{\mathrm{T}}]^{\mathrm{T}}.$$

The Kronecker (or direct) product of two matrices, $\mathbf{A}$ and $\mathbf{B}$, is defined as

$$\mathbf{A} \otimes \mathbf{B} = \begin{bmatrix} a_{1,1}\mathbf{B} & a_{1,2}\mathbf{B} & \cdots & a_{1,K}\mathbf{B} \\ a_{2,1}\mathbf{B} & a_{2,2}\mathbf{B} & \cdots & a_{2,K}\mathbf{B} \\ & & & \\ a_{N,1}\mathbf{B} & a_{N,2}\mathbf{B} & \cdots & a_{N,K}\mathbf{B} \end{bmatrix}. \tag{3.6}$$

Some basic properties of the Kronecker product are as follows.

K1: $(\mathbf{A} \otimes \mathbf{B})^{\mathrm{H}} = \mathbf{A}^{\mathrm{H}} \otimes \mathbf{B}^{\mathrm{H}}$
K2: $\mathbf{A} \otimes (\mathbf{B} \otimes \mathbf{C}) = (\mathbf{A} \otimes \mathbf{B}) \otimes \mathbf{C}$
K3: $(\mathbf{A} \otimes \mathbf{B})(\mathbf{C} \otimes \mathbf{D}) = \mathbf{AC} \otimes \mathbf{BD}$, if $\mathbf{AC}$ and $\mathbf{BD}$ are defined.
K4: $\mathrm{tr}(\mathbf{A} \otimes \mathbf{B}) = \mathrm{tr}(\mathbf{A})\mathrm{tr}(\mathbf{B})$, if $\mathbf{A}$ and $\mathbf{B}$ are square matrices.
K5: $\mathrm{vec}(\mathbf{ABC}) = (\mathbf{C}^{\mathrm{T}} \otimes \mathbf{A})\mathrm{vec}(\mathbf{B})$.

Appendix 4 Lattice theory

A4.1 Definitions and properties of lattices

Suppose that there are K real-valued linearly independent basis vectors, $\{\mathbf{b}_1, \mathbf{b}_2, \ldots, \mathbf{b}_K\}$, where $\mathbf{b}_k \in \mathbb{R}^N$. The *lattice*, denoted by $\mathbf{\Lambda}$, generated by $\{\mathbf{b}_1, \mathbf{b}_2, \ldots, \mathbf{b}_K\}$ is the set of linear combinations of $\{\mathbf{b}_1, \mathbf{b}_2, \ldots, \mathbf{b}_K\}$ with integer coefficients. That is,

$$\mathbf{\Lambda} = \left\{ \mathbf{b} \mid \mathbf{b} = \sum_{k=1}^{K} \mathbf{b}_k u_k,\ u_k \in \mathbb{Z} \right\}, \tag{4.1}$$

where $\mathbb{Z}$ is the set of integer numbers. Let

$$\mathbf{B} = [\mathbf{b}_1\ \mathbf{b}_2\ \ldots\ \mathbf{b}_K].$$

Then, a vector in the lattice $\mathbf{\Lambda}$ can be expressed as $\mathbf{b} = \mathbf{B}\mathbf{u}$, where $\mathbf{u} \in \mathbb{Z}^K$. The basis vectors, $\{\mathbf{b}_k\}$, are called *generators* and $\mathbf{B}$ is called *generator matrix*. Note that since any vector in $\mathbf{\Lambda}$ lies in a K-dimensional subspace of $\mathbb{R}^N$, we simply assume that $N = K$ throughout this appendix.

The dimension of $\mathbf{\Lambda}$ is the number of basis vectors, K. For a given lattice, there could be more than one basis. Suppose that the lattices generated by $\mathbf{B}_1$ and $\mathbf{B}_2$ are the same. This means that any column vector of $\mathbf{B}_1$, which is a vector in the lattice, should be a linear combination of the column vectors of $\mathbf{B}_2$ with integer coefficients. Denote by $\mathbf{b}_{i,k}$ the kth column vector of $\mathbf{B}_i$. Then, we have

$$\begin{aligned}
\mathbf{b}_{1,1} &= u_{1,1}\mathbf{b}_{2,1} + u_{2,1}\mathbf{b}_{2,2} + \cdots + u_{K,1}\mathbf{b}_{2,K} \\
\mathbf{b}_{1,2} &= u_{1,2}\mathbf{b}_{2,1} + u_{2,2}\mathbf{b}_{2,2} + \cdots + u_{K,2}\mathbf{b}_{2,K} \\
\vdots\ &\quad \vdots \\
\mathbf{b}_{K,2} &= u_{1,K}\mathbf{b}_{2,1} + u_{2,K}\mathbf{b}_{2,2} + \cdots + u_{K,K}\mathbf{b}_{2,K},
\end{aligned} \tag{4.2}$$

where the coefficients, $u_{m,n}$, are integers. This leads to the following relation between the two bases:

$$\mathbf{B}_1 = \mathbf{B}_2\mathbf{U}. \tag{4.3}$$

Furthermore, since $\mathbf{B}_2 = \mathbf{B}_1\mathbf{U}^{-1}$, $\mathbf{U}^{-1}$ should also be a $K \times K$ matrix whose elements are integers. Since the elements of $\mathbf{U}$ and $\mathbf{U}^{-1}$ are integers, $\det(\mathbf{U})$ and $\det(\mathbf{U}^{-1})$ should also be integers. Since

$$1 = \det(\mathbf{U}\mathbf{U}^{-1}) = \det(\mathbf{U})\det(\mathbf{U}^{-1}),$$

we can conclude that $\mathbf{U}$ is an integer unimodular, i.e. $\det(\mathbf{U}) = \pm 1$.[1] We can summarize the above result as follows.

Lemma A4.1.1 *Any two bases that generate the same lattice are related by an integer unimodular matrix.*

Consider a lattice, $\mathbf{\Lambda}$, generated by a basis $\{\mathbf{b}_1, \mathbf{b}_2, \ldots, \mathbf{b}_K\}$. The *fundamental domain* (or fundamental parallelepiped) for $\mathbf{\Lambda}$ corresponding to this basis is the following set:

$$\mathcal{F}(\mathbf{b}_1, \mathbf{b}_2, \ldots, \mathbf{b}_K) = \left\{ \mathbf{b} \mid \mathbf{b} = w_1 \mathbf{b}_1 + w_2 \mathbf{b}_2 + \cdots + w_K \mathbf{b}_K, 0 \leq w_k < 1 \right\}. \quad (4.4)$$

It is noteworthy that there is no unique fundamental domain for a lattice as there are different bases that generate the same lattice. Using the notion of the fundamental domain, we can show that any vector in $\mathbb{R}^K$ can be decomposed into two vectors, where one vector is a lattice and the other vector is the fundamental domain for this lattice.

Lemma A4.1.2 *Suppose that* $\mathbf{\Lambda}$ *is a lattice in* $\mathbb{R}^K$*, which is generated by* $\{\mathbf{b}_1, \mathbf{b}_2, \ldots, \mathbf{b}_K\}$*. Let* $\mathbf{a}$ *be any vector in* $\mathbb{R}^K$*. Then,*

$$\mathbf{a} = \mathbf{b} + \mathbf{c}, \quad (4.5)$$

where $\mathbf{b}$ *is a unique vector in* $\mathbf{\Lambda}$ *and* $\mathbf{c}$ *is a unique vector in* $\mathcal{F}(\mathbf{b}_1, \mathbf{b}_2, \ldots, \mathbf{b}_K)$.

The volume of the fundamental domain, $\mathcal{F}(\mathbf{b}_1, \mathbf{b}_2, \ldots, \mathbf{b}_K)$, is given by

$$\begin{aligned} \mathrm{Vol}\left(\mathcal{F}(\mathbf{b}_1, \mathbf{b}_2, \ldots, \mathbf{b}_K)\right) &= \int_{\mathcal{F}} \mathrm{d}x_1 \mathrm{d}x_2 \ldots \mathrm{d}x_K \\ &= |\det(\mathbf{B})| \\ &= \sqrt{\det(\mathbf{B}^{\mathrm{T}}\mathbf{B})}. \end{aligned} \quad (4.6)$$

Lemma A4.1.3 *The fundamental domains for* $\mathbf{\Lambda}$ *have the same volume.*

Proof: Suppose that $\mathbf{B}_1$ and $\mathbf{B}_2$ generate the same lattice, $\mathbf{\Lambda}$. Then, we have

$$\mathbf{B}_1 = \mathbf{B}_2 \mathbf{U},$$

where $\mathbf{U}$ is an integer unimodular matrix. Since

$$\begin{aligned} |\det(\mathbf{B}_1)| &= \sqrt{\det(\mathbf{B}_1^{\mathrm{T}}\mathbf{B}_1)} \\ &= \sqrt{\det(\mathbf{U}^{\mathrm{T}}\mathbf{B}_2^{\mathrm{T}}\mathbf{B}_2\mathbf{U})} \\ &= \sqrt{\det(\mathbf{B}_2^{\mathrm{T}}\mathbf{B}_2)} \\ &= |\det(\mathbf{B}_2)|, \end{aligned} \quad (4.7)$$

we can show that the volumes of the fundamental domains from $\mathbf{B}_1$ and $\mathbf{B}_2$ are the same. This completes the proof. □

[1] A square matrix whose determinant is ± 1 is called *unimodular*. Furthermore, a unimodular matrix whose elements are integer is called an integer unimodular matrix.

Now, the volume of the fundamental domain for $\mathbf{\Lambda}$ can be denoted by Vol($\mathbf{\Lambda}$), which is

$$\mathrm{Vol}(\mathbf{\Lambda}) = \mathrm{Vol}\left(\mathcal{F}(\mathbf{b}_1, \mathbf{b}_2, \ldots, \mathbf{b}_K)\right) = |\det(\mathbf{B})|.$$

For convenience, we also write

$$\det(\mathbf{\Lambda}) = \mathrm{Vol}(\mathbf{\Lambda}),$$

where det($\mathbf{\Lambda}$) denotes the determinant of lattice $\mathbf{\Lambda}$.

The volume of a lattice can also be expressed by a product of the norms of orthogonalized basis vecors. To see this, consider the Gram–Schmidt orthogonalization of $\{\mathbf{b}_1, \mathbf{b}_2, \ldots, \mathbf{b}_K\}$. Let $\tilde{\mathbf{b}}_1 = \mathbf{b}_1$. Then, the other orthogonal basis vectors are given by

$$\tilde{\mathbf{b}}_k = \mathbf{b}_k - \sum_{q=1}^{k-1} \mu_{k,q} \tilde{\mathbf{b}}_q, \ k = 2, 3, \ldots, K, \tag{4.8}$$

where

$$\mu_{k,q} = \frac{\tilde{\mathbf{b}}_q^{\mathrm{T}} \mathbf{b}_k}{||\tilde{\mathbf{b}}_q||^2}.$$

The coefficients $\mu_{k,q}$ are determined to ensure the following orthogonality:

$$\tilde{\mathbf{b}}_l^{\mathrm{T}} \tilde{\mathbf{b}}_k = 0, \ l = 1, 2, \ldots, k-1; \ k = 1, 2, \ldots, K.$$

As a result, we can see that the column vectors $\tilde{\mathbf{b}}_k$'s are mutually orthogonal. From the $\tilde{\mathbf{b}}_k$'s, the original vectors $\mathbf{b}_k$'s can also be found. From (4.8), it follows that

$$\begin{aligned}\mathbf{b}_k &= \tilde{\mathbf{b}}_k + \sum_{q=1}^{k-1} \mu_{k,q} \tilde{\mathbf{b}}_q \\ &= \sum_{q=1}^{k} \mu_{k,q} \tilde{\mathbf{b}}_q, \ k = 1, 2, \ldots, K,\end{aligned} \tag{4.9}$$

where it is assumed that $\mu_{k,k} = 1$ for all k. For convenience, we can adopt a matrix representation as follows:

$$\begin{aligned}\mathbf{B} &= [\mathbf{b}_1 \ \mathbf{b}_2 \ \ldots \ \mathbf{b}_K] \\ &= \tilde{\mathbf{B}}[\boldsymbol{\mu}],\end{aligned} \tag{4.10}$$

where

$$\tilde{\mathbf{B}} = [\tilde{\mathbf{b}}_1 \ \tilde{\mathbf{b}}_2 \ \ldots \ \tilde{\mathbf{b}}_K];$$

$$[\boldsymbol{\mu}] = \begin{bmatrix} \mu_{1,1} & \mu_{2,1} & \cdots & \mu_{K,1} \\ 0 & \mu_{2,2} & \cdots & \mu_{K,2} \\ \vdots & \vdots & \ddots & \vdots \\ 0 & 0 & \cdots & \mu_{K,K} \end{bmatrix}. \tag{4.11}$$

Since $\mu_{k,k} = 1$ for all k, $\det([\boldsymbol{\mu}]) = 1$. From this, we can show that

$$\begin{aligned}\det(\mathbf{B}^{\mathrm{T}}\mathbf{B}) &= \det((\tilde{\mathbf{B}}[\boldsymbol{\mu}])^{\mathrm{T}}\tilde{\mathbf{B}}[\boldsymbol{\mu}]) \\ &= \det(\tilde{\mathbf{B}}^{\mathrm{T}}\tilde{\mathbf{B}}) \\ &= ||\tilde{\mathbf{b}}_1||^2||\tilde{\mathbf{b}}_2||^2 \cdots ||\tilde{\mathbf{b}}_K||^2 \end{aligned} \tag{4.12}$$

and finally

$$\mathrm{Vol}(\boldsymbol{\Lambda}) = \prod_{k=1}^{K} ||\tilde{\mathbf{b}}_k||. \tag{4.13}$$

Lattices are considered for channel coding due to their algebraic properties. For a detailed account, the reader is referred to (Forney, Jr. 1988) and (Conway & Sloane 1991).

A4.2 Key problems and bounds

A4.2.1 Key problems

There are two key problems with a lattice.

Shortest vector problem (SVP): For a given lattice $\boldsymbol{\Lambda}$, find a non-zero shortest vector (that minimizes its Euclidean norm) in $\boldsymbol{\Lambda}$ or find a $\mathbf{b} \in \boldsymbol{\Lambda} \setminus \{\mathbf{0}\}$ that minimizes $||\mathbf{b}||$.
Closest vector problem (CVP): For a given $\mathbf{x} \in \mathbb{R}^K$, find a closest vector to $\mathbf{x}$ in $\boldsymbol{\Lambda}$ or find a $\mathbf{b} \in \boldsymbol{\Lambda}$ that minimizes $||\mathbf{b} - \mathbf{x}||$.

A variation of the SVP is the shortest basis problem (SBP). In a SBP, for a given lattice, a basis, $\{\mathbf{b}_1, \mathbf{b}_2, \ldots, \mathbf{b}_K\}$, that minimizes the maximum norm that is given by

$$\max_{1 \le k \le K} ||\mathbf{b}_k||$$

is to be found. The SBP can be modified if a different criterion to measure the maximum norm is used. Using the SVP, we can solve a SBP. For a given $\boldsymbol{\Lambda}$, let

$$\begin{aligned}\mathbf{b}_1 &= \text{shortest vector in } \boldsymbol{\Lambda} \\ \mathbf{b}_2 &= \text{shortest vector in } \boldsymbol{\Lambda} \text{ linearly independent of } \mathbf{b}_1 \\ &\vdots \\ \mathbf{b}_K &= \text{shortest vector in } \boldsymbol{\Lambda} \text{ linearly independent of} \\ &\quad \{\mathbf{b}_1, \mathbf{b}_2, \ldots, \mathbf{b}_{K-1}\}. \end{aligned} \tag{4.14}$$

Each subproblem to find $\mathbf{b}_k$ is a SVP.

Although these problems are NP-hard problems which require exceedingly high complexity, they play a key role in a number of applications, including MIMO detection as demonstrated in Chapter 9. In general, if we can find a basis that is nearly orthogonal, the solutions (or their approximates) of these problems can be easily found. For example, consider the SVP. For a given lattice, suppose that there exists an orthogonal basis whose basis vectors are short, denoted by $\{\mathbf{b}_1, \mathbf{b}_2, \ldots, \mathbf{b}_K\}$, the SVP reduces to find a basis vector that corresponds to the minimum norm:

$$\text{(SVP)} \qquad \min_k ||\mathbf{b}_k||.$$

In addition, the solution of the CVP becomes

$$\text{(CVP)} \qquad \arg\min_{\mathbf{b}\in\mathbf{\Lambda}} ||\mathbf{b} - \mathbf{x}|| = \sum_{k=1}^{K} a_k \mathbf{b}_k,$$

where $a_k = \lceil (\mathbf{b}_k^{\mathrm{T}}\mathbf{x})/||\mathbf{b}_k||^2 \rfloor$. Therefore, it is crucial to find a basis that is nearly orthogonal and whose vectors are short.

A4.2.2 Bounds

To see how much the basis vectors are close to orthogonal, the *Hadamard inequality* becomes useful:

$$\det(\mathbf{\Lambda}) \leq ||\mathbf{b}_1||\, ||\mathbf{b}_2|| \cdots ||\mathbf{b}_K||, \tag{4.15}$$

where the equality holds if and only if the basis vectors are orthogonal to each other. Thus, the degree of orthogonality can be measured by the *Hadamard ratio* of the basis $\mathbf{B}$ that is defined by

$$\mathcal{H}(\mathbf{B}) = \left(\frac{\det(\mathbf{B})}{||\mathbf{b}_1||\, ||\mathbf{b}_2|| \cdots ||\mathbf{b}_K||} \right)^{1/K}. \tag{4.16}$$

Due to the Hadamard inequality, we have $0 < \mathcal{H}(\mathbf{B}) \leq 1$. As $\mathcal{H}(\mathbf{B})$ approaches 1, the basis vectors are more orthogonal.

Since there are multiple bases for a given lattice, there could be a basis that is more orthogonal than the others. Using bounds, we can see how a basis could be close to orthogonal for a given lattice.

Theorem A4.2.1 (Hermite's theorem) *For a given lattice, $\mathbf{\Lambda}$, there is a non-zero vector $\mathbf{b} \in \mathbf{\Lambda}$ satisfying*

$$||\mathbf{b}|| \leq \gamma_K^{1/2} \det^{\frac{1}{K}}(\mathbf{\Lambda}), \tag{4.17}$$

where γ_n is called Hermite's constant. Furthermore,

$$||\mathbf{b}_1||\, ||\mathbf{b}_2|| \cdots ||\mathbf{b}_K|| \leq \gamma_K^{K/2} \det(\mathbf{\Lambda}). \tag{4.18}$$

From (4.15) and (4.18), we can have bounds as follows:

$$\det(\mathbf{\Lambda}) \leq ||\mathbf{b}_1||\, ||\mathbf{b}_2|| \cdots ||\mathbf{b}_K|| \leq \gamma_K^{K/2} \det(\mathbf{\Lambda}).$$

To see how much the basis vectors could be close to orthogonal, it is crucial to find Hermite's constant, because, according to Hermite's theorem, there exists a basis for a given $\mathbf{\Lambda}$ satisfying

$$\mathcal{H}(\mathbf{B}) \geq \frac{1}{\sqrt{\gamma_K}}. \tag{4.19}$$

It is noteworthy that this bound is independent of $\mathbf{\Lambda}$. To determine Hermite's constant, we need the following famous theorem.

Theorem A4.2.2 **(Minkowski's theorem)** *Let $S \subset \mathbb{R}^K$ be a symmetric convex set and $\mathbf{\Lambda}$ be a lattice in $\mathbb{R}^K$. If*

$$\mathrm{Vol}(S) > 2^K \det(\mathbf{\Lambda}), \tag{4.20}$$

S contains a non-zero lattice vector in $\mathbf{\Lambda}$. If S is closed, $>$ in (4.20) *is replaced by $\geq$.*

Using Minkowski's theorem, we can have an estimate of Hermite's constant. Consider the following hyper-cube in $\mathbb{R}^K$:

$$S = \{\mathbf{x} = [x_1\ x_2\ \ldots\ x_K]^{\mathrm{T}} \mid -A \leq x_k \leq A\},$$

where $A > 0$ is a constant. This set is symmetric, convex, and closed. The volume of S is

$$\mathrm{Vol}(S) = (2A)^K = 2^K A^K.$$

For a given lattice, $\mathbf{\Lambda}$, let

$$A = (\det(\mathbf{\Lambda}))^{\frac{1}{K}}.$$

According to Minkowski's theorem, there is a non-zero lattice vector, $\mathbf{b}$, in S. This implies that

$$\begin{aligned} ||\mathbf{b}|| &= \sqrt{b_1^2 + b_2^2 + \cdots + b_K^2} \\ &\leq \sqrt{K} A \\ &= \sqrt{K}\,(\det(\mathbf{\Lambda}))^{\frac{1}{K}}. \end{aligned} \tag{4.21}$$

Thus, from (4.17) and (4.21), we have

$$\gamma_K \leq K. \tag{4.22}$$

Another estimate of γ_K can be obtained using a ball. Consider the following ball in $\mathbb{R}^K$:

$$S = \{\mathbf{x} \mid ||\mathbf{x}|| \leq R\},$$

where R is the radius. The volume of the ball is given by

$$\begin{aligned} \mathrm{Vol}(S) &= V_K(R) \\ &\triangleq \frac{\pi^{\frac{K}{2}} R^K}{\Gamma\left(\frac{K}{2}+1\right)}, \end{aligned} \tag{4.23}$$

where $\Gamma(x)$ is the Gamma function and $V_K(R)$ denotes the volume of a ball of radius R in $\mathbb{R}^K$. Let $C_K = \frac{\pi^{\frac{K}{2}}}{\Gamma(\frac{K}{2}+1)}$. Then,

$$R = \left(\frac{V_K(R)}{C_K}\right)^{\frac{1}{K}}.$$

As with a hyper-cube above, if we let

$$\text{Vol}(S) = 2^K \det(\mathbf{\Lambda}),$$

we have

$$R = \left(\frac{2^K \det(\mathbf{\Lambda})}{C_K}\right)^{\frac{1}{K}}.$$

From Minkowski's theorem, it follows

$$\begin{aligned} ||\mathbf{b}|| &\le R \\ &= \left(\frac{2^K \det(\mathbf{\Lambda})}{C_K}\right)^{\frac{1}{K}} \\ &= \left(\frac{2^K}{C_K}\right)^{\frac{1}{K}} \det^{\frac{1}{K}}(\mathbf{\Lambda}). \end{aligned} \tag{4.24}$$

Using Stirling's formula

$$\Gamma(n+1) = n! \simeq \sqrt{2\pi n} n^n \mathrm{e}^{-n} \quad (n \gg 1),$$

we have

$$\left(\frac{2^K}{C_K}\right)^{\frac{1}{K}} \simeq \sqrt{\frac{2K}{\pi \mathrm{e}}} (\pi K)^{\frac{1}{K}} \simeq \sqrt{\frac{2K}{\pi \mathrm{e}}}, \ K \gg 1. \tag{4.25}$$

Finally, for a large K,

$$||\mathbf{b}|| \le \sqrt{\frac{2K}{\pi \mathrm{e}}} \det^{\frac{1}{K}}(\mathbf{\Lambda}) \tag{4.26}$$

and an estimate of Hermite's constant is given by

$$\gamma_K \simeq \frac{2K}{\pi \mathrm{e}}. \tag{4.27}$$

Using Hermite's constant, we can find bounds for the SVP and SBP. In particular, for the SBP in (4.14),

$$\begin{aligned} \prod_{k=1}^{K} ||\mathbf{b}_k||^2 &\le \gamma_K^K \det^2(\mathbf{\Lambda}) \\ &\le \left(\frac{2K}{\pi \mathrm{e}}\right)^K \det^2(\mathbf{\Lambda}). \end{aligned}$$

A4.3 Reduced basis

There are various definitions for a reduced basis. To describe basis reduction conditions, the Gram–Schmidt orthogonalization in (4.10) for a given basis $\mathbf{B}$ is necessary:

$$\begin{aligned}\mathbf{B} &= [\mathbf{b}_1\ \mathbf{b}_2\ \dots\ \mathbf{b}_K] \\ &= \tilde{\mathbf{B}}[\mu].\end{aligned}$$

Three well-known definitions for a reduced basis are given below.

- **Minkowski reduced basis**: For a given lattice $\mathbf{\Lambda}$ generated by $\mathbf{B}$, the basis $\mathbf{B}$ is *Minkowski reduced* if $\mathbf{b}_1$ is a shortest vector in $\mathbf{\Lambda}$, $\mathbf{b}_k$ is a shortest vector independent of $\mathbf{b}_1, \mathbf{b}_2, \dots, \mathbf{b}_{k-1}$, for $k = 2, 3, \dots, K$.
- **Korkin–Zolotarev (KZ) reduced basis**: For a given lattice $\mathbf{\Lambda}$ generated by $\mathbf{B}$, the basis $\mathbf{B}$ is *KZ reduced* if $\mathbf{b}_1$ is a shortest vector in $\mathbf{\Lambda}$, $\tilde{\mathbf{b}}_k$ (not $\mathbf{b}_k$) is a shortest vector independent of $\mathbf{b}_1, \mathbf{b}_2, \dots, \mathbf{b}_{k-1}$, for $k = 2, 3, \dots, K$. It is known that if a basis is KZ reduced, it is also LLL reduced (Kannan 1987).
- **LLL reduced basis**: For a given basis $\mathbf{B}$, if

$$|\mu_{k,l}| \le \frac{1}{2}, \quad 1 \le l < k \le K; \tag{4.28}$$

$$||\tilde{\mathbf{b}}_k||^2 \le \omega ||\mu_{k+1,k}\tilde{\mathbf{b}}_k + \tilde{\mathbf{b}}_{k+1}||^2, \ k = 1, 2, \dots, K-1, \tag{4.29}$$

  where $1 \le \omega \le 4$, $\mathbf{B}$ is LLL reduced.

While the LLL algorithm has a polynomial time complexity to find a LLL reduced basis, there is no algorithm to find a KZ or Minkowski reduced basis with a polynomial time complexity yet. A review of lattice reduction algorithms can be found in (Schnorr & Euchner 1994).

References

Agrell, E., Eriksson, T., Vardy, A. & Zeger, K. (2002), "Closest point search in lattices," *IEEE Trans. Inform. Theory* **48**(8), 2201–2214.

Alamouti, S. M. (1998), "A simple transmit diversity technique for wireless communications," *IEEE J. Selec. Areas in Commun.* **16**(8), 1451–1458.

Albert, A. E. & Gardner, L. A. (1966), *Stochastic Approximation and Nonlinear Regression*, MIT Press.

Anderson, B. & Moore, J. (1979), *Optimal Filtering*, Prentice-Hall.

Babai, L. (1986), "On Lovasz' lattice reduction and the nearest lattice point problem," *Combinatorica* **6**(1), 1–13.

Bienvenu, G. & Kopp, L. (1983), "Optimality of high resolution array processing using the eigensystem approach," *IEEE Trans. Acoust., Speech Signal Process.* **31**(5), 1234–1248.

Biglieri, E., Taricco, G. & Tulino, A. (2002), "Performance of space-time codes for a large number of antennas," *IEEE Trans. Inform. Theory* **48**(7), 1794–1803.

Cary, M. C. (2002), Lattice basis reduction algorithms and applications. unpublished manuscript. URL: *www.cs.washington.edu/homes/cary*

Chiani, M., Win, M. Z., Zanella, A., Mallik, R. K. & Winters, J. H. (2003), "Bounds and approximations for optimal combining of signals in the presence of multiple cochannel interferers and thermal noise," *IEEE Trans. Commun.* **51**(2), 296–307.

Choi, J. (2005*a*), "Nulling and cancellation detector for MIMO channels and its application to multistage receiver for coded signals: performance and optimization," *IEEE Trans. Wireless Commun.* **5**(5), 1207–1216.

Choi, J. (2005*b*), "On the partial MAP detection with applications to MIMO channels," *IEEE Trans. Signal Proc.* **53**(1), 158–167.

Choi, J. (2006), *Adaptive and Iterative Signal Processing in Communications*, Cambridge University Press.

Cohen, H. (1993), *A Course in Computational Algebraic Number Theory*, Springer-Verlag.

Conway, J. H. & Sloane, N. J. A. (1991), *Sphere Packings, Lattices and Groups*, Springer-Verlag.

Cover, T. M. & Thomas, J. A. (1991), *Elements of Information Theory*, Wiley.

Cox, D. R. & Hinkley, D. V. (1974), *Theoretical Statistics*, Chapman and Hall.

Damen, M. O., El Gamal, H. & Caire, G. (2003), "On maximum-likelihood detection and the search for the closest lattice point," *IEEE Trans. Inform. Theory* **49**(10), 2389–2402.

Edelman, A. (1989), Eigenvalues and condition numbers of random matrices, PhD thesis, MIT. URL: *www-math.mit.edu/edelman/comprehensive.html*

Er, M. E. & Cantoni, A. (1983), "Derivative constraints for broad-band element space antenna array processors," *IEEE Trans. Acoust., Speech, Signal Process.* **31**(6), 1378–1393.

Ertel, R., Cardieri, P., Sowerby, K., Rappaport, T. & Reed, J. (1998), "Overview of spatial channel models for antenna array communication systems," *IEEE Personal Commun.* pp. 10–22.

Fincke, U. & Pohst, M. (1985), "Improved methods for calculating vectors of short length in a lattice, including a complexity analysis," *Math. Comput.* **44**, 463–471.

Fischer, R. F. H. (2002), *Precoding and Signal Shaping for Digital Transmission*, Wiley-IEEE Press.

Forney, Jr., G. D. (1988), "Coset codes. i. introduction and geometrical classification," *IEEE Trans. Inform. Theory* **34**(5), 1123–1151.

Foschini, G. (1996), "Layered space-time architecture for wireless communication in a fading environment when using multiple-element antennas," *Bell Labs Tech. Jour.* 41–59.

Foschini, G., Chizhik, D., Gans, M., Papadias, C. & Valenzuela, R. (2003), "Analysis and performance of some basic space-time architectures," *IEEE J. Selected Areas Commun.* **21**(4), 303–320.

Foschini, G. & Gans, M. (1998), "On limits of wireless communications in a fading environment when using multiple antennas," *Wireless Personal Commun.* **6**, 311–335.

Foschini, G., Golden, G., Valenzuela, R. & Wolniansky, P. (1999), "Simplified processing for high spectral efficiency wireless communication employing multi-element arrays," *IEEE J. Selected Areas Commun.* **17**(11), 1841–1852.

Frost, III, O. L. (1972), "An algorithm for linearly constrained adaptive array processing," *Proc. IEEE* **60**(8), 926–935.

Gallager, R. G. (1994), "An inequality on the capacity region of multiple access multipath channels". In R. E. Blahut, D. J. Costello, U. Maurer & T. Mittelholzer, eds., *Communications and Cryptography: Two Sides of One Tapestry*, Kluwar Academic Publishers.

Gan, Y. H. & Mow, W. H. (2005), "Complex lattice reduction algorithms for low-complexity MIMO detection," in *Proc. IEEE Global Telecommunications Conf.*, vol. 5, pp. 2953–2957.

Gan, Y. H., Ling, C. & Mow, W. H. (2006), *Complex lattice reduction algorithm for low-complexity MIMO detection*, Reprint.

Gesbert, D., Shafi, M., Shiu, D., Smith, P. J. & Naguib, A. (2003), "From theory to practice: An overview of mimo space-time coded wireless systems," *IEEE J. Selec. Areas Commun.* **21**(3), 281–302.

Goldsmith, A. (2005), *Wireless Communications*, Cambridge University Press.

Golub, G. H. & Loan, C. F. V. (1983), *Matrix Computations*, The Johns Hopkins University Press.

Guey, J.-C., Fitz, M. P., Bell, M. R. & Kuo, W.-Y. (1999), "Signal design for transmitter diversity wireless communication systems over rayleigh fading channels," *IEEE Trans. Commun.* **47**(4), 527–537.

Guo, D., Shamai, S. & Verdu, S. (2005), "Mutual information and minimum mean-square error in gaussian channels," *IEEE Trans. Inform. Theory* **51**(4), 1261–1282.

Hagenauer, J., Offer, E. & Papke, L. (1996), "Iterative decoding of binary block and convolutional codes," *IEEE Trans. Inform. Theory* **42**(2), 429–445.

Hassibi, B. (2000), "An efficient square-root algorithm for BLAST," in *Acoustics, Speech, and Signal Processing, 2000. ICASSP '00. Proceedings, 2000 IEEE International Conference*, vol. 2, pp. II737–II740.

Haykin, S. & Veen, B. V. (1999), *Signals and Systems*, John Wiley & Sons.

Hochwald, B. M., Peel, C. B. & Swindlehurst, A. L. (2005), "A vector-perturbation technique for near-capacity multichannel multiuser communications – part ii: Perturbation," *IEEE Trans. Commun.* **53**(3), 537–544.

Hochwald, B. & ten Brink, S. (2003), "Achieving near-capacity on a multiple-antenna channel," *IEEE Trans. Commun.* **51**(3), 389–399.

Hoffstein, J., Pipher, J. & Silverman, J. H. (2002), *An Introduction to Mathematical Cryptography*, Springer Science+Business Media.

Hong, S.-C., Choi, J., Jung, Y.-H. & Kim, S. R. (2004), "Constrained MMSE receivers for CDMA systems in frequency-selective fading channels," *IEEE Trans. Wireless Commun.* **3**(5), 1393–1398.

Jafarkhani, H. (2005), *Space-Time Coding*, Cambridge University Press.

Jalden, J., Seethaler, D. & Matz, G. (2008), "Worst- and average-case complexity of LLL lattice reduction in MIMO wireless systems," in *Proc. IEEE ICASSP*, pp. 2685–2688.

Johnson, D. H. & Dudgeon, D. E. (1993), *Array Signal Processing: Concepts and Techniques*, Prentice-Hall.

Kannan, R. (1987), "Algorithmic geometry of numbers," *Annual Review of Comp. Sci.* **2**, 231–267.

Kaveh, M. & Barabell, A. J. (1986), "The statistical performance of the MUSIC and the minimum-norm algorithms in resolving plane waves in noise," *IEEE Trans. Acoust., Speech, Signal Process.* **34**(2), 331–341.

Kay, S. (1993), *Fundamentals of Statistical Signal Processing, Estimation Theory*, Prentice-Hall.

Kay, S. (1998), *Fundamentals of Statistical Signal Processing, Detection Theory*, Prentice-Hall.

Kim, K. J., Yue, J., Iltis, R. A. & Gibson, J. D. (2005), "A QRD-M/Kalman filter based detection and channel estimation algorithm for MIMO-OFDM systems," *IEEE Trans. Wireless Commun.* **4**(2), 710–721.

Kong, N. & Milstein, L. B. (1999), "Average SNR of a generalized diversity selection combining scheme," *IEEE Commun. Lett.* **3**(3), 57–59.

Kumaresan, R. & Tufts, D. W. (1983), "Estimating the angles of arrival of multiple plane waves," *IEEE Trans. Aerospace Electron. Syst.* **19**(1), 134–139.

Lee, H. B. & Wengrovitz, M. S. (1991), "Statistical characterization of the music nukk spectrum," *IEEE Trans. Signal Proc.* **39**(6), 1333–1347.

Lee, W. C. Y. (1982), *Mobile Communication Engineering*, McGraw-Hill.

Lehmann, E. L. (1983), *Theory of Point Estimation*, John Wiley & Sons.

Lenstra, A. K., Lenstra, J. H. W. & Lovasz, L. (1982), "Factorizing polynomials with rational coefficients," *Math. Ann.* **216**(4), 515–534.

Leon-Garcia, A. (1994), *Probability and Random Processes for Electrical Engineering*, 2nd edn, Addison-Wesley.

Li, P., Paul, D., Narasimhan, R. & Cioffi, J. (2006), "On the distribution of SINR for the MMSE MIMO receiver and performance analysis," *IEEE Trans. Inform. Theory* **52**(1), 271–286.

Liberti, Jr., J. C. & Rappaport, T. S. (1999), *Smart Antennas for Wireless Communications: IS-95 and Third Generation CDMA Applications*, Prentice-Hall.

Ling, C. (2006), On the proximity factors of lattice reduction-aided decoding. Unpublished manuscript.

Luk, F. T. & Tracy, D. M. (2008), "An improved LLL algorithm," *Linear Algebra Appl.* **248**(2–3), 441–452.

Madhow, U. & Honig, M. (1994), "MMSE interference suppression for direct sequence spread spectrum CDMA," *IEEE Trans. Commun.* **42**(12), 3178–3188.

Monzingo, R. A. & Miller, T. W. (1980), *Introduction to Adaptive Arrays*, John Wiley & Sons.

Mow, W. H. (2003), 'Universal lattice decoding: principle and recent advances', *Wireless Commun. Mob. Comput.* **3**, 553–569.

Orfanidis, S. J. (1988), *Optimum Signal Processing: An Introduction*, 2nd edn, Macmillan.

Papoulis, A. (1984), *Probability, Random Variables, and Stochastic Processes*, 2nd edn, McGraw-Hill.

Paulraj, A., Gore, D. & Nabar, R. (2003), *Introduction to Space-Time Wireless Communication*, Cambridge University Press.

Paulraj, A. J., Gore, D. A., Nabar, R. U. & Bolcskei, H. (2004), "An overview of MIMO communications – a key to gigabit wireless," *Proc. IEEE* **92**(2), 198–218.

Paulraj, A. J. & Ng, B. C. (1998), "Space-time modems for wireless personal communications," *IEEE Person. Commun.* **5**, 36–48.

Peel, C. B., Hochwald, B. M. & Swindlehurst, A. L. (2005), "A vector-perturbation technique for near-capacity multichannel multiuser communications – part i: channel inversion and regularization," *IEEE Trans. Commun.* **53**(1), 195–202.

Pohst, M. (1993), *Computational Algebraic Number Theory*, Birkhauser Verlag. DMV Seminar.

Porat, B. (1994), *Digital Processing of Random Signals: Theory and Methods*, Prentice-Hall.

Proakis, J. G. (1995), *Digital Communications*, 3rd edn, McGraw-Hill.

Rapajic, P. B. & Vucetic, B. S. (1994), "Adaptive receiver structures for asynchronous CDMA systems," *IEEE J. Selec. Areas in Commun.* **12**(4), 685–697.

Rappaport, T. (1996), *Wireless Communications: Principles and Practice*, IEEE Press.

Reed, J. H. (2002), *Software Radio: A Modern Approach to Radio Engineering*, Prentice-Hall.

Richardson, T. & Urbanke, R. (2008), *Modern Coding Theory*, Cambridge University Press.

Sage, A. P. & Melsa, J. L. (1971), *Estimation Theory with Applications to Communications and Control*, McGraw-Hill.

Salz, J. (1985), "Digital transmission over cross-coupled linear channels," *AT & T Tech. Jour.* **64**, 1147–1159.

Sandell, M., Lillie, A., McNamara, D., Ponnampalam, V. & Miforf, D. (2007), "Complexity study of lattice reduction for mimo detection," in *Proc. IEEE Wireless Communications Networking Conf.*, pp. 1088–1092.

Scharf, L. L. (1991), *Statistical Signal Processing: Detection, Estimation, and Time Series Analysis*, Addison-Wesley.

Schmidt, R. O. (1986), "Multiple emitter location and signal parameter estimation," *IEEE Trans. Antennas Propag.* **34**(3), 276–280.

Schnorr, C. P. & Euchner, M. (1994), "Lattice basis reduction: improved practical algorithms and solving subset sum problems," *Math. Program.* **66**, 181–191.

Schubert, M. & Boche, H. (2004), "Solution of the multiuser downlink beamforming problem with individual sinr constraints," *IEEE Trans. Veh. Tech.* **53**(1), 18–28.

Schwartz, M., Bennet, W. R. & Stein, S. (1966), *Communication Systems and Techniques*, McGraw-Hill.

Shin, H. & Lee, J. (2003), "Capacity of multiple-antenna fading channels: spatial fading correlation, double scattering, and keyhole," *IEEE Trans. Inform. Theory* **49**(10), 2636–2647.

Silverstein, J. W. & Bai, Z. D. (1995), "On the empirical distribution of eigenvalues of a class of large dimensional random matrices," *J. Multivariate Analysis* **54**, 175–192.

Solo, V. & Kong, X. (1995), *Adaptive Signal Processing Algorithms: Stability and Performance*, Prentice-Hall.

Taherzadeh, M., Mobasher, A. & Khandani, A. K. (2007*a*), "Communication over MIMO broadcast channels using lattice-basis reduction," *IEEE Trans. Inform. Theory* **53**(12), 4567–4582.

Taherzadeh, M., Mobasher, A. & Khandani, A. K. (2007*b*), "LLL reduction achieves the receive diversity in MIMO decoding," *IEEE Trans. Inform. Theory* **53**(12), 4801–4805.

Tarokh, V., Seshadri, N. & Calderbank, A. (1998), "Space-time codes for high data rate wireless communication: performance criterion and code construction," *IEEE Trans. Inform. Theory* **44**(2), 744–765.

Telatar, I. (1999), "Capacity of multiple-antenna gaussian channels," *Europ. Trans. Telecomm.* **10**, 585–595.

Tse, D. N. C. & Hanly, S. V. (1999), "Linear multiuser receivers: effective interferences, effective bandwidth and user capacity," *IEEE Trans. Inform. Theory* **45**(2), 641–657.

Tse, D. & Viswanath, P. (2005), *Fundamentals of Wireless Communication*, Cambridge University Press.

Tulino, A. M. & Verdu, S. (2004), *Random Matrix Theory and Wireless Communications (Foundations and Trends in Commun. and Inform. Theory)*, Vol. 1, Now Publishers Inc.

Van Trees, H. L. (2002), *Optimum Array Processing*, John Wiley & Sons.

Veen, B. V. & Buckley, K. (1988), "Beamforming: A versatile approach to spatial filtering," *IEEE ASSP Mag.* **5**, 4–24.

Ventura-Traveset, J., Caire, G., Biglieri, E. & Taricco, G. (1997), "Impact of diversity reception on fading channels with coded modulation – part i: coherent detection," *IEEE Trans. Commun.* **45**(5), 563–572.

Verdu, S. (1998), *Multiuser Detection*, Cambridge University Press.

Vikalo, H., Hassibi, B. & Kailath, T. (2004), "Iterative decoding for MIMO channels via modified sphere decoding," *IEEE Trans. Wireless Commun.* **3**(6), 2299–2311.

Viterbi, A. J. & Omura, J. K. (1979), *Principles of Digital Communication and Coding*, McGraw-Hill.

Viterbo, E. & Boutros, J. (1999), "A universal lattice code decoder for fading channels," *IEEE Trans. Inform. Theory* **45**(5), 1639–1642.

Whalen, A. (1971), *Detection of Signals in Noise*, Academic Press.

Wiener, N. (1949), *Extrapolation, Interpolation, and Smoothing of Stationary Time Series*, MIT Press.

Wilson, S. G. (1996), *Digital Modulation and Coding*, Prentice-Hall.

Windpassinger, C., Fischer, R. F. H. & Huber, J. B. (2004), "Lattice-reduction-aided broadcast precoding," *IEEE Trans. Commun.* **52**(12), 2057–2060.

Winters, J. H. (1998), "Smart antennas for wireless systems," *IEEE Person. Commun.* **5**(1), 23–27.

Winters, J. H., Salz, J. & Gitlin, R. D. (1994), "The impact of antenna diversity on the capacity of wireless communication systems," *IEEE Trans. Commun.* **42**(2/3/4), 1740–1751.

Wolniansky, P., Foschini, G., Golden, G. & Valenzuela, R. (1998), "V-BLAST: an architecture for realizing very high data rates over the rich-scattering wireless channel," in *IEEE Proc. ISSSE-98*, pp. 295–300.

Wubben, D., Bohnke, R., Kuhn, V. & Kammeyer, K. D. (2004), "Near-maximum-likelihood detection of MIMO systems using MMSE-based lattice reduction," in *Proc. IEEE International Conf. Communications*, Paris, pp. 798–802.

Yao, H. (2003), Efficient Signal, Code, and Receiver Designs for MIMO Communication Systems, PhD thesis, MIT.
URL: *www.rle.mit.edu/dspg/documents/EfficientSignalCode.pdf*

Yao, H. & Wornell, G. W. (2002), "Lattice-reduction-aided detectors for MIMO communication systems," in *Proc. IEEE Global Telecommunications Conf.*, Taiwan, pp. 424–428.

Zhang, Q. T. (1995), "Probability of resolution of the MUSIC algorithm," *IEEE Trans. Signal Proc.* **43**(4), 978–987.

Zheng, L. & Tse, D. N. C. (2003), "Diversity and multiplexing: a fundamental trade-off in multiple-antenna channels," *IEEE Trans. Inform. Theory* **49**(5), 1073–1096.

Index